Biological Data Mining in Protein Interaction Networks

Xiao-Li Li
Institute for Infocomm Research, Singapore

See-Kiong Ng
Institute for Infocomm Research, Singapore

Medical Information Science REFERENCE

MEDICAL INFORMATION SCIENCE REFERENCE

Hershey · New York

Director of Editorial Content: Kristin Klinger
Managing Editor: Jamie Snavely
Assistant Managing Editor: Carole Coulson
Typesetter: Amanda Appicello
Cover Design: Lisa Tosheff
Printed at: Yurchak Printing Inc.

Published in the United States of America by
Information Science Reference (an imprint of IGI Global)
701 E. Chocolate Avenue, Suite 200
Hershey PA 17033
Tel: 717-533-8845
Fax: 717-533-8661
E-mail: cust@igi-global.com
Web site: http://www.igi-global.com/reference

and in the United Kingdom by
Information Science Reference (an imprint of IGI Global)
3 Henrietta Street
Covent Garden
London WC2E 8LU
Tel: 44 20 7240 0856
Fax: 44 20 7379 0609
Web site: http://www.eurospanbookstore.com

Library of Congress Cataloging-in-Publication Data

Biological data mining in protein interaction networks / Xiao-Li Li and See-Kiong Ng, editors.
p. ; cm.
Includes bibliographical references and index.
Summary: "The goal of this book is to disseminate research results and best practices from cross-disciplinary researchers and practitioners interested in, and working on bioinformatics, data mining, and proteomics"--Provided by publisher.
ISBN 978-1-60566-398-2 (h/c)
1. Protein-protein interactions--Data processing. 2. Proteomics. 3. Data mining. I. Li, Xiao-Li, 1969- II. Ng, See-Kiong.
[DNLM: 1. Protein Interaction Mapping--methods. 2. Computational Biology--methods. 3. Databases, Protein. QU 55 B61535 2009]
QP551.5.B56 2009
572'.640285--dc22

2008041607

British Cataloguing in Publication Data
A Cataloguing in Publication record for this book is available from the British Library.

Dedication

This book is dedicated to our beloved parents.

Editorial Advisory Board

List of Reviewers

Zeyar Ang, *Institute for Infocomm Research, Singapore*
Tero Aittokallio, *University of Turku, Finland*
Tatsuya Akutsu, *Kyoto University, Japan*
Christopher Baker, *Institute for Infocomm Research, Singapore*
Svetlana Bulashevska, *German Cancer Research Centre, German*
Jin Chen, *Stanford University, USA*
Honnian Chua, *Institute for Infocomm Research, Singapore*
Joaquin Dopazo, *Centro de Investigación Príncipe Felipe (CIPF), Spain*
Valeria Fionda, *Università della Calabria, Italy*
Chuan Sheng Foo, *Stanford University, USA*
Justin Chookhar Heng, *Institute for Infocomm Research, Singapore*
Charlie Hodgman, *University of Nottingham, UK*
Willy Hugo, *National University of Singapore, Singapore*
Daisuke Kihara, *Purdue University, USA*
Jung-Jae Kim, *European Bioinformatics Institute, UK*
Martin Krallinger, *Campus Universidad Autónoma, Spain*

Hai Quan Li, *Noble Foundation, USA*
Xiao-Li Li, *Institute for Infocomm Research, Singapore*
Tzia Liang Mah, *Institute for Infocomm Research, Singapore*
Takashi Makino, *University of Dublin, Ireland*
Hiroshi Mamitsuka, *Kyoto University, Japan*
Ali Masoudi-Nejad, *University of Tehran, Iran*
See-Kiong Ng, *Institute for Infocomm Research, Singapore*
Martin Samuel Rao Paradesi, *Kansas State University, USA*
Clara Pizzuti, *ICAR, Consiglio Nazionale delle Ricerche, Italy*
Menaka Priyadharsani Rajapakse, *Institute for Infocomm Research, Singapore*
Rune Sætre, *Tokyo University, Japan*
Kelvin Sianhui Sim, *Institute for Infocomm Research, Singapore*
Narayanaswamy Srinivasan, *Indian Institute of Science, India*
Harukazu Suzuki, *RIKEN Genomic Sciences Center, Japan*
Sui Sheng Tang, *Institute for Infocomm Research, Singapore*
Kar Leong Tew, *Institute for Infocomm Research, Singapore*
Jerzy Tiuryn, *Warsaw University, Poland*
Victor Joo Chuan Tong, *Institute for Infocomm Research, Singapore*
Anna Tramontano, *Sapienza University, Italy*
Raymond Wan, *Kyoto University, Japan*
Min Wu, *Nanyang Technological University, Singapore*
Ghimeng Yap, *Institute for Infocomm Research, Singapore*
Aidong Zhang, *University at Buffalo, The State University of New York, USA*
Xiaoyue Zhao, *Bionovo Inc., USA*

Table of Contents

Detailed Table of Contents

Section I
Introduction

In Section I of the book, we will provide the two introductory chapters. Chapter I aims to provide a biological primer on protein-protein interactions for those readers who are computer scientists and who may not have adequate knowledge in biology. Chapter II is a primer on the key concepts and pattern mining algorithms in data mining useful for the readers who are biologists.

Advances in protein-protein interaction (PPI) detection technology and computational analysis methods have produced numerous PPI networks, whose completeness appears to depend on the extent of data derived from different PPI assay methods and the complexity of the studied organism. Despite the partial nature of human PPI networks, computational data integration and analyses helped to elucidate new interactions and disease pathways. The success of computational analyses considerably depends on PPI data understanding. Exploration of the data and verification of their quality requires basic knowledge of the molecular biology of PPIs and familiarity with the assay methods used to detect PPIs. Both topics are reviewed in this chapter. After introducing various types of PPIs the principles of selected PPI assays are explained and their limitations discussed. Case studies of the Wnt signaling pathway and splice regulation demonstrate some of the challenges and opportunities that arise from assaying and analyzing PPIs. The chapter is concluded with an extrapolation to human systems biology that offers a glimpse into the future of PPI networks.

In this tutorial chapter, the author reviews basics about frequent pattern mining algorithms, including itemset mining, association rule mining, and graph mining. These algorithms can find frequently appearing substructures in discrete data. They can discover structural motifs, for example, from mutation data, protein structures, and chemical compounds. As they have been primarily used for business data, biological applications are not so common yet, but their potential impact would be large. Recent advances in computers including multicore machines and ever increasing memory capacity support the application of such methods to larger datasets. The author explains technical aspects of the algorithms, but do not go into details. Current biological applications are summarized and possible future directions are given.

Section II
PPI Network Construction and Cleansing

We will describe the methods for constructing and cleansing PPI networks. Section II of this book will describe key bioinformatics methods for predicting/validating protein interactions, as well as network cleansing techniques to detect reliable protein interaction networks.

Many methods have been proposed for inference of protein-protein interactions from protein sequence data. This chapter focuses on methods based on domain-domain interactions, where a domain is defined as a region within a protein that either performs a specific function or constitutes a stable structural unit. In these methods, the probabilities of domain-domain interactions are inferred from known protein-protein interaction data and protein domain data, and then prediction of interactions is performed based on these probabilities and contents of domains of given proteins. This chapter overviews several fundamental methods, which include association method, expectation maximization-based method, support vector machine-based method, and linear programming-based method. This chapter also reviews a simple evolutionary model of protein domains, which yields a scale-free distribution of protein domains. By combining with a domain-based protein interaction model, a scale-free distribution of protein-protein interaction networks is also derived.

This chapter presents applications of machine learning to predicting protein-protein interactions (PPI) in Saccharomyces cerevisiae. Several supervised inductive learning methods have been developed that treat this task as a classification problem over candidate links in a PPI network – a graph whose nodes represent proteins and whose arcs represent interactions. Most such methods use feature extraction from

reader in the choice of the most convenient methods, experiments and integrative data and to underline the most common biases and errors to obtain a portrait of PINs which is not only reliable but as well able to correctly retrieve the biological information contained in such data.

Section III
Knowledge Discovery from PPI Networks

Section III of the book will focus on knowledge discovery – that is, how do we computationally interrogate the PPI networks to discover new useful biological knowledge such as interaction motifs, network motifs, and protein complexes using sophisticated graph data mining approaches previously inaccessible to the bench biologists. We will also describe the evolutionary analyses of protein interaction networks for gaining insights on molecular evolution and comparative genomics.

Recent breakthroughs in high throughput experiments to determine protein-protein interaction have generated a vast amount of protein interaction data. However, most of the experiments could only answer the question of whether two proteins interact but not the question on the mechanisms by which proteins interact. Such understanding is crucial for understanding the protein interaction of an organism as a whole (the interactome) and even predicting novel protein interactions. Protein interaction usually occurs at some specific sites on the proteins and, given their importance, they are usually well conserved throughout the evolution of the proteins of the same family. Based on this observation, a number of works on finding protein patterns/motifs conserved in interacting proteins have emerged in the last few years. Such motifs are collectively termed as the interaction motifs. This chapter provides a review on the different approaches on finding interaction motifs with a discussion on their implications, potentials and possible areas of improvements in the future.

This chapter examines some of the available techniques for analyzing a protein interaction network (PIN) when depicted as an undirected graph. Within this graph, algorithms have been developed which identify "notable" smaller building blocks called network motifs. The authors examine these algorithms by dividing them into two broad categories based on two definitions of "notable": (a) statistically-based methods and (b) frequency-based methods. They describe how these two classes of algorithms differ not only in terms of efficiency, but also in terms of the type of results that they report. Some publicly-available programs are demonstrated as part of their comparison. While most of the techniques are generic and were originally proposed for other types of networks, the focus of this chapter is on the application of these methods and software tools to PINs.

In this chapter a survey on the main graph-based clustering techniques proposed in the literature to mine protein-protein interaction networks (PINs) is presented. The detection of putative protein complexes is an important research problem in systems biology. In fact it may help in understanding the mechanisms regulating cell life, in deriving conservations across species, in predicting the biological functions of uncharacterized proteins, and, more importantly, for therapeutic purposes. Different kind of approaches are described and classified. Furthermore, some validation techniques commonly exploited in this context are illustrated. The goal of the chapter is to provide a useful guide and reference for both computer scientists and biologists. Computer scientists may have a complete vision of what has already been made and which are the new challenges about PINs clustering, taking them as a starting point for further researches and new proposals; on the other hand, biologists may find in the chapter the necessary material to select the most appropriate methods to apply for their specific purposes.

This chapter introduces evolutionary analyses of protein interaction networks and of proteins as components of the networks. The authors show relationships between proteins in the networks and their evolutionary rates. For understanding protein-protein interaction (PPI) divergence, duplicated genes are often compared because they are derived from a common ancestral gene. In order to reveal evolutionary mechanisms acting on the interactome it is necessary to compare PPIs across species. Investigation of co-localization of interacting genes in a genome shows that PPIs have an important role in the maintenance of a physical link between neighboring genes. The purpose of this chapter is to introduce methodologies for analyzing PPI data and to describe molecular evolution and comparative genomics insights gained from such studies.

Section IV
Biological Applications Using PPI Analysis

Section IV of the book will present the various potential biological applications of protein interaction networks, e.g. the identification of lethal proteins, the functional annotation of unknown proteins, and the enrichment of functional genomics analyses. We will discuss how to use the protein interaction network and other biological resources for drug discovery, how to uncover the underlying disease pathways, and how to exploit the PPI networks to understand the complex dynamics of cellular processes.

Genetic factors play a major role in the etiology of many human diseases. Genome-wide experimental methods produce an increasing number of genes associated with such diseases. This chapter introduces data sources, bioinformatics tools, and computational methods for prioritizing disease candidate genes and identifying disease pathways. The main strategy is to examine the similarity among the candidate genes and known disease genes at the functional level. The authors review different similarity measures and prevailing methods for integrating results from different functional aspects. They hope this chapter will help advocate many useful resources that the researchers can use to investigate diseases of their interest.

Integration of organism-wide protein interactome data with information on expression of genes, cellular localization of proteins and their functions has proved extremely useful in developing biologically intuitive interaction networks. This chapter highlights the dynamics in protein interaction network across different stages in the lifecycle of Plasmodium falciparum, a malarial parasite, and the implication of the network dynamics in different physiological processes. The main focus of the chapter is the integration of information on experimentally derived interactions of P.falciparum proteins with expression data and analysis of the implications of interactions in different cellular processes. Extensive analysis has been made to quantify the interaction dynamics across various stages, as well as correlating it with the dynamics of the cellular pathways involving the interacting proteins. The authors' analysis demonstrates the power of strategic integration of genome-wide datasets in extracting information on dynamics of biological pathways and processes.

Section V
Tools for Analysis of PPI Networks

We will conclude the book by three chapters that discuss state-of-the-art software tools that provide visualization and analysis for protein interaction networks.

Software for the visualisation and analysis of protein-protein interaction (PPI) networks can enable general exploration, as well as providing graph-theoretic algorithms for specific tasks. Analyses can include reduction of complexity or the scope of the network in order to make it more manageable, or increase in complexity by integration with other datasets, to represent biology more accurately. Two software approaches are outlined in this chapter: desktop applications and web services. Desktop applications have attractive user interfaces with a wide range of analysis tools, and often capabilities for integration of other bio-molecular data. Web services provide a newer approach to network analysis. They have the advantages of a broader range of potential functionalities and a more extensible framework than standalone desktop tools. However, their relative infancy means that they are not as well developed. This chapter provides an evaluation of some common desktop applications, compared to and contrasted with several examples of web services.

The aim of this chapter is that of analyzing and comparing network querying techniques as applied to protein interaction networks. In the last few years, several automatic tools supporting knowledge discovery from available biological interaction data have been developed. In particular, network querying tools search a whole biological network to identify conserved occurrences of a query network module. The goal of such techniques is that of transferring biological knowledge. Indeed, the query subnetwork generally encodes a well-characterized functional module, and its occurrences in the queried network probably denote that this function is featured by the associated organism. The proposed analysis is intended to be useful to understand problems and research issues, state of the art and opportunities for researchers working in this research area.

This chapter provides an overview of the computational approaches developed for exploring the modular organization of protein interaction networks. A special emphasis is placed on the module finding tools implemented in three freely available software packages, VisANT, Cytoscape and MATISSE, as well as on their biomedical applications. The selected methods are presented in the broader context of module discovery options, ranging from approaches that rely merely on topological properties of the underlying network to those that take into account also other complementary data sources, such as the mRNA levels of the proteins. The author will also highlight some current limitations in the measured network data that should be understood when developing and applying module finding methodology, and discuss some key future trends and promising research directions with potential implications for clinical research.

Preface

The time has come for modern biology to move from the study of single molecules to networked systems. The recent technological advances in genomics and proteomics have resulted in the identification of many genes and proteins in various living organisms. However, simply knowing the existence of the genes and proteins in organisms does not provide insights about the molecular functions and biological processes in which they participate. Biologically, the functions of the bio-molecules and their participation in various cellular processes are mediated by networks of protein interactions and gene regulation. The new challenge for bioinformatics is to interrogate the biological data from the network perspectives, using new network-based data mining methodologies to uncover useful biological knowledge.

Recently, high-throughput methods (e.g. yeast-two-hybrid) for detecting protein-protein interactions (PPIs) *en masse* have enabled the construction of PPI networks on a genomic scale. A graphical map of an organism's interactome can be constructed from such experiments by considering individual proteins as the nodes, and the existence of a physical interaction between a pair of proteins as a link between two corresponding nodes. The growing flood of molecular interaction data, as biologists step up their efforts in mapping the interactomes of various species, is expected to surpass the data flux resulted from the development of genome sequencing in the past decade.

Such developments offer unprecedented opportunities to develop new bioinformatics methods for data mining the PPI networks. The study of protein interaction networks will provide invaluable insights about the inner working mechanisms of cells, which can lead to the uncovering of the underlying disease pathways and even the discovery of new drugs to benefit human beings. The subject area is therefore of interest to both biologists and computer scientists — to the biologists, the unraveling of PPI networks can unlock the many secrets of life; to the computer scientists, the PPI networks, being large-scale real-world graphical data, provide the ideal computational challenge for data mining research.

The objective of this book is to disseminate the research results and best practice from cross-disciplinary researchers and practitioners interested in, and working on bioinformatics, data mining, and proteomics. We aim to present the various methodologies in an accessible way, and we hope to bring better awareness of this interesting and challenging problem to inspire new computational solutions. The book is aimed at three overlapping audiences: (1) Researchers in the areas of bioinformatics, data mining, machine learning and data structure; (2) practitioners in the industry in these areas; (3) instructors and postgraduate students in colleges and universities. Upon reading the book, we hope that the molecular biologists would have acquired the necessary information to select the most appropriate methods and useful tools to apply, while the computer scientists would have a complete view of what have already been done and what are the new computational challenges that can serve as impactful starting points for furthering their data mining research.

Two main principles have been used to guide the writing of the book. Firstly, the chapters in the book are written in a comprehensive manner instead of focusing solely on a specific work. Each of the chapters serves as a tutorial for network-based computational analysis of PPI networks. This book hopes to provide the reader a good understanding on how to mine the PPI networks by using effective computational algorithms as well as insights into the associated research challenges in data mining the PPI networks.

Secondly, the chapters are written with a balance of theory and practice to meet the needs of different categories of readers. We aim to present the various methodologies in an understandable way to our inter-disciplinary audience. Theoretical descriptions of the computational challenges and methods are provided in details for the computationally inclined readers. For the more practical minded readers, each chapter discusses and compares the experimental results and scientific findings obtained by using the methods. Where possible, each chapter lists and briefly describes online tools and database resources related to the topic at hand. Each chapter also provides a section of discussion on future trends from which researchers and postgraduate students can shape their research topics for further study.

As far as we know, this is the first book that is focused primarily on applying sophisticated data mining and graph mining techniques on protein interaction networks. We hope this book will be a useful resource for professionals and researchers who wish to learn how to apply advanced data mining techniques in protein interaction network. In addition, it can be used as a supplementary text book for Masters and PhD students studying bioinformatics.

This book is focused on presenting the current bioinformatics methods to interrogate the PPI networks. It is important to bear in mind that the PPI network is only one of the many components in the complex machinery of life. To fully understand the interplay of bio-molecules in carrying out critical life processes at the network, it is necessary to integrate and analyze other biological data.

The ultimate success of the PPI network analysis will depend on the parallel improvements both in the biological experimental techniques from the biologists which provide rich biological datasets for data mining community, as well as in the data mining techniques from the computer scientists which provide efficient ways to exploit the protein interaction data to help biologists better understand the life processes. We hope this book will play a role in helping to realize the wonderful magic of what a perfect marriage of biology with computer science can bring.

THE CHALLENGES

It is a well-known fact that protein interactions play a central role at virtually every level of cellular activities. Analysis of the PPI networks is necessary to decipher the underlying cellular mechanisms as well as the behavior of various biological systems. We will discuss the computational challenges for mining PPI networks.

Noisy PPI Data: Alarmingly High False Positive and False Negative Rates

While high-throughput methods such as yeast-two-hybrid (Y2H) (Fields & Song, 1989) and tandem affinity purification-mass spectrometry (TAP-MS) (Puig et al., 2001; Rigaut et al., 1999) have enabled comprehensive detection of protein interactions, the quality of detected protein interactions is far from satisfactory. On the one hand, the experimental conditions in which the detection methods are carried out may cause a bias towards detecting interactions that do not occur under physiological conditions, resulting in false positive detection rates that could be alarmingly high. In other words, the experimental data may not be very accurate and contain interaction data that do not occur in the cell. On the other hand, the high-throughput methods can also fail to detect various types of interactions, for example, loss of weak transient interactions, loss of post-translational modification, and bias against soluble or membrane proteins (Lalonde et al., 2008; Tarassov et al., 2008). This results in false negative detection and low experimental coverage of the interactomes.

How serious is the situation — specifically, what are the false positive and false negative rates in current protein interaction data? Many researchers have attempted to answer this question recently (G. D. Bader & Hogue, 2002; J. S. Bader, Chaudhuri, Rothberg, & Chant, 2004; Gentleman & Huber, 2007; Hart, Ramani, & Marcotte, 2006; Von Mering et al., 2002), by comparing the overlapping between the collected protein interactions from different large-scale biological experiments, or checking the consistency of the function or location information between the interaction partners to estimate the accuracy and coverage of the PPI data. The results of all these research consistently showed that the quality of the protein interaction data is indeed very problematic, with accuracy values ranging from 10% to 50% (Gentleman & Huber, 2007; Sprinzak, Sattath, & Margalit, 2003; Von Mering et al., 2002) and coverage values lower than 50%, even for the most studied and curated interactome of *Saccharomyces cerevisiae* (yeast) (Hart et al., 2006). For many other species, one can expect the accuracy and coverage of the PPIs to be even lower.

How to deal with the false positive and false negative interactions? Computational methods can help to validate the existing protein interactions (to address false positive interaction issue) and predict novel protein interactions (to address false negative interaction issue). Our book provides bioinformatics solutions to predict/validate protein interactions by using various biological features/properties, such as protein domains, protein sequences (e.g., amino acids composition etc), protein functions, biological processes, cellular locations, structural information, and topological features extracted from the PPI network. Network cleansing techniques to improve the quality of the entire protein interaction networks are also surveyed in this book.

Mining PPI Network: A Computationally Challenging Graph Mining Problem

As mentioned earlier, a PPI network is typically modelled as an undirected graph (sometimes with weighted links) where the nodes represent unique proteins and the links denote interactions between two proteins. Such a network is a very large graph with thousands of vertices and tens of thousands of edges, even for a simple model organism such as yeast. One can only imagine the insurmountable complexity of the PPI networks for the more complicated species such as the human being. If we wish to investigate the evolution of various protein interaction networks or to alignment the PPI networks, the computational challenge is even more overwhelming since we will need to handle multiple humongous networks.

Graph theory is an important tool to facilitate the efficient analyses of the large scale interaction networks (Barabasi & Oltvai, 2004). In our book, we have surveyed several state-of-the-art graph mining techniques and their applications in discovering such useful biological objects from PPI networks as interaction motifs, network motifs, lethal proteins, protein complexes/functional modules, as well as in the evolutionary analyses PPI network and also network alignment/querying tasks. Although mining these objects from protein interaction networks are computationally challenging problems, it is possible to reduce the search space and time complexity and obtain better mining results by exploiting biological knowledge coupled with the development of novel efficient graph mining techniques.

Integrating Various Biological Evidences for System-Level Understanding

Recently, high throughput experimental technologies have been developed and as a result, an increasing number of large datasets at the various biological levels from genomics, proteomics to metabolomics have been generated. Many of the datasets are deposited in centralized databases that are publicly accessible by all the researchers. At the same time, there are increasing efforts by biologists to provide annotations of the biological knowledge. Projects such as the Gene Ontology have been initiated to enable collective and systematic functional annotations of genes and proteins. All these efforts are motivated by the need to integrate various biological evidences for system-level understanding of the inherently complex cellular processes.

The benefits of integrating various biological evidences are two-fold. Firstly, as mentioned, the PPI data generated from the high-throughput methods are inevitably noisy and incomplete, so it is important to weight the links (i.e. the protein interactions) in the PPI networks by using appropriate confidence measures. For example, we can employ metrics from biological evidences such as reproducibility of the interactions from multiple experimental methods, support from such other non-interaction data as co-expression, co-localization and shared functions, as well as the conservation of the protein interactions across other genomes, and so forth. For example, a Bayesian network model had been developed to predict protein interactions by integrating noisy experimental interaction data with the weighted genomic features which are only weakly associated with interactions, for example, messenger RNA coexpression, coessentiality, and colocalization, and so forth (Jansen et al., 2003). The results were positive, indicating that it is useful to integrate biological evidences in addressing the limitations in the current quality of PPI data. Similarly, it is also possible to use machine learning methods, such as kernel methods (Ben-Hur & Noble, 2005), to integrate different biological resources into high-dimensional vector space to do classification and thus better predict protein interactions.

Secondly, integrating the PPI networks with additional biological evidences enables a systems-level understanding of biological processes and human diseases (Ideker & Sharan, 2008) by utilizing the rich information and strengths from different sources at different biological levels (Ghazalpour, Doss, Zhang, Wang, & Plaisier, 2006; Kelley & Ideker, 2005; Mootha et al., 2003). Obtaining quantitative and dynamic PPI data across different tissue cells and their integration with gene expression, functional, structural, and metabolic pathway data is the key towards successful development of diagnostics and therapeutics that target disease-relevant protein interactions. Numerous chapters in this book (e.g. the two book chapters that focus on disease related research) therefore describe methods that involved integration of various biological evidences.

On the other hand, one should be mindful that integration may not necessarily result high-quality data and results—the intrinsic noisiness of data and integration methods does really matter. Computational methods, such as machine learning approaches, probabilistic and statistical tools, should be carefully designed to maximally exploit the knowledge from various biological evidences and minimize the side effect from the irrelevant and noisy sources. At the same time, the users of the computational methods should be aware of the limitations and assumptions made by the designers of the computational methods. We hope such a message is clear in this book to both the computer scientists (designers of computational methods) and the biologists (users of the computational methods).

ORGANIZATION OF THE BOOK

The book is divided into five major sections, described below, covering the topics of: introduction (Section I), PPI network construction and cleansing (Section II), knowledge discovery from PPI networks (Section III), biological applications using PPI analysis (Section IV) and tools for analysis of PPI networks (Section V).

Section I: Introduction

In Sectino I, we will provide two introductory chapters, namely Chapter I "*Molecular Biology of Protein-Protein Interactions for Computer Scientists*" and Chapter II "*Data Mining for Biologists*".

Chapter I is intended for the readers from computer science who may not have adequate background about protein-protein interactions in biology. This introductory chapter will provide a brief tour of protein interaction types, experimental detection methods and their limitations. Biologically interesting examples of the Wnt signaling pathway and splice regulation will be used to demonstrate the challenges and opportunities that arise from assaying and analyzing protein interactions.

Chapter II is useful for the biologists who may not have adequate background knowledge on data mining algorithms. This chapter reviews the basics about various relevant data mining algorithms and their applications to biology. The chapter will focus on frequent pattern mining algorithms, including itemset mining, association rule mining, and graph mining. Originally developed for applications in totally different domains, these pattern mining algorithms can be used to automatically detect frequently appearing patterns/substructures from biological data. The chapter will summarize data mining's current biological applications and discuss the new directions of applying data mining techniques in biology.

Section II: PPI Network Construction and Cleansing

Chapter III, "*Domain-Based Prediction and Analysis of Protein Interaction Network*", will review domain-based models both for prediction of protein-protein interactions and for explaining the scale-freeness of protein-protein interaction networks. The chapter will describe the use computational methods such as the association method, EM method, SVM-based method and LP-based method, to infer domain-domain interactions from known protein-protein interaction data. These domain-domain interactions can then be used to predict new protein interactions based on the domains in the given proteins. This chapter will also review an evolutionary model of protein domains to explain how to derive a scale-free distribution of protein-protein interaction networks.

Chapter IV, "*Incorporating Graph Features for Prediction of Protein-Protein Interactions*", will review the machine learning methods which have been designed to predict protein interactions. Many of the current techniques either used the features extracted from protein sequences or other biological properties, or incorporated relational and structural features extracted from the PPI network. This chapter will describe predicting protein interactions using the graph features extracted from a PPI network along with other available biological features of the proteins and their interactions. The future trends in this area of predicting protein interactions based on information from PPI network will be highlighted.

Chapter V, "*Discovering Protein-Protein Interaction Sites from Sequence and Structure*", will first introduce both sequence and structural features which characterize protein-protein interaction sites and then review and compare the methodologies for predicting protein-protein binding site. The chapter will show how protein-protein interaction data can serve as a critical platform for advanced molecular recognition research and experimental design.

Chapter VI, "*Network Cleansing: Reliable Protein Interaction Networks*", will provide an overview of the existing methods for network cleansing. Different classes of cleansing algorithms will be described and their results are compared. The chapter will provide the information to guide the readers in the choice of the most appropriate methods, experiments and integrative data to obtain a portrait of reliable protein interaction network, at the same time highlighting the common biases and errors.

Section III: Knowledge Discovery from PPI Networks

Chapter VII, "*Discovering Interaction Motifs from Protein Interaction Networks*", will describe how to discover protein interaction motifs that are conserved in interacting proteins. Such knowledge discovery can help understand the mechanisms of protein interactions. It is based on the observation that protein interactions usually occur at some specific sites/motifs on the proteins and that these motifs are well conserved throughout the evolution among the proteins of the same family. The chapter will provide a review on the different approaches on mining for interaction motifs in PPI data, together with their implications, potentials, and possible areas of improvements in the future.

Chapter VIII, "*Discovering Network Motifs in Protein Interaction Networks*", will examine the methods to mine network motifs from PPI networks. Network motifs are potentially the basic building blocks of the protein interaction networks. By discovering the network motifs, researchers can gain insights on the general structure of the overall network, categorize different PPI networks into so-called "super-families", and even formulate hypothesis on how the network was formed through evolution. Both statistically-based methods and frequency-based methods will be described and the experimental results are compared in this chapter.

Chapter IX, "*Discovering Protein Complexes in Protein Interaction Networks*", will survey different state-of-the-art graph-based clustering techniques for detecting putative protein complexes from PPI networks. The discovered putative protein complexes are useful biological knowledge. For example, the knowledge of the complexes may help in understanding the mechanisms regulating cell life, in deriving conservations across species, in predicting the biological functions of uncharacterized proteins, and, more importantly, for therapeutic purposes.

Chapter X, "*Evolutionary Analyses of Protein Interaction Networks*", will describe the methodologies for analyzing PPI data to understand molecular evolution and gain comparative genomics insights from such studies. In order to reveal the evolutionary mechanisms acting on the interactomes, it is necessary to compare protein interactions across species. This chapter will show that the evolution of proteins as the components of protein interaction networks can be understood through the evolutionary rates of the PPI networks. It will also show that protein interactions can influence the genomic locations of genes during evolution.

Section IV: Biological Applications Using PPI Analysis

Chapter XI, "*Discovering Lethal Proteins in Protein Interaction Networks*", will describe current methods for detecting lethal proteins from protein interaction networks. These methods exploit either the network properties alone or integrate with biological various biological information and properties. The study of lethal proteins is useful for understanding the minimal condition for cellular development and survival.

Chapter XII, "*Predicting Protein Functions from Protein Interaction Networks*", will investigate in detail the popular methods of using protein-protein interactions to predict protein functions. In particular, both local prediction methods and global optimization methods will be described and compared.

Chapter XIII, "*Protein Interactions for Functional Genomics*", will review the use of protein-protein interactions for the interpretation of genomic experiments. As an example, this chapter will describe the available resources and methodologies which are used to create a curated compilation of protein interactions and a novel approach to filter interactions.

Chapter XIV, "*Prioritizing Disease Genes and Understanding the Disease Pathways*", will describe a variety of useful data sources and bioinformatics tools and methods that can help prioritize disease genes and identify disease pathways. The main strategy is to examine the similarity among the candidate genes and known disease genes at the functional level. The chapter will review different similarity measures and prevailing methods for integrating results from different functional aspects, and advocate many useful resources that the readers can use to investigate various diseases of their interest.

Chapter XV, "*Dynamics of Protein-Protein Interaction Network in Plasmodium Falciparum*", will present an application to integrate gene expression data with the protein interaction network acting in the different cellular process for *Plasmodium falciparum*, a malarial parasite, to reveal the dynamics in protein interaction network across different stages in the lifecycle of *Plasmodium falciparum*. The analysis hopes to demonstrate the power of strategic integration of genome-wide datasets for unraveling the dynamic complexity of biological pathways and processes.

Section V: Tools for Analysis of PPI Networks

Chapter XVI, "*Graphical Analysis and Visualization Tools for Protein Interaction Networks*", will present several visualisation and analysis software tools for protein interaction networks. The major advantages and disadvantages of each tool will be given, along with the analyses offered, and the capability for integration of other types of data. This chapter will also outline and evaluate two typical software approaches, namely desktop applications and web services.

Chapter XVII, "*Network Querying Techniques for PPI Network Comparison*", will describe and compare network querying techniques and tools applied to protein interaction networks. Network querying tools can be used to search a whole biological network to identify conserved occurrences of a query network module.

Chapter XVIII, "*Module Finding Approaches for Protein Interaction Networks*", will review the computational approaches for finding functional modules from protein interaction networks. The chapter will focus on those module finding tools implemented in freely available software packages. As in the other chapters in this book, this chapter will also discuss the key future trends and promising research directions with potential implications for clinical research.

REFERENCES

Bader, G. D., & Hogue, C. W. (2002). Analyzing yeast protein-protein interaction data obtained from different sources. *Nat Biotechnol, 20*(10), 991-997.

Bader, J. S., Chaudhuri, A., Rothberg, J. M., & Chant, J. (2004). Gaining confidence in high-throughput protein interaction networks. *Nat Biotechnol, 22*(1), 78-85.

Barabasi, A. L., & Oltvai, Z. N. (2004). Network biology: understanding the cell's functional organization. *Nat Rev Genet, 5*(2), 101-113.

Ben-Hur, A., & Noble, W. (2005). Kernel methods for predicting protein-protein interactions. *Bioinformatics, 21 Suppl.*

Fields, S., & Song, O. (1989). A novel genetic system to detect protein-protein interactions. *Nature, 340*(6230), 245-246.

Gentleman, R., & Huber, W. (2007). Making the most of high-throughput protein-interaction data. *Genome Biol, 8*(10), 112.

Ghazalpour, A., Doss, S., Zhang, B., Wang, S., & Plaisier, C. (2006). *Integrating Genetic and Network Analysis to Characterize Genes Related to Mouse Weight.*

Hart, G. T., Ramani, A. K., & Marcotte, E. M. (2006). How complete are current yeast and human protein-interaction networks? *Genome Biol, 7*(11), 120.

Ideker, T., & Sharan, R. (2008). Protein networks in disease. *Genome Res, 18*(4), 644-652.

Jansen, R., Yu, H., Greenbaum, D., Kluger, Y., Krogan, N. J., Chung, S., et al. (2003). A Bayesian networks approach for predicting protein-protein interactions from genomic data. *Science, 302*(5644), 449-453.

Kelley, R., & Ideker, T. (2005). Systematic interpretation of genetic interactions using protein networks. *Nature Biotechnology, 23*(5), 561-566.

Lalonde, S., Ehrhardt, D. W., Loque, D., Chen, J., Rhee, S. Y., & Frommer, W. B. (2008). Molecular and cellular approaches for the detection of protein-protein interactions: latest techniques and current limitations. *Plant J, 53*(4), 610-635.

Mootha, V. K., Lepage, P., Miller, K., Bunkenborg, J., Reich, M., & Hjerrild, M. (2003). Identification of a gene causing human cytochrome c oxidase deficiency by integrative genomics. *Proc Natl Acad Sci U S A, 100*(2), 605-610.

Puig, O., Caspary, F., Rigaut, G., Rutz, B., Bouveret, E., & Bragado-Nilsson, E. (2001). The tandem affinity purification (TAP) method: A general procedure of protein complex purification. *Methods, 24*(3), 218-229.

Rigaut, G., Shevchenko, A., Rutz, B., Wilm, M., Mann, M., & Séraphin, B. (1999). A generic protein purification method for protein complex characterization and proteome exploration. *Nat Biotechnol, 17*(10), 1030-1032.

Sprinzak, E., Sattath, S., & Margalit, H. (2003). How reliable are experimental protein-protein interaction data? *J Mol Biol, 327*(5), 919-923.

Tarassov, K., Messier, V., Landry, C. R., Radinovic, S., Molina, M. M., & Shames, I. (2008). An in vivo map of the yeast protein interactome. *Science, 320*(5882), 1465-1470.

Von Mering, C., Krause, R., Snel, B., Cornell, M., Oliver, S. G., & Fields, S. (2002). Comparative assessment of large-scale data sets of protein-protein interactions. *Nature, 417*(6887), 399-403.

Acknowledgment

The editors would like to acknowledge the help of all involved in the collation and review process of the book, without whose support the project could not have been satisfactorily completed. Deep appreciation and gratitude is due to Institute for Infocomm Research, Singapore, for ongoing sponsorship in terms of generous allocation of Internet, hardware and software resources and other editorial support services for coordination of this year-long project.

Most of the authors of chapters included in this book also served as referees for chapters written by other authors. Additionally, some domain experts also offered the help to conduct the constructive and comprehensive double-blind review process — we required at least three reviewers for each chapter to provide a thorough review of all submitted material. This plays a very significant role in choosing the best materials for the final publication. Our heartfelt thanks go to all those who provided the rigorous reviews. Some of the reviewers must be mentioned as their reviews have clearly set the benchmark. Reviewers who provided the most comprehensive, critical and constructive comments include: Aidong Zhang of the State University of New York; Anna Tramontano of Sapienza University; Charlie Hodgman of University of Nottingham; Clara Pizzuti of ICAR Consiglio Nazionale delle Ricerche; Daisuke Kihara of Purdue University; Hai Quan Li of Noble Foundation; Hiroshi Mamitsuka of Kyoto University; Jerzy Tiuryn of Warsaw University; Jin Chen of Stanford University; Joaquin Dopazo of Centro de Investigación Príncipe Felipe; Jung-jae Kim of European Bioinformatics Institute; Raymond Wan of Kyoto University; Tatsuya Akutsu of Kyoto University; Tero Aittokallio of University of Turku; Valeria Fionda of Università della Calabria; and Xiaoyue Zhao of Cold Spring Harbor Laboratory. We also want to acknowledge the great support from many of our colleagues in Data Mining Department in the review process.

We gratefully appreciate our Editorial Advisory Board (EAB) members who have provided us useful suggestions and comments. They also helped recommend suitable leading experts to contribute various key chapters in this book.

Special thanks also go to the publishing team at IGI Global, whose contributions throughout the whole process from inception of the initial idea to final publication have been invaluable. In particular to Julia Mosemann, who continuously prodded via e-mail for keeping the project on schedule and clear our doubts and questions promptly.

Last but not least, Xiao-Li would like to thank his dear daughter Li Qi and wife Liu Futao, for their unfailing support and encouragement during the months it took to give birth to this book. See-Kiong would like to thank Prof. Masaru Tomita, his former advisor from Carnegie Mellon University, who has introduced him to the wonderful world of Bioinformatics.

In closing, we wish to thank all of the authors for their insights and excellent contributions to this book.

Editor,
Xiao-Li Li and See-Kiong Ng
Institute for Infocomm Research, Singapore
September 2008

Section I
Introduction

In Section I of the book, we will provide the two introductory chapters. Chapter I aims to provide a biological primer on protein-protein interactions for those readers who are computer scientists and who may not have adequate knowledge in biology. Chapter II is a primer on the key concepts and pattern mining algorithms in data mining useful for the readers who are biologists.

Chapter I
Molecular Biology of Protein–Protein Interactions for Computer Scientists

Christian Schönbach
Nanyang Technological University, Singapore

ABSTRACT

Advances in protein-protein interaction (PPI) detection technology and computational analysis methods have produced numerous PPI networks, whose completeness appears to depend on the extent of data derived from different PPI assay methods and the complexity of the studied organism. Despite the partial nature of human PPI networks, computational data integration and analyses helped to elucidate new interactions and disease pathways. The success of computational analyses considerably depends on PPI data understanding. Exploration of the data and verification of their quality requires basic knowledge of the molecular biology of PPIs and familiarity with the assay methods used to detect PPIs. Both topics are reviewed in this chapter. After introducing various types of PPIs the principles of selected PPI assays are explained and their limitations discussed. Case studies of the Wnt signaling pathway and splice regulation demonstrate some of the challenges and opportunities that arise from assaying and analyzing PPIs. The chapter is concluded with an extrapolation to human systems biology that offers a glimpse into the future of PPI networks.

INTRODUCTION

The function and functioning of a cell from differentiation to death depends on information flow among its molecular and cellular components. Processes enabling intra- and extracellular information flow include PPIs, signal transduction, and protein transport, among others. Methods to identify PPIs (reviewed by Berggård et al., 2007; Lalonde et al., 2008) *in vitro* and *in vivo* have produced thousands of interaction pairs that are presented in mostly static interactomes. For the most part, full understanding of the biological significance of interactions in cellular context, particularly in human and mammalian model organisms remains a challenge. Owing to the intrinsic properties of PPI assays none of the interactomes are complete. The best sampled and curated (Reguly et al., 2006) interactome of yeast is thought to represent about half of the complete PPI set (Hart et al., 2006).

Each PPI assay has different biases for soluble or membrane proteins, protein abundance, affinity range, and post-translational modifications that define stable or transient interactions (Charbonnier et al., 2008). Coordination and standardization (Hermjakob et al., 2004; Kerrien et al., 2007) of protein interaction data has improved analyses and integration, but a reference human interactome that can be used for benchmarking is not yet available. In the coming five years results of the Human Interactome Project (2008) and the Human Proteome Organization (HUPO) initiative to map the human proteome (Pearson, 2008) are expected to improve the molecular network-based understanding of biological processes and human diseases. However, successful development of diagnostics and therapeutics that target disease-relevant PPIs will depend on the progress of obtaining quantitative and dynamic PPI data across different tissue cells and their integration with gene expression, structural, and metabolic data. In view of the following chapters this review will give a brief tour of PPI interaction types and detection methods illustrated by biological interesting examples. Caveats and aspects of biological properties that are of interest for computational analyses and predictions of PPIs are highlighted.

BACKGROUND: PPI TYPES

PPI as any other molecular interaction must meet at least two conditions: sufficient concentration of interactants and a favorable difference of free energy upon interaction. Both parameters are dynamic and depend on the cellular environment. The concentration of a protein may depend on the level of gene expression, translation efficacy, its stability or turnover, and availability in the same cellular compartment as the potential interacting partner. The level of free of energy difference depends on the sum free energy differences derived from atomic level interactions such as hydrogen bond formation, van der Waals interactions and ionic bond formation. Corresponding to the difference of total free energy the resulting interactions are broadly classified into stable, strong transient, or weak transient interactions. Stable interactions are often formed by homodimers, for example RNA polymerase II. Many heterodimeric interactions, for instance the binding of cell division cycle 42 to Wiskott-Aldrich syndrome protein (Rudolph et al., 1998) are transient, but relatively strong. Interactions of cell surface receptors or cargo proteins are weak and transient (e.g., carbohydrate mediated binding of lectin mannose binding 1 to cathepsin C (Nyfeler et al., 2005)).

The detection of PPIs depends on the intrinsic properties and sensitivities of the assays which are reviewed in detail by Berggård (2007) and Lalonde (2008). Briefly, medium throughput surface plasmon resonance assays (e.g., BIACore™) are among the most sensitive *in vitro* assay systems that capture also the kinetics of PPIs. The popular *in vivo* yeast two-hybrid (Y2H) assay is high throughput, but not quantitative and biased towards the detection of stronger binary interactions between protein domains.

PPI DETECTION

Principles of Y2H Method

Appropriate use of wet lab data in computational analyses requires at least a basic understanding of the method(s) used to generate the data. Here, the principles of two commonly used large-scale PPI detection methods, Y2H and tandem affinity purification-mass spectrometry (TAP-MS) are introduced and their limitations discussed.

Almost 20 years ago Field and Song (1989) published a genetic system to detect PPIs that is now termed Y2H. The authors utilized the transcriptional activation properties of GAL4, which depend on a DNA binding domain (DBD) and an activation domain (AD). GAL4 binds with its DBD to specific upstream sites (UAS promoter) of genes encoding enzymes of the galactose metabolism (Fig. 1A). At the same time, AD facilitates the transcription of the downstream genes by RNA polymerase II. When DBD and AD were physically separated transcriptional activation was abolished, whereas reconstitution of DBD and AD enabled transcription. Indeed, when Fields and Song fused DBD to SNF1 and AD to SNF4, the interaction of SNF1 and SNF4 reconstituted the transcriptional activity of GAL4. In Y2H assays (Fig. 1B) the hybrid protein DBD-X is called bait. The fusion of AD with target Y is termed prey. The construct expressing the prey is introduced into the bait protein-expressing yeast strain by co-transfection or mating. In the latter case bait and prey are expressed in two different haploid yeast strains. Upon mating the cells become diploid. If protein X of the bait interacts with protein Y of the prey in the diploid cells transcription of the downstream reporter gene will be activated. Contemporary Y2H assays use multiple reporter genes that allow for both growth and color-based read-outs. When reporter genes for growth are activated and transcribed the yeast cell will grow on agar plates lacking histidine and uracil, among other nutrients. The color-based read-out of interactions utilizes the classical *lacZ* reporter system which results in blue colonies upon incubation with the beta-galactosidase substrate X-Gal (5-bromo-4-chloro-3-indolyl-beta-D-galactopyranoside).

Figure 1. Principle of yeast GAL4 transcription activation (A) and Y2H assay (B). Fig. 1A. Binding of GAL4 DBD to the target promoter enables GAL4 AD to activate transcription of the GAL4-regulated gene by RNA polymerase II (left side). Separation of the DBD and AD domains as shown on the right side prevents transcription activation. Fig. 1B. PPI detection using the Y2H method. The bait plasmid expresses upon transfection into haploid yeast the hybrid protein DBD-protein X. Hybrid AD-protein Y is expressed by the prey plasmid. After mating of the haploid yeast strains, the bait is used to sample interactions of protein X with other proteins (e.g., protein Y). If X interacts with Y, transcription of the reporter gene is activated. Here, the read-out product of the reporter is associated with yeast growth. If protein X does not interact with protein Y the reporter gene remains silent and the yeast cell will die.

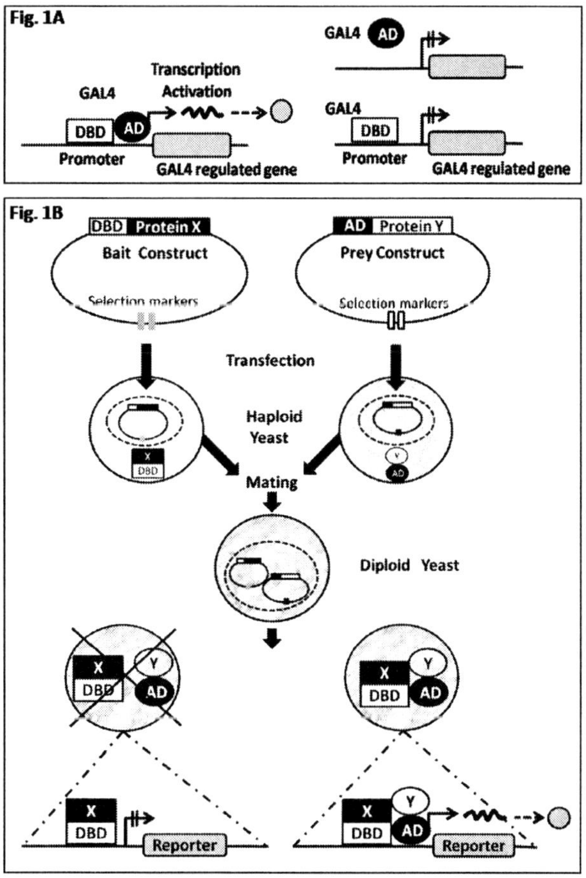

The identity of the interacting proteins is established by isolating plasmid DNA from the yeast colonies followed by sequencing of their inserts. The entire procedure of bait and prey library construction, screening until establishing the identity of the interacting protein takes at least three weeks.

Commercially available Y2H assays include various modifications that aid large-scale library screening and aim to reduce false positives. Despite improved reliabilities of Y2H-derived PPIs the legacy of older high throughput dataset remains. A study by Sprinzak et al. (2003) indicated 50% reliability of PPIs when considering known functions of the interacting proteins and their subcellular locations.

The detection of binary interactions without direct quantitative and subcellular location information is a methodological limitation common to Y2H assays that necessitates the use of complementary methods. For technical details and potential problems of prey and bait library screening, interested readers are advised to refer to the Y2H assay units in Current Protocols in Molecular Biology (Golemis et al., 2008).

Principles of TAP-MS Method

TAP-MS (Fig. 2A) was developed by Seraphin (Rigaut et al., 1999; Puig et al., 2001) at EMBL (European Molecular Biology Laboratory). The method extends the detection of binary PPIs to multimeric interactions of protein complexes. The isolation of *in vivo* protein complexes is assisted by tagged target proteins (Fig. 2B). Since

Figure 2A. TAP-MS method. B. TAP tag and TAP-tagged protein complex purification

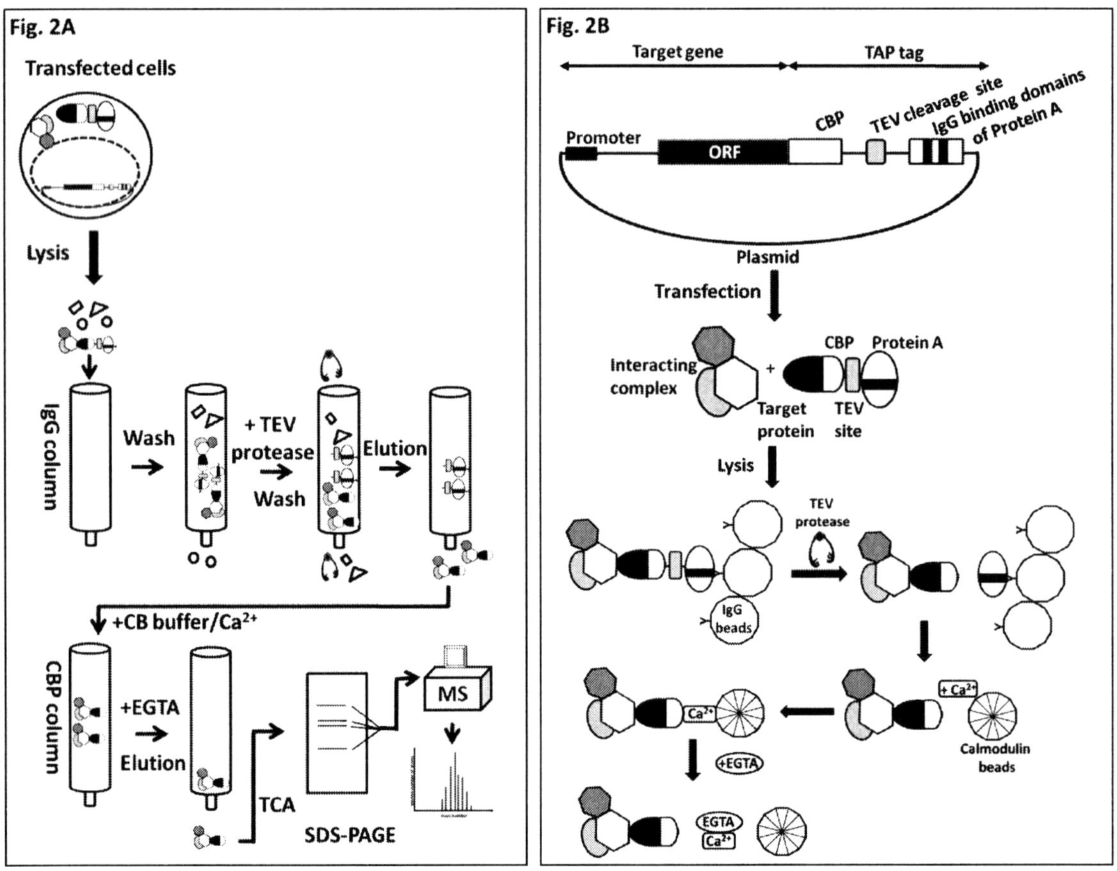

TAP-MS can resolve the question whether binary interactions of the type A-B and A-C constitute independent PPI pairs or form a complex A-B-C, the method is ideally suited to complement Y2H assays. The following overview draws from the technical details published at a dedicated home page (The TAP method home page, 2002) and two method articles (Rigaut et al., 1999; Puig et al., 2001).

A typical TAP tag consists of the calmodulin-binding protein (CBP), a tobacco etch virus (TEV) protease cleavage site, and the immunoglobulin G-binding (IgG) units of protein A (ProtA) of *Staphylococcus aureus* (Fig. 2B). The binding properties of ProtA to IgG-coupled and CBP to calmodulin-coupled sepharose beads permit the capture and purification of low abundant fusion protein complexes by affinity column chromatography.

The TAP-tagged target protein is constructed by ligating the open reading frame (ORF) to the tag construct. The plasmid is then transfected into yeast which expresses the tagged target protein under control of its native target gene promoter. Alternatively, the TAP tag fusion can be constructed by genomic-based PCR (polymerase chain reaction). Latter method is applied when hundreds or thousands of TAP-tagged target proteins are employed to sample PPIs. Most target proteins are tagged at the C-terminus, using CBP-TEV-ProtA (Fig 2B). However, target proteins that contain at the C-terminus a functional site (e.g., peroxisome targeting signal 1) require N-terminal tagging. In this case, the order of the tag components would be ProtA-TEV-CBP. Modified TAP tags for application in mammalian cells are biotinylated to allow high-affinity purification using streptavidin (Drakas et al., 2005)

The transfected cells are lysed under mild conditions to increase the probability of retaining a protein complex in its native state. When the lysate is applied to the IgG sepharose column, protein complexes (dimers and multimers) containing the TAP-tagged target are specifically bound by IgG (Fig 2A and B). Unspecific bound proteins and any contaminating material are removed during the washing steps. Upon adding TEV protease the retained protein complexes are specifically cleaved at the TEV recognition site and released from the IgG-bound ProtA tag component. The eluate is then transferred to calmodulin-coated beads in the presence of calcium (Ca^{2+}) containing calmodulin binding (CB) buffer. Calmodulin is an EF-hand helix-turn-helix domain-containing Ca^{2+}-binding protein that binds with high affinity to the CPB-tagged target protein complexes. During the second affinity chromatography the TEV protease and other contaminants are removed. After washing, the calmodulin-bound CBP-tagged protein complexes are released by chelating free Ca^{2+} with ethylene glycol tetra-acetic acid (EGTA).

Depending on the objective of the investigators, the next steps may include direct submission of the eluted protein complexes to MS analysis or functional characterization of a specific complex. In the former case, the identity of the protein complex components can be established, but not quantified. To determine the stoichiometry of the protein complex subunits the eluted material is concentrated in a trichloroacetic acid (TCA) precipitation step and separated by using sodium dodecyl sulfate polyacrylamide gel electrophoresis (SDS-PAGE). The separated subunits are visualized by staining and isolated from the gel. After a digestion step the material is subjected to MS. Matrix-assisted laser desorption/ionisation-time of flight (MALDI-TOF) MS (Vestal & Campbell, 2005) and nanoelectrospray ionization (ESI) MS (Valaskovic & Kelleher, 2002) allow to establish protein identities from sample concentrations in picomol (10^{-12}) to femtomol (10^{-15}) ranges.

Although TAP-MS appears to detect in yeast three times as many known interactions as Y2H assays (Ho et al., 2002) with an error rate of 15% (Gingras et al., 2005), the method is not without problems. The multiple washing and purification steps are prone to eliminate transient low affinity protein complexes.

No Assay is Perfect

In yeast, various high throughput methods using Y2H (Ito et al., 2001) and TAP-MS (Ho et al., 2002; Gavin et al., 2002) made it possible to generate comprehensive PPI and multiprotein complex data that led upon integration with existing microarray (Spellman et al., 1998; Cho et al., 1998) and protein location data (Christie et al., 2004) to the first dynamic protein complex network model of the eukaryotic cell cycle (de Lichtenberg et al., 2005). Interestingly, two missed protein complexes (anaphase-promoting complex and Skp1p/Cullin/F-box protein complex) were added back for network analysis purposes. An estimate by Gavin (2002) indicated that TAP-MS captures at least 43.5% of known yeast protein complex subunits. The estimate excludes homodimeric (self-interacting) protein interactions, which TAP-MS is unable to detect. Despite the good fit of the model with known yeast cell cycle biology, Y2H and TAP-MS are not well suited to detect very weak transient protein complexes. In addition, in TAP-MS the subcellular location of complexes is lost due to the *in vitro* purification of whole-cell lysates. If the strategy is applied to a less-studied cellular process without time-consuming preparation of subcellular fractionated lysates, false negatives may not be rescued by integrating external biological information.

Recently, the authors of the first yeast *in vivo* interactome map (Tarassov et al., 2008) have circumvented caveats of traditional Y2H and *in vitro* high throughput PPI detection: 1) loss of weak transient interactions, 2) loss of protein location information, 3) loss of post-translational modification, and 4) bias against membrane proteins. The high throughput method called protein-fragment complementation assay (PCA) is based on detecting reconstituted reporter activity upon interaction of two proteins that are fused to mouse dihydrofolate reductase (DHFR) reporter fragments. The majority of 2,770 high-confidence interactions were reported as new, whereas the overlap with previously reported interactions was 16-43%, if detected by both PCA and Y2H, or TAP-MS. Alternative methods (e.g., split-TEV) using other reporter systems were reviewed by Kerppola (2006). Although the PPIs are detected *in vivo*, the PCA system does not resemble *in situ* conditions. The transfection of cells with expression constructs affects the abundance of proteins. Indeed, expression of DHFR PCA proteins was about twice the level of control proteins. Higher abundance of proteins will increase the sensitivity of PPI detection, but may also cause a bias towards interactions that do not occur under physiological conditions.

Elusive PPIs

On the other hand, weak dynamic interactions and interactions of inducible proteins (e.g., proteases) are easily missed, if they are below the sensitivity threshold of assays. For example the serine-cysteine protease Tysnd1 localizes from the cytosol to peroxisomes where it specifically recognizes and cleaves enzymes of the peroxisomal beta-oxidation pathway (Kurochkin et al., 2007). The resulting protein fragments may interact with each other to form a beta-oxidation enhancing complex. A general peroxisome enzyme-processing protein like Tysnd1 was proposed for decades, but protein-based methods were unable to detect and characterize Tysnd1, unless it was overexpressed or drug-induced. The endogenous protein was only detected with a specific antibody or by using the highly sensitive multiplexed isobaric tagging tandem MS (iTRAQ™ MS/MS) (Islinger et al., 2007). There will be many more proteins, including splice variants, falling into the category of inducible, low abundance proteins with transient interactions.

In-cell nuclear magnetic resonance (NMR) or structural interactions (STINT) NMR (Burz et al., 2006a; Selenko & Wagner, 2006) may have the potential to detect elusive PPIs under almost natural conditions, once the problems with expression in eukaryotic cells have been solved. The method allows to measure structural changes of the interaction surface, but is currently only applicable for proteins overexpressed in *E. coli.* (Burz et al., 2006b).

PPI or not PPI?

As there is no perfect PPI assay, it is important to know an assay's limitations and perform independent confirmation experiments for novel interactions. Two widely applied methods to confirm direct PPIs, detected by high throughput assays, are *in vivo* co-immunoprecipitation (co-IP) and *in vitro* glutathione-S transferase (GST) pull-down assays. Depending on the affinity of the interaction these assays may not be able to distinguish between non-specific interactions and weak interactions, as pointed out by Mackay et al. (2007). Weak interactions are of biological significance and can be confirmed by other methods. Isothermal titration microcalorimetry (Doyle, 2001) is one alternative method. Another option, proximity ligation *in situ* assay, (Söderberg et al., 2006) allows the detection of endogenous PPI in cells and tissues, provided a specific antibody is available. In consideration of potential contradictory findings, the authors suggested to include "negative interactions" (by co-IP and GST-pulldown) as comments in PPI databases. Other discrepancies between *in vivo* and *in vitro* PPI assays will occur when a protein binds to a complex, whose formation is dependent on binding to DNA (e.g. transcriptional activators) (McNabb & Pinto, 2005) or regulated by phosphorylation.

BIOLOGY OF PPI

What Connects PPI and Splicing?

Reversible phosphorylation mediated PPIs of splice regulatory (SR) proteins are an interesting example of signal transduction-dependent splicing. Kinases phosphorylate SR proteins in the cytoplasm. Depending on the phosphorylation status of SR proteins their subcellular location changes and binding to other proteins (e.g. U1 snRNP) involved in spliceosome formation increases or decreases (reviewed by Stamm, 2008; Hagopian et al., 2008). The mechanism is linked to splice site selection and regulation of alternative splicing, as demonstrated for exon inclusion in protein kinase C beta 1 upon insulin induced phosphorylation of SR proteins via AKT1 (Ghosh et al., 2007). Alterations in splicing patterns were also observed in gene expression profiling studies of various cancer tissue cells (Gardina et al., 2006; Soreq et al., 2008), but were so far not analysed in context of PPIs. The parallel application of a phosphorylation sensitive PPI assay (e.g., split-TEV) and RNA-Seq short-read high-throughput sequencing (Mortazavi et al., 2008) in sampling normal cells and cells of different cancer stages would allow the identification of relationships between cancer-specific splice variants, splice regulatory proteins, phosphatases and kinases on both expression and protein network level. The significance of detected PPIs towards cancer splicing patterns and their potential as drug targets could be tested by specific inhibition of SR-protein binding or phosphorylation.

Making Sense of PPIs in Wnt Signaling

Many projects that aim to identify disease-associated pathways generate PPI data around one or two disease proteins to create after integration of gene expression and other complementary data subnetworks. For example the TNF-alpha/NF-kappa B network of Bouwmeester and co-workers (2004) represents a subnetwork of an inflammation process. Various computational approaches rely on global draft human PPI networks (Stelzl et al., 2005; Rual et al., 2005) to predict disease associations utilizing existing pathway information and microarray data to filter out PPIs, that are unlikely to occur at a given time or condition. Since PPIs in disease networks are subject of Chapter XIV and covered by two recent reviews (Ideker & Sharan, 2008; Pache et al., 2008) only three PPIs, that are part of the Wnt signaling pathway (Fig. 3A) are discussed to demonstrate surprises and opportunities of reanalyzing PPIs of a published network.

The potential tumor suppressor ANP32A is one of seven disease proteins with high confidence interactions in the Wnt signaling pathway that were reported by Stelzl et al. (2005). Of interest are the interaction pairs ANP32A-AXIN1 (new), AXIN1-PPP2CA (known) and ANP32-PPP2CA (known). The latter PPI was extracted by the authors from a previous version of HPRD database (Mishra et al., 2006). Interestingly, the binding of ANP32 to PPP2CA is not anymore listed in the current release 7 of HPRD and cannot be found in MINT (Chatr-aryamontri et al., 2007), IntAct (Kerrien et al., 2007) or BIND (Alfarano et al., 2005) databases (Fig. 3B). However, literature supports binding of ANP32A to PPP2CA, which results in the inhibition of PPP2CA (Li et al., 1996). The interaction is mediated by the N-terminal LRR domain of ANP32A and C subunit of PPP2CA (Chen et al., 2008; de Chiara et al., 2008). Mouse Anp32a also binds to Atx1 and Clip170 proteins, which are associated with spinocerebellar ataxia type 1 and Hodgkin's disease, respectively. Both diseases are known for the involvement of Wnt signaling.

Fig. 3A shows PPIs of the cytoplasmic located beta-catenin scaffold. Wnt signaling activates DVL1 through CSNK1E which increases binding of FRAT1 to GSK3B, thereby decreasing AXIN1-PPP2CA affinity and phophorylation of CTNNB1. As a result CTNNB1 accumulates and locates to the nucleus where it activates transcription factors (Luo & Lin, 2004). Suppression of Wnt signaling was observed when the acetyltransferase inhibitory domain-containing fragment of ANP32A bound to AXIN1 (Stelzl et al., 2005). On the other hand, binding of PPP2CA to AXIN1 enhances the degradation of beta-catenin in absence of Wnt signaling. Since the binding of ANP32A to PPP2CA appears to cause non-competitive inhibition of PPP2CA (de Chiara et al., 2008), it is tempting to propose that ANP32A binding of PPP2CA modulates AXIN1-mediated beta-catenin phosphorylation and

Figure 3A. Scaffold for beta-catenin degradation in Wnt signaling pathway (modified after KEGG PATHWAY Wnt signaling pathway (2008)). Details of Wnt signaling are explained in the text. B. Twenty PPIs reported for ANP32A. A.The proteins shown in beta-catenin scaffold are CSNK1E: casein kinase 1, epsilon; DVL1: dishevelled, dsh homolog 1 (Drosophila); SENP2: SUMO1/sentrin/SMT3 specific peptidase 2; FRAT1: frequently rearranged in advanced T-cell lymphomas; GSK3B: glycogen synthase kinase 3 beta; PPP2CA: protein phosphatase 2 (formerly 2A), catalytic subunit, alpha; APC2: adenomatosis polyposis coli 2; CSNK1A1L: casein kinase 1, alpha 1-like isoform; AXIN1: axin 1; ANP32A: Acidic leucine-rich nuclear phosphoprotein 32 family member A; CTNNB1: catenin (cadherin-associated protein), beta 1. +p indicates increased phosphorylation. B. Symbols in grey-shaded boxes indicate PPIs detected by Stelzl et al. (2005). Nineteen ANP32A interacting proteins that were retrieved from PPI databases are: APEX1: APEX nuclease (multifunctional DNA repair enzyme) 1; ATX1: ataxin 1; CENTA1: centaurin alpha 1; ELAVL1: embryonic lethal abnormal vision Drosophila-like 1; ESR1: estrogen receptor 1; HSPB3: heat shock 27kDa protein 3; KPNA1: karyopherin alpha 1 (importin alpha 1); KPNA5: karyopherin alpha 5 (importin alpha 6); KPNA6: karyopherin alpha 6 (importin alpha 7); MAP1B: microtubule-associated protein 1B; NSFL1C: NSFL1 (p97) cofactor (p47); PTPRC: protein tyrosine phosphatase receptor type C; FLJ46889: highly similar to Homo sapiens likely ortholog of rat p47; RARA: retinoic acid receptor alpha; SET: SET nuclear oncogene; THRB: thyroid hormone receptor beta (avian erythroblastic leukemia viral (v-erb-a) oncogene homolog 2); TRAF1: TNF receptor-associated factor 1; XPO1: exportin 1.

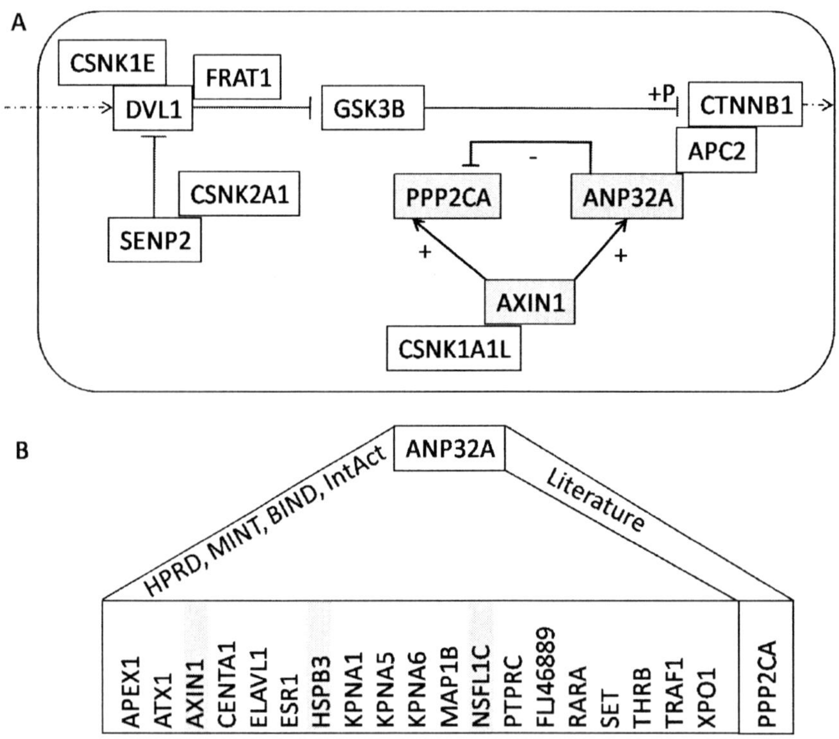

degradation in both presence and absence of Wnt signaling. The mechanism of differential interactions, which may depend on conformational changes in both ANP32A and PPP2CA remains to be elucidated.

The fortuitous reanalysis of two ANP32A interactions while preparing this chapter yielded one new hypothesis and three problems that will be amplified by larger analyses of potential disease-associated PPIs.

1. The Y2H assay detected only three, presumably highly stable domain-domain interaction out of twenty known interactions for ANP32A (Fig.3 B). While the missing interactions relevant for a chosen pathway can easily be extracted from PPI databases, one should keep in mind that databases have their own limitations.
2. The vanishing of the ANP32A-PPP2CA interaction from HPRD might be related to curation or confusing naming of the protein in combination with a name change. However integration of data upon reading full-text articles that were extracted by PubMed searching with "PPP2CA interaction binding " and "ANP32A interaction binding" led to the new hypothesis of ANP32A as potential modulator of the axin scaffold for beta-catenin.
3. The FLJ46889-ANP32A and mouse Clip170-Anp32a interactions will require further computational and biological experiments to determine their function and the role of CLIP170 in human Wnt signaling pathway.

WHAT'S NEXT? FROM PPI TO A VIRTUAL HUMAN

Protein networks are dynamic and not isolated entities in and across cellular systems. The Human Interactome Project (2008) and Human Proteome Map initiative (Pearson, 2008) are expected to deliver a reference interactome and proteome that will spawn further studies to capture the dynamics of direct and indirect interactions of proteins with lipids, RNA, DNA or metabolites. Linking of the PPI network with other networks, that regulate as a whole the function and fate of cells will eventually enable a systems-level understanding of cell biology. A coordinated long term strategy to generate, test and simulate networks, as proposed at the JST (Japan Science and Technology Agency) and BBSRC (Biotechnology and Biological Sciences Research Council, UK) sponsored workshop Future Challenges for Systems Biology (FCSB 2008; 2008), would be an excellent framework to overcome current limitations in the field of protein-protein interactions. The workshop concluded with the announcement of a grand challenge (The Tokyo Declaration; 2008) to establish over the next 30 years a molecule-based, multi-scale, computational model of the human ('virtual human'). Appropriate for the scope of the challenge, it was featured in News of the journal Nature (Systems biologists; 2008).

The 'virtual human' is supposed to provide a framework to delineate and simulate the networks of molecular interactions on DNA, RNA, protein and metabolite levels associated with biological processes in both healthy and disease states (Jones, 2008). If the initiative is to succeed, it will be vital to secure broad interdisciplinary support and perhaps a dedicated international funding structure that is amenable to the challenges of technology development, numerous global and local-scale experiments up to clinical trials. A 'virtual human' should be a boon for the pharmaceutical industry where drug candidate to market attrition rates are estimated to be as high as 89% (Kola & Landis, 2004).

ACKNOWLEDGMENT

I thank an anonymous referee for helpful feedback. This work was supported in part by a grant from Nanyang Technological University.

REFERENCES

Alfarano, C., Andrade, C. E., Anthony, K., Bahroos, N., Bajec, M., Bantoft, K., et al. (2005). The Biomolecular Interaction Network Database and related tools 2005 update. *Nucleic Acids Res, 33*(Database issue), D418 424.

Berggård, T., Linse, S., & James, P. (2007). Methods for the detection and analysis of protein-protein interactions. *Proteomics, 7*(16), 2833-2842.

Bouwmeester, T., Bauch, A., Ruffner, H., Angrand, P. O., Bergamini, G., Croughton, K., et al. (2004). A physical and functional map of the human TNF-alpha/NF-kappa B signal transduction pathway. *Nat Cell Biol, 6*(2), 97-105.

Burz, D. S., Dutta, K., Cowburn, D., & Shekhtman, A. (2006a). Mapping structural interactions using in-cell NMR spectroscopy (STINT-NMR). *Nat Methods, 3*(2), 91-93.

Burz, D. S., Dutta, K., Cowburn, D., & Shekhtman, A. (2006b). In-cell NMR for protein-protein interactions (STINT-NMR). *Nat Protoc, 1*(1), 146-152.

Chatr-aryamontri, A., Ceol, A., Palazzi, L. M., Nardelli, G., Schneider, M. V., Castagnoli, L., et al. (2007). MINT: The Molecular INTeraction database. *Nucleic Acids Res, 35*(Database issue), D572-574.

Chen, S., Li, B., Grundke-Iqbal, I., & Iqbal, K. (2008). I1PP2A affects tau phosphorylation via association with the catalytic subunit of protein phosphatase 2A. *J Biol Chem, 283*(16), 10513-10521.

Cho, R. J., Campbell, M. J., Winzeler, E. A., Steinmetz, L., Conway, A., Wodicka, L., et al. (1998). A genome-wide transcriptional analysis of the mitotic cell cycle. *Mol Cell, 2*(1), 65-73.

Christie, K. R., Weng, S., Balakrishnan, R., Costanzo, M. C., Dolinski, K., Dwight, S. S., et al. (2004). Saccharomyces Genome Database (SGD) provides tools to identify and analyze sequences from Saccharomyces cerevisiae and related sequences from other organisms. *Nucleic Acids Res, 32*(Database issue), D311-314.

de Chiara, C., Menon, R. P., & Pastore, A. (2008). Structural bases for recognition of Anp32/LANP proteins. *FEBS J, 275*(10), 2548-2560.

de Lichtenberg, U., Jensen, L. J., Brunak, S., & Bork, P. (2005). Dynamic complex formation during the yeast cell cycle. *Science, 307*(5710), 724-727.

Doyle, M. L. (2001). Titration microcalorimetry. *Curr Protoc Protein Sci, Chapter 20*, Unit20 24.

Drakas, R., Prisco, M., & Baserga, R. (2005). A modified tandem affinity purification tag technique for the purification of protein complexes in mammalian cells. *Proteomics 5*(1), 132-137.

FCSB 2008. Retrieved August 15, 2008, from http://www.systems-biology.org/~myukiko/FCSB2008/doku.php.

Fields, S., & Song, O. (1989). A novel genetic system to detect protein-protein interactions. *Nature 340*(6230), 245-246.

Gardina, P. J., Clark, T. A., Shimada, B., Staples, M. K., Yang, Q., Veitch, J., et al. (2006). Alternative splicing and differential gene expression in colon cancer detected by a whole genome exon array. *BMC Genomics, 7*, 325.

Gavin, A. C., Bosche, M., Krause, R., Grandi, P., Marzioch, M., Bauer, A., et al. (2002). Functional organization of the yeast proteome by systematic analysis of protein complexes. *Nature, 415*(6868), 141-147.

Ghosh, N., Patel, N., Jiang, K., Watson, J. E., Cheng, J., Chalfant, C. E., et al. (2007). Ceramide-activated protein phosphatase involvement in insulin resistance via Akt, serine/arginine-rich protein 40, and ribonucleic acid splicing in L6 skeletal muscle cells. *Endocrinology, 148*(3), 1359-1366.

Gingras, A. C., Gstaiger, M., Raught, B., & Aebersold, R. (2005). Analysis of protein complexes using mass spectrometry. *Nat Rev Mol Cell Biol., 8*(8), 645-654.

Golemis, G. A., Serebriiskii, I., Finley Jr., R. L., Kolonin, M. G., Gyuris, J., & Brent R. (2008). Interaction trap/two-hybrid system to identify interacting proteins. In F.M. Ausubel, R. Brent, R. E. Kingston, D. D. Moore, J. G. Seidman, J. A. Smith, et al. *Current Protocols in Molecular Biology* (pp. 20.1.1-20.1.35). New York: John Wiley & Sons, Inc.

Hagopian, J. C., Ma, C. T., Meade, B. R., Albuquerque, C. P., Ngo, J. C., Ghosh, G., et al. (2008). Adaptable Molecular Interactions Guide Phosphorylation of the SR Protein ASF/SF2 by SRPK1. *J Mol Bio.*

Hart, G. T., Ramani, A. K., & Marcotte, E. M. (2006). How complete are current yeast and human protein-interaction networks? *Genome Biol, 7*(11), 120.

Hermjakob, H., Montecchi-Palazzi, L., Bader, G., Wojcik, J., Salwinski, L., Ceol, A., et al. (2004). The HUPO PSI's molecular interaction format--a community standard for the representation of protein interaction data. *Nat Biotechnol, 22*(2), 177-183.

Ho, Y., Gruhler, A., Heilbut, A., Bader, G. D., Moore, L., Adams, S. L., et al. (2002). Systematic identification of protein complexes in Saccharomyces cerevisiae by mass spectrometry. *Nature, 415*(6868), 180-183.

Human Interactome Project. (2008). Retrieved August 15, 2008, from http://interactome.dfci.harvard.edu/H_sapiens/

Ideker, T., & Sharan, R. (2008). Protein networks in disease. *Genome Res, 18*(4), 644-652.

Islinger, M., Lüers, G.H., Li, K.W., Loos, M.,& Völkl, A. (2007). Rat liver peroxisomes after fibrate treatment. A survey using quantitative mass spectrometry. *J Biol Chem. 282*(32), 23055-23069.

Ito, T., Chiba, T., Ozawa, R., Yoshida, M., Hattori, M., & Sakaki, Y. (2001). A comprehensive two-hybrid analysis to explore the yeast protein interactome. *Proc Natl Acad Sci U S A, 98*(8), 4569-4574.

KEGG PATHWAY: Wnt signaling pathway – Homo sapiens (human). (2008). Retrieved August 15, 2008, from http://www.genome.ad.jp/kegg/pathway/hsa/hsa04310.html

Kerppola, T. K. (2006). Complementary methods for studies of protein interactions in living cells. *Nat Methods, 3*(12), 969-971.

Kerrien, S., Alam-Faruque, Y., Aranda, B., Bancarz, I., Bridge, A., Derow, C., et al. (2007). IntAct--open source resource for molecular interaction data. *Nucleic Acids Res, 35*(Database issue), D561-565.

Kerrien, S., Orchard, S., Montecchi-Palazzi, L., Aranda, B., Quinn, A. F., Vinod, N., et al. (2007). Broadening the horizon--level 2.5 of the HUPO-PSI format for molecular interactions. *BMC Biol, 5*, 44.

Kola, I. & Landis, J. (2004). Can the pharmaceutical industry reduce attrition rates? *Nat Rev Drug Discov 3*(8), 711-716.

Kurochkin, I. V., Mizuno, Y., Konagaya, A., Sakaki, Y., Schönbach, C., & Okazaki, Y. (2007). Novel peroxisomal protease Tysnd1 processes PTS1- and PTS2-containing enzymes involved in beta-oxidation of fatty acids. *EMBO J, 26*(3), 835-845.

Jones, D. (2008). All systems go. *Nat Rev Drug Discov, 7*(4), 278-279.

Lalonde, S., Ehrhardt, D. W., Loque, D., Chen, J., Rhee, S. Y., & Frommer, W. B. (2008). Molecular and cellular approaches for the detection of protein-protein interactions: latest techniques and current limitations. *Plant J, 53*(4), 610-635.

Li, M., Makkinje, A., & Damuni, Z. (1996). Molecular identification of I1PP2A, a novel potent heat-stable inhibitor protein of protein phosphatase 2A. *Biochemistry, 35*(22), 6998-7002.

Luo, W., & Lin, S. C. (2004). Axin: a master scaffold for multiple signaling pathways. *Neurosignals, 13*(3), 99-113.

Mackay, J. P., Sunde, M., Lowry, J. A., Crossley, M., & Matthews, J. M. (2007). Protein interactions: is seeing believing? *Trends Biochem Sci, 32*(12), 530-531.

McNabb, D. S., & Pinto, I. (2005). Assembly of the Hap2p/Hap3p/Hap4p/Hap5p-DNA complex in Saccharomyces cerevisiae. *Eukaryot Cell, 4*(11), 1829-1839.

Mishra, G. R., Suresh, M., Kumaran, K., Kannabiran, N., Suresh, S., Bala, P., et al. (2006). Human protein reference database--2006 update. *Nucleic Acids Res, 34*(Database issue), D411-414.

Mortazavi, A., Williams, B. A., McCue, K., Schaeffer, L., & Wold, B. (2008). Mapping and quantifying mammalian transcriptomes by RNA-Seq. *Nat Methods, 5*(7), 621-628.

Nash, R., Weng, S., Hitz, B., Balakrishnan, R., Christie, K. R., Costanzo, M. C., et al. (2007). Expanded protein information at SGD: new pages and proteome browser. *Nucleic Acids Res, 35*(Database issue), D468-471.

Nyfeler, B., Michnick, S.W., & Hauri, H.P. (2005). Capturing protein interactions in the secretory pathway of living cells. *Proc Natl Acad Sci U S A. 102*(18), 6350-6355.

Pache, R. A., Zanzoni, A., Naval, J., Mas, J. M., & Aloy, P. (2008). Towards a molecular characterisation of pathological pathways. *FEBS Lett, 582*(8), 1259-1265.

Pearson, H. (2008). Biologists initiate plan to map human proteome. (2008). *Nature, 452*(7190), 920-921.

Puig, O., Caspary, F., Rigaut, G., Rutz, B., Bouveret, E., Bragado-Nilsson, E., et al. (2001). The tandem affinity purification (TAP) method: a general procedure of protein complex purification. *Methods 24*(3), 218-229.

Reguly, T., Breitkreutz, A., Boucher, L., Breitkreutz, B. J., Hon, G. C., Myers, C. L., et al. (2006). Comprehensive curation and analysis of global interaction networks in Saccharomyces cerevisiae. *J Biol, 5*(4), 11.

Rigaut, G., Shevchenko, A., Rutz, B., Wilm, M., Mann, M., & Séraphin, B. (1999). A generic protein purification method for protein complex characterization and proteome exploration. *Nat Biotechnol. 17*(10), 1030-1032.

Rual, J. F., Venkatesan, K., Hao, T., Hirozane-Kishikawa, T., Dricot, A., Li, N., et al. (2005). Towards a proteome-scale map of the human protein-protein interaction network. *Nature, 437*(7062), 1173-1178.

Rudolph, M.G., Bayer, P., Abo, A., Kuhlmann, J., Vetter I.R., & Wittinghofer, A. (1998). The Cdc42/Rac interactive binding region motif of the Wiskott Aldrich syndrome protein (WASP) is necessary but not sufficient for tight binding to Cdc42 and structure formation. *J. Biol. Chem. 273* (29), 18067-18076.

Selenko, P., & Wagner, G. (2006). NMR mapping of protein interactions in living cells. *Nat Methods, 3*(2), 80-81.

Soderberg, O., Gullberg, M., Jarvius, M., Ridderstrale, K., Leuchowius, K. J., Jarvius, J., et al. (2006). Direct observation of individual endogenous protein complexes in situ by proximity ligation. *Nat Methods, 3*(12), 995-1000.

Soreq, L., Gilboa-Geffen, A., Berrih-Aknin, S., Lacoste, P., Darvasi, A., Soreq, E., et al. (2008). Identifying alternative hyper-splicing signatures in MG-thymoma by exon arrays. *PLoS ONE, 3*(6), e2392.

Spellman, P. T., Sherlock, G., Zhang, M. Q., Iyer, V. R., Anders, K., Eisen, M. B., et al. (1998). Comprehensive identification of cell cycle-regulated genes of the yeast Saccharomyces cerevisiae by microarray hybridization. *Mol Biol Cell, 9*(12), 3273-3297.

Sprinzak, E., Sattath, S., & Margalit, H. (2003). How reliable are experimental protein-protein interaction data? *J Mol Biol. 327*(5), 919-923.

Stamm, S. (2008). Regulation of alternative splicing by reversible protein phosphorylation. *J Biol Chem, 283*(3), 1223-1227.

Stelzl, U., Worm, U., Lalowski, M., Haenig, C., Brembeck, F. H., Goehler, H., et al. (2005). A human protein-protein interaction network: a resource for annotating the proteome. *Cell, 122*(6), 957-968.

Systems biologists hatch plan for virtual human. (2008, February 21) *Nature 451*(7181), 879.

Tarassov, K., Messier, V., Landry, C. R., Radinovic, S., Molina, M. M., Shames, I., et al. (2008). An in vivo map of the yeast protein interactome. *Science, 320*(5882), 1465-1470.

The TAP method home page (2002). Retrieved August 30, 2008, from http://www-db.embl-heidelberg.de/jss/servlet/de.embl.bk.wwwTools.GroupLeftEMBL/ExternalInfo/seraphin/TAP.html.

The Tokyo Declaration. (2008). Retrieved August 15, 2008, from http://www.systems-biology.org/~myukiko/FCSB2008/lib/exe/fetch.php?id=workshop%3Astatement&cache=cache&media=workshop:fcsb_2008-tokyo-declaration.pdf.

Tong, A. H., Lesage, G., Bader, G. D., Ding, H., Xu, H., Xin, X., et al. (2004). Global mapping of the yeast genetic interaction network. *Science, 303*(5659), 808-813.

Valaskovic, G. A. & Kelleher, N. L. (2002). Miniaturized formats for efficient mass spectrometry-based proteomics and therapeutic development. *Curr Top Med Chem., 2*(1), 1-12.

Vestal, M. L. & Campbell, J. M. (2005). Tandem time-of-flight mass spectrometry. *Methods Enzymol.,* 402, 79-108.

Chapter II
Data Mining for Biologists

Koji Tsuda
Max Planck Institute for Biological Cybernetics, Germany

ABSTRACT

In this tutorial chapter, the author reviews basics about frequent pattern mining algorithms, including itemset mining, association rule mining, and graph mining. These algorithms can find frequently appearing substructures in discrete data. They can discover structural motifs, for example, from mutation data, protein structures, and chemical compounds. As they have been primarily used for business data, biological applications are not so common yet, but their potential impact would be large. Recent advances in computers including multicore machines and ever increasing memory capacity support the application of such methods to larger datasets. The author explains technical aspects of the algorithms, but do not go into details. Current biological applications are summarized and possible future directions are given.

INTRODUCTION

As new techniques for obtaining biological data continue to be introduced and a huge amount of data is accumulated in a large number of databases, the importance of data mining is ever increasing (Han & Kamber 2000). Modern high throughput technologies create comprehensive information about genome sequences, gene expression, polymorphism, protein interactions etc. However, extracting important information is not a trivial task. It is not only because of the huge amount of data. It is hard to define what is the important information in a current context mathematically. In collaboration among computer scientists and biologists, clear definition of the task is the key to success. It is often the case that computer scientists require biologists to define the mathematical definition of what they want from data. For example, it might be a statistical classifier predicting some phenotype from SNP data. But when biologists do not have sufficient knowledge about data mining methods, or computer scientists do not understand the biologists' motivation well, such collaboration might end up in failure. One typical failure is to define an impossible task which cannot be solved by any reasonable data mining method. This tutorial chapter is supposed to give computational biologists some background knowledge about data mining methods.

There are many different classes of methods in data mining. Most popular ones include clustering, supervised classification, regression, sequence alignment etc. There are good introductory books about data mining methods in biology, for example, (Schoelkopf et al., 2004). In this chapter, we cannot focus on all techniques, but concentrate on frequent pattern mining algorithms and their possible applications to biological data.

The most popular method of pattern mining algorithms is itemset mining (Agrawal & Srikant, 1994), which has been extensively applied to market basket analysis. Suppose a supermarket welcoming many customers every day. Each customer buys more than one item. When each item is indexed by an integer, each transaction (i.e., the items bought by a customer) is summarized as a set of integers (Figure 1). Then, the goal of itemset mining is to enumerate the subsets of items which occurred in more than m transactions. The frequency threshold m is called the minimum support parameter. Found subsets are called patterns. In a popular example of market basket analysis, it is discovered that beer and diaper are likely to be sold simultaneously. The frequent co-occurrence of two items is considered as very strong signal of association between them. Biological objects are often represented as an itemset. For example, a promoter sequence can be represented as a set of sequence motifs. A patient is a set of genotypes at several quantitative loci. Applications to gene expression data can be seen, e.g., in (Creighton & Hanash, 2003; Tamura and D'haeseleer, 2008).

The idea of frequently appearing substructure is commonly exploited in sequence motif analysis (Bailey et al., 2006). Their objective is to find a frequently appearing substring from a set of strings. One point which discriminates motif analysis from frequent pattern mining is that the motif analysis finds only one motif at a time, whereas frequent pattern mining enumerates all frequently appearing patterns. For example, MEME (Bailey et al., 2006) can derive a probabilistic motif from a set of k-mers. Even though there are many possible motifs, it can extract one motif because it is based on a mixture model of two components, one of which is the background model. It could be extended to deal with multiple motifs by increasing the number of components, but then the optimization of parameters gets more difficult (Blekas et al., 2003).

Pattern mining algorithms are combinatorial methods that enumerate all solutions satisfying a set of pre-specified criteria (e.g., frequency). We call the other methods which create only one optimal solution "optimization methods". Often, the enumerative nature of itemset mining is misunderstood as confusing. Itemset mining creates a lot of frequent patterns which have more or less the same quality. On the other hand, a conventional optimization method show only one solution, which is conceived as "clear". However, in many situations in biological data analysis, the amount of data is far too short for determining unique solution with absolute confidence. In my opinion, multiple solutions by pattern mining methods are a natural consequence of the reality. An advantage is that biologists can select good ones from the solutions by their own biological knowledge. Those who anticipate a magic box that provides one reliable solution which is both statistically significant and biologically meaningful might find enumerative methods useless. However, the only one solution by optimization methods is easily perturbed by noise and premature finish of parameter optimization (i.e., local minima).

Itemset mining has been extended to more structured data, such as transaction sequences (Pei et al., 2004), trees (Asai et al., 2002) and labeled graphs (Yan & Han, 2002). In this tutorial, we focus on graph mining, because other techniques are considered as specialization of graph mining. Much of real data is described as labeled graphs. For example, chemical compounds are represented as graphs where nodes represent atoms, and edges represent bonds among them (Kazius et al., 2006). Graph mining can be applied to the 3D structure of proteins as well

Figure 1. Examples of frequent itemsets

A: 1,2,5,6,7,9
B: 2,3,4,5
C: 1,2,7,8,9
D: 1,7,9
E: 2,3,7,9

Frequent itemsets included in at least 3 transactions

{1} {2} {7} {9}
{1,7} {1,9}
{2,7} {2,9} {7,9}

(Borgwardt et al., 2006). The purpose of graph mining is to find subgraph patterns which occur frequently in the database. In other words, they could be called graph motifs. Currently, the most prominent application field of graph mining is chemoinformatics (Gasteiger & Engel, 2003). Especially, in quantitative structure-activity relationship (QSAR) analysis, the biochemical activity of a drug-candidate is predicted based on the molecular structure. Since the number of atoms is very small (e.g. C,O,H,S,N), the connectivity between them has crucial meaning. Graph mining can prepare a good set of subgraph features which are further used by prediction algorithms.

The rest of this chapter is organized as follows. In Itemset Mining, we explain itemset mining and its applications to biology. In Graph Mining, finding frequent subgraphs from chemical compounds is the main topic. Based on basic knowledge about mining algorithms, possible new directions such as statistical graph mining and pair mining are presented in Future Trends. Finally, Conclusion summarizes the chapter.

ITEMSET MINING

As mentioned in the previous section, itemset mining aims to enumerate all frequent subsets of items in the set of transactions. More concretely, definition of itemsets and their properties are summarized as follows:

1. Given a set S of items, any nonempty subset of S is called an itemset.
2. Given an itemset I and a set T of transactions, the support of I, denoted as support(I), is the number of transactions that contain all the items in I.
3. Given a positive integer α, I is a frequent itemset, if support(I)$\geq\alpha$. We refer to α as the minimum support parameter.

There are many implementations of itemset mining freely available on the web. For example, LCM is one of the most efficient algorithm (Uno et al., 2005), available at http://research.nii.ac.jp/~uno/codes.htm. Weka, an integrated Java-based environment for machine learning, has a function for itemset mining (http://www.cs.waikato.ac.nz/ml/weka). The computational time of itemset mining depends on the number of solutions (i.e., the number of frequent itemsets), which is unknown in advance. If α is too low, it takes a long time to finish. So a practical advice is to set α to a large value initially (e.g., 50% of the number of transactions), and decrease it gradually until you have a moderate number of solutions (e.g., 10000).

Association Rule Mining

Using frequent itemsets, one can come up with useful association rules in the form LHS→RHS, where LHS and RHS are itemsets, the RHS set being likely to occur whenever the LHS occurs. In market analysis, one obtains association rules like {cereal}→{milk,juice}, indicating that if a customer buys cereal, he or she is likely to buy milk and juice as well. The mathematical definition of association rules is as follows:

1. An association rule is a pair of disjoint itemsets. If LHS and RHS denote the two disjoint itemsets, the association rule is written as LHS→RHS.
2. The support of the association rule LHS→RHS is the support of the union of itemsets LHS and RHS.
3. The confidence of the rule LHS→RHS is the ratio support(LHS+RHS)/support(LHS)

In association rule mining, users define thresholds for support and confidence, and the algorithm enumerates all rules. First, the itemset mining is applied and enumerate all frequent itemsets. They correspond to LHS+RHS, for example {A,B,C,D}. Then, all partitions of the itemsets are tried such as {A}→{B,C,D} and {A,D}→{B,C}, and those with high confidence are recorded. Codes for association rule mining are available at http://fimi.cs.helsinki.fi/fimi04/.

In the analysis of yeast gene expression data by (Creighton & Hanash, 2003), they derived an association rule:

{YHM1}→{ARG1, ARG4, ARO3, CTF13, HIS5, LYS1, RIB5, SNO1, SNZ1, YHR029C, YOL118C}

The support was 11% and the confidence was 81%. It implies that if YHM1 is expressed, then the genes in RHS are likely to be expressed as well. Such rules could reflect the underlying regulatory network. Since it is an enumeration approach, the user receives many of such association rules which describe different pieces of the network. The integration of such rules to a network is also a difficult combinatorial problem which has not been solved to a sufficient level.

The Apriori Algorithm

There are several families of algorithms of itemset mining, but the Apriori algorithm is the earliest and the simplest of all. Given 4 items, all itemsets form a lattice depicted in Figure 2.

If a transaction t includes itemset I, then t includes any subset of I. This property, called anti-monotonicity, plays a very important role. It means that, frequent itemsets whose support is above a threshold appears as a connected region in the lattice (i.e., the highlighted region in the figure).

Therefore, we can enumerate frequent itemsets by starting from the empty set and traversing the tree by adding items. Whenever an infrequent itemset is found, we do not need to traverse itemsets in its downstream (tree pruning). One can choose either breadth-first search or depth-first search in traversing the lattice. In the former case, the method is called Apriori algorithm (Agrawal & Srikant, 1994), otherwise the backtrack algorithm (Zaki et al., 1997). Here, we present a summary of the Apriori algorithm.

Apriori
1. D1 = all frequent itemsets of size 1, k = 1
2. while Dk is not empty
3. Take the union of two itemsets in Dk. If their size is k+1, add to Dk+1
4. Remove all infrequent itemsets from Dk+1

Figure 2. Search tree of Apriori algorithm. The highlighted region indicates frequent patterns.

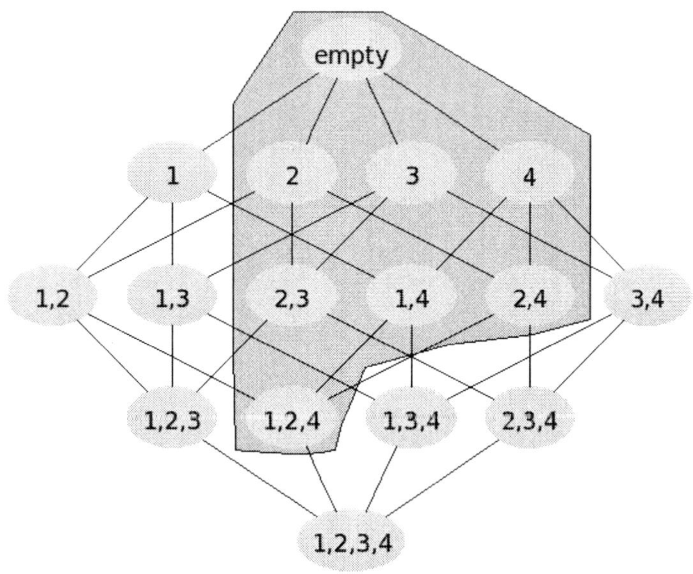

5. k = k + 1
6. end while

Computational Cost

Evaluation of computational cost of enumeration algorithms like itemset mining is tricky. Usually, the computational time is measured as a function of input size like O(n). In our case, the input size corresponds to the number of transactions. However, the efficiency of itemset mining crucially depends on the number of solutions. Namely, if the number of frequent itemsets is small, the algorithm finishes quickly, but if there are many frequent itemsets, it takes long time. In worst case evaluation, itemset mining is NP-hard with respect to the number of transactions. However, the 'worst case' means that all transactions have the same set of items. When there are large overlaps in transactions, the dataset is called dense. For dense datasets, itemset mining can be extremely slow. However, in market basket analysis, it is not usually the case, because each customer buys only a small fraction of items. For such sparse data, itemset mining can scale to millions of transactions. For biological applications, it is important to evaluate the density of transactions in advance. If it is too dense, itemset mining might not be a viable choice.

For enumeration algorithms, a reasonable cost evaluation can be done by measuring the delay, i.e., the computational time spent between two solutions. If the delay is polynomial to the input size, it is called polynomial delay. Unfortunately, the Apriori algorithm is not polynomial delay, but there are faster algorithms like LCM (Uno et al., 2005), which is linear delay.

Combination with Regression

In machine learning terminology, frequent itemset mining and association graph mining are unsupervised learning methods, because they are not to predict an external target variable. On the other hand, in many biological problems, prediction is the central subject. For example, it is important to predict if a tumor is benign or malignant based on expression profiles. In this case, the target variable is binary (either 1 or -1), and it is called a classification problem. When the remaining life of a patient is predicted, the target variable is real-valued, which is a regression problem.

Once a tumor is represented by an itemset of expressed genes, one of the most popular way to predict the target variable is to apply linear regression, using statistical software such as R (http://www.r-project.org). Here, the itemset is encoded as a binary vector, where expressed genes are described as 1, the other genes 0. After learning the regression coefficients, one can claim that the genes with large coefficients are associated with the phenotype (benign or malignant). However, this analysis makes sense only when the target variable is a linear function of the features. It is assumed that the phenotype is determined by linear sum of individual effects posed by the genes. Considering complex regulatory mechanisms of the cell, this assumption sounds too naïve. To introduce nonlinearity, it makes sense to perform frequent itemset mining first, and then apply linear regression. The number of features will increase as the number of frequent itemsets is usually much larger than the number of items. Features are constructed such that, if all members of the itemset are expressed, the feature is 1, otherwise 0. After learning the coefficients, one can obtain the set of gene sets, which contributes to determine the phenotype.

Recently, (Saigo *et al.*, 2007) applied this approach for the prediction of the drug resistance of HIV-1. They predicted the phenotype (drug resistance) from the mutation set in the amino acid sequence of the reverse transcriptase. The notation 77L means that the 77th amino acid of the reverse transcriptase has changed to L. After itemset mining, we obtain the set of associated mutations like {77L,151M}. Then, they performed linear regression on the itemsets and obtained coefficients as shown in Figure 3. This figure shows the results for 6 FDA approved drugs. It is known that the drug resistance cannot be induced by only one mutation, but require several mutations at once. It turns out that the chosen itemsets correspond well to known mutation associations (Thyrimidine-associated mutations and the Q151M complex). See the paper for details. Notice that they proposed a cleverer algorithm called itemset boosting to quickly collect necessary itemsets in an iterative way, but we do not explain the algorithm in detail here.

Figure 3. Regression coefficients for predicting the drug resistance based on itemsets of amino acid mutations.

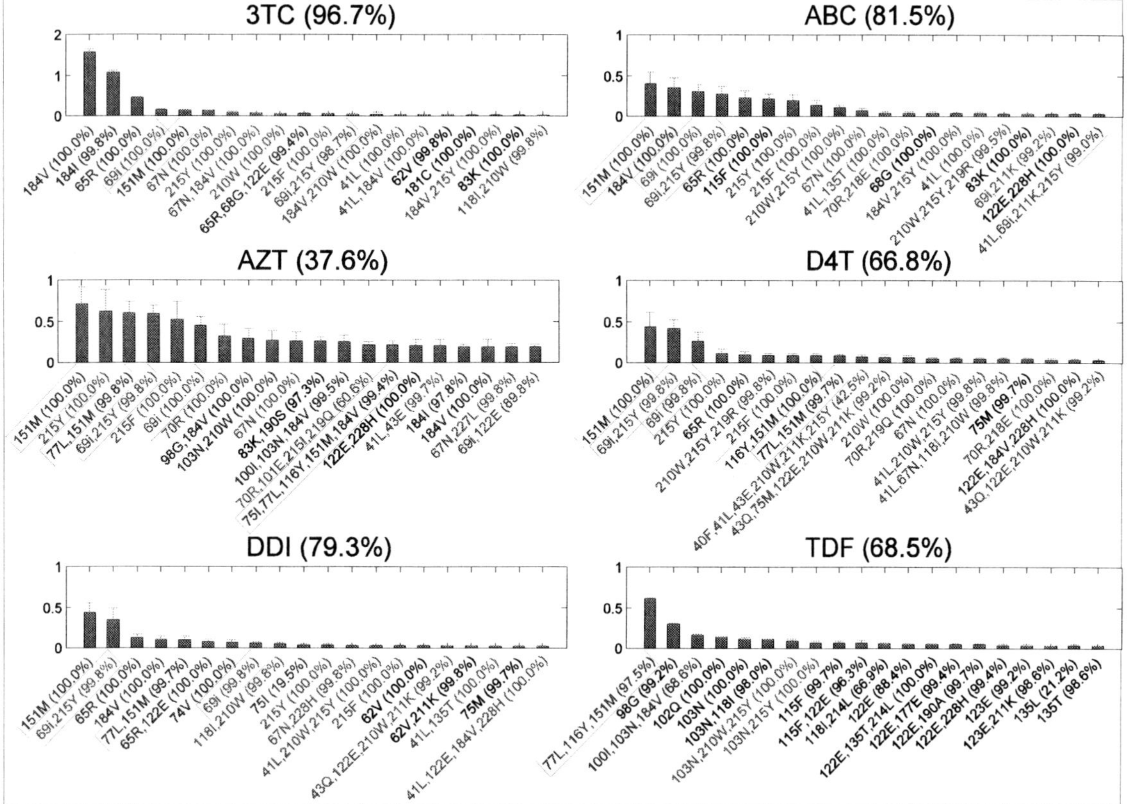

In this section, we have shown that itemset mining can be utilized in combination with a machine learning algorithm. Association rule mining has been combined with rule-based classification (Liu et al., 1998; Cong et al., 2005) and support vector machines (Sun et al., 2007) as well. Itemset mining has inherent weakness that it could give too many patterns for biologists to comprehend. Statistical methods are expected to help in choosing important itemsets and prioritize them. Since itemset mining algorithms themselves are very well studied, it is not expected that the scalability improves significantly. However, there are a lot of new research possibilities when combined with statistical methods.

GRAPH MINING

Graphs are general and powerful data structures that can be used to represent diverse kind of objects. As mentioned in Introduction, graph-based methods are prevalently used in computational biology. In this section, we are interested in the analysis of graph databases that contain tens of thousands of labeled graphs. We assume that each graph is relatively small, say, less than 100 nodes. In computational biology, there are much larger graphs such as protein-protein interaction networks. The number of nodes is typically more than 10000, but the number of graphs is only one typically. If one considers evolution of networks, it could be more than one, but not so many.

Figure 4. Examples of chemical compounds with and without mutagenicity

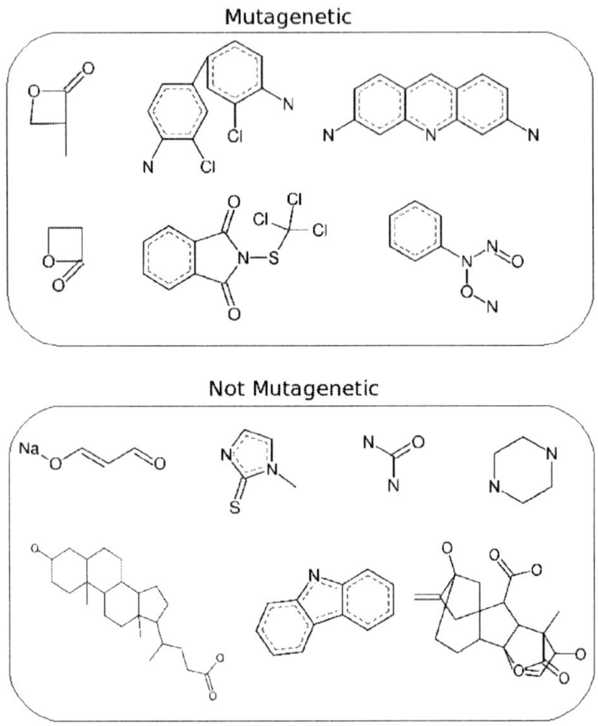

Extension of graph mining to such a one-big-graph scenario is also considered recently (Fiedler & Borgelt, 2007), but in this chapter, we concentrate on discovering frequent pattterns in relatively small graphs.

Chemoinformatics, especially drug discovery, is deeply associated with graph mining problems. In Figure 4, we have shown examples of a chemoinformatics dataset which contains both mutagenetic and non-mutagenetic drug candidates. In drug development, one has to make sure that the drug is not mutagenetic, otherwise it causes cancer in patients. Therefore, computational prediction of mutagenicity or other biochemical activities is of great interest in chemistry. As seen in the figure, it is not a simple prediction task. The graphs in the same category differ significantly in size, and there are no visible features discriminating two categories. To extract features from such data, graph mining has shown to be very powerful, because it enumerates frequently occurring substructures which should be related to bioactivities (Kazius et al., 2006). There are several graph mining methods available such as AGM (Inokuchi, 2004), gSpan (Yan & Han, 2002) and Gaston (Nijssen and Kok, 2004). In the following, we pick up gSpan and explain it in detail.

Search Tree of Graphs

Our subgraph enumeration strategy requires a canonical search space in which a whole set of patterns are traversed without duplication. As the search space, we adopt the DFS (depth first search) code tree. The basic idea

Figure 5. Schematic figure of the DFS code tree

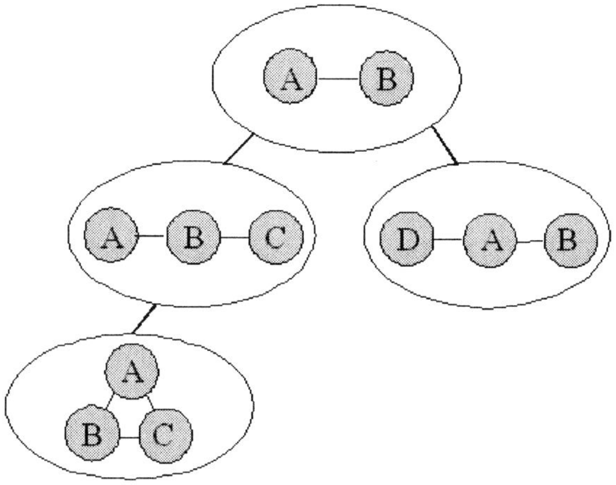

Figure 6. Depth first search and DFS code of graph. (a) A graph example. (b), (c) Two different depth-first-searches of the same graph. Red numbers represent the DFS indices. Bold edges and dashed edges represent the forward edges and the backward edges respectively.

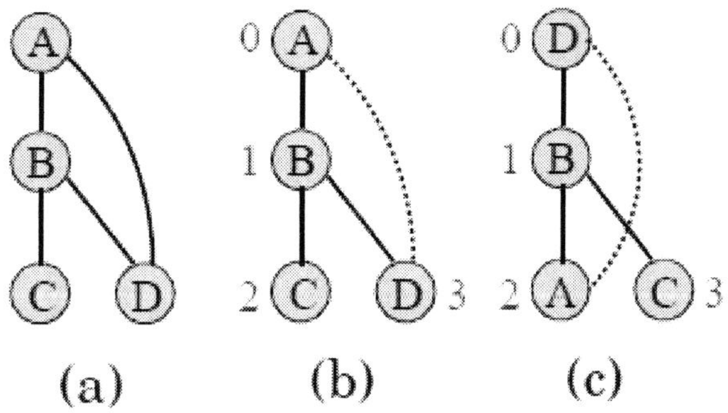

of the DFS code tree is to organize patterns as a tree, where a child node has a supergraph of the parent's pattern (Figure 5). In the tree, a pattern is represented as a text string called the DFS code, which is made by traversing the graph by depth first search. Each node is indexed from 0 to n-1 according to the discovery time in the DFS. All the edges traversed in the DFS are called forward edges and the rest is called backward edges. One important fact is that, according to the starting node, there are several DFS codes for the same graph (Figure 6). The canonical representation is determined as the minimum code according to the lexicographical order.

The patterns are enumerated by generating the tree from the root to leaves using a recursive algorithm. Node generation is systematically done by rightmost extensions. Still, it is often the case that the same DFS code is generated through different paths. To avoid the duplication, whenever a new node is made, the associated DFS code has to be minimum. Yan & Han (2002) have proven that, by assuring the minimality of the DFS code in each extension step, the whole set of patterns can be enumerated without duplication. As in the itemset mining, we adopt tree pruning according to the support. If the support of a pattern is found to be smaller than the minimum support threshold, the search tree extension is stopped immediately.

All embeddings of a pattern in the graphs have to be maintained in each node to calculate its support. If a pattern matches a graph in different ways, all such embeddings are stored. When a new pattern is created by adding an edge, it is not necessary to perform full isomorphism checks with respect to all graphs in the database. A new list of embeddings is made by extending the embeddings of the parent. Technically, it is necessary to devise a data structure such that the embeddings are stored incrementally, because it takes a prohibitive amount of memory to keep all embeddings independently in each node.

The most time consuming part of gSpan is the minimality check of the DFS code. It is as expensive as automorphism checking (Yan & Han, 2002). Accordingly, gSpan is not polynomial-delay. However, for sparse graphs like chemical compounds, it can scale up to tens of thousands of graphs. As in itemset mining, gSpan might not be feasible when graphs are dense with many edges and similar to each other. An C++ implementation of gSpan can be obtained in gboost soft ware package (http://www.kyb.mpg.de/bs/people/nowozin/gboost)

QSAR Analysis

After frequent subgraph mining, one can represent each graph as binary vectors as in Figure 7. Now it is straightforward to apply linear regression algorithms to predict the chemical activity. Figure 8 shows the top 20 patterns required to predict mutagenicity in a chemical dataset (Saigo et al., 2008). It is observed that the chosen patterns are relatively large (up to 10 nodes). To predict the bioactivity, it is necessary to extract reasonably large patterns. It would be possible to use simpler patterns like paths or trees to save computational cost. However, the obtained patterns are not as interpretable as shown here. Notice that the results are obtained by graph boosting algorithm, which can save computational cost substantially in comparison to the naïve application of frequent substructure mining.

Apart from linear regression, rule-based methods and support vector machines can be applied to graph data in a similar way. For more possibilities, see http://www.cs.unc.edu/~weiwang/, for example.

RNA Common Secondary Structure Discovery

Let us introduce another example of biological applications of graph mining by (Hamada et al., 2006). Multiple alignment of RNA sequences are difficult, because one has to predict their secondary structure to come up with

Figure 7. Feature vector based on subgraph patterns

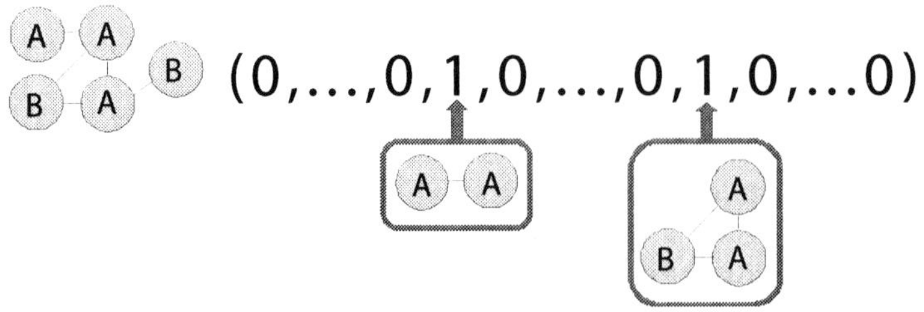

Figure 8. Top 20 discriminative subgraph patterns extracted from a mutagenicity dataset. The digits under each pattern indicate the regression coefficients.

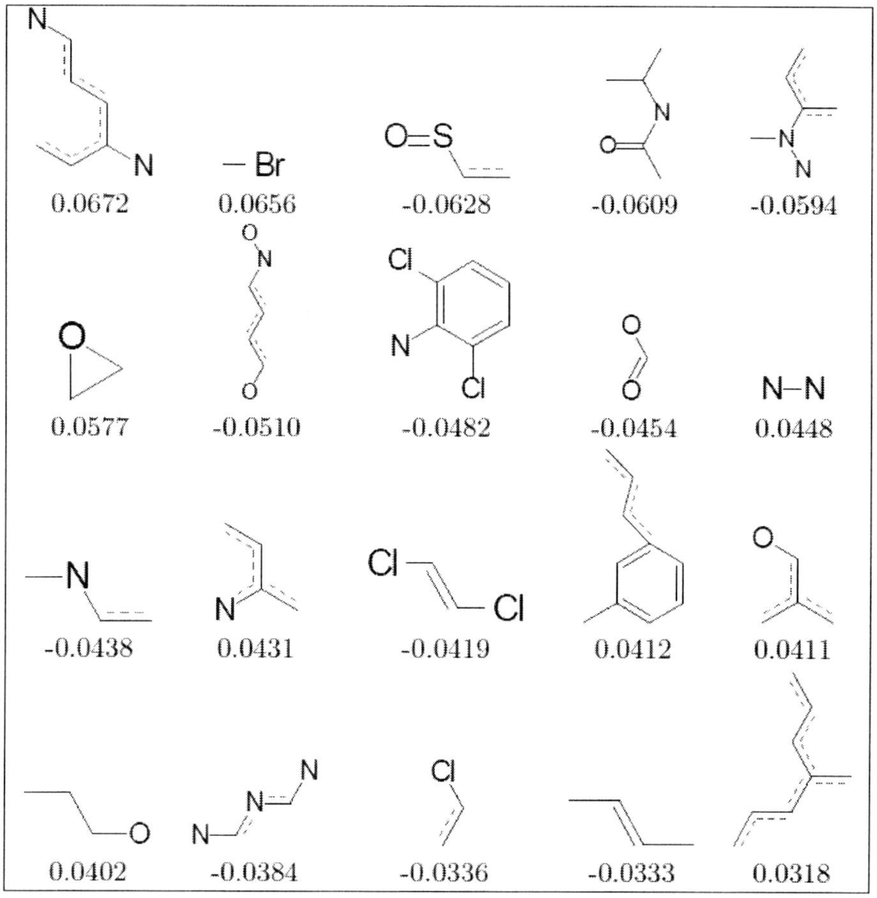

correct alignment. Given a set of unaligned RNA sequences, we need to find parts of the sequences that share the common secondary structure (Figure 9). It is possible to enumerate stem candidates using McCaskill's algorithm (McCaskill, 1990), but one has to choose right ones to construct the true secondary structure. For each sequence, a labeled graph is created by representing stem candidates as nodes and their position relationships as edges. Using a modified version of gSpan, the common secondary structure is obtained as a subgraph pattern. It is also possible to map the pattern back to the sequences to obtain multiple alignment. Due to the enumerative nature of gSpan, one can obtain several possible suggestions for multiple alignments. The accuracy of obtained common structures is competitive among state-of-the-art methods such as CMFinder (Yao et al., 2005). This example illustrates that graph mining algorithms could be applied to the objects which are not graphs per se.

Figure 9. Discovering common secondary structures from unaligned RNA sequences

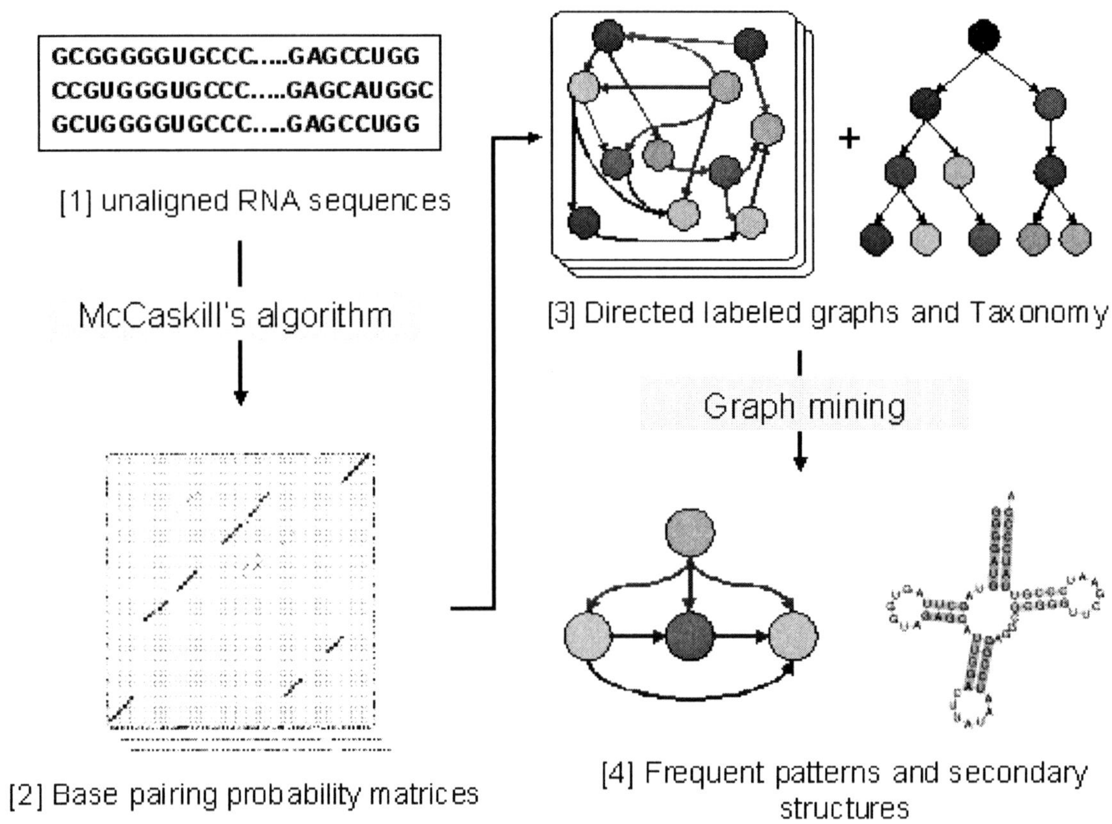

FUTURE TRENDS

Network motifs (Milo et el., 2002; Mittendorf et al., 2005) are frequently appearing graph patterns in a regulatory network of genes and proteins. Typically, the regulatory network is directed, thus network motifs are also directed. It is claimed that the motifs can help determining unknown functions of proteins and understanding the evolution from a network perspective. They are very close to what we call subgraph patterns, but there are two major differences.

1. Network motifs are frequently occurring subgraphs in one large graph, whereas the subgraph patterns are occurring in many graphs in the database
2. Network motifs are purely topological, namely they do not have node and edge labels

When graphs do not have any labels, all possible patterns of small size (say 5 nodes) can easily be enumerated. So identifying network motifs is not a difficult algorithmic problem. However, since each node has relevant information as nucleotide sequences, 3D structures of proteins, GO functional annotations and so forth, it seems interesting to include them as node labels. Also, edge labels can be diverse, for example, protein interaction, genetic interaction, co-occurrence in literature etc. To detect network motifs with labels, one needs an algorithm like gSpan. The algorithms presented in this chapter are all exact, and obviously not fast enough to detect all such motifs in a genome-wide regulatory network. It poses an interesting challenge for computer scientists how to develop an efficient method to discover labeled network motifs. Probably some approximations or early pruning by statistical criteria would be necessary.

Most of biological processes are carried out by molecular interaction, for example, the binding of ligands to proteins, proteins to DNA, proteins to proteins, RNA to RNA etc. In the QSAR analysis described in the previous section often focus on finding inhibitor ligands of a particular protein. However, one protein can bind to several ligands and vice versa. Therefore, it makes sense to consider pair patterns. When we have a database of protein-ligand bindings, a pair pattern refers to the combination of two graph patterns coming from protein and ligand, which occurs frequently in binding pairs. It poses a more difficult enumeration problem than conventional ones focusing on only one object, because the search space is defined as the product of two spaces.

In traditional data mining, obtained patterns are assumed to be viewed by humans, but it is increasingly popular to use the patterns as an input for a machine learning algorithms. We have introduced several application examples of regression techniques. Humans would like to capture a general overview of the database, therefore the minimum support threshold could be set to a high value. However, in regression, it is often the case that infrequent pattterns play an important role in classifying rare cases. Therefore, to achieve the highest accuracy, minimum support threshold has to be low (Saigo et al., 2008), leading to efficiency problems. An effective solution comes from the fact that machine learning algorithms typically do not need all features. In particular, sparse methods such as LASSO (Tibshirani, 1996) are designed to reduce the number of used features. Statistical substructure mining methods can find the patterns necessary for a given machine learning algorithms, avoiding the generation of unnecessary branches in the search tree (Saigo et al., 2008; Saigo et al., 2007; Nowozin et al., 2007; Tsuda, 2007). The mining algorithm is called repeatedly with different selection criterion to progressively collect necessary patterns. In classification cases, the example weights are updated to focus on the samples still misclassified. In my opinion, the integration of statistics and pattern mining is the key step to achieve both high prediction accuracy and good efficiency.

CONCLUSION

We have presented the essence of pattern mining algorithms with focus on their implications rather than technical subtleties. Convincing someone to use pattern mining algorithms is sometimes hard, because we cannot tell him or her about the computational time in advance. It totally depends on the denseness of the dataset and the minimum support threshold. To see if your problem can be solved by mining, the best way is to download software introduced in this chapter and try it out. For sparse data, it is remarkably fast in most cases.

Enumeration algorithms are not new, as theoretical studies exist since 70s. However, with modest computational power in 80s or 90s, such algorithms were out of question for a reasonable amount of data. Therefore, we had to resort to optimization algorithms. Now we can run the algorithms comfortably with a common PC.

Given scarcity of data, it is extremely hard to determine the unique solution in most biological questions. As application of computational methods cannot be avoided in modern biology, one has to be careful not to make false statements. Using optimization methods, one tries to convince people by showing the best-looking aspect of the data, intentionally or non-intentionally ignoring the aspects which do not match one's claim. In my opinion, enumeration algorithms can provide multiple perspectives to view biological data and has smaller room for human (malicious) intervention because it shows all good solutions exhaustively.

ACKNOWLEDGMENT

KT would like to thank Takeaki Uno and Hiroki Arimura for providing presentation materials.

REFERENCES

Agrawal, R., & Srikant, R. (1994). Fast algorithms for mining association rules in large databases. *Proc. VLDB 1994* (pp. 487–499).

Asai, T., Abe, K., Kawasoe, S., Arimura, H., Sakamoto, H., & Arikawa, S. (2002). Efficient substructure discovery from large semi-structured data. *Proc 2nd SIAM Data Mining Conference (SDM)* (pp. 158–174).

Bailey, T. L., Williams, N., Misleh, C., & Li, W. W. (2006). Meme: discovering and analyzing dna and protein sequence motifs. *Nucleic Acids Res, 34*, W369–W373.

Blekas, K., Fotiadis, D. I., & Likas, A. (2003). Greedy mixture learning for multiple motif discovery in biological sequences. *Bioinformatics, 19*, 607–617.

Borgwardt, K. M., Ong, C. S., Schoenauer, S., Vishwanathan, S. V. N., Smola, A. J., & Kriegel, H.-P. (2006). Protein function prediction via graph kernels. *Bioinformatics, 21*(suppl. 1), i47–i56.

Cong, G., Tan, K.-L., Tung, A., & Xu, X. (2005). Mining top-k covering rule groups for gene expression data. *Proceedings of the 2005 ACM SIGMOD International Conference on Management of Data* (pp. 670–681).

Creighton, C., & Hanash, S. (2003). Mining gene expression databases for association rules. *Bioinformatics, 19*, 79–86.

Fiedler, M., & Borgelt, C. (2007). Support computation for mining frequent subgraphs in a single graph. *Mining and Learning with Graphs (MLG).*

Gasteiger, J., & Engel, T. (2003). *Chemoinformatics: A textbook*. Wiley-VCH.

Hamada, M., Tsuda, K., Kudo, T., Kin, T., & Asai, K. (2006). Mining frequent stem patterns from unaligned RNA sequences. *Bioinformatics, 22,* 2480–2487.

Han, J., & Kamber, M. (2000). *Data mining: Concepts and techniques*. Morgan Kaufmann.

Inokuchi, A. (2005). Mining generalized substructures from a set of labeled graphs. *Proceedings of the 4th IEEE Internatinal Conference on Data Mining* (pp. 415–418). IEEE Computer Society.

Kazius, J., Nijssen, S., Kok, J., & Ijzerman, T. B. A. (2006). Substructure mining using elaborate chemical representation. *J. Chem. Inf. Model., 46*, 597–605.

Liu, B., Hsu, W., & Ma, Y. (1998). Integrating classification and association rule mining. *Proceedings of the Fourth International Conference on Knowledge Discovery and Data Mining (KDD-98)* (pp. 80–86).

McCaskill, J. S. (1990). The equilibrium partition function and base pair binding probabilities for RNA secondary structure. *Biopolymers, 29*, 1105–1119.

Middendorf, M., Ziv, E., & Wiggins, C. H. (2005). Inferring network mechanisms: the drosophila melanogaster protein interaction network. *Proc Natl Acad Sci U S A, 102*, 3192–3197.

Milo, R., Shen-Orr, S., Itzkovitz, S., Kashtan, N., Chklovskii, D., & Alon, U. (2002). Network motifs: simple building blocks of complex networks. *Science, 298*, 824–827.

Nijssen, S., & Kok, J. (2004). A quickstart in frequent structure mining can make a difference. *Proceedings of the 10th ACM SIGKDD International Conference on Know ledge Discovery and Data Mining* (pp. 647–652). ACM Press.

Nowozin, S., Bakir, G., & Tsuda, K. (2007). Discriminative subsequence mining for action classification. *Proceedings of the 11th IEEE International Conference on Computer Vision (ICCV 2007).* IEEE Computer Society.

Pei, J., Han, J., Mortazavi-Asl, B., Wang, J., Pinto, H., Chen, Q., Dayal, U., & Hsu, M.-C. (2004). Mining sequential patterns by pattern-growth: the prefixspan approach. *IEEE Trans. Knowledge and Data Engineering, 16*, 1424–1440.

Saigo, H., Nowozin, S., Kadowaki, T., Kudo, T., & Tsuda, K. (2008). gBoost: a mathematical programming approach to graph classification and regression. *Machine Learning*. to appear.

Saigo, H., Uno, T., & Tsuda, K. (2007). Mining complex genotypic features for predicting HIV-1 drug resistance. *Bioinformatics, 23,* 2455–2462.

Schoelkopf, B., Tsuda, K., & Vert, J. (Eds.). (2004). *Kernel methods for computational biology*. MIT Press.

Sun, G., Liu, X., Cong, G., Zhou, M., Xiong, Z., Lin, C.-Y., & Lee, J. (2007). Detecting erroneous sentences using automatically mined sequential patterns. *ACL 2007, Proceedings of the 45th Annual Meeting of the Association for Computational Linguistics.*

Tamura, M., & D'haeseleer, P. (2008). Microbial genotype-phenotype mapping by class association rule mining. *Bioinformatics*. to appear.

Tibshirani, R. (1996). Regression shrinkage and selection via LASSO. J. R. *Stat. Soc. Ser. B Stat. Methodol., 58,* 267–288.

Tsuda, K. (2007). Entire regularization paths for graph data. *Proceedings of the 24th International Conference on Machine Learning* (pp. 919–926).

Uno, T., Kiyomi, M., & Arimura, H. (2005). LCM ver.3: collaboration of array, bitmap and prefix tree for freque nt itemset mining. *OSDM '05: Proceedings of the 1st international workshop on open so urce data mining* (pp. 77–86). Chicago, Illinois.

Yan, X., & Han, J. (2002). gSpan: graph-based substructure pattern mining. *Proceedings of the 2002 IEEE International Conference on Data Mining* (pp. 721–724). IEEE Computer Society.

Yao, Z., Weinberg, Z., & WL, W. R. (2005). CMfinder–a covariance model based RNA motif finding algorithm. *Bioinformatics, 22*, 445–452.

Zaki, M., Parthasarathy, S., Ogihara, M., & Li, W. (1997). New algorithms for fast discovery of association rules. *Proc. KDD 1997* (pp. 283–286).

Section II
PPI Network Construction and Cleansing

We will describe the methods for constructing and cleansing PPI networks. Section II of this book will describe key bioinformatics methods for predicting/validating protein interactions, as well as network cleansing techniques to detect reliable protein interaction networks.

Chapter III
Domain-Based Prediction and Analysis of Protein-Protein Interactions

Tatsuya Akutsu
Kyoto University, Japan

Morihiro Hayashida
Kyoto University, Japan

ABSTRACT

Many methods have been proposed for inference of protein-protein interactions from protein sequence data. This chapter focuses on methods based on domain-domain interactions, where a domain is defined as a region within a protein that either performs a specific function or constitutes a stable structural unit. In these methods, the probabilities of domain-domain interactions are inferred from known protein-protein interaction data and protein domain data, and then prediction of interactions is performed based on these probabilities and contents of domains of given proteins. This chapter overviews several fundamental methods, which include association method, expectation maximization-based method, support vector machine-based method, and linear programming-based method. This chapter also reviews a simple evolutionary model of protein domains, which yields a scale-free distribution of protein domains. By combining with a domain-based protein interaction model, a scale-free distribution of protein-protein interaction networks is also derived.

INTRODUCTION

Understanding of functions of genes and proteins is important in post-genomic era. Information on protein-protein interactions is useful for understanding protein functions because protein-protein interactions play a key role in many cellular processes. Since the end of the last century, some experimental techniques have been developed for comprehensive analysis of protein-protein interactions, which include two-hybrid systems and proteomics methods. Though these experimental methods revealed many unknown interactions, there were large gaps between results done by different groups (Ito et al., 2001; Uetz et al., 2000). Therefore, computational methods should be developed for inference of protein-protein interactions. For that purpose, various approaches have been proposed. Since other approaches and aspects will be covered in other chapters in this book, this chapter focuses on computational and mathematical aspects of domain-based approaches.

A protein consists of one or multiple domains, where a *domain* is defined as a region within a protein that either performs a specific function or constitutes a stable structural unit. Examples of structural domains are illustrated in Fig. 1 though domains are sometimes defined not based on structures but based on sequence/functional similarities. In a word, domains are considered as parts of a protein. Though there is no exact or mathematical definition of protein domains, several hundreds of protein domains are currently known. In order to classify domains, several database systems have been constructed, which include Pfam (Finn et al., 2005), InterPro (Nicola et al., 2007) and ProDOM (Bru et al., 2005). Furthermore, most of these databases provide facilities to identify protein domains from a given protein sequence. In Pfam, each domain is represented by an HMM (Hidden-Markov Model) and protein domains contained in a given protein sequence are identified by using these HMMs.

Utilizing information of domain organizations of proteins, several methods have been proposed for prediction of protein-protein interactions. In these methods, scores or probabilities of *domain-domain interactions* are first derived from known protein-protein interactions and then these are utilized for calculating the score or probability of protein-protein interaction for given protein sequences. Sprinzak and Margalit (2001) proposed the *association method* for computing the score of each domain pair. Kim et al. (2002) proposed similar scores and applied the scores to inference of protein-protein interactions. Deng et al. (2002) proposed an *EM (Expectation-Maximization) algorithm* for estimating the probability of interaction for each domain pair.

In these methods, it is assumed that protein-protein interaction data are given as binary data (i.e., whether or not each protein pair interacts is given). However, multiple experiments are performed for the same protein pairs in practice and thus the ratio of the number of observed interactions to the number of experiments is available for each protein pair. For example, Ito et al. (2001) performed multiple experiments for each protein pair. But, the results are not always the same. Therefore, it is reasonable to use the ratio of the number of observed interactions to the number of experiments as input data. We developed a method utilizing these ratios (Hayashida et al., 2003; Hayashida et al., 2004), which was further improved by Chen et al. (2006).

Independent of our work, various approaches have also been proposed for improving the prediction accuracy of the domain-based approach. Li et al. (2006) developed probabilistic model and method that make active use of negative data (i.e., non-interacting proteins pairs). Chen et al. (2006) developed methods using decision

Figure 1. Example of protein domains. Protein P_1 consists of domains D_1 and D_2, whereas protein P_2 consists of domains D_3, D_4 and D_5. In domain-based models, it is assumed that P_1 and P_2 interact with each other if at least one domain pair interacts.

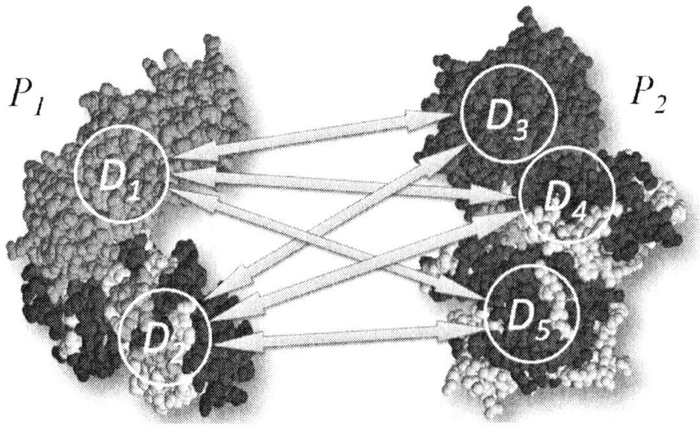

trees and neural networks. Singhal and Resat (2007) developed a genetic algorithm-based method. Kim et al. (2007) developed Bayesian methods and showed that the Bayesian approach is superior to the Likelihood-based approach by Deng et al. (2002). In addition to prediction of pairwise interactions, some studies have been done on prediction of protein complexes based on domain-domain interactions. Ng et al. (2003) developed a database of interacting protein domains, which are useful for prediction of both pairwise protein-protein interactions and protein complexes. Sprinzak et al. (2006) used logistic regression to predict both intra-complex interactions and inter-complex interactions. It is also important to verify the role of domain-domain interactions in protein-protein interactions. For that purpose, some detailed studies have been done on assessing the conservation of domain-domain interactions in protein-protein interactions (Itzhaki et al., 2006; Schuster-Böckler and Bateman, 2007).

Domain-based models are also useful for understanding structures of protein-protein interaction networks. Recent studies suggest that protein-protein interaction networks are *scale-free* (Barabási et al., 2004, Jeong et al., 2001). That is, the degree distribution in protein-protein interaction networks follows a power-law. Several models have been proposed for explaining the scale-free property of protein-protein interaction networks (Berg et al., 2004; Evlampiev & Isambert, 2007; Ispolativ et al., 2005; Vázquez et al., 2003). However, these models are complicated or biologically unreasonable. Thus, we developed a simple model for explaining the scale-free property of protein-protein interaction networks (Nacher et al., 2006; Nacher et al., 2007). In this model, a simple evolutionary model of protein domains was combined with the above mentioned domain-based model of protein-protein interactions.

In this chapter, we first review fundamental domain-based methods for inference of protein-protein interactions. Then, we review domain-based models of protein-protein interaction networks. Finally, we conclude with discussions on future directions. It is to be noted that it is impossible to give a comprehensive review because various models and methods have been proposed for both problems. Thus, in this chapter, we focus on fundamental and simple models and methods and aim to give basic ideas in them from computational and mathematical viewpoints.

DOMAIN-BASED MODELS FOR PROTEIN-PROTEIN INTERACTIONS

Domain-based models of protein-protein interactions are based on a very simple idea: two proteins interact with each other if and only if there exists an interacting domain pair. We explain this idea using a simple example in Fig. 2. In this chapter, P_1, P_2, ... denote proteins, and D_1, D_2, ... denote *domains*. We also use P_i to denote the set

Figure 2. Example of the domain-based model of protein-protein interactions. In this case, P_1 and P_2 interact with each other because D_1 in P_1 and D_2 in P_2 interact with each other. On the other hand, P_2 and P_3 do not interact with each other because neither (D_2, D_2) nor (D_2, D_3) interacts. We can also infer domain-domain interactions from known protein-protein interactions.

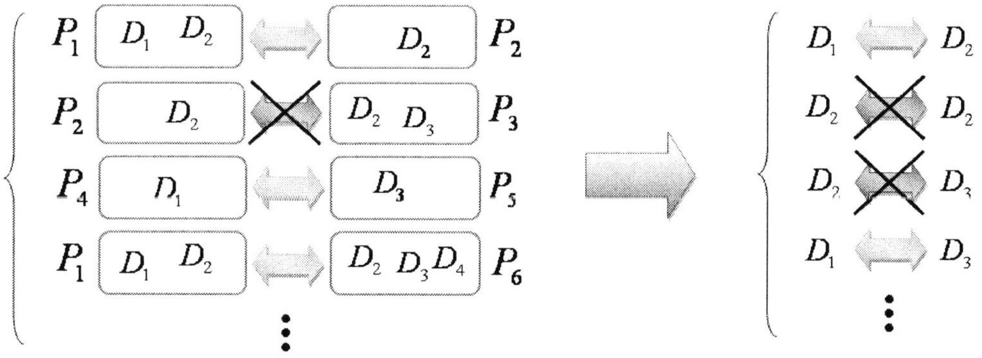

of domains contained in P_i. In the example of Fig. 2, P_1 consists of D_1 and D_2, P_2 consists of D_2, P_3 consists of D_2 and D_3, P_4 consists of D_1, P_5 consists of D_3, and P_6 consists of D_2, D_3, and D_4. That is, P_1={ D_1, D_2},..., P_6={ D_2, D_3, D_4}. Suppose that domains D_1 and D_2 interact with each other, domains D_1 and D_3 interact with each other, and the other domain pairs do not interact. Then, P_1 and P_2 interact with each other because D_1 in P_1 and D_2 in P_2 interact with each other. Similarly, P_4 and P_5, and P_1 and P_6 interact with each other, respectively. On the other hand, P_2 and P_3 do not interact with each other because neither (D_2, D_2) nor (D_2, D_3) interacts.

Though we assumed that interacting domain pairs are already known, we can consider the inverse direction: inference of *domain-domain interactions* from protein-protein interactions. In this case, the problem is to infer domain-domain interactions from domain organization data of proteins and protein-protein interaction data. In the example in Fig. 2, we infer that (D_1, D_2) and/or (D_2, D_2) interact since P_1 and P_2 interact. However, we see that (D_2, D_2) should not interact since P_2 and P_3 do not interact. Therefore, we can conclude (D_1, D_2) should be an interacting domain pair. Similarly, we can conclude that (D_1, D_3) should be an interacting domain pair. In this way, we can infer interacting domain pairs from domain organization data and known interacting protein pairs. This corresponds to a learning or training phase. Here, suppose that a new protein pair (P_7,P_8) is given where P_7 = {D_2, D_4} and P_8={D_1, D_3, D_4}. Then, we can infer that P_7 and P_8 interact since (D_2, D_1) is an interacting domain pair. This corresponds to a prediction or testing phase. Though the above example explains the basic idea of the domain-based approach, more elaborated models are required in practice. In particular, it is much better to introduce probabilities or scores because real data contain noise and protein-protein interactions may not be governed by deterministic processes. In the following, we review such models.

Association Method

Sprinzak and Margalit (2001) proposed a simple method called the *association method.* In this method, a score is assigned to each domain pair (D_m,D_n). Let N_{mn} be the number of protein pairs (in the training data set) containing a domain pair (D_m,D_n). Let I_{mn} be the number of interacting protein pairs (in the training data set) containing a domain pair (D_m,D_n). Then, the score for (D_m,D_n) is given by

$$ASSOC(D_m, D_n) = \frac{I_{mn}}{N_{mn}}.$$

EM Method

Deng et al. (2002) introduced a probabilistic model of protein-protein interactions. In their model, protein-protein interactions and domain-domain interactions are represented by random variables: P_{ij}=1 if P_i and P_j interact with each other, and D_{mn}=1 if D_m and D_n interact with each other. It is also assumed that domain-domain interactions are independent and two proteins interact if and only if at least one domain pairs from the two proteins interact. Under this assumption, the probability that P_i and P_j interact with each other is given by

$$\Pr(P_{ij} = 1) = 1 - \prod_{D_{mn} \in P_{ij}} (1 - \lambda_{mn}),$$

where $D_{mn} \in P_{ij}$ means that $D_m \in P_i$ and $D_n \in P_j$, and λ_{mn} denotes the probability that D_m and D_n interact with each other (i.e., $\lambda_{mn} = \Pr(D_{mn} = 1)$). Deng et al. (2002) considered two types of experimental errors: false positives, in which two proteins do not really interact but were observed to interact in the experiments, and false negatives, in which two proteins really interact but were not observed to interact in the experiments. Let *fp* and *fn* denote the false positive rate and the false negative rate, respectively. Let O_{ij} be the variable for the observed interaction result for P_i and P_j, where O_{ij}=1 if the interaction is observed, otherwise O_{ij}=0. Using P_{ij} and O_{ij}, *fp* and *fn* are given as below.

$$fp = \Pr(O_{ij} = 1 \mid P_{ij} = 0),$$
$$fn = \Pr(O_{ij} = 0 \mid P_{ij} = 1).$$

Then, $\Pr(O_{ij})$ is given by

$$\begin{aligned}\Pr(O_{ij} = 1) \\ &= \Pr(O_{ij} = 1, P_{ij} = 1) + \Pr(O_{ij} = 1, P_{ij} = 0) \\ &= \Pr(P_{ij} = 1)(1 - fn) + (1 - \Pr(P_{ij} = 1))fp.\end{aligned}$$

Deng et al. defined the likelihood function (i.e., the probability of the observed whole interaction data) by

$$L = \prod_{i,j} (\Pr(O_{ij} = 1))^{O_{ij}} (1 - \Pr(O_{ij} = 1))^{1-O_{ij}}.$$

Then, likelihood L is a function of (λ_{mn}, fp, fn). Since it is quite difficult to directly compute (λ_{mn}, fp, fn) maximizing L, they employed an *EM (Expectation-Maximization) algorithm* in order to locally maximize L where fp and fn were fixed to certain values.

SVM-Based Method

Support vector machines (SVMs) have been successfully applied to various problems in bioinformatics. Indeed, various approaches have been proposed for application of SVMs to inference of protein-protein interactions. Since we are focusing on domain-based methods here, we review application of SVMs to domain-based prediction of protein-protein interactions (Hayashida et al., 2003).

In order to apply SVMs, we need positive examples and negative examples. In this case, we treat observed interacting protein pairs as positive examples and non-observed protein pairs as negative examples. For each protein pair (P_i, P_j), we define the feature vector $\mathbf{f}_{ij}$ by

$$(\mathbf{f}_{ij})_{mn} = 1 \text{ if } D_{mn} \in P_{ij},$$
$$(\mathbf{f}_{ij})_{mn} = 0 \text{ otherwise},$$

where $(\mathbf{f}_{ij})_{mn}$ denotes the mn-th element of the vector $\mathbf{f}_{ij}$.

The experimental results in (Hayshida et al., 2003) suggest that among all above method and LPBN, the EM method has the best performance for binary data.

INFERENCE OF STRENGTHS OF PROTEIN-PROTEIN INTERACTIONS

In the previous section, it is assumed that protein-protein interaction data are given as binary data. That is, whether or not each protein pair interacts is given. However, multiple experiments are performed for the same protein pairs in practice and thus the ratio of the number of observed interactions to the number of experiments is available for each protein pair (Ito et al., 2001). We call this ratio the *strength of interaction*. By utilizing strength data of interactions, we may be able to derive better probabilities for domain-domain interactions (see also Fig. 3). In this section, we review such an approach.

Figure 3. Inference of the probabilities of domain-domain interactions from strengths of protein-protein interactions. The probabilities of domain-domain interactions can also be used for inference of strengths of protein-protein interactions.

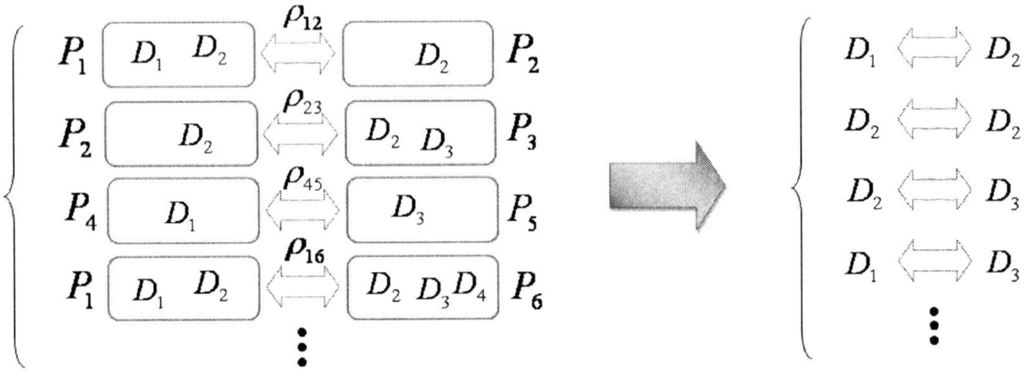

LPBN: LP-Based Method for Binary Interaction Data

Before explaining an LP (*linear programming*)-based method for numerical interaction data (i.e., strength data), we explain LPBN (Hayashida et al., 2003) because it is easier to understand.

The probabilistic model proposed for the EM method is also employed in the LPBN method. By introducing some threshold Θ, we can infer interactions by using the following rule:

$$P_i \text{ and } P_j \text{ interact} \Leftrightarrow 1 - \prod_{D_{mn} \in P_{ij}} (1 - \lambda_{mn}) \geq \Theta .$$

This condition can be transformed as follows.

$$\prod_{D_{mn} \in P_{ij}} (1 - \lambda_{mn}) \leq 1 - \Theta ,$$

$$\ln\left(\prod_{D_{mn} \in P_{ij}} (1 - \lambda_{mn}) \right) \leq \ln(1 - \Theta),$$

$$\sum_{D_{mn} \in P_{ij}} \ln(1 - \lambda_{mn}) \leq \ln(1 - \Theta),$$

where 'ln' denotes the natural logarithm. Here, we let $\gamma_{mn} = \ln(1 - \lambda_{mn})$ and $\beta = \ln(1 - \Theta)$. Then, the above condition can be written as a linear inequality:

$$\sum_{D_{mn} \in P_{ij}} \gamma_{mn} \leq \beta .$$

Thus, if we can find γ_{mn} satisfying

$$\sum_{D_{mn} \in P_{ij}} \gamma_{mn} \le \beta \quad \text{for all } i,j \text{ such that } O_{ij} = 1,$$

$$\sum_{D_{mn} \in P_{ij}} \gamma_{mn} > \beta \quad \text{for all } i,j \text{ such that } O_{ij} = 0,$$

$$\gamma_{mn} \le 0 \quad \text{for all } m,n,$$

we can obtain the parameters consistent with all observed data, where $\gamma_{mn} \le 0$ means $\lambda_{mn} \ge 0$ However, it is usually impossible to satisfy all constraints. Thus, as in SVMs, we introduce slack variables in order to cope with violated constraints and obtain the following *linear program*:

$$\text{minimize } \sum_{i,j} \xi_{ij}$$

subject to

$$\sum_{D_{mn} \in P_{ij}} \gamma_{mn} \le \beta - c + \xi_{ij} \quad \text{for all } i,j \text{ such that } O_{ij} = 1,$$

$$\sum_{D_{mn} \in P_{ij}} \gamma_{mn} > \beta + c - \xi_{ij} \quad \text{for all } i,j \text{ such that } O_{ij} = 0,$$

$$\gamma_{mn} \le 0 \quad \text{for all } \gamma_{mn},$$

$$\xi_{ij} \ge 0 \quad \text{for all } \xi_{ij},$$

$$\beta < 0,$$

where c is an appropriate small constant (e.g., c=0.01). Once γ_{mn} and β are determined, we can determine λ_{mn} and Θ by $\lambda_{mn} = 1 - \exp(\gamma_{mn})$ and $\Theta = 1 - \exp(\beta)$.

LPNM: LP-Based Method for Numerical Interaction Data

LPBN can be extended for coping with numerical interaction data (i.e., *strength of interactions*), where the extended method is called LPNM (Hayashida et al., 2003).

Recall that threshold Θ was employed in LPBN for making binary predictions. Instead, in LPNM, we use ρ_{ij} which is set to be the ratio of interactions between proteins P_i and P_j in a series of experiments:

$$\rho_{ij} = \frac{N(O_{ij})}{Z},$$

where $N(O_{ij})$ denotes the number of times that an interaction between P_i and P_j is observed in the experiments and Z is the total number of experiments.

Since ρ_{ij} is the ratio of interactions between P_i and P_j, it is reasonable to try to minimize the difference between $\Pr(P_{ij} = 1)$ and ρ_{ij}. However, it seems not computationally easy to minimize this difference because inference of domain-domain interactions that are most consistent with given protein-protein interactions is proven to be NP-hard (Hayashida et al., 2004). Thus, instead of minimizing the difference, we minimize the following sum:

$$\sum_{i,j} \left| \left(\sum_{D_{mn} \in P_{ij}} \ln(1 - \lambda_{mn}) \right) - \ln(1 - \rho_{ij}) \right| = \sum_{i,j} \left| \left(\sum_{D_{mn} \in P_{ij}} \gamma_{mn} \right) - \beta_{ij} \right|$$

By combining with other constraints, we obtain the following *linear program*.

$$\text{minimize} \sum_{i,j} \alpha_{ij}$$

subject to

$$\left(\sum_{D_{mn} \in P_{ij}} \gamma_{mn} \right) - \beta_{ij} \le \alpha_{ij} \,,$$

$$\beta_{ij} - \left(\sum_{D_{mn} \in P_{ij}} \gamma_{mn} \right) \le \alpha_{ij} \,,$$

$$\gamma_{mn} \le 0 \quad \text{for all } \gamma_{mn} \,,$$

$$\alpha_{ij} \ge 0 \quad \text{for all } \alpha_{ij} \,,$$

$$\beta_{ij} < 0.$$

ASNM: Association Method for Numerical Interaction Data

We can also adapt the *association method* for numerical interaction data (Hayashida et al., 2004). Instead of counting the number of interacting protein pairs that contain domain pair (D_m, D_n), we use the sum of strengths of interacting protein pairs that contain domain pair (D_m, D_n). That is, we use the following score

$$ASNM(D_m, D_n) = \frac{\sum_{(i,j):D_{mn} \in P_{ij}} \rho_{ij}}{N_{mn}} \,.$$

For example, suppose that $\rho_{12} = 0.9$, $\rho_{23} = 0.1$, $\rho_{45} = 0.75$, $\rho_{16} = 0.8$ in Fig. 3. Then, $ASNM(D_1, D_2)$ will be $(0.9 + 0.8)/2 = 0.85$ since (P_1, P_2) and (P_1, P_6) contain (D_1, D_2), whereas (P_2, P_3) or (P_4, P_5) does not contain (D_1, D_2). It is to be noted that ASNM is equivalent to the association method if each ρ_{ij} takes a binary value.

APM: Association Probabilistic Method

In the definition of the ASNM score, all protein pairs are treated equally. However, some protein pairs may contain (D_1, D_2) only whereas other protein pairs may contain (D_1, D_2) and other domain pairs. For example, in Fig. 3, (P_1, P_2) contains 2 domain pairs, whereas (P_1, P_6) contains 6 domain pairs. In ASNM, contributions of these two protein pairs to (D_1, D_2) are equally treated. However, it is reasonable to give more weight to (P_1, P_2) than to (P_1, P_6). Chen et al. (2006) noticed this point and developed an alternative scoring scheme, which is called APM and is defined as below:

$$APM(D_m, D_n) = \frac{\sum_{(i,j):D_{mn} \in P_{ij}} \left[1 - (1 - \rho_{ij})^{1/|P_{ij}|} \right]}{N_{mn}} \,.$$

The results of computational experiments showed that APM has better accuracy than ASNM and LPNM and runs much faster than LPNM (Chen et al., 2006). Thus, APM is considered as one of the best existing methods for inference of protein-protein interactions from numerical interaction data (i.e., strengths of known interactions). The scoring schemes are summarized in Fig. 4.

Figure 4. Summary of scoring schemes on domain-based methods for prediction of protein-protein interactions

$ASSOC(D_m, D_n)$	$ASNM(D_m, D_n)$	$APM(D_m, D_n)$
$\frac{I_{mn}}{N_{mn}}$	$\frac{\sum_{(i,j):D_{mn}\in P_{ij}} \rho_{ij}}{N_{mn}}$	$\frac{\sum_{(i,j):D_{mn}\in P_{ij}} \left[1-(1-\rho_{ij})^{1/\lvert P_{ij}\rvert}\right]}{N_{mn}}$
LPBN		*LPNM*
minimize $\sum_{i,j} \xi_{ij}$ subject to $\sum_{D_{mn}\in P_{ij}} \gamma_{mn} \le \beta - c + \xi_{ij}$ for all i,j such that $O_{ij} = 1$, $\sum_{D_{mn}\in P_{ij}} \gamma_{mn} > \beta + c - \xi_{ij}$ for all i,j such that $O_{ij} = 0$, $\gamma_{mn} \le 0$ for all γ_{mn}, $\xi_{ij} \ge 0$ for all ξ_{ij}, $\beta < 0$		minimize $\sum_{i,j} \alpha_{ij}$ subject to $\left(\sum_{D_{mn}\in P_{ij}} \gamma_{mn}\right) - \beta_{ij} \le \alpha_{ij}$, $\beta_{ij} - \left(\sum_{D_{mn}\in P_{ij}} \gamma_{mn}\right) \le \alpha_{ij}$, $\gamma_{mn} \le 0$ for all γ_{mn} , $\alpha_{ij} \ge 0$ for all α_{ij} , $\beta_{ij} < 0$

A DOMAIN-BASED MODEL OF PROTEIN-PROTEIN INTERACTION NETWORKS

We have overviewed application of domain-based models to inference of protein-protein interactions. However, application of domain-based models is not limited to inference of protein-protein interactions. In this section, we show that a domain-based model is useful for explaining the scale-free distribution of protein-protein interaction networks.

Scale-Free Network

Before explaining domain-based models of protein-protein interaction networks, we briefly review the *scale-free* network and *preferential attachment*, where the latter is the most well-known mechanism for artificially generating scale-free networks (Barabási & Albert, 1999; Barabási et al., 1999).

We regard each network as an undirected graph. Each node v in a graph is said to have degree k if it is connected to k other nodes. Let $P(k)$ denote the frequency of nodes that have degree k (see also Fig. 5). Then, a graph (or a network) is called *scale-free* if $P(k)$ approximately follows a *power-law*

$$P(k) \propto k^{-\gamma} ,$$

where γ is a positive constant depending on networks and γ usually takes a value between 2 and 3. It is observed that this power-law emerges in many biological networks and other networks, which include metabolic networks,

Figure 5. Explanation of degree distribution. In this example, P(1)=1/10, P(2)=5/10, P(3)=3/10, P(4)=1/10, P(5)=P(6)=...=0.

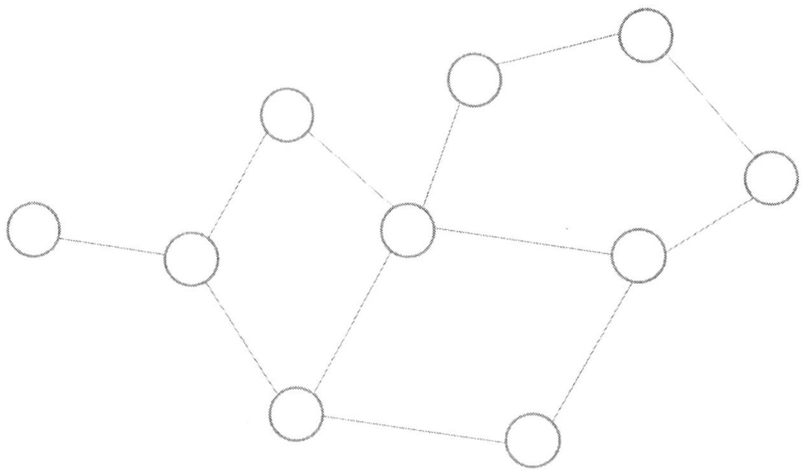

protein-protein interaction networks, the World Wide Web, the Internet, and the movie star network (Barabási & Oltvai, 2004).

It is known that this scale-free distribution is very different from the degree distribution of random graphs, which follows the Poisson distribution. Thus, it motivated many researchers to develop various mathematical models of generating scale-free networks. Among these, the model proposed by Barabási and Albert (1999) is well-known, which is described as follows.

1. Create a complete graph with N_0 nodes, where N_0 is a constant greater than 1.
2. Create a new node v and connect it to m nodes in the current graph, where each node is selected with the probability proportional to its degree, and m is a constant greater than 1.
3. Repeat (2) until the number of nodes reaches the specified number.

They showed that this procedure generates networks following the power-law of $P(k) \propto k^{-3}$, regardless of N_0 and m.

They used mean-field theory to show this power-law. Here, we briefly review their analysis (Barabási et al., 1999). Let $k_i(t)$ denote the degree of a node i at step t. Then, the average $k_i(t)$ should approximately satisfy:

$$\frac{d\,k_i(t)}{d\,t} = \frac{mk_i(t)}{2mt} = \left(\frac{1}{2t}\right)k_i(t)$$

because the sum of degrees at step t is around $2mt$ and thus the probability that node i is connected to a new node at step t is

$$\frac{mk_i(t)}{2mt}.$$

Based on this differential equation, they obtained

$$P(k) \propto k^{-3}.$$

The above mechanism of choosing a node according to the degree distribution is called preferential attachment. Though preferential attachment is simple and interesting, relations to biological networks are unclear. It looks rather artificial from a biological point of view, as well as many other mathematical models. Thus, we should develop more biologically appropriate models.

Duplication-Divergence Model

Since preferential attachment is not adequate for explaining scale-freeness of protein-protein interaction networks, several models have been proposed. Among those, extensive studies have been done on the *duplication-divergence model* (Berg et al., 2004; Ispolatov et al., 2005; Vásquetz et al., 2003). Though there are several variants, this model is defined as follows (see also Fig. 6).

Duplication: A node (protein) *i* is selected at random. A new node *i'* with links to all neighbors of *i* is created. That is, the replica of the original node is introduced and connected to each neighbor of the original node.

Divergence: Each link emanating from the replica *i'* is activated with retention probability σ.This means that a new link disappears with probability 1-σ by means of mutations. If at least one link is established, the replica is preserved. Otherwise, the replica is discarded.

It is analytically shown that the degree distributions of networks constructed by this model approximately follow a *power-law*. It seems that this model is more biologically appropriate than the preferential attachment model. However, mathematical analysis of this model is somewhat involved. Furthermore, though this model is based on evolution of networks, what evolve in organisms are not networks but are genes or protein sequences. Therefore, it is worthy to consider other models that are based on evolution of protein sequences.

Figure 6. Duplication and divergence model. In this model, some protein i is randomly selected and the replica i' of i is created. Links emanating from i are copied but some of copied links disappear as a result of divergence.

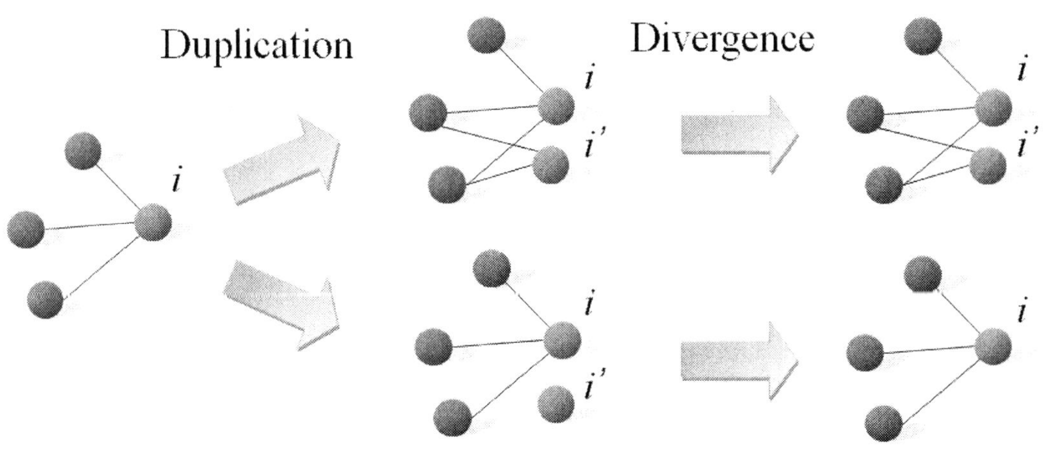

A Simple Evolutionary Model for Protein-Protein Interaction Networks

It is well-known that mutations and gene duplications play an important role in evolution of genomes. Based on this fact, we developed a simple evolutionary model of protein *domains* (Nacher et al., 2006). The model is defined as follows (see also Fig. 7).

1. At first, there exist N_0 different types of domains.
2. With probability 1-*a*, a new type of a domain is created (it corresponds to a mutation). Otherwise, one protein is selected at uniformly random and a copy of the protein is created (it corresponds to a duplication).
3. Step (2) is repeated until the number of proteins reaches a specified number.

It should be noted that every protein consists of a single domain in this model. That is, we do not consider fusion or shuffle of domains.

This simple model yields a *scale-free* distribution of protein *domains*. Here, we briefly review the analysis done in Nacher et al (2006). Let $P_D(k)$ denote the number of domains for each of which there exist exactly *k* copies of the domain. For example, suppose that there are 2,2,2,3,3,5 copies of proteins of domain type A, B, C, D, E, F, respectively. Then, $P_D(1) = P_D(4) = P_D(6) = \cdots = 0$, $P_D(2) = \frac{3}{6}$, $P_D(3) = \frac{2}{6}$, and $P_D(5) = \frac{1}{6}$. Let *i* and $k_i(t)$ denote a type of a domain and the number of proteins consisting of domain *i* at step *t*, respectively. Then, using the mean-field approximation, we have

$$\frac{d\,k_i(t)}{dt} = a \cdot \frac{k_i(t)}{t}$$

because a copy of a protein is created with probability *a* at each step and there are *t* proteins in total at step *t*. Since this equation is very similar to that for *preferential attachment*, we can use the same analysis method as in Barabási et al. (1999) and thus we have

Figure 7. Duplication and mutation model of evolution of protein domains. With probability 1-a, a new type of a domain is created (mutation). With probability a, one protein (domain) is selected at uniformly random and a copy of the protein is created (duplication).

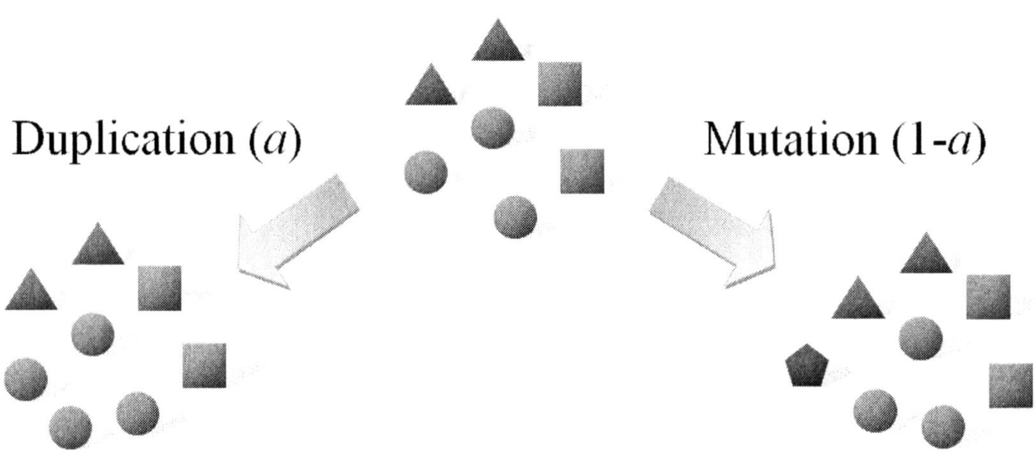

$$P_D(k) \propto k^{-(1+(1/a))}.$$

It is to be noted that the exponent in the case of preferential attachment is always 3, whereas the exponent can be controlled by parameter a in this case. Since it is usually considered that the mutation rate is small compared to the duplication rate, a should be close to 1 and thus the exponent of the power-law should be close to $-(1 + (1/a)) = -2$. This scale-free distribution of domains was confirmed by computational analysis of real protein sequences obtained from publicly available databases.

Combination of Evolutionary Model and Domain-Based Interaction Model

In the domain-based model for prediction of protein-protein interactions, we assumed that two proteins interact with each other if some domains contained in these proteins interact with each other. Here, we further pose a simple assumption that interacting domain pairs are determined uniformly at random. We combine this assumption with the evolutionary model of protein domains and the domain-based model of protein-protein interactions (Nacher et al., 2007). The combined model is summarized as below.

1. Proteins are generated according to the simple evolutionary model of domains, where we assume that each protein consists of a single domain,
2. Interacting domain pairs are determined uniformly at random.
3. Two proteins are defined to interact if the domains in these proteins interact.

Here, we briefly review analysis of this model. From (1), we can assume that *domain* distribution follows $P_D(k) \propto k^{-(1+(1/a))}$. From (2), we assume that among all possible domain pairs, N domain pairs are selected as interacting domain pairs at uniformly random. First, we consider the case N=1. Let n_X be the number of copies of domain X. Suppose that domains A and B are selected as interacting domains. The probability that such a pair is selected is proportional to $x^{-\gamma} \cdot y^{-\gamma}$ if $n_A = x$ and $n_B = y$. Then, we will have x proteins with degree y, and y proteins with degree x. Thus, the expected number of proteins having degree y is approximated by

$$\Pr(n_B = y) \cdot E[n_A] \propto y^{-\gamma} \int_{x=1}^{\infty} x^{-\gamma} dx = \frac{1}{\gamma - 2} y^{-\gamma} ,$$

where $E[Z]$ means the expected value of Z. We need to repeat this procedure for N times. We can also show that the degree distribution is approximately proportional to $y^{-\gamma}$ if N is not too large, and thus we can obtain the *scale-free* degree distribution of protein-protein interaction networks. This theoretical result was confirmed by computer simulation and database analysis (Nacher et al., 2007).

DISCUSSIONS AND FUTURE TRENDS

We have seen that domain-based models are useful both for prediction of protein-protein interactions and for explaining the scale-freeness of protein-protein interaction networks. It is quite interesting that these models are based on a simple assumption: two proteins interact with each other if two domains contained in these proteins interact with each other. The results of computational experiments on both problems suggest that this simple assumption captures some features of the complex mechanisms of real protein-protein interactions. However, the models reviewed in this paper are too simple. In order to make better predictions and more detailed analysis, we need to improve models and methods. In the following, we discuss some of possible improvements for each problem separately.

Inference of Protein-Protein Interactions

The domain-based prediction methods are simple and useful. As shown by Hayashida et al. (2003) and Chen et al. (2006), the domain-based prediction methods showed quite high accuracy for training data whereas they did not show very high accuracy for test data. This suggests that there exist some overfitting problems. Thus, it might be able to improve the prediction accuracy if some technique for avoiding overfit can be incorporated into the domain-based methods.

However, it may be still difficult to significantly improve the prediction accuracy for test data by using domain-based methods only because domain-based models capture some partial aspects of protein-protein interactions. In order to significantly improve the prediction accuracy, we should analyze more detailed patterns of protein-protein interactions. Even if two pairs of proteins have the same domain organization, it is possible that one pair interacts whereas the other pair does not. We should carefully collect and analyze such cases. Then, we may find other important factors affecting protein-protein interactions. Therefore, detailed analysis of correlation between domain organizations and protein-protein interactions should be performed.

In all the models reviewed in this chapter, independence of domain-domain interactions was assumed. However, it is not independent. For example, suppose that protein A consists of domain D_1 and protein B consists of domains D_2 and D_3. If D_1 interacts with D_2, D_1 cannot interact with D_3. Thus, interaction between D_1 and D_2 is not independent from interaction between D_1 and D_3. Therefore, domain-based models should be modified so that correlations between domain-domain interactions are taken into account.

Analysis of Protein-Protein Interaction Networks

We have reviewed the duplication-divergence model and the domain-based model of protein-protein interaction networks. Though both models are simple and seem to be more biologically appropriate than the preferential attachment model, multi-domain proteins are not taken into account. Since there exist a lot of multi-domain proteins, it is necessary to consider multi-domain proteins. In particular, it is important to consider fusion and shuffle of domains. We performed simulations on a domain-based model with fusion and shuffle of domains, and observed that the scale-free degree distributions of domains are also reproduced in this case (Nacher et al., 2006). However, we were not successful in making mathematical analysis. Therefore, mathematical analysis of such a model is left as future work. For a multi-domain variant of the duplication-divergence model, some mathematical analyses have been done (Evlampiev & Isambert, 2007; Karev et al., 2002). However, these are complicated. It would be better if much simpler analyses could be done.

In the models reviewed in this chapter, information of interface sites on proteins was not taken into account. However, Kim et al. (2006) pointed out the importance of considering interface sites and distinguishing overlapping interfaces from non-overlapping interfaces. That is, it is important to distinguish multi-interface hubs from singlish-interface hubs. Therefore, development and analysis of mathematical models considering interface sites are important future works.

CONCLUDING REMARKS

In this chapter, we have reviewed domain-based models both for prediction of protein-protein interactions and for explaining the scale-freeness of protein-protein interaction networks. Though the reviewed models are quite simple, it seems that these models capture important factors of the complex mechanisms of protein-protein interactions.

Of course, as reviewed in the other chapters, there exist other important factors in protein-protein interactions. In particular, such factors as three-dimensional structures, sequence and structural motifs, physico-chemical properties of interface sites and evolutionary conserved regions should be taken into account. Therefore, integration of important factors should be studied and would be useful for both prediction and analysis of protein-protein interactions.

REFERENCES

Barabási, A-L., & Oltvai, Z. N. (2004). Network biology: understanding the cells' functional organization. *Nature Reviews Genetics*, *5*, 101-113.

Barabási, A-L., Albert, R., & Jeong, H. (1999). Mean-field theory for scale-free random networks. *Physica A*, *272*, 174-187.

Barabási, A-L., & Albert, R. (1999). Emergence of scaling in random networks. *Science, 286*, 509-512.

Berg, J., Lässig, M., & Wagner, A. (2004). Structure and evolution of protein interaction networks: a statistical model for link dynamics and gene duplications. *BMC Evolutionary Biology*, *4*, 51.

Bru, C., Courcelle, E., Carrère, S., Beausse, Y., Dalmar, S., & Kahn, D. (2005). The ProDom database of protein domain families: more emphasis on 3D. *Nucleic Acids Research*, *33*, D212-D215.

Chen, L., Wu, L-Y., Wang, Y., & Zhang, X-S. (2006). Inferring protein interactions from experimental data by association probabilistic method. *PROTEINS: Structure, Function, and Bioinformatics*, *62*, 833-827.

Chen, X-W., & Liu, M. (2006). Domain-based predictive models for protein-protein interaction predictions. *EURASIP Journal on Applied Signal Processing*, *2006*, 32767.

Deng, M., Mehta, S., Sun, F., & Chen, T. (2002). Inferring domain-domain interactions from protein-protein interactions. *Genome Research*, *12*, 1540-1548.

Evlampiev, K., & Isambert, H. (2007). Modeling protein network evolution under genome duplication and domain shuffling. *BMC Systems Biology*, *1*, 49.

Finn, R. D., et al. (2006). Pfam: clans, web tools and services. *Nucleic Acids Research*, *34*, D247-D251.

Hayashida, M., Ueda, N., & Akutsu, T. (2003). Inferring strengths of protein-protein interactions from experimental data using linear programming. *Bioinformatics*, *19*, ii58-ii65.

Hayashida, M., Ueda, N., & Akutsu, T. (2004). A simple method for inferring strengths of protein-protein interactions. *Genome Informatics*, *15-1*, 56-68.

Ispolatov, I., Krapivsky, P. L., & Yuryev, A. (2005). Duplication-divergence model of protein interaction network. *Physical Review E*, *71*, 061911.

Ito, T., Chiba, T., Ozawa, R., Yoshida, M., Hattori, M., & Sakaki, Y. (2001). A comprehensive two-hybrid analysis to explore the yeast protein interactome. *Proceedings of the National Academy of Sciences USA*, *98*, 4569-4574.

Itzhaki, Z., Akiva, E., Altuvia, Y., & Margalit, H. (2006). Evolutionary conservation of domain-domain interactions. *Genome Biology*, *7*, R125.

Jeong, H., Tombor, B., Albert, R., Oltvai, Z. N., & Barabási, A-L. (2001). Lethality and centrality in protein networks. *Nature*, *411*, 41-42.

Karev, G. P., Wolf, Y. I, Rzhetsky, A. Y., Berezovskaya, F. S., & Koonon, E. V. (2002). Birth and death of protein domains: A simple model of evolution explains power law behavior. *BMC Evolutionary Biology*, *2*, 18.

Kim, I., Liu, Y., & Zhao, H. (2007). Bayesian methods for predicting interacting protein pairs using domain information. *Biometrics*, *63*, 824-833.

Kim, P. M., Lu, L. J., Xia, Y., & Gerstein, M. B. (2006). Relating three-dimensional structures to protein networks provides evolutionary insights. *Science*, *314*, 1938-1941.

Kim, W. K., Park, J., & Suh, J. K. (2002). Large scale statistical prediction of protein-protein interaction by potentially interacting domain (PID) pair. *Genome Informatics, 13*, 42-40.

Li, X-L., Tan, S-H., & Ng, S-K. (2006). Improving domain-based protein interaction prediction using biologically-significant negative dataset. *International Journal of Data Mining and Bioinformatics, 1*, 138-149.

Nacher, J. C., Hayashida, M., & Akutsu, T. (2006). Protein domain networks: Scale-free mixing of positive and negative exponents. *Physica A, 367*, 538-552.

Nacher, J. C., Hayashida, M., & Akutsu, T. (2007). Topological aspects of protein networks. *Studies in Computational Intelligence, 56*, 147-158.

Ng, S-K., Zhang, Z., Tan, S-H., & Lin, K. (2003). InterDom: a database of putative interacting protein domains for validating predicted protein interactions and complexes. *Nucleic Acids Research, 31*, 251-254.

Nicola, J., et al. (2007). New developments in the InterPro database. *Nucleic Acids Research, 35*, D224-D228.

Schuster-Böckler, B., & Bateman, A. (2007). Reuse of structural domain-domain interactions in protein networks. *BMC Bioinformatics, 8*, 259.

Singhal, M., & Resat, H. (2007). A domain-based approach to predict protein-protein interactions. *BMC Bioinformatics, 8*, 199.

Sprinzak, E., & Margalit, H. (2001). Correlated sequence-signatures as markets of protein-protein interaction. *Journal of Molecular Biology, 311*, 681-692.

Sprinzak, E., Altuvia, Y., & Margalit, H. (2006). Characterization and prediction of protein-protein interactions within and between complexes. *Proceedings of the National Academy of Sciences USA, 103*, 14718-14723.

Uetz, P. *et al.* (2000). A comprehensive analysis of protein-protein interactions in *Saccharomyces cerevisiae. Nature, 403*, 623-627.

Vázquez, A., Flammini, A., Maritan, A., & Vespignani, A. (2003). Modeling of protein interaction networks. *ComPlexUs, 1*, 38-44.

Chapter IV
Incorporating Graph Features for Predicting Protein-Protein Interactions

Martin S. R. Paradesi
Kansas State University, USA

Doina Caragea
Kansas State University, USA

William H. Hsu
Kansas State University, USA

ABSTRACT

This chapter presents applications of machine learning to predicting protein-protein interactions (PPI) in Saccharomyces cerevisiae. Several supervised inductive learning methods have been developed that treat this task as a classification problem over candidate links in a PPI network – a graph whose nodes represent proteins and whose arcs represent interactions. Most such methods use feature extraction from protein sequences (e.g., amino acid composition) or associated with protein sequences directly (e.g., GO annotation). Others use relational and structural features extracted from the PPI network, along with the features related to the protein sequence. Topological features of nodes and node pairs can be extracted directly from the underlying graph. This chapter presents two approaches from the literature (Qi et al., 2006; Licamele & Getoor, 2006) that construct features on the basis of background knowledge, an approach that extracts purely topological graph features (Paradesi et al., 2007), and one that combines knowledge-based and topological features (Paradesi, 2008). Specific graph features that help in predicting protein interactions are reviewed. This study uses two previously published datasets (Chen & Liu, 2005; Qi et al., 2006) and a third dataset (Paradesi, 2008) that was created by combining and augmenting three existing PPI databases. The chapter includes a comparative study of the impact of each type of feature (topological, protein sequence-based, etc.) on the sensitivity and specificity of classifiers trained using specific types of features. The results indicate gains in the area under the sensitivity-specificity curve for certain algorithms when topological graph features are combined with other biological features such as protein sequence-based features.

INTRODUCTION

Protein-Protein Interaction Prediction Problem

The term *protein-protein interaction (PPI)* refers to associations between proteins as manifested through biochemical processes such as formation of structures, signal transduction, transport, and phosphorylation. PPI plays an important role in the study of biological processes. Many PPIs have been discovered over the years and several databases have been created to store the information about these interactions such as BIND (Bader *et al.*, 2003), DIP (Salwinski *et al.*, 2004), MIPS (Mewes *et al.*, 2002), IntAct (Kerrien *et al.*, 2007) and MINT (Chatr-aryamontri *et al.*, 2007). In particular, more than 80,000 interactions between yeast proteins are available from various high-throughput interaction detection methods (von Mering *et al.*, 2002). These methods can detect if the interaction is either a physical binding between proteins or a functional association between proteins. Often, the functional association between two proteins leads to physical binding among them. Determining PPI using high-throughput methods is expensive and time-consuming. Furthermore, a high number of false positives and false negatives can be generated. Therefore, there is a need for computational approaches that can help in the process of identifying real protein-protein interactions.

Several methods have been designed to address the task of predicting protein-protein interactions using machine learning. Most of them use features from protein sequences (e.g., amino acids composition) or associated with protein sequences directly (e.g., GO annotation). However, the PPI network can be used to design node and topological features from the associated graph. Several methods use such relational and structural features extracted from the PPI network, along with the features related to the protein sequence. This chapter provides an overview of several machine learning methods for predicting PPI using the graph information extracted from a PPI network along with other available biological features of the proteins and their interactions, and shows the importance of the graph features for accurate predictions.

Overview of PPI Databases

Several PPI databases have been used to extract examples of PPIs for machine learning algorithms. We review the main PPI databases in what follows.

The Biomolecular Interaction Network Database (BIND)

BIND (Bader *et al.*, 2003) stores information about interactions, complexes and pathways. It also contains a number of large scale interaction and complex mapping experiments using yeast two-hybrid, mass spectrometry, genetic interactions and phage display. The group that maintains BIND has also developed a graphical analysis tool that provides users an understanding of functional domains in protein interactions. They have also developed a clustering tool that allows users to divide the protein interaction network into specific regions of interest. BIND assumes that interactions can occur between two biological 'objects', which could be proteins, RNA or DNA sequences, genes, molecular complexes, small molecules, or photons (light).

The Database of Interacting Proteins (DIP)

DIP (Salwinski *et al.*, 2004) is a database containing 18,343 interactions between 4,923 proteins validated from 23,366 experiments of the *Saccharomyces cerevisiae* organism. A few of the experiments from which they validate protein interactions are co-immunoprecipitation, yeast two-hybrid and in vitro binding assays. The group that maintains DIP has developed several quality assessment methods and uses them to identify the most reliable subset of the interactions that are inferred from high-throughput experiments. They also provide an online implementation of their evaluation methods that can be used to evaluate the reliability of new experimental and predicted interactions.

IntAct

IntAct (Kerrien *et al.*, 2007) contains data such as experimental methods, conditions and interacting domains that is extracted entirely from publications and is manually annotated by curators. It also formalizes the data by using a comprehensive set of controlled vocabularies in order to ensure data integrity. It is thus far the only published database that contains negative examples of protein interactions, *i.e.* explicitly identifies pairs of proteins that do not interact. The database contains 169,792 interactions between 63,427 proteins. These interactions were obtained from 8,477 experiments that were performed on several organisms. The web site provides tools allowing users to search, visualize and download data from the repository.

A Molecular INTeraction database (MINT)

MINT (Chatr-aryamontri *et al.*, 2007) stores molecular interaction data extracted from several publications. Most of its curation work is focused on physical interactions, direct interactions and colocalizations between proteins. Genetic or computationally inferred interactions are not included in the database. It contains 42,044 interactions between 5,256 proteins of the *Saccharomyces cerevisiae* organism. An online graph visualization and editing tool called "MINT Viewer" is available that allows users to view the interaction network and delete edges that are not of interest to the user.

The Munich Information Center for Protein Sequences (MIPS)

MIPS (Mewes *et al.*, 2002) provides information on Open Reading Frames (ORFs), RNA genes and other genetic elements. The research group that maintains MIPS has also applied techniques such as gene disruption in conjunction with powerful expression analysis and two-hybrid techniques as part of a systematic functional genome analysis. These methods generate information on how proteins cooperate in complexes, pathways and cellular networks. In addition, detailed information on transcription factors and their binding sites, transport proteins and metabolic pathways are being included or interlinked to the core data. The database also provides information on the molecular structure and the functional network of the yeast genome.

Introduction to Machine Learning

Machine learning algorithms (Mitchell, 1997) offer some of the most cost-effective approaches to automated knowledge discovery and data mining (discovery of features, correlations, and other complex relationships and hypotheses that describe potentially interesting regularities) from large data sets. In particular, machine learning algorithms have proven to be very successful for many bioinformatics problems, including protein-protein interaction prediction.

In this chapter, we formulate the problem of PPI prediction as a classification task, i.e. a task where the learning algorithm is provided with experience in the form of labeled examples (a.k.a., training data set or data source) and is asked to classify new unlabeled examples in one of several possible classes. In our case, the training examples consist of existing information about protein interactions extracted from PPI databases. Each example is encoded using a set of variables called attributes or features. A special attribute, called the class label, is used to represent the class to which that particular example belongs. The class label in a protein interaction prediction problem indicates whether the proteins in a candidate pair interact with each other (i.e., it takes two values: yes and no). The output of a learning algorithm for a classification task is called a classifier. Several strategies can be used to estimate the true error of a classifier. The simplest one is to divide the labeled data into a training set and a test set. The classifier is learned from the training set and its error is estimated using the test set. More commonly, the error is estimated by using a method called k-fold cross-validation. To use this method, the labeled data is divided into k folds. A classifier is learned from a training set consisting of k - 1 folds and tested on the remaining k^{th} fold. The estimate for the true error is obtained by taking the average of the error of the k possible classifiers learned by leaving out one fold at a time.

We will review several learning algorithms that have been used to predict PPI interactions in the next few paragraphs.

Decision Trees

Decision tree algorithms (Quinlan, 1986; Breiman *et al.*, 1984) are among some of the most widely used machine learning algorithms for building pattern classifiers from data. Their popularity is due in part to their ability to: select from all attributes used to describe the data, a subset of attributes that are relevant for classification; identify complex predictive relations among attributes; and produce classifiers that are easy to comprehend for humans. The ID3 (Iterative Dichotomizer 3) algorithm proposed by Quinlan (1986) and its more recent variants such as C4.5 (Quinlan, 1993) are representative for a widely used family of decision tree learning algorithms. The ID3 algorithm searches in a greedy fashion, for attributes that yield the maximum amount of information for determining the class membership of instances in a training set *D* of labeled instances. The result is a decision tree that correctly assigns each instance in *D* to its respective class. The construction of the decision tree is accomplished by recursively partitioning *D* into subsets based on values of the chosen attribute until each resulting subset has instances that belong to exactly one of the *m* classes. The selection of an attribute at each stage of construction of the decision tree maximizes the estimated expected information gained from knowing the value of the attribute in question. C4.5 (Quinlan, 1993) is the most popular variant of the ID3 algorithm that has been implemented as the J48 classifier in *WEKA*, the *Waikato Environment for Knowledge Analysis* (Witten & Frank, 2005), a popular machine learning toolkit. Some of the improvements that C4.5 has made over ID3 algorithm are: dealing with missing data, pruning the tree after creation and dealing with attributes of different costs.

Random Forests

Random Forest classifiers are seen to produce highly accurate results for many supervised classification problems (Breiman, 2001). This algorithm involves the construction of multiple trees from the data. Each tree votes for the class of a new instance and the class with the maximum number of votes is chosen. The method of constructing each tree is described by Breiman (2001) in the following steps:

1. If there are *N* examples in the training set, the tree will be built by sampling *N* examples at random with replacement,
2. If there are *M* input variables, a small subset of these examples m is chosen at each node to find the best split of the data at that node, and
3. There is no pruning of the trees that are constructed at each stage.

Naïve Bayes

Naïve Bayes is a highly practical learning algorithm (Mitchell, 1997), comparable to more powerful algorithms such as decision trees or neural networks in terms of performance in some domains. In the Naïve Bayes framework, each example x is described by a conjunction of attribute values, i.e. $x = < a_1, a_2, \ldots, a_n >$. The class label of an example can take any value from a finite set $C = \{c_1, c_2, \ldots, c_m\}$. The attribute values are assumed to be conditionally independent given the class label. A training set of labeled examples, $D = \{< x_1, y_1 >, < x_2, y_2 >, \ldots, < x_t, y_t >\}$, is presented to the algorithm. During the learning phase, a hypothesis *h* consisting of conditional and prior probabilities is learned from the training set. During the evaluation phase, the trained Naïve Bayes classifier predicts the class label of new instances *x* as follows:

$$c_{NB}(x) = \arg\max_{c_j \in C} \prod_{i=1}^{n} P(c_j)P(a_i \mid c_j)$$

Support Vector Machine

The Support Vector Machine (SVM) algorithm (Vapnik, 1998; Cortes & Vapnik, 1995; Scholkopf *et al.*, 1997; Cristianini & Shawe-Taylor, 2000) is a binary classification algorithm. If the data are linearly separable, it outputs a separating hyperplane, which maximizes the "margin" between classes. If data are not linearly separable, the algorithm works by implicitly mapping the data to a higher dimensional space, where the data become separable. A maximum margin separating hyperplane is found in this space. This hyperplane in the high dimensional space corresponds to a nonlinear surface in the original space. SVM classifiers are sometimes called "large margin classifiers" because they find a maximum margin separation. Large margin classifiers are very popular due to theoretical results that show that a large margin ensures a small generalization error bound (Vapnik, 1998) and also because they proved to be very effective in practice.

K-Nearest Neighbors

The K-Nearest Neighbors classifier (Cover & Hart, 1967; Mitchell, 1997) is a simple example of instance-based learning, also known as lazy learning. In the K-Nearest Neighbors algorithm, the nearest neighbors are defined in terms of a metric (e.g., Euclidean distance) D between instances. The class label for a new instance x is given by the most common class label among the k training examples nearest to x (according to the distance D).

Bagging

Bootstrap aggregating or bagging (Breiman, 1996) is an algorithm that helps improve the accuracy of a classifier. Bagging works by sampling examples from the training dataset D with replacement to create subsets of the training data, which are called bootstrap samples. Classifiers are learned from the different bootstrap samples. To predict the class label of a new example, the outputs of the resulting classifiers are averaged or the classifiers are allowed to vote for the class of this new example. Bagging has been shown to avoid overfitting and to reduce the variance of learning algorithms in several domains (Breiman, 1996).

REPTree

REPTree is a supervised inductive learning algorithm implemented in WEKA (Witten & Frank, 2005) that builds a decision/regression tree using information gain/variance reduction (Quinlan, 1986) as the splitting criterion. It then prunes the tree using reduced-error pruning with backfitting (Quinlan, 1993; Mitchell, 1997). Missing values are dealt with by using fractional instances as in C4.5 (Quinlan, 1993).

BACKGROUND AND SIGNIFICANCE

Several graph-based approaches have been used to address the problem of predicting PPIs. These approaches represent the PPI network as a graph and extract relational and structural features from it. The training dataset provided to the learning algorithms consists of examples represented by such graph-based features, sometimes together with other features (e.g., amino acid composition, GO functions) associated with the protein pairs. The test dataset presented to the classifier produced by the learning algorithm contains protein pairs that are not present in the training dataset. Statistical measures such as accuracy, sensitivity, specificity and AUC score can be calculated to evaluate the performance of the learning algorithms. Several approaches to PPI prediction, including approaches based on graph-based features, are described and compared below.

Qi, Bar-Joseph & Klein-Seetharaman (2006)

Qi *et al.* (2006) divide the protein interaction prediction task into three sub-tasks: (1) prediction of physical (or actual) interaction among proteins, (2) prediction of proteins belonging to the same complex and (3) prediction of proteins belonging to the same pathway. They use different data sources for different subtasks: data from the MIPS database (Mewes *et al.*, 2002) for the first subtask, data from the DIP database (Salwinski *et al.*, 2004) for the second subtask and data from the KEGG database (Kanehisa & Goto, 2000) for the third subtask. One hundred sixty two features were constructed, grouped into seventeen distinct categories and studied to understand their effect on the protein interaction prediction subtasks. The categories that the 162 features were grouped into are:

- *Gene expression*: This category includes 20 features (each being a Pearson's correlation coefficient) calculated on 20 gene expression datasets that were recorded under more than 500 conditions (Bar-Joseph *et al.*, 2003).
- *Gene Ontology (Molecular Function, Biological Process & Cellular Component):* These three categories contain information on how many times a pair of proteins occurs in the trees (Ashburner *et al.*, 2000; Christie *et al.*, 2004).
- *Protein Expression:* Features in this category capture the difference in the expression levels for the candidate pair of proteins (Ghaemmaghami *et al.*, 2003).
- *Essentiality:* An essential protein is a protein for which deletion of the encoding gene results in a lethal phenotype, which is usually measured under laboratory conditions. The singleton feature in this category records whether the members of a pair of proteins are essential.
- *High-throughput PPI datasets (HMS_PCI, TAP & Y2H):* These three categories contain information extracted from several high-throughput protein interaction methods (Bader *et al.*, 2003; Gavin *et al.*, 2002; Ho *et al.*, 2002; Ito *et al.*, 2001; Uetz *et al.*, 2000).
- *Synthetic Lethal*: Synthetic interactions are identified if mutations in two separate genes produce a different phenotype from either gene alone, and indicate a functional association between the two genes. Two genes have a synthetic lethal relationship if mutants in either gene are viable but the double mutation is lethal. The single feature in this category was extracted by taking the union of lethality indicators from Tong *et al.* (2001) and MIPS (Mewes *et al.*, 2002).
- *Gene neighborhood/Gene Fusion/Gene Co-occurrence:* The single feature in this category is the disjunction of indicators from the three datasets described by von Mering *et al.* (2002).
- *Sequence Similarity*: The single feature in this category is a BLAST hit indicator for the query protein on the *Saccharomyces* Genome Database or SGD (Christie *et al.*, 2004).
- *Homology-based PPI*: Sequence similarity information is used to identify homology pairs. These pairs are then "BLASTed" against NCBI's non-redundant protein database and the count of their interactions extracted, resulting in four features in this category.
- *Domain-Domain Interaction*: Deng *et al.* (2002) identify domain interactions based on sequence analysis. The value of the single feature in this category is the probability of interaction of a candidate protein pair.
- *Protein-DNA Transcription Factor (TF) group binding*: Qi *et al.* (2006) group the TFs based on the MIPS protein class catalog into 16 TF groups. For each TF group, the number of TFs that bind to both genes is found and used as one of the 16 attributes in this category.
- *MIPS features (Protein Class and Mutant Phenotype):* These 2 categories contain features that identify if the protein pair belongs to the same protein class (among 25) and mutant phenotype (among 11).

The inductive learning algorithms used in (Qi *et al.*, 2006) are: Random Forests (RF), RF similarity-based k-Nearest-Neighbor, Naive Bayes, Decision Trees (J48), Logistic Regression, and Support Vector Machines (SVM). R50, a partial AUC score, was used to evaluate the performance of the resulting classifiers. R50 is defined as the area under the ROC curve with up to 50 negative predictions. In addition to R50 being a commonly used metric in the machine learning literature, the justification provided by the authors for using this score is based on the fact that the observed frequency of interacting proteins is 1:600, resulting in an estimate of 50 protein interactions among the 30,000 selected pairs. When comparing the classifiers learned from the data using the algorithms listed above, the best relative AUC scores were obtained using RandomForest (RF) and RandomForest similarity-based k-Nearest-Neighbor (kRF) for all tasks. The maximum R50 AUC scores were 0.67 for the DIP-based direct PPI task, 0.25 for the MIPS-based co-complex PPI task, and 0.25 for the KEGG-based co-pathway PPI task. The study by Qi *et al.* also showed that the feature with the highest coverage for all three PPI subtasks was the gene coexpression, followed by the process, component, and function categories extracted from the Gene Ontology (Ashburner *et al.*, 2000).

Licamele & Getoor (2006)

Licamele and Getoor (2006) combine the link structure of the PPI graph with the information about proteins in order to predict the interactions in a yeast dataset. More specifically, they look at the shared neighborhood among proteins and calculate the clustering coefficient among the neighborhoods for the first-order and second-order protein relations. The Gene Ontology distance between proteins is also considered. However, no distinction is made between direct (physical interaction) and indirect (proteins belonging to the same complex) interactions in the Licamele & Getoor (2006) study. The training set is assembled from multiple data sources such as MIPS (Mewes *et al.*, 1999), BIND (Bader *et al.*, 2001), DIP (Xenarios *et al.*, 2002), yeast two-hybrid (Ito *et al.*, 2001; Uetz *et al.*, 2000) and In vivo pull-down (Gavin *et al.*, 2002; Ho *et al.*, 2002). Classifiers such as Naive Bayes, kNN, Logistic Regression, C4.5, SVM, JRIP and Bagging with REPTrees were used on the resulting dataset. Licamele and Getoor report that the highest accuracy (81.7%) among the classifiers learned was achieved using Bagged REPTrees and that the corresponding AUC score was 0.8967, when predicting new links from noisy high throughput data.

Paradesi, Caragea & Hsu (2007)

The approaches by Qi *et al.* (2006) and by Licamele and Getoor (2006) use relational data of the PPI network along with other biologically relevant information (such as, sequence, gene expression data, GO terms, etc.) to predict the protein interactions. Paradesi *et al.* (2007) address the problem of predicting protein-protein interactions based solely on the graph features of the PPI network. They identify nine structural features for the *Saccharomyces cerevisiae* protein interaction network. These features include indegree, outdegree, number of mutual proteins and backward distance between proteins, among others.

Two datasets were used in the Paradesi *et al.* study: the DIP dataset (Salwinski *et al.*, 2004) and also the dataset generated by Qi *et al.* (2006). Similar to the other two approaches described above, Paradesi *et al.* (2007) learn several classifiers, such as Bagged Random Forest, Bagged REPTree, Random Tree, J48 and Classification via Regression from the training data. The results show that the method developed by Paradesi *et al.* (2007) compares well with the methods by Qi *et al.* and Licamele and Getoor, although no sequence information is used (as in the other two approaches), but only the relational features of the network data.

However, please note that the comparison between the approach by Paradesi *et al.* (2007) and the one by Licamele and Getoor (2006) was done using different datasets, for which the final results obtained with each method (reported in published work) were compared. Also, the method that we used for generating negative examples from the dataset provided by Qi *et al.* (2006) does not produce the same negative examples as those used in the study by Qi *et al.* (2006). Thus, the comparisons reported suffer from some data bias. Nevertheless, the complete comparisons based on previously published results are shown below:

Chen & Liu (2005)

Protein interaction sites refer to the locations on the protein structures where one protein physically interacts with another protein. A protein domain is a functionally defined protein region. Chen & Liu (2005) predict PPI using protein domain information. Many domain-based models for protein interaction prediction have been developed, and preliminary results have demonstrated their feasibility (Chen & Liu, 2005). Most of the existing domain-based

Figure 1. Comparison of results by Licamele and Getoor (2006) and Paradesi et al. (2007)

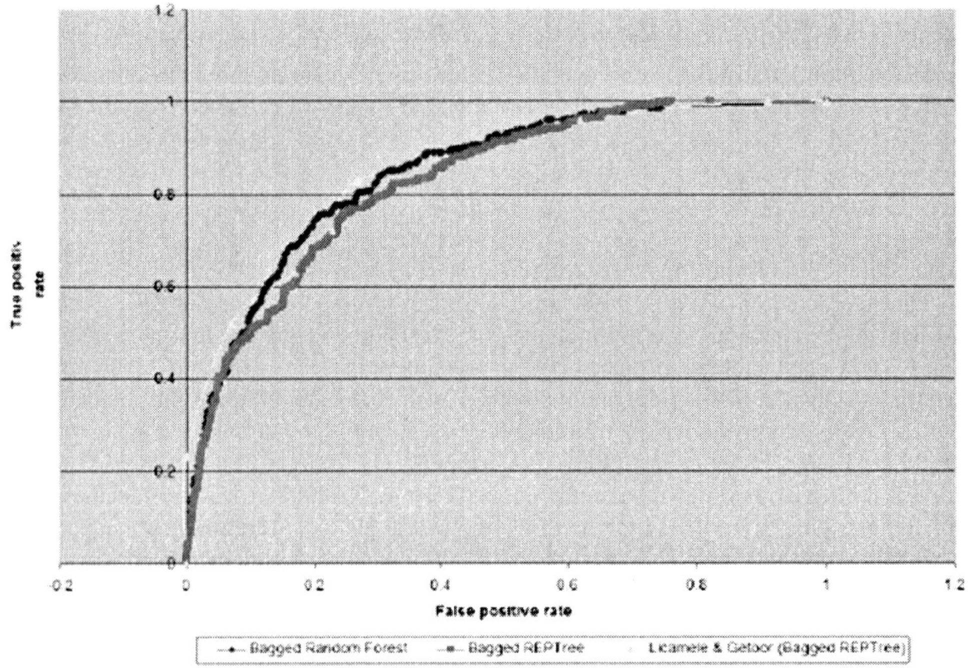

Figure 2. Comparison of the results by Qi et al. (2006) and those by Paradesi et al. (2007). The first dataset is DIP (Salwinski et al., 2006), while the second dataset is generated by Qi et al. (2006).

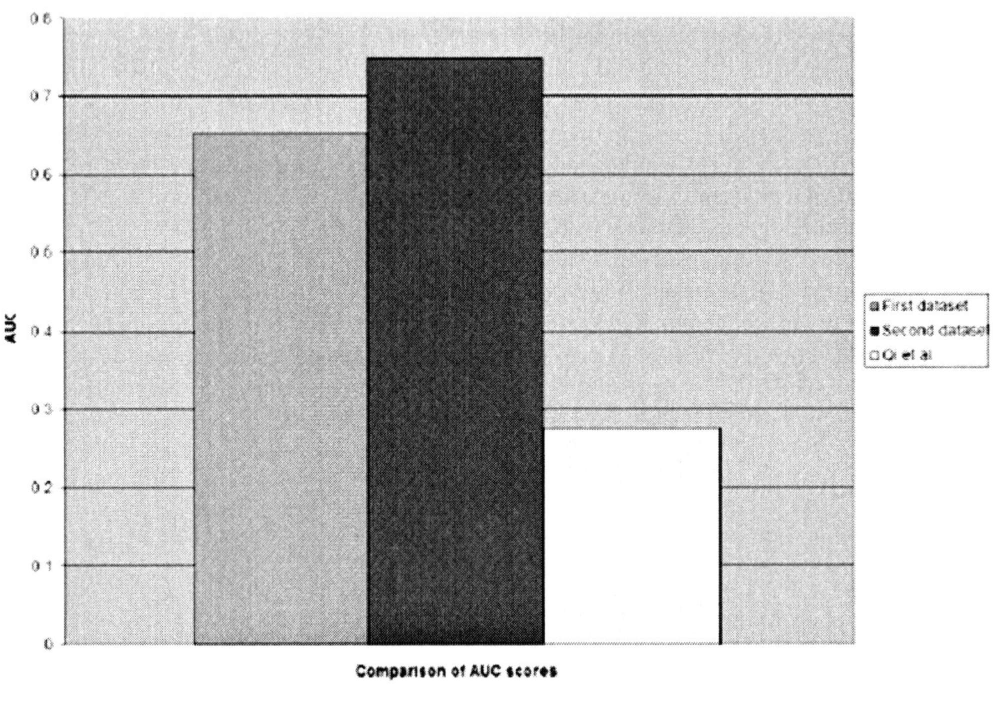

methods, however, consider only single-domain pairs (one domain from one protein) and assume independence between domain–domain interactions. Chen & Liu (2005) introduced a new framework based on random forest for PPI prediction, which explores the contributions of all the possible domain combinations to predicting protein interactions. Furthermore, their model does not assume that domain pairs are independent of each other. They obtained the PPI data from DIP (Salwinski *et al.*, 2004; Deng *et al.*, 2002; Schwikowski *et al.*, 2000; Xenarios *et al.*, 2001). Chen & Liu (2005) extract the domain information for each protein and build a vector of the domain list of each candidate protein pair. The values in the vector are given by the number of occurrences of the domain in both proteins. They obtain a sensitivity of 79.78% and specificity of 64.38%, a better result than that achieved using the MLE method of Deng *et al.* (2002), which yields a sensitivity of 78.30% and a specificity of 37.53%.

Paradesi (2008)

Paradesi (2008) attempts to perform a fair comparison between methods that use only biological information and methods that use both graph features and biological information. Thus, two new sets of experiments are performed on the datasets provided by Chen and Liu (2005) and by Qi *et al.* (2006), respectively. Various features, including those originally used by Chen and Liu and by Qi *et al.*, but also the graph features used in Paradesi *et al.* (2007), are studied to indentify their importance for the accuracy of prediction. The results are presented in

Table 1. Results obtained from experiments using Chen & Liu (2005) dataset (5-fold cross-validation)

	J48		NB		CVR		SVM	
	Se %	Sp %	Se %	Sp %	Se %	Sp %	Se %	Sp %
Domain	73.3	62.1	73.44	63.08	55.1	0	74.3	73.56
Degree	86.62	85.74	89.78	60.66	85.78	86.98	87.2	76.02
MutualProtein	96.68	59.3	97.5	57.72	55.1	0	98.96	55.94
BackwardDistance	99.52	65.8	99.52	65.8	99.52	65.8	99.52	65.8
Domain + Degree	86.3	86.08	88.94	62.52	85.92	86.7	80.7	77.76
Domain + MutualProtein	85.14	68.4	95.64	59.08	55.1	0	78.62	75.26
Domain + BackwardDistance	91.14	72.46	83.16	71.82	99.52	65.8	86.06	82.68
Degree + MutualProtein	87.6	86.56	93.96	65.24	86.72	87.14	89.26	77.44
Degree + BackwardDistance	92.56	91.54	93.1	71.88	92.02	92.14	93.3	83.54
MutualProtein + BackwardDistance	97.86	80.04	98.18	74.42	99.54	70.88	99.26	76.16
Domain + Degree + MutualProtein	87.8	85.88	93.16	65.94	86.86	86.84	83.18	78.76
Domain + MutualProtein + BackwardDistance	96.94	81.5	97.82	76.42	99.54	70.88	90.7	85
Domain + Degree + BackwardDistance	91.86	91.96	92.3	73.06	91.96	92.24	88.7	85.58
Degree + MutualProtein + BackwardDistance	93.04	93.1	94.68	69.4	92.94	93.14	95.74	85.68
Domain + Degree + MutualProtein + BackwardDistance	93.6	92.4	94.28	70.54	92.76	93.2	91.14	87.4

Table 2. Results obtained from experiments using Qi et al. (2006) dataset (5-fold cross-validation)

	J48	RF	NB	CVR	SVM
Feature	0.504	0.7052	0.7244	0.7466	0.504
Degree	0.5	0.7394	0.9442	0.9798	0.5
MutualProtein	0.61	0.806	0.826	0.5	0.622
BackwardDistance	0.79	0.79	0.79	0.5	0.73
Feature + Degree	0.57	0.7276	0.7378	0.9526	0.506
Feature + MutualProtein	0.5944	0.8172	0.737	0.8347	0.626
Feature + BackwardDistance	0.796	0.8438	0.7374	0.8634	0.736
Degree + MutualProtein	0.7052	0.8604	0.9632	0.9828	0.624
Degree + BackwardDistance	0.79	0.833	0.9646	0.9896	0.73
MutualProtein + BackwardDistance	0.79	0.94	0.948	0.5	0.758
Feature + Degree + MutualProtein	0.5784	0.8358	0.7428	0.9496	0.626
Feature + MutualProtein + BackwardDistance	0.796	0.9408	0.7438	0.9052	0.758
Feature + Degree + BackwardDistance	0.796	0.8678	0.7428	0.9768	0.736
Degree + MutualProtein + BackwardDistance	0.79	0.944	0.9774	0.9926	0.758
Feature + Degree + MutualProtein + BackwardDistance	0.796	0.927	0.7468	0.98	0.756

Tables 1 and 2, respectively, show that the graph features alone can result in very good predictive results, while the sequence features by themselves are less predictive.

ISSUES AND PROBLEMS

There are many methods of inferring protein interactions, some of them described above. The goal of this chapter was to show that methods that use graph-based features from the PPI network, sometimes along with the biological features of the interacting proteins, can produce results better than those using the biological features alone. However, the task of predicting protein interactions using machine learning approaches is far from being solved completely. The machine learning algorithms are obviously not 100% accurate. In particular, the graph-based approaches advocated in the chapter are faced with several problems and issues discussed in this section.

Combining Data from Multiple Databases

There are many databases that provide information about protein interactions. However, different databases store protein interactions from different high-throughput interaction detection methods. This results in very few protein interactions that are contained in several databases. Qi *et al.* (2006) states that there are only 293 protein interactions present in all DIP (Salwinski *et al.*, 2004), MIPS (Mewes *et al.*, 2002) and KEGG (Kanehisa & Goto, 2000) databases. Moreover, different databases store different attributes for protein interactions. Let us consider the data available in all databases as published by their respective research groups. The BIND (Bader *et al.*, 2003) database stores information about interactions, complexes and pathways. BIND (Bader *et al.*, 2003) contains a number of large-scale interactions and complex mapping experiments using yeast two-hybrid, mass spectrometry, genetic interactions and phage display. DIP (Salwinski *et al.*, 2004) develops methods of quality assessment and uses them to identify the most reliable subset of the interactions that are inferred from high-throughput experiments. MIPS (Mewes *et al.*, 2002) provides information on Open Reading Frames (ORFs), RNA genes and other genetic elements. It also contains information on how proteins cooperate in complexes, pathways and cellular networks. In addition, detailed information on transcription factors and their binding sites, transport proteins and metabolic pathways are being included or interlinked to the core data. IntAct (Kerrien *et al.*, 2007) manually annotates published manuscripts reporting molecular interaction data and formalizing it by using a comprehensive set of controlled vocabularies in order to ensure data integrity. IntAct (Kerrien *et al.*, 2007) is probably the only database that contains negative examples of protein interactions. MINT (Chatr-aryamontri *et al.*, 2007) focuses on the curation work on physical interactions between proteins. Genetic or computationally inferred interactions are not included in the database. However, von Mering *et al.*(2002) states that there are very few protein interactions that are supported by one of the high-throughput methods. Given the diversity of the protein interactions in different databases and also the diversity of the features, it is very difficult to assemble a comprehensive training dataset that contains all the known protein interactions, as well as all the biological and structural features that can be defined for a pair of interactions, thus making it difficult to use all the existing information to learn more accurate prediction models.

Negative Protein Interactions

One of the most common problems that researchers face while predicting protein interactions using computational methods is to deal with the large number of negative examples. Let us assume that there are around 6,000 proteins in a database and around 80,000 interactions between them are known. This means that there are still $6000^2 - 80{,}000 = 35{,}920{,}000$ interactions that are yet to be classified as true positive or true negative. If we provide these 80,000 positive examples and 35,920,000 negative examples to a random algorithm that predicts that any two given proteins do not interact, we will still obtain an accuracy of 99.78%. Although, this is a good result, it is important that a machine learning algorithm must predict the true positives accurately. We are more interested in the problem of predicting which proteins interact than in that of predicting which proteins do not interact. In order to solve our desired problem, we must find ways of reducing the number of negative examples that will be provided to the algorithms. Qi *et al.* (2006) randomly select protein pairs that do not interact as negative examples. Licamele and Getoor (2006) randomly sample negative protein pairs without replacement. They choose an equal number of positive protein pairs and negative protein pairs. Chen & Liu (2005) also sample negative protein pairs that are roughly equal to the number of positive protein pairs. Paradesi *et al.* (2007) perform a depth-limited Breadth First Search (with depth 2) from each protein and generate protein pairs. By using this technique, they obtain all positive interactions and several negative interactions. All of the above-mentioned techniques of generating negative examples are not accurate because they assume that any protein pair that is not present in the positive interaction list constitutes a negative example. A protein pair must be labeled as a negative interaction if and only if it is experimentally determined that those two proteins do not interact in any high-throughput methods.

False Positives in High-Throughput Methods

There are several high-throughput methods for detecting protein-protein interactions, including yeast two-hybrid method (Ito *et al.*, 2001; Uetz *et al.*, 2000), analysis of protein complexes using mass spectrometry (Gavin *et al.*, 2002; Ho *et al.*, 2002), co-immunoprecipitation (Sambrook *et al.*, 2006), etc. Some of the interactions that are identified from different high-throughput methods may be false positives. Many researchers have tried to assess the quality of the high-throughput data (von Mering *et al.*, 2002; Mrowka *et al.*, 2001; Deane *et al.*, 2002). However, they claim that sometimes a subset of interactions that were identified by using one method could not be identified by another method. These interactions are called false positives because they might not have been interactions but were wrongly labeled as interactions by the high-throughput method (Salwinski *et al.*, 2003).

Representation of Data

There is a need to handle appropriately data from different interaction-detection methods. For example, the yeast two-hybrid screening method, which is one of the most popular methods of detecting protein interactions, uses a bait-and-prey approach to find interactions in yeast (Young, 1998). Furthermore, several databases provide the protein interaction data in the form of different interacting "bait" proteins and "prey" proteins. However, some researchers still treat the protein interaction network from the yeast two-hybrid screening method as an undirected network. In general, different high-throughput techniques have different interacting relationships (either directed or undirected) between interacting protein pairs.

Validity of Predicted Protein Interaction

It might be easier and faster to predict or identify new protein interactions using the graph-based features of the PPI network. However, the newly discovered protein interaction is accurate only from a relational learning (or graph-mining) perspective. There is no guarantee that the new interacting protein pair is biologically valid. The new protein interaction could either be a true positive interaction or a false positive interaction. Moreover, there are protein pairs that are identified as interacting pairs because of the graph-based prediction and a single high-throughput method. These protein pairs are most likely false positive protein interactions because they have been identified only through a single high-throughput protein interaction detection method. Therefore, there is a need to verify the validity of the interactions discovered through the graph-based prediction approach.

Identification of Useful Graph-Based Features

There are many features that can be extracted from a PPI network as explained by the previous approaches of predicting protein interactions. However, not all features are useful in the prediction task. In fact, some features may actually hurt the learning process of the classifier and thereby lower the accuracy of the results. Qi *et al.* (2006) have discovered that features that might be useful for solving an interaction subtask may not be useful for solving another interaction subtask. The authors have also observed that different combinations of various features affect the performance of the learning algorithms.

SUGGESTIONS AND RECOMMENDATIONS

In this section, we provide possible solutions and recommendations to the problems and issues presented above.

Combining Data from Multiple Databases

Gathering more data and features can provide a more tractable representation of the training data, increasing the achievable accuracy, sensitivity, and specificity of the learning system. Given the nature of the data in the PPI databases, there are two different methods we could use to combine data from multiple sources. The first method is to take the intersection across all databases. This would result in a small number of protein interactions but many features about those interactions. The second method is to take the union across all databases. This would result in a dataset with many protein interactions but also many missing attributes. Thus, there is a tradeoff between data with a small number of protein interactions but good quality information about these interactions and data with a large number of protein interactions but a lot of missing information about the interactions. This tradeoff suggests that there is a need to efficiently unify information across all databases by the research groups that maintain these databases. To address this need, the research groups that maintain BIND (Bader *et al.*, 2003), MINT (Chatr-aryamontri *et al.*, 2007), DIP (Salwinski *et al.*, 2004), MPact (Güldener *et al.*, 2006) and IntAct (Kerrien *et al.*, 2007) have formed the IMEx consortium to build a large, consistent and non-redundant repository of protein interactions and information about the interactions. According to the IMEx consortium, the data gathered by them will be broader in scope and deeper in information than any individual effort (Kerrien *et al.*, 2007). By integrating multiple data sources, the PPI network will be more complete than before and will enable researchers to achieve better quality graph-based feature extraction.

Negative Protein Interactions

A better method of selecting non-interacting protein pairs (i.e., negative examples) needs to be developed. One possible method for generating negative protein pairs would be to mark protein pairs that do not belong to the same complex or the same pathway as negative, if there is no known interaction between them. Another method for generating negative examples is similar to the previous method, but could be achieved by extracting information from a PPI network. If the proteins in a pair belong to different cliques that are apart by a very large distance in the network and they do not have any known interaction between them, they could be considered to form a negative protein pair. Also, if we cluster the proteins in a network based on complexes and choose a protein pair from different clusters, that pair could be considered to be a negative example. It would help to use the above-mentioned techniques on protein interaction visualization tools such as ProViz (Iragne *et al.*, 2005), iPfam (Finn *et al.*, 2005), VisANT (Hu *et al.*, 2007), etc. and querying tools such as PathBLAST (Kelley *et al.*, 2004), APID (Prieto & De Las Rivas, 2006), etc. to detect negative protein interactions.

False Positives in High-Throughput Methods

DIP provides several tests, such as Expression Profile Reliability Index (EPR Index) (Deane *et al.*, 2002), Paralogous Verification (PVM) (Deane *et al.*, 2002) and Domain Pair Verification (DPV) (Deng *et al.*, 2002) as online services, to ensure that the number of false positives is reduced in any given dataset. A graph-based approach to reduce the number of false positives obtained from computational techniques would be to build separate graphs for each of the high-throughput methods. By taking the intersection of all the graphs we could obtain the true positive protein interactions in the simplest case. In a more complex case, we could use protein visualization tools such as ProViz (Iragne *et al.*, 2005), iPfam (Finn *et al.*, 2005), VisANT (Hu *et al.*, 2007), etc. to identify proteins that frequently occur in most of the graphs and infer actual protein interactions among them.

Representation of Data

The protein interaction network must be treated as a directed network especially when using yeast two-hybrid screening data in order to enable the classification algorithm to learn using bait and prey proteins. More importantly, there needs to be a way of dealing with representation of data collected from several high-throughput methods. In other words, if the experiment proves that the interaction occurs in one way from one protein to another, the interaction must be treated as a directed edge; otherwise, the interaction must be treated as an undirected edge. In a few cases, there might be protein interactions that require the above-mentioned data representation. In order to solve a protein interaction prediction problem that involves data with different representations, the problems must be divided into minimal sub-problems that can be solved using the desired representation and learning algorithm. The results from the sub-problems must be combined using either a weighting system or a voting technique, and the protein pair must be classified as interacting or not.

Validity of Predicted Protein Interaction

We could label each positive interacting protein pair identified by the graph-based interaction prediction method as a true positive example. However, most probably the results from the machine learning graph-based approach are not 100% accurate when the learned classifier is used to classify new data in the test dataset. One of the techniques of validating newly discovered protein interactions is to perform a high-throughput protein interaction detection method to experimentally validate if the protein pair interacts or not. If a protein pair has been identified as an interacting pair by only one high-throughput method, then the result of the graph-based prediction can help confirm the protein pair as a true positive interaction. In other words, the results of graph-based PPI prediction algorithms can be used along with the high-throughput interaction detection methods to identify true positive interactions. Therefore, graph-based interaction prediction methods can serve a two-fold purpose – to identify new interactions and to strengthen our confidence in the results of existing high-throughput methods.

Identification of Useful Graph-Based Features

It is important to use only features that help in the learning process. One method by which this can be achieved is by running classification algorithms on each feature individually and selecting only those features that provide higher accuracy. Another method of identifying important features is by applying dection on the input data. There are several statistical measures to identify the importance of features such as Entropy, Gini index, etc. It is also important to identify the right combination of features to better predict protein interactions. This can be achieved by performing several experiments with various combinations of features. It is often computationally difficult to perform many experiments with thousands of features to detect the right combinations of features. A better solution would be to first calculate the most useful features, thereby reducing the size of the feature list, and then perform experiments by varying the combinations of the selected features. It has been observed by the authors that by choosing the important features and reducing the size of the feature list, one may not achieve the highest accuracy possible. In other words, a comprehensive examination of all features will provide the highest accuracy for some combination of features, while an examination of a reduced set of features will be computationally less-intensive. Thus, there is a trade-off between the number of features chosen and the desired accuracy.

FUTURE TRENDS

The future trends in the field of predicting protein interactions using graph-based features look very promising due to the following changes in this area:

Increase in Quality and Quantity of Data

There is a rapid increase in the discovery of new proteins and interactions based on high-throughput methods. There is also a growth in techniques to identify actual protein interactions and eliminate false positive interactions. The IMEx consortium, as mentioned previously in this chapter, is allowing individual researchers and research groups to submit protein interaction information. This newly submitted data is run through several tests and manually curated to ensure that the interaction is a true positive interaction. As the quality and quantity of data increases, the PPI network becomes more complete, thereby allowing more complete features to be extracted from the network. There is scope for researchers to work on increasing the quality and quantity of protein interactions by developing new computational techniques.

Improvement in Classification Algorithms

There are also rapid advances in the machine learning and data mining algorithms. Most of the supervised and unsupervised learning algorithms used in the prediction of protein interactions were developed for tasks other than that. Although it has been observed that these algorithms have worked well for the protein interaction prediction task, there is a need for developing custom algorithms that can handle protein interaction data even better.

Use of Protein Interaction Network Analysis Tools

There are many protein interaction visualization tools such as ProViz (Iragne *et al.*, 2005), iPfam (Finn *et al.*, 2005), VisANT (Hu *et al.*, 2007), etc. and querying tools such as PathBLAST (Kelley *et al.*, 2004), APID (Prieto

& De Las Rivas, 2006), etc. available. These tools allow users to view and search for proteins in any PPI network. They can be exploited to gather several graph-based features from the PPI network. The visualization and querying tools can also be used to split the PPI network into several overlapping sub-graphs. Interactions can be predicted at the sub-graph level and these predictions can be combined to identify protein interactions at the original graph level. A protein pair can be labeled as interacting if it is observed that the interaction between the protein pair appears in more than one sub-graph.

Development of Different Approaches

Although there is an increase in data and improvement of algorithms and tools, attention must be paid to improve the approaches of solving the protein interactions prediction problem. There is no one-solution-solves-all-problems approach anymore. Instead, there is a need for developing approaches that solve the problem by applying an ensemble of various machine learning algorithms for different subgraphs of the PPI network. In other words, one could extract different graph-based features from different subsets of the PPI network and run different machine learning algorithms on the features, depending on the data. The different machine learning classifiers could "vote" on the class, and the weighted average of the output could be assigned as the actual class.

CONCLUSION

In this chapter, we provide an overview of several methods for predicting protein interactions using biological and graph features extracted from a PPI network. Thus, machine learning algorithms that can be used to predict protein interactions based on biological and graph-based features are described, along with some specific methods that make use of such algorithms and features. Furthermore, several issues and open problems in the PPI prediction area are presented together with suggestions and recommendations for how to deal with them. At last, important future trends are highlighted.

REFERENCES

Altman, D. G., & Bland, J. M. (1994). Diagnostic tests. 1: Sensitivity and specificity. *BMJ, 308*(6943):1552.

Albert, I., & Albert, R. (2004). Conserved network motifs allow protein-protein interaction prediction. *Bioinformatics, 20*(18):3346-52.

Ashburner, M., Ball, C. A., Blake, J. A., Botstein, D., Butler, H., Cherry, J. M., Davis, A. P., Dolinski, K., Dwight, S. S., Eppig, J. T., Harris, M. A., Hill, D. P., Issel-Tarver, L., Kasarskis, A., Lewis, S., Matese, J. C., Richardson, J. E., Ringwald, M., Rubin, G. M., & Sherlock, G. (2000). Gene ontology: Tool for the unification of biology. The Gene Ontology Consortium. *Nat Genet., 25*(1), 25-9.

Bader, G.D., Donaldson, I., Wolting, C., Ouellette, B. F., Pawson, T., & Hogue, C. W. (2001). BIND--The Biomolecular Interaction Network Database. *Nucleic Acids Res, 29*(1), 242-5.

Bader, G. D., & Hogue, C. W. (2002). Analyzing yeast protein-protein interaction data obtained from different sources. *Nat Biotechnol., 20*(10), 991-7.

Bader, G. D., Betel, D., & Hogue, C. W. (2003). BIND: the Biomolecular Interaction Network Database. *Nucleic Acids Res, 31*(1), 248-50.

Bar-Joseph, Z., Gerber, G. K., Lee, T. I., Rinaldi, N. J., Yoo, J. Y., Robert, F., Gordon, D. B., Fraenkel, E., Jaakkola, T. S., Young, R. A., & Gifford, D. K. (2003). Computational discovery of gene modules and regulatory networks. *Nat Biotechnol., 21*(11), 1337-42.

Bateman, A., Coin, L., Durbin, R., Finn, R. D., Hollich, V., Griffiths-Jones, S., Khanna, A., Marshall, M., Moxon, S., Sonnhammer, E. L., Studholme, D. J., Yeats, C., Eddy, S. R. (2004). The Pfam protein families database. *Nucleic Acids Res.*, *32*(Database issue), D138-41.

Breiman, L., Freidman, J. H., Olshen, R. A., & Stone, C. J. (1984). *Classification and Regression Trees.* Wadsworth.

Breiman, L. (1996). Bagging predictors. *Machine Learning, 24*(2), 123-40.

Breiman, L. (2001). Random Forests. *Machine Learning, 45*(1), 5-32.

Bu, D., Zhao, Y., Cai, L., Xue, H., Zhu, X., Lu, H., Zhang, J., Sun, S., Ling, L., Zhang, N., Li, G., & Chen, R. (2003). Topological structure analysis of the protein-protein interaction network in budding yeast. *Nucleic Acids Res, 31*(9), 2443-50.

Chatr-aryamontri, A., Ceol, A., Palazzi, L. M., Nardelli, G., Schneider, M. V., Castagnoli, L., & Cesareni, G. (2007). MINT: The Molecular INTeraction database. *Nucleic Acids Res, 35*(Database issue): D572–D574.

Chen, X. W., & Liu, M. (2005). Prediction of protein-protein interactions using random decision forest framework. *Bioinformatics, 21*(24), 4394-4400.

Christie, K. R., Weng, S., Balakrishnan, R., Costanzo, M. C., Dolinski, K., Dwight, S. S., Engel, S. R., Feierbach, B., Fisk, D. G., Hirschman, J. E., Hong, E. L., Issel-Tarver, L., Nash, R., Sethuraman, A., Starr, B., Theesfeld, C. L., Andrada, R., Binkley, G., Dong, Q., Lane, C., Schroeder, M., Botstein, D., & Cherry, J. M. (2004). Saccharomyces Genome Database (SGD) provides tools to identify and analyze sequences from *Saccharomyces cerevisiae* and related sequences from other organisms. *Nucleic Acids Res.*, *32*(Database issue), D311-4.

Cortes, C., & Vapnik, V. (1995). Support-Vector Networks. *Machine Learning, 20*(3), 273-297.

Cover, T., & Hart, P. (1967). Nearest neighbor pattern classification. *Information Theory, IEEE Transactions on, 13*(1), 21-27.

Cristianini, N., & Shawe-Taylor, J. (2000). *An Introduction to Support Vector Machines.* Cambridge University.

Deane, C.M., Salwinski, L., Xenarios, I., & Eisenberg, D. (2002). Protein interactions: two methods for assessment of the reliability of high throughput observations. *Mol Cell Proteomics*, *1*(5), 349-56.

Deng, M., Mehta, S., Sun, F., & Chen, T. (2002). Inferring domain-domain interactions from protein-protein interactions. *Genome Res*, *12*(10), 1540-8.

Fawcett, T. (2004). *ROC Graphs: Notes and Practical Considerations for Researchers.* Technical report, Palo Alto, USA: HP Laboratories.

Fields, S., & Song, O. (1989). A novel genetic system to detect protein-protein interactions. *Nature*, *340*(6230), 245-6.

Finn, R. D., Marshall, M., & Bateman, A. (2005). iPfam: visualization of protein-protein interactions in PDB at domain and amino acid resolutions. *Bioinformatics*, *21*(3), 410-2.

Frank, E., Wang, Y., Holmes, G., & Witten, I. H. (1998). Using model trees for classification. *Machine Learning, 32*(1), 63-76.

Gavin, A. C., Bösche, M., Krause, R., Grandi, P., Marzioch, M., Bauer, A., Schultz, J., Rick, J. M., Michon, A. M., Cruciat, C. M., Remor, M., Höfert, C., Schelder, M., Brajenovic, M., Ruffner, H., Merino, A., Klein, K., Hudak, M., Dickson, D., Rudi, T., Gnau, V., Bauch, A., Bastuck, S., Huhse, B., Leutwein, C., Heurtier, M. A., Copley, R. R., Edelmann, A., Querfurth, E., Rybin, V., Drewes, G., Raida, M., Bouwmeester, T., Bork, P., Seraphin, B., Kuster, B., Neubauer, G., & Superti-Furga, G. (2002). Functional organization of the yeast proteome by systematic analysis of protein complexes. *Nature*, *415*(6868), 141-7.

Ghaemmaghami, S., Huh, W. K., Bower, K., Howson, R. W., Belle, A., Dephoure, N., O'Shea, E.K., Weissman, J. S. (2003). Global analysis of protein expression in yeast. *Nature, 425*(6959), 737-41.

Güldener, U., Münsterkötter, M., Oesterheld, M., Pagel, P., Ruepp, A., Mewes, H. W., Stümpflen, V. (2006). MPact: the MIPS protein interaction resource on yeast. *Nucleic Acids Res, 34*(Database issue), D436-41.

Harbison, C. T., Gordon, D. B., Lee, T. I., Rinaldi, N. J., Macisaac, K. D., Danford, T. W., Hannett, N. M., Tagne, J. B., Reynolds, D. B., Yoo, J., Jennings, E. G., Zeitlinger, J., Pokholok, D. K., Kellis, M., Rolfe, P. A., Takusagawa, K. T., Lander, E. S., Gifford, D. K., Fraenkel, E., Young, R. A. (2004). Transcriptional regulatory code of a eukaryotic genome. *Nature, 431*(7004), 99-104.

Ho, Y., Gruhler, A., Heilbut, A., Bader, G. D., Moore, L., Adams, S. L., Millar, A., Taylor, P., Bennett, K., Boutilier, K., Yang, L., Wolting, C., Donaldson, I., Schandorff, S., Shewnarane, J., Vo, M., Taggart, J., Goudreault, M., Muskat, B., Alfarano, C., Dewar, D., Lin, Z., Michalickova, K., Willems, A. R., Sassi, H., Nielsen, P. A., Rasmussen, K. J., Andersen, J. R., Johansen, L. E., Hansen, L. H., Jespersen, H., Podtelejnikov, A., Nielsen, E., Crawford, J., Poulsen, V., Sørensen, B. D., Matthiesen, J., Hendrickson, R. C., Gleeson, F., Pawson, T., Moran, M. F., Durocher, D., Mann, M., Hogue, C. W., Figeys, D., & Tyers, M. (2002). Systematic identification of protein complexes in *Saccharomyces cerevisiae* by mass spectrometry. *Nature, 415*(6868), 180-3.

Hu, Z., Ng, D. M., Yamada, T., Chen, C., Kawashima, S., Mellor, J., Linghu, B., Kanehisa, M., Stuart, J. M., & DeLisi, C. (2007). VisANT 3.0: New modules for pathway visualization, editing, prediction and construction. *Nucleic Acids Res, 35*(Web Server issue), W625-32.

Huh, W. K., Falvo, J. V., Gerke, L. C., Carroll, A. S., Howson, R. W., Weissman, J. S., O'Shea, E. K. (2003). Global analysis of protein localization in budding yeast. *Nature, 425*(6959), 686-91.

Hsu, W. H., King, A. L., Paradesi, M. S. R., Pydimarri, T., & Weninger, T. (2006). Collaborative and Structural Recommendation of Friends using Weblog-based Social Network Analysis. *Proc. of Computational Approaches to Analyzing Weblogs* - AAAI 2006 Technical Report SS-06-03, (pp. 55-60).

Iragne, F., Nikolski, M., Mathieu, B., Auber, D., & Sherman, D. (2005). ProViz: Protein interaction visualization and exploration. *Bioinformatics, 21*(2), 272-4.

Ito, T., Chiba, T., Ozawa, R., Yoshida, M., Hattori, M., & Sakaki, Y. (2001). A comprehensive two-hybrid analysis to explore the yeast protein interactome. *Proc Natl Acad Sci U S A., 98*(8), 4569-74.

Kanehisa, M., & Goto, S. (2000). KEGG: Kyoto Encyclopedia of Genes and Genomes. *Nucleic Acids Res, 28*, 27-30.

Kelley, B. P., Yuan, B., Lewitter, F., Sharan, R., Stockwell, B. R., & Ideker, T. (2004). PathBLAST: a tool for alignment of protein interaction networks. *Nucleic Acids Res, 32*(Web Server issue), W83-8.

Kerrien, S., Alam-Faruque, Y., Aranda, B., Bancarz, I., Bridge, A., Derow, C., Dimmer, E., Feuermann, M., Friedrichsen, A., Huntley, R., Kohler, C., Khadake, J., Leroy, C., Liban, A., Lieftink, C., Montecchi-Palazzi, L., Orchard, S., Risse, J., Robbe, K., Roechert, B., Thorneycroft, D., Zhang, Y., Apweiler, R., & Hermjakob, H. (2007). IntAct--open source resource for molecular interaction data. *Nucleic Acids Res, 35*(Database issue), D561-5.

Licamele, K., & Getoor, L. (2006). Predicting Protein-Protein Interactions Using Relational Features. *Proc. of ICML Workshop on Statistical Network Analysis.*

Liu, H., & Motoda, H. (1998). *Feature Selection for Knowledge Discovery and Data Mining.* Kluwer.

MacBeath, G., & Schreiber, S. L. (2000). Printing proteins as microarrays for high-throughput function determination. *Science; 289*(5485), 1760-3.

Maslov, S., & Sneppen, K. (2002). Specificity and stability in topology of protein networks. *Science, 296*(5569), 910-3.

Mewes, H. W., Heumann, K., Kaps, A., Mayer, K., Pfeiffer, F., Stocker, S., & Frishman, D. (1999). MIPS: A database for genomes and protein sequences. *Nucleic Acids Res*, *27*, 44–48.

Mewes, H. W., Frishman, D., Güldener, U., Mannhaupt, G., Mayer, K., Mokrejs, M., Morgenstern, B., Münsterkötter, M., Rudd, S., & Weil, B. (2002). MIPS: A database for genomes and protein sequences. *Nucleic Acids Res*, *30*(1), 31-4.

Mitchell, T. (1997). *Machine Learning*. McGraw Hill.

Mrowka, R., Patzak, A., & Herzel, H. (2001). Is there a bias in proteome research? *Genome Res.*, *11*, 1971-73.

Najafabadi, H. S, & Salavati, R. (2008). Sequence-based prediction of protein-protein interactions by means of codon usage. *Genome Biology*, *9*, R87.

Paradesi, M. S. R. (2008). Graph-Based Protein-Protein Interaction Prediction in *Saccharomyces Cerevisiae*. Master's thesis, Kansas State University, Department of Computing and Information Sciences.

Paradesi, M. S. R., Caragea, D., & Hsu, W. H. (2007). Structural Prediction of Protein-Protein Interactions in *Saccharomyces cerevisiae. Proc. of IEEE 7th International Symposium on BioInformatics and BioEngineering*, 2, 1270-1274.

Paradesi, M. S. R., Wang, L., Brown, S. J., & Hsu, W. H. (2006). Mining Domain Association Rules From Protein-Protein Interaction data. *Intelligent Engineering Systems through Artificial Neural Networks*, *16*, 213-218.

Prieto, C., & De Las Rivas, J. (2006). APID: Agile Protein Interaction DataAnalyzer. *Nucleic Acids Res*, *34*(Web Server issue), W298-302.

Qi, Y., Bar-Joseph, Z., & Klein-Seetharaman, J., (2006). Evaluation of different biological data and computational classification methods for use in protein interaction prediction. *Proteins: Structure, Function, and Bioinformatics*, *63*(3), 490-500.

Quinlan, J. R. (1986). Induction of decision trees. *Machine Learning*, *1*, 81-106.

Quinlan, J. R. (1993). *C4.5: Programs for Machine Learning*. Morgan Kaufmann.

Salwinski, L., & Eisenberg, D. (2003). Computational methods of analysis of protein-protein interactions. *Curr. Opin. Struct. Biol.*, *13*, 377-382.

Salwinski, L., Miller, C. S., Smith, A. J., Pettit, F. K., Bowie, J. U., & Eisenberg, D. (2004). The Database of Interacting Proteins: 2004 update. *Nucleic Acids Res*, *32*(Database issue), D449-51.

Sambrook, J., & Russell, D. W. (2006). Identification of Associated Proteins by Coimmunoprecipitation, *CSH Protocols*, doi:10.1101/pdb.prot3898.

Schwikowski, B., Uetz, P., & Fields, S. (2000). A network of protein-protein interactions in yeast. *Nat Biotechnol*, *18*(12), 1257-61.

Scholkopf, B., Sung, K.-K., Burges, C. J. C., Girosi, F., Niyogi, P., Poggio, T., & Vapnik, V. (1997). Comparing support vector machines with gaussian kernels to radial basis function classiers. *IEEE Trans. on Signal Processing*, *45*(11), 2758-65.

Shoemaker, B. A., & Panchenko, A. R. (2007). Deciphering Protein–Protein Interactions. Part I. Experimental Techniques and Databases. *PLoS Comput Biol.*, *3*(3), e42.

Shoemaker, B. A., & Panchenko, A. R. (2007). Deciphering Protein–Protein Interactions. Part II. Computational Methods to Predict Protein and Domain Interaction Partners. *PLoS Comput Biol* *3*(4), e43.

Tong, A. H., Evangelista, M., Parsons, A. B., Xu, H., Bader, G. D., Pagé, N., Robinson, M., Raghibizadeh, S., Hogue, C.W., Bussey, H., Andrews, B., Tyers, M., & Boone, C. (2001). Systematic genetic analysis with ordered arrays of yeast deletion mutants. *Science*, *294*(5550), 2364-8.

Tong, A. H., Lesage, G., Bader, G. D., Ding, H., Xu, H., Xin, X., Young, J., Berriz, G. F., Brost, R. L., Chang, M., Chen, Y., Cheng, X., Chua, G., Friesen, H., Goldberg, D. S., Haynes, J., Humphries, C., He, G., Hussein, S., Ke, L., Krogan, N., Li, Z., Levinson, J.N., Lu, H., Ménard, P., Munyana, C., Parsons, A. B., Ryan, O., Tonikian, R., Roberts, T., Sdicu, A.M., Shapiro, J., Sheikh, B., Suter, B., Wong, S. L., Zhang, L.V., Zhu, H., Burd, C. G., Munro, S., Sander, C., Rine, J., Greenblatt, J., Peter, M., Bretscher, A., Bell, G., Roth, F. P., Brown, G. W., Andrews, B., Bussey, H., Boone, C. (2004). Global mapping of the yeast genetic interaction network. *Science*, *303*(5659), 808-13.

Uetz, P., Giot, L., Cagney, G., Mansfield, T. A., Judson, R. S., Knight, J. R., Lockshon, D., Narayan, V., Srinivasan, M., Pochart, P., Qureshi-Emili, A., Li, Y., Godwin, B., Conover, D., Kalbfleisch, T., Vijayadamodar, G., Yang, M., Johnston, M., Fields, S., & Rothberg, J. M. (2000). A comprehensive analysis of protein-protein interactions in *Saccharomyces cerevisiae*. *Nature*, *403*(6770), 623-7.

Vapnik, V. (1998). *Statistical Learning Theory*. Wiley.

von Mering, C., Krause, R., Snel, B., Cornell, M., Oliver, S. G., Fields, S., & Bork, P. (2002). Comparative assessment of large-scale data sets of protein-protein interactions. *Nature*, *417*, 399-403.

Witten, I. H., & Frank, E. (2005). *Data Mining: Practical machine learning tools and techniques*. 2nd Edition. Morgan Kaufmann.

Xenarios, I., Salwnski, L., Duan, X. J., Higney, P., Kim, S. M., & Eisenberg, D. (2002). DIP, the database of interacting proteins: a research tool for studying cellular networks of protein interactions. *Nucleic Acids Res*, *30*, 303–5.

Young, K. H. (1998). Yeast two-hybrid: so many interactions, (in) so little time. *Biol Reprod*, *58*(2), 302-11.

Chapter V
Discovering Protein–Protein Interaction Sites from Sequence and Structure

David La
Dept. of Biological Sciences, Purdue University, USA

Daisuke Kihara
Dept. of Biological Sciences, Dept. of Computer Science, Purdue University, USA

ABSTRACT

This chapter gives a comprehensive introduction of the sequence/structural features that are characteristic of protein-protein interaction sites and reviews state-of-the-art methodologies for protein-protein binding site prediction. Protein-protein interaction residues are largely responsible for mediating physical binding processes such as inhibitory effects through enzyme-inhibitor interaction, initiating immune response by an antibody-antigen interaction, and regulation of cell signaling proteins. Currently, various methods are available for predicting protein-protein interaction sites, which allow a residue-level understanding of the physical protein binding phenomena presented by the global construction protein-protein interaction networks. The overview of the discussed protein-protein binding site prediction strategies and detailed comparison of their weaknesses and strengths is aimed towards assisting protein researchers in gaining more insight to protein-protein interaction networks.

INTRODUCTION

In the recent years, the biology community has been fascinated by snapshots of complex inner workings of cellular activities from various different angles enabled by the advancement of high throughput experiments. The construction of proteome-wide protein-protein interaction maps provided us for the first time a vivid visualization of networks of physical protein binding which maintain cellular events involving cell signaling, gene regulation, and metabolism (Giot *et al.* 2003;Uetz *et al.* 2000;Ito *et al.* 2001). On the other hand, genome sequencing projects and structural genomics projects continue to accumulate sequence and structure information of individual proteins. Therefore, the urgent need in the post-genomic era is to capitalize these two types of enormous amount of data to provide useful and practical information to the biology community. Effectively, protein-protein interaction sites plays a crucial role in proteomics; a greater understanding of protein-protein interaction sites ultimately bridges the gap between abstract information of protein-protein interaction (PPI) network and physical entity of protein structures, substantiating protein-protein interaction data as a critical platform for advanced molecular recognition research and experimental design. For example, knowing (or predicting) binding residues in a PPI site enables designing point mutation experiments to verify the role of the residues in the PPI network. In this chapter, we review current computational methods for predicting PPI sites from sequence and structure and extend discussion to the future direction of this field.

Physical aspects of protein-protein interactions are well studied in the context of protein-protein docking prediction. In the protein-protein docking area, at a molecular basis, these physical protein-protein binding events are commonly classified into three major groups in the context of their function: (1) enzyme-inhibitor complexes, (2) antibody-antigen interactions, and (3) other types of interactions (Hwang *et al.* 2008;Mintseris *et al.* 2005;Chen *et al.* 2003). This classification has been proved quite useful to atomic-level analysis employed for protein-protein binding site prediction and protein-protein docking. A genome-scale PPI network would include interactions of proteins of much broader range of functions, but here we mention these three classes to illustrate variation of the nature of protein-protein interaction.

Figure 1. Examples of the different groups of protein-protein interactions: (A) PDBID: 1BRS, a Barnase-barstar complex (B) PDBID: 1KXQ, an alpha-amylase to antibody interaction, and (C) PDBID: 1B6C, an example of the other types of interactions such as those transient that involve kinase to isomerase binding.

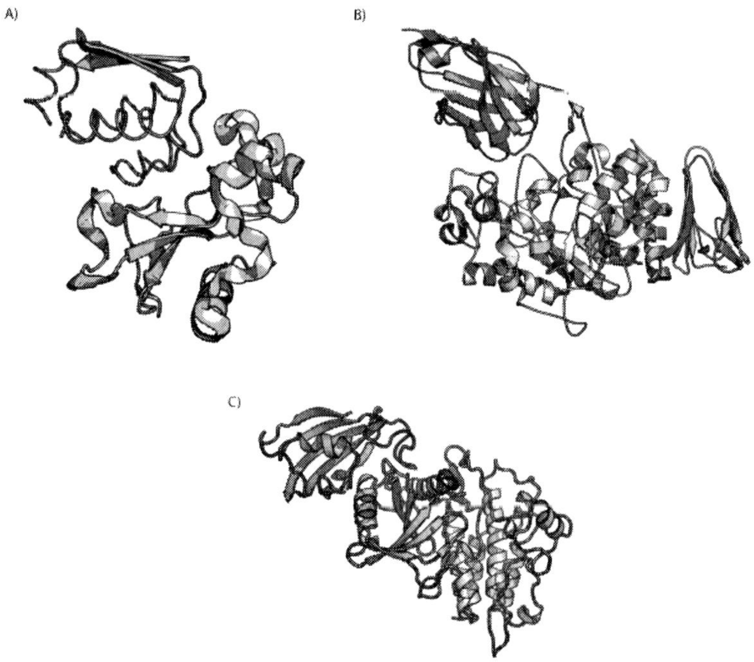

An example structural complex for each interaction class is shown in Figure 1. The class of enzyme-inhibitor interactions involves the inhibition or regulation of an enzyme function by physical binding. It has been shown that there is a general tendency for enzyme-inhibitor interfaces to be more evolutionary conserved than other surface residues (de Vries *et al.* 2006;Nooren, Thornton 2003a;Nooren, Thornton 2003b). Further, antibody-antigen interactions are generally established through hydrophobic contacts, where regions of the hyper-variable loops are generated through rapid mutation resulting in the effective recognition of complementary antigenic proteins. Interactions classified as *other*, include obligate homo- or hetero-oligomers (permanently bound) and other transient (temporarily bound) complexes involved in cell signaling, *e.g.* protein kinase interactions and G-coupled receptors. Classification of permanent and transient interaction is also important as the binding sites of the two classes exhibit notable differences.

IMPORTANT PROTEIN FEATURES FOR DISCOVERING PROTEIN-PROTEIN INTERACTION SITES

There are a broad range of sequence-based and structure-based features which can be used to predict PPI sites of a protein. A PPI site prediction method aims to predict amino acid residues which bind to another protein (*i.e.* closer than a threshold value to the binding partner) from the primary sequence or a set of the sequence and the tertiary structure of the protein of interest. Residues at the PPI sites which make significant energetic contribution to binding constitute *hot spots* (Bogan, Thorn 1998). In principle hot spots on a protein surface are to be predicted by biophysical energetic consideration (Fernandez-Recio *et al.* 2005;Shulman-Peleg *et al.* 2007;Liang *et al.* 2004;Liang *et al.* 2006;Murga *et al.* 2008) but they can be also predicted the other sequence-based and structure-based features.

Features useful for predicting PPI sites can be roughly classified into three different categories: (1) Sequence-based features: information that are directly taken from sequences without structural information, (2) Low-resolution structural features: sequence data that make use low resolution spatial data to find neighboring residue so that surface clusters or patches can be readily identified, and (3) High resolution structural features: structural properties that require high resolution atomic detail. In the following sections, we will review various critical protein features that fall into these general categories. Some of the features are used by itself but it is more common to combine several features to construct a composite score.

Sequence-Based Features

As the whole genomes continue to become available, protein sequence information is almost always available and can be readily applied for a target protein even when its structure has not experimentally solved or reliably predicted.

Residue Propensity

It is clear that certain residues are more preferred at PPI sites than other residues (Lo Conte *et al.* 1999;Jones, Thornton 1996) and thus should be incorporated as an important feature in PPI site prediction methods (Liang *et al.* 2006;Neuvirth *et al.* 2004;Negi, Braun 2007). Studies have shown that obligate complexes have more aromatic and hydrophobic residues at the PPI sites compared to the site of all other surfaces (Jones, Thornton 1997a). On the other hand, transient complexes generally have more polar residues at regions that interact with other proteins (Mintseris, Weng 2005;Block et al. 2006;Janin et al. 2007). Protein surface clustered with aromatic residues is indicative of π-π stacking interactions between two complexes or cation-π interactions (Pletneva *et al.* 2001). Further, a rare observation of a cysteine residue on a protein surface is energetically unfavorable and thus must be involved in some functional role. Cysteine residues generally form a disulfide bridge on another cysteine residue of an adjacent protein helping stabilize a resulting complex.

Sequence Conservation

When a set of homologous sequences are available, a multiple sequence alignment of the sequences will provide additional value in a sequence analysis. Residues are conserved if they have structural and/or functional importance, and residues in PPI site show some degree of conservation (Mihalek *et al.* 2003). Sequence conservation is commonly represented in the form of *sequence profile*. Sequence profiles represents the amino acid occurrences at individual sequence positions in a given protein family alignment. Specifically, common sequence profiles take form of a position specific scoring matrix (PSSM), which describes the likelihood of each of the standard 20 amino acids at each position of a sequence in question. Given a query sequence, PSSMs can be constructed by using PSI-BLAST (Altschul et al. 1997), where the significant search results are used to generate the profile. Machine learning techniques, *i.e.* neural network and Support Vector machines, are commonly applied for predicting PPI sites by using PSSM as input (Zhou, Shan 2001;Tjong *et al.* 2007;Porollo, Meller 2007).

Shannon Entropy (Shannon 2001) is alternative approach for measuring sequence conservation among protein families (Valdar 2002). The form of an entropy score involves the following: a given K number of amino acid types (typically 20 from the number standard amino acids), N total number of residues in the column, na number of type a residues, and the fa frequency of type a residue within all N residues in a given column ($fa=na/N$), we use the equation to define Shannon's entropy as:

$$S = -\sum_{a}^{K} fa \log_2 fa \quad (1)$$

Larger values of S indicate more entropy or residue-type diversity and lower values indicate less residue character diversity and more evolutionary sequence conservation. Further, residues may be classified into their chemically similar groups. A common classification would be to specify $K=6$, where residues are grouped into aliphatic, aromatic, polar, positively charged, negatively charged, and unusual conformations (Mirny, Shakhnovich 1999). Such a classification would intuitively give more chemically and biologically reasonable measures of sequence entropy.

However, various studies have shown that sequence conservation alone is not very discriminative for PPI sites, because residues of several other classes, such as those locating at protein cores and small chemical ligand binding sites of enzymes, are equally well or more conserved than PPI sites (Capra, Singh 2007;Caffrey *et al.* 2004;Lichtarge *et al.* 1996a;Lichtarge *et al.* 1996b;Glaser *et al.* 2003;Pupko *et al.* 2002;Landau *et al.* 2005;Armon *et al.* 2001;Glaser *et al.* 2005;Mintseris, Weng 2005). Therefore, sequence conservation information should be used in combination with other features.

Correlated Mutation Analysis

Correlated mutation analysis (CMA) is a technique for detecting compensatory sequence variations among protein family sequence alignments. The assumption is that interacting residues share co-evolutionary mutation patterns. When a mutation occurs in a sequence position, its spatially neighboring residues will also change to compensate for this mutation in a complementary manner. CMA was first used to measure the relationship between these mutations and intra-residue contact pairs in proteins structures (Goebel *et al.* 1994), where residue positions that have high correlated mutations are likely to be also spatially close to each other. Similarly, correlated mutations are also used for identifying inter-residue contacts where they correspond to PPI sites (Halperin *et al.* 2006;Pazos *et al.* 1997). Pazos and Valencia further applied CMA to detect interacting proteins from a large set of protein pairs (Pazos, Valencia 2002).

Given an all-to-all pair-wise comparison of sequence k and l in a multiple sequence alignment, the mutational relationship between two alignment positions i and j are evaluated by a correlation coefficient:

$$r_{ij} = \frac{1}{N^2} \sum_{kl} \frac{(s_{ikl} - \bar{s}_i)(s_{jkl} - \bar{s}_j)}{\sigma_i \sigma_j} \quad (2)$$

Where N is the number of sequences in the multiple sequence alignment, and $\bar{s}_i$ and $\bar{s}_j$ are mean values of N^2 mutation scores in the position i and j respectively, while σ_i and σ_j are standard deviations of the N^2 similarity scores in the position i and j respectively. s_{ikl} is the similarity score of position i in sequence k and l, and s_{jkl} is the similarity score of position j in sequence k and l. A similarity matrix provides the individual similarity scores for the pair-wise mutations and represents the level in which the physiochemical properties between mutations are conserved.

Low-Resolution Structural Features

Low-resolution structural features involve the minimal detailed use of backbone information and their corresponding residue identities. Those low-resolution structures, which don't have atomic details, are still very useful to identify surface residues which are spatially clustered. Low-resolution structures can be obtained also by protein structure prediction (Kihara *et al.* 2002;Kihara, Skolnick 2004;Kihara *et al.* 2001).

When a low-resolution structure for a target protein is available, PPI site prediction can be made by a *patch*-basis. A patch is defined either as (1) a spatial region of a given surface residue and its neighboring residues with a defined cut-off distance value, essentially capturing residues into a sphere, or (2) a fixed number of residues that form a cluster of residues around a single site of interest. The first patch definition extracts a uniform elliptical shape and simple in implementation. When PPI site residues may cluster into shapes that are non-elliptical, the second patch definition is more appropriate.

Hydrophobic Patch

A typical example of using patch is to consider average hydrophobicity of patches, because PPI sites, particularly obligate PPI sites, generally consist of hydrophobic residues. Of a given patch in question, all residues in the patch are assigned a hydrophobicity scale value (Kyte, Doolittle 1982). Taking the average hydrophobicity scale value of all the residues in the patch will result in a value that is represented as the patch hydrophobicity. All possible patches are then evaluated for a given structure and the patch with the largest hydrophobicity values would be designated as the PPI site prediction.

Secondary Structure

Some preferences in the secondary structure at PPI sites are observed for certain classes of proteins. Protein binding site regions of transient homo-complexes are found to favor β-sheets (Neuvirth *et al.* 2004). For making complexes stable, a β-sheet of one protein may bridge a β-sheet of another, forming an extended β-sheet interfacial structure. To a lesser extent, there are also PPI sites that mainly consist of α-helical regions. For example, coiled-coil regions largely consist of the interaction between the hydrophobic regions in α-helices, where they are common in transcription factors that form complexes to regulate gene expression. Lastly, intrinsically disordered regions are commonly found to be involved in the transient interactions of hub proteins (Dosztanyi et al. 2006;Dunker et al. 2005;Oldfield et al. 2008;Ekman et al. 2006;Singh et al. 2007;Higurashi et al. 2008).

High-Resolution Structural Features

Atomic detailed structure enables consideration of biophysical properties and detailed shapes of surface regions. In principle, computing these features needs a protein crystal structure solved by X-ray crystallography.

Atom Group Propensity

Similar to the amino acid residue propensity discussed in the sections of sequence-based features, propensity of atom groups to be on a PPI site can be considered. Kufareva *et al.* classified heavy atoms in amino acids into 32 types according to their chemical features including charge and sp-, sp2-, or sp3-hybridization and computed the probability to find each atom group on the interface. Using the probability, surface patches are predicted to be at the PPI site or not (Kufareva *et al.* 2007).

Desolvation Energy

PPI sites are generally excluded from solvent where more hydrophobic residues are located. Rather than evaluating hydrophobicity on a per residue basis, atom based desolvation could more precisely discriminate between PPI sites and non-PPI sites because desolvation features are not limited to only hydrophobic residues. Therefore, atomic based desolvation parameters (ASP), which evaluate the free energy of transferring from water to a protein-protein interface, are optimized for protein-protein docking application (Fernandez-Recio *et al.* 2005). These parameters are originally determined from water/octanol transfer experiments (Eisenberg, McLachlan 1986).

Desolvation energies are particularly discriminative for predicting interacting regions of antibodies (Burgoyne, Jackson 2006). However, prediction performance on antigens compared to antibodies is considerably less discriminative. Nevertheless, use of desolvation energies still results in prediction accuracy that is noticeably better than the alternative types of features which includes sequence conservation (Burgoyne, Jackson 2006).

Relative Accessible Surface Area

The average relative accessible surface area (ASA) of residues in a local patch is found to have discriminating power for patches at PPI site from other surface patches (Jones, Thornton 1997a), *i.e.* PPI site patches tend to have larger relative ASA. The relative ASA of a residue in a protein structure is defined by the ratio of the observed ASA of the residue relative to the reference ASA. And the reference ASA of an amino acid residue X is obtained by calculating the ASA of Gly-X-Gly peptide in extended conformations (Miller *et al.* 1987). This interesting observation indicates that a patch with residues which are more exposed than usual is preferred to be covered by binding to another protein.

Molecular Surface Shape

Protein molecular surfaces are generally observed in terms of cavities, protrusions, and flatness. These abstract molecular surface shapes are important for identifying unique and functionally important regions (Li *et al.* 2008). In the same paper mentioned in the previous section of relative ASA (Jones, Thornton 1997a;Jones, Thornton 1997b) it was found that PPI sites are geometrically flat in general. The flatness of a surface patch is defined as the least square fitness of a plane to atoms of residues in the patch. Jones and Thornton examined another geometrical measure, named protrusion. The protrusion is defined as the average protrusion of each residue in the patch from the surface of the protein. Thus, intuitively, the flatness indicates an overall geometrical shape of the surface of a patch and the protrusion indicates local ruggedness of the patch.

Side Chain Energy

Liang *et al.* used a composite score for evaluating the energy of side chain conformation (*i.e.* rotamer) of surface residues, which they called the side chain energy, and showed that residues at PPI sites have higher side-chain energy than other surface residues (Liang *et al.* 2004). The side chain energy is calculated as a linear combination of several energetic terms including side-chain atom surface contact area ($S_{contact}$), overlap volume ($V_{overlap}$) hydrogen bonding energy (E_{hbond}), electrostatic interaction energy (E_{elec}), hydrophobic solvent accessible surface (S_{pho}), hydrophilic solvent accessible surface (S_{phi}), difference between the fraction buried hydrophilic atom surfaces from rotamers in protein environment and in isolation (ΔF_{phi}), solvent exclusion volume around charged atoms ($V_{exclusion}$), frequency of observed rotamers (f_1), frequency of conformations of a given backbone residue (f_2), the presence of a disulfide bridge (N_{ssbond}), and the difference between the rotamer free energy in solvent and the denatured state (ΔG_{ref}). The side chain energy for a given residue (R_i) is given in the following equation:

$$
\begin{aligned}
E_{sidechain}(R) = & -0.143 S_{contact} + 0.724 V_{overlap} + 1.72 E_{hbond} + 28.6 E_{elec} - 0.0467 \Delta S_{pho} \\
& + 0.0042 \Delta S_{phi} + 1.14 \Delta (F_{phi})^{10} + 7.95 V_{exclusion} - 0.919 \ln(f_1 f_3) \\
& - 4.3 N_{unhbond} - \Delta G_{ref}
\end{aligned} \tag{3}
$$

The weights are optimized so that each term with reference values best fit the observed energetically favorable residue conformations. The higher side-chain energy of PPI sites for unbound monomer implies that the energy of two interacting proteins declines significantly upon association.

B-Factor

B-factors or temperature-factors of an X-ray crystal structure quantify the atomic positional variation to the average displacement of the scattered electron density. Higher B-factors are generally found at regions of proteins that are flexible (*i.e.* loops or disordered regions), whereas lower B-factors usually correspond to regions that are less flexible (*i.e.* the compact core of the protein). Hydrophobic protein cores are tightly packed and exhibit less flexibility and hence lower B-Factors. Similarly, PPI sites of bound complexes also have low B-factors because they are not exposed to bulk solvent and exhibit less thermo-motion. Interestingly, it was found that the B-factors of unbound PPI sites also exhibit lower B-factor values compared to non-protein binding regions (Neuvirth *et al.* 2004). The authors discussed that the preference for a low B-factor might be due to the potential of binding sites to be involved in crystal contacts. Nevertheless, as the formation of crystal contacts at a certain position would suggest physicochemical differences of the involved surface patches, the B-factor would be still useful for predicting PPI sites of an X-ray crystal structure.

Water Molecule

It is common that a X-ray crystal structure of a protein-protein bound complex contains water molecules at its PPI site, forming water-mediated polar interactions (Lo Conte *et al.* 1999). Surface shape complementarity is often better established when water facilitates close atomic packing to fill voids and spaces at PPI sites. Neuvirth *et al.* further found that PPI sites of unbound monomers have a higher water content compared to other surface regions (Neuvirth *et al.* 2004). This observation may be rationalized by polar residues found at PPI sites, which provide the coordination points for water. In addition, recent molecular dynamic simulations indicate that residues at PPI site exhibit less water molecule traffic than other sites, showing that the hydrogen bonds at interfaces to water are more stable than those that are located away from the binding interface. (Mihalek *et al.* 2007).

Electrostatic Potential

Burgoyne and Jackson tested peak and average values of electrostatic potential for local surface regions as a predictor of PPI site among other predictors (Burgoyne, Jackson 2006). The electrostatic potential is computed by the finite difference Poisson–Boltzmann calculations. It was found that electrostatics showed some predictive capacity for enzymes but did not show general trends in the prediction of protein–protein interfaces. The results imply that electrostatic complementarity play an important role in on orienting enzymes and inhibitor complexes.

EXISTING METHODS FOR PREDICTING PPI SITES

In the previous section we have seen various features which can be used for predicting PPI sites. In most of the prediction methods, several features are combined to capture different characteristics of PPI sites to improve the prediction accuracy. As examples of how features are combined in actual prediction methods, in this section we overview currently available PPI site prediction methods. These methods and their websites are listed in Table 1.

These methods fall into three general classes. The first class involve the use of the combination of sequence- and structure-based information through machine learning, such as support vector machines (SVM) (Bradford, Westhead 2005;Res *et al.* 2005) and neural networks (NN) (Tjong *et al.* 2007;Porollo, Meller 2007;Pettit *et al.* 2007). In the second class of approaches, we will detail the methods that only make use structural information

Table 1. Web servers available for predicting PPI sites

Available Methods	URL
Class 1 (Use sequence and structural information)	
Promate	http://bioportal.weizmann.ac.il/promate/
PPI-Pred	http://www.bioinformatics.leeds.ac.uk/ppi_pred/
Cons-PPISP	http://pipe.scs.fsu.edu/ppisp.html
SPPIDER	http://sppider.cchmc.org/
Sharp²	http://www.bioinformatics.sussex.ac.uk/SHARP2
PINUP	http://sparks.informatics.iupui.edu/PINUP/
InterProSurf	http://curie.utmb.edu/prosurf.html
WHISCY	http://nmr.chem.uu.nl/Software/whiscy/index.html
Class 2 (Use only structural information)	
Optimal Docking Area	http://www.molsoft.com/oda.cgi
HotPatch	http://hotpatch.mbi.ucla.edu/
Class 3 (Integrate predictions from various servers)	
Meta-PPISP	http://pipe.scs.fsu.edu/meta-ppisp.html

Figure 2. Comparison between methods and the sequence- or/and structure-based features they use

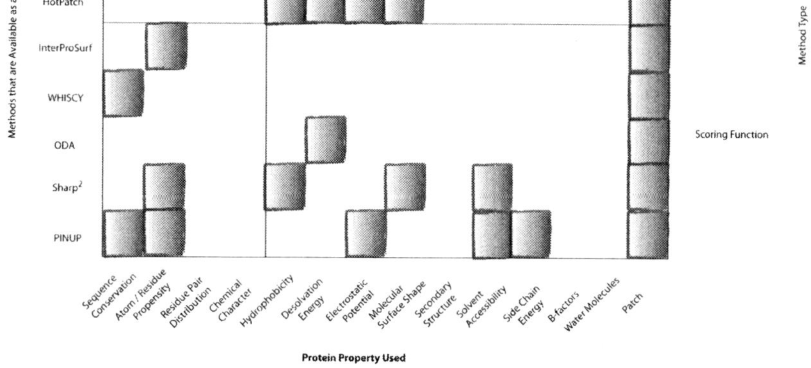

to predict PPI sites. Finally, in the third class, recent developments integrate predictions from various state-of-the-art methods (from classes previously mentioned) through a *meta-server* to achieve improved performance (Qin, Zhou 2007). The features and the strategy used to combine those features by the methods listed in Table 1 are summarized in Figure 2.

Promate utilizes a naïve Bayesian to integrate the following features (Neuvirth *et al.* 2004): sequence level features include evolutionary sequence conservation, single residue frequency, paired-residue frequency, and residue chemical character (aromatic, positively charged, negatively charged, hydrophobic). Structure-based information includes desolvation energy, secondary structure elements, B-factors, and water molecules. All of such sequence- and structure-based information is derived from a defined patch encompassed by a radius of 10Å drawn from a center of a given protein surface residue. Compared to other methods for PPI site prediction, this approach uses the largest number of features. An extended server, called *Promateus,* allows users to upload custom features to the naïve Bayesian approach to add to the set of features of used in *Promate* (Neuvirth *et al.* 2007).

PPI-Pred uses SVM to integrate sequence-based features, including sequence conservation, sequence profile, the chemical character of a given residue, and structure-based features, which include electrostatic potential and molecular surface shape data (Bradford, Westhead 2005). This approach uses features similar to that found in *Promate* except it does not consider B-factors and position of water molecules.

PPISP (Protein-Protein Interaction Site Predictor) uses a position specific sequence profile derived from PSI-BLAST and solvent accessibility of known PPI sites as input for NN training (Zhou, Shan 2001). To further improve the prediction performance, multiple PPISP NNs are trained and combined into a consensus-scoring framework called *cons-PPISP.*

Porollo and Meller found that surface residues that are incorrectly predicted to be part of the core of the protein from their sequence characteristics frequently correspond to PPI sites. They have implemented this interesting idea into a PPI site prediction program, *SPPIDER* (Solvent accessibility-based Protein-Protein Interaction site Identification and Recognition) (Porollo, Meller 2007). SPPIDER basically inputs the difference between the predicted residue solvent accessibility from their sequence-based features and actual residue solvent accessibility computed from the structure of the input protein into NN architecture. The sequence-based features used include the measure of expected number of neighboring residue contacts, sequence conservation (entropy), residue frequencies and sequence profiles from PSI-BLAST, sequence conservation of chemical property, size, and hydrophobicity.

InterProSurf solely relies on residue PPI site propensity. It identifies clusters of residues on a protein surface that are observed at high frequency in a representative dataset of PPI sites (Negi *et al.* 2007;Negi, Braun 2007).

The *Sharp²* (Solvation potential, Hydrophobicity, Accessible surface area, Residue interface propensity, and Planarity and Protrusion) method measures three geometric features, *i.e.* surface planarity, the average protrusion of residues, and the average relative accessible surface area of residues, along with three other residue-based features, namely, solvation potential, hydrophobicity, and interface propensity (Murakami, Jones 2006). These six features in relation to PPI sites are examined in their previous work. (Jones, Thornton 1997a). The three geometrical features are explained in the previous section. Each parameter is normalized so that resulting values ranges from 0 to 100 and an average value for all parameter values will be the representative score for a given patch in question.

PINUP (Protein Interface residUe Prediction) utilizes the side chain energy explained in the previous section, linearly combined with a residue conservation score and residue PPI site propensity (Liang *et al.* 2006). It was reported that the residue PPI site propensity score performed the best among the three individual terms and the combination of the three showed improvement on top of it.

The *Optimal Docking Area* method computes the docking desolvation energy of a given surface patch (Fernandez-Recio *et al.* 2005). The desolvation energy of a patch is a sum of parameterized per-atomic contributions (σ_i) in proportion to the solvent accessible surface area (ASA):

$$E_{desolv} = -\sum_i \sigma_i ASA_i \qquad (4)$$

where i is an atom included in the patch. The atomic solvation parameter, σ_i, is determined through linear fitting of octanol/water transfer energies from experiment and further optimized for protein-protein docking. For each surface point distributed on a protein surface, the desolvation energy of a series of patches with different radii is computed. Then the lowest energy among the patches is assigned to the surface point. Finally a patch of a surface point that gives the lowest desolvation energy will be considered a putative PPI site.

Hotpatch is aimed to identify various functional sites on a protein surface, including catalytic sites of particular enzyme classes, metal ion binding sites, DNA/RNA interaction sites, and PPI sites (Pettit *et al.* 2007). It integrates several features in a NN architecture depending on which sites are going to be predicted. When specified to predict PPI sites, it integrates residue-based hydrophobicity with other structure-based features, including electrostatic potential, charge, surface shape, residue-based hydrophobicity and atomic-based desolvation energies.

WHISCY (WHat Information does Sequence Conservation Yield) uses sequence conservation alone to predict PPI sites (de Vries *et al.* 2006). The conservation scoring function used by WHISCY is a pair-wise comparison of sequences in a multiple sequence alignment which uses a modified Dayhoff amino acid substitution matrix (Dayhoff *et al.* 1983). Further, the WHISCYMATE option provided by the server combines the conservation sores with *Promate* predictions. The resulting predictions are then used to guide protein-protein docking strategies (Dominguez *et al.* 2003).

Meta-PPISP takes a consensus approach (Qin, Zhou 2007;Zhou, Qin 2007); it combines raw scores from Promate, PINUP and Cons-PPSIP through linear regression to produce a final score. In general meta-server approaches can combine strengths of component methods thus successful in many prediction methods in bioinformatics. Likewise Meta-PPISP showed a better performance over the three component methods in their paper.

PREDICTION PERFORMANCE OF CURRENT METHODS

Up to this point, we overviewed useful features for predicting PPI sites and how existing methods combine those features. Next, to understand the current status of PPI methods, we will discuss the prediciton accuracy. We will review two recent benchmark studies, because they assessed the performance of most of the methods discussed in the previous section (Zhou, Qin 2007;Huang, Schroeder 2008).The prediction performance is commonly measured in terms of the sensitivity (the proportion of corectly predicted residues at PPI sites relative to the acutal residues at PPI sites) and the specificity (the proportion of the corectly predicted PPI site residues relative to the total number of predicted PPI site residues). Since the sensitivity and the specificty is tradeoff, usually the specificity is plotted relative to the sensitivity.

Zhou and Qin prepared two benchmark datasets; one dataset consists of 35 proteins in the enzyme and inhibitor category from ZDOCK protein-protein docking benchmark dataset (Mintseris *et al.* 2005) and another one consists of 25 docking prediction targets in CAPRI (Critical Assessment of Prediction of Interactions) prediction contest (http://www.ebi.ac.uk/msd-srv/capri/). The CAPRI dataset contains eight proteins which are not in the enzyme/inhibitor category, *i.e.* antibody/antigen or other immune system complexes. Six webservers in Table 1, namely, cons-PPISP, Promate, PINUP, PPI-Pred, SPPIDER, and Meta-PPISP are benchmarked. On the 35 enzyme dataset, PPI-Pred, SPIDDER, cons-PPISP, Promate, PINUP, and meta-PPISP showed the specificity of 27, 33, 36, 36, 48, and 50%, respectively, at the sensitivity of 50%. These methods peformed uniformly worse on another dataset of 25 CAPRI targets, probably because proteins in the CAPRI dataset is more diverse. At the sensitivity of 30%, the methods showed the specificity of 23, 25, 26, 26, 28, and 31%, respectively (the methods are placed in the same order). At the sensitivity of 50%, the specificity of the methods ranges roughly between 15 to 28%. Note that the purpose of showing the actual numbers is to have a general idea of the performace of currently avaialble methods but not to rank webservers. Acutally a fair comparison of existing methods is difficult, since some of the tested proteins may be used in tuning settings of the methods. What we can conclude from this benchmark study is that current prediction methods show the specificity of around 20 to 50% at the sensitivity of 50% and that a better performance can be achieved by taking a meta-server approach. A comparison of the benchmark performances (measured by sensitivity verses specificity curves) for each PPI site prediction method can be found in the review by Zhou and Qin.

In attempt to formulate an alternative meta-serer solution, Huang and Schroeder developed MetaPPI which combines five individual PPI site prediction methods, namely, PPI-Pred, SPIDDER, Promate, PINUP, and PPISP. They reported success rates for these five methods in comparison to their MetaPPI method on 62 protein-protein complexes. The protein complexes are calssified into two datasets, one consists of 20 complexes classified as enzyme-inhibitors and another one consists of the remaining 42 complexes that were classified as other types complexes. In their work, the success rate for a dataset is defined as the percentage of protein complexes whose PPI sites are predicted with a sensitivity of over 50% (note that this is different from the average specificity at the sensitivity of 50% , which are reported in the review by Zhou and Qin).

For the dataset of enzyme-inhibitors, the success rate of PPI-Pred, SPIDDER, Promate, PINUP, PPISP, and MetaPPI were 45, 23, 36, 52, 55, and 70%, respectively. While for the category of other types of complexes, the perfomance of the methods were performed with 28, 10, 13, 15, 25, and 44%, respectively. Consistent to Zhou and Qin's benchmark, enzyme-ihibitor complexes can be generally predicted with greater success than other types of complexes.

The performance shown here might not seem very good. However, encouragingly, Zhou and Qin's work showed that a PPI prediction with 40-45% of the sensitivity and the specificity can improve ranking of near-native docking decoys compared to the original scores of the ZDOCK docking program (Chen *et al.* 2003), thus useful for protein docking prediction.

DISCUSSION

The current methods use a combination of reasonable sequence-based structure-based features but still leave room for improvement. A part of the difficulty of PPI site prediction comes from the fact that different classes of protein-protein interaction have different characteristics. For example, PPI sites of enzymes are known to have relatively well conserved residues, while sequence conservation information is less useful in predicting PPI sites of antigen in antibody/antigen interaction. Sequence variation at antigenic sites may not exhibit any biological pressure in terms of their sequence evolution (because the antibody is evolving to bind to the antigen rather than the reverse), it is important to understand the biophysical properties that allow antibodies to be specific in recognizing antigenic proteins. Similarly, PPI sites of obligate complexes and those of proteins which undergo transient interaction with several different proteins may use different biophysical mechanism for establishing binding. Therefore, practically, developing class specific prediction methods separately for different class of PPI will certainly improve prediction performance. Another potential reason for the limited accuracy of current PPI methods is that usually prediction of a protein is made without using information from its binding partner. That will largely prevent a PPI method from considering detailed energy of protein binding. Of course it is not trivial to specify PPI sites of a docking partner (expect for a case of a PPI site of an antibody), but incorporating some information of the partner may make significant improvement, if implemented in an intelligent way. Finally, the effect of water at the PPI site, which may play a critical role in protein-protein binding, is not well understood biophysically and still limited in use for PPI site prediction. As mentioned above, water molecules are frequently found in PPI sites of crystal structures of binding proteins. General understanding of interplay between water molecules and a PPI site would provide very important information for predicting PPI sites.

Although current PPI site prediction methods have limitations, with more structure and sequence data of protein complexes becoming available, the performance of prediction methods would only improve in the future.

CONCLUSION

In this chapter, we discussed PPI site prediction methods from several angles including features they use, the accuracy of current methods, and intrinsic limitation of PPI prediction and potential directions for improvement. As PPI networks have drawn much attention in understanding complex nature of cellular activities, the importance of PPI site prediction has been increased because it can bridge PPI networks determined by high throughput experiments and protein structure information solved by structural genomic projects. With the increasing availability of new protein structures and sequences and understanding of biophysics of protein-protein interaction, we expect to see advanced PPI site prediction methods to be applicable on a more diverse dataset of interacting proteins in PPI networks.

ACKNOWLEDGMENT

This work is partially supported by grants from National Institute of General Medical Sciences of the National Institutes of Health (R01GM075004, U24 GM077905) and National Science Foundation (DMS 0604776).

REFERENCES

Altschul, S. F., Madden, T. L., Schaffer, A. A., Zhang, J., Zhang, Z., Miller, W., & Lipman, D. J. (1997). Gapped BLAST and PSI-BLAST: A new generation of protein database search programs. *Nucleic Acids Res, 25*, 3389-3402.

Armon, A., Graur, D., & Ben-Tal, N. (2001). ConSurf: an algorithmic tool for the identification of functional regions in proteins by surface mapping of phylogenetic information. *J Mol Biol, 307*, 447-463.

Block, P., Paern, J., Hullermeier, E., Sanschagrin, P., Sotriffer, C. A., & Klebe, G. (2006). Physicochemical descriptors to discriminate protein-protein interactions in permanent and transient complexes selected by means of machine learning algorithms. *Proteins, 65*, 607-622.

Bogan, A. A., & Thorn, K. S. (1998). Anatomy of hot spots in protein interfaces. *J Mol Biol, 280*, 1-9.

Bradford, J. R., & Westhead, D. R. (2005). Improved prediction of protein-protein binding sites using a support vector machines approach. *Bioinformatics, 21*, 1487-1494.

Burgoyne, N. J., & Jackson, R. M. (2006). Predicting protein interaction sites: binding hot-spots in protein-protein and protein-ligand interfaces. *Bioinformatics, 22*, 1335-1342.

Caffrey, D. R., Somaroo, S., Hughes, J. D., Mintseris, J., & Huang, E. S. (2004). Are protein-protein interfaces more conserved in sequence than the rest of the protein surface? *Protein Sci, 13*, 190-202.

Capra, J. A., & Singh, M. (2007). Predicting functionally important residues from sequence conservation. *Bioinformatics, 23*, 1875.

Chen, R., Mintseris, J., Janin, J., & Weng, Z. (2003). A protein-protein docking benchmark. *Proteins, 52*, 88-91.

Dayhoff, M. O., Barker, W. C., & Hunt, L. T. (1983). Establishing homologies in protein sequences. *Methods Enzymol., 91*, 524-545.

de Vries, S. J., van Dijk, A. D., & Bonvin, A. M. (2006). WHISCY: what information does surface conservation yield? Application to data-driven docking. *Proteins, 63*, 479-489.

Dominguez, C., Boelens, R., & Bonvin, A. M. (2003). HADDOCK: A protein-protein docking approach based on biochemical or biophysical information. *J Am Chem Soc, 125*, 1731-1737.

Dosztanyi, Z., Chen, J., Dunker, A. K., Simon, I., & Tompa, P. (2006). Disorder and sequence repeats in hub proteins and their implications for network evolution. *J Proteome Res, 5*, 2985-2995.

Dunker, A. K., Cortese, M. S., Romero, P., Iakoucheva, L. M., & Uversky, V. N. (2005). Flexible nets. The roles of intrinsic disorder in protein interaction networks. *FEBS J, 272*, 5129-5148.

Eisenberg, D., & McLachlan, A. D. (1986). Solvation energy in protein folding and binding. *Nature, 319*, 199-203.

Ekman, D., Light, S., Bjorklund, A. K., & Elofsson, A. 2006. What properties characterize the hub proteins of the protein-protein interaction network of Saccharomyces cerevisiae? *Genome Biology, 2007 8, R192* 7:R45.

Fernandez-Recio, J., Totrov, M., Skorodumov, C., & Abagyan, R. 2005. Optimal docking area: a new method for predicting protein-protein interaction sites. *Proteins, 58*, 134-143.

Giot, L., Bader, J. S., Brouwer, C., Chaudhuri, A., Kuang, B., Li, Y., et al. (2003). A protein interaction map of Drosophila melanogaster. *Science, 302*, 1727-1736.

Glaser, F., Pupko, T., Paz,I ., Bell, R. E., Bechor-Shental, D., Martz, E., & Ben-Tal, N. 2003. ConSurf: identification of functional regions in proteins by surface-mapping of phylogenetic information. *Bioinformatics, 19*, 163-164.

Glaser, F., Rosenberg, Y., Kessel, A., Pupko, T., and Ben-Tal, N. (2005). The ConSurf-HSSP database: The mapping of evolutionary conservation among homologs onto PDB structures. *Proteins, 58*, 610-617.

Goebel, U., Sander, C., Schneider, R., & Valencia, A. (1994). Correlated mutations and residue contacts in proteins. *Proteins, 18*, 309-317.

Halperin, I., Wolfson, H., & Nussinov, R. (2006). Correlated mutations: advances and limitations. A study on fusion proteins and on the Cohesin-Dockerin families. *Proteins 63*, 832-845.

Higurashi, M., Ishida, T., & Kinoshita, K. (2008). Identification of transient hub proteins and the possible structural basis for their multiple interactions. *Protein Sci, 17*, 72-78.

Huang, B., & Schroeder, M. (2008). Using protein binding site prediction to improve protein docking. *Gene, 422*, 14-21.

Hwang, H., Pierce, B., Mintseris, J., Janin, J., & Weng, Z. (2008). Protein-protein docking benchmark version 3.0. *Proteins*.

Ito, T., Chiba, T., Ozawa, R., Yoshida, M., Hattori, M., & Sakaki,Y. (2001). A comprehensive two-hybrid analysis to explore the yeast protein interactome. *Proc Natl Acad Sci USA, 98*, 4569-4574.

Janin, J., Rodier, F., Chakrabarti, P., & Bahadur, R. P. (2007). Macromolecular recognition in the Protein Data Bank. *Acta Crystallogr D Biol Crystallogr, 63*, 1-8.

Jones, S., & Thornton, J. M. (1996). Principles of protein-protein interactions. *Proc Natl Acad Sci USA, 93*, 13-20.

Jones, S., & Thornton, J. M. (1997a). Analysis of protein-protein interaction sites using surface patches. *J Mol Biol, 272*, 121-132.

Jones, S. & Thornton, J. M. (1997b). Prediction of protein-protein interaction sites using patch analysis. *J Mol Biol, 272*, 133-143.

Kihara, D., Lu, H., Kolinski, A., & Skolnick, J. (2001). TOUCHSTONE: An ab initio protein structure prediction method that uses threading-based tertiary restraints. *Proc. Natl. Acad. Sci U. S. A, 98*, 10125-10130.

Kihara, D., & Skolnick, J. (2004). Microbial genomes have over 72% structure assignment by the threading algorithm PROSPECTOR_Q. *Proteins, 55*, 464-473.

Kihara, D., Zhang, Y., Lu, H., Kolinski, A., & Skolnick, J. (2002). Ab initio protein structure prediction on a genomic scale: application to the Mycoplasma genitalium genome. *Proc Natl Acad Sci USA, 99*, 5993-5998.

Kufareva, I., Budagyan, L., Raush, E., Totrov, M., & Abagyan, R. (2007). PIER: Protein interface recognition for structural proteomics. *Proteins, 67*, 400-417.

Kyte, J., & Doolittle, R. F. (1982). A simple method for displaying the hydropathic character of a protein. *J Mol Biol, 157*, 105-132.

Landau, M., Mayrose, I., Rosenberg, Y., Glaser, F., Martz, E., Pupko, T., & Ben-Tal, N. (2005). ConSurf 2005: The projection of evolutionary conservation scores of residues on protein structures. *Nucleic Acids Res, 33*, W299-W302.

Li, B., Turuvekere, S., Agrawal, M., La, D., Ramani, K., & Kihara, D. (2008). Characterization of local geometry of protein surfaces with the visibility criterion. *Proteins, 71*, 670-683.

Liang, S., Zhang, C., Liu, S., & Zhou, Y. (2006). Protein binding site prediction using an empirical scoring function. *Nucleic Acids Res, 34*, 3698-3707.

Liang, S., Zhang, J., Zhang, S., & Guo, H. (2004). Prediction of the interaction site on the surface of an isolated protein structure by analysis of side chain energy scores. *Proteins 57*, 548-557.

Lichtarge, O., Bourne, H. R., & Cohen, F. E. (1996a). An evolutionary trace method defines binding surfaces common to protein families. *J Mol Biol, 257*, 342-358.

Lichtarge, O., Bourne, H. R., & Cohen, F. E. (1996b). Evolutionarily conserved Galphabetagamma binding surfaces support a model of the G protein-receptor complex. *Proc Natl Acad Sci USA, 93*, 7507-7511.

Lo Conte, L., Chothia, C., & Janin, J. (1999). The atomic structure of protein-protein recognition sites. *J Mol Biol, 285*, 2177-2198.

Mihalek, I., Res, I., & Lichtarge, O. (2007). On itinerant water molecules and detectability of protein-protein interfaces through comparative analysis of homologues. *J Mol Biol, 369*, 584-595.

Mihalek, I., Res, I., Yao, H., & Lichtarge, O. (2003). Combining inference from evolution and geometric probability in protein structure evaluation. *J Mol Biol, 331*, 263-279.

Miller, S., Janin, J., Lesk, A. M., & Chothia, C. (1987). Interior and surface of monomeric proteins. *J Mol Biol, 196*, 641-656.

Mintseris, J., & Weng, Z. (2005). Structure, function, and evolution of transient and obligate protein-protein interactions. *Proc Natl Acad Sci USA, 102*, 10930-10935.

Mintseris, J., Wiehe, K., Pierce, B., Anderson, R., Chen, R., Janin, J., & Weng, Z. (2005). Protein-Protein Docking Benchmark 2.0: an update. *Proteins, 60*, 214-216.

Mirny, L. A., & Shakhnovich, E. I. (1999). Universally conserved positions in protein folds: reading evolutionary signals about stability, folding kinetics and function. *J Mol Biol, 291*, 177-196.

Murakami, Y., & Jones, S. (2006). SHARP2: protein-protein interaction predictions using patch analysis. *Bioinformatics, 22*, 1794-1795.

Murga, L. F., Ondrechen, M. J., & Ringe, D. (2008). Prediction of interaction sites from apo 3D structures when the holo conformation is different. *Proteins.*

Negi, S. S., & Braun, W. (2007). Statistical analysis of physical-chemical properties and prediction of protein-protein interfaces. *Journal of molecular modeling, 13*, 1157-1167.

Negi, S. S., Schein, C. H., Oezguen, N., Power, T. D., & Braun, W. (2007). InterProSurf: a Web server for predicting interacting sites on protein surfaces. *Bioinformatics, 23*, 3397-3399.

Neuvirth, H., Heinemann, U., Birnbaum, D., Tishby, N., & Schreiber, G. (2007). ProMateus--an open research approach to protein-binding sites analysis. *Nucleic Acids Res, 35*, W543-W548.

Neuvirth, H., Raz, R., & Schreiber, G. (2004). ProMate: a structure based prediction program to identify the location of protein-protein binding sites. *J Mol Biol, 338*, 181-199.

Nooren, I. M., & Thornton, J. M. (2003a). Diversity of protein-protein interactions. *EMBO J, 22*, 3486-3492.

Nooren, I. M., & Thornton, J. M. (2003b). Structural characterisation and functional significance of transient protein-protein interactions. *J Mol Biol, 325*, 991-1018.

Oldfield, C. J., Meng, J., Yang, J. Y., Yang, M. Q., Uversky, V. N., & Dunker, A. K. (2008). Flexible nets: disorder and induced fit in the associations of p53 and 14-3-3 with their partners. *BMC genomics, 9*(Suppl 1), S1.

Pazos, F., Helmer-Citterich, M., Ausiello, G., & Valencia, A. (1997). Correlated mutations contain information about protein-protein interaction. *J Mol Biol, 271*, 511-523.

Pazos, F., & Valencia, A. (2002). In silico two-hybrid system for the selection of physically interacting protein pairs. *Proteins, 47*, 219-227.

Pettit, F. K., Bare, E., Tsai, A., & Bowie, J. U. (2007). HotPatch: a statistical approach to finding biologically relevant features on protein surfaces. *J Mol Biol, 369*, 863-879.

Pletneva, E. V., Laederach, A. T., Fulton, D. B., & Kostic, N. M. (2001). The role of cation-pi interactions in biomolecular association. Design of peptides favoring interactions between cationic and aromatic amino acid side chains. *J Am Chem Soc, 123*, 6232-6245.

Porollo, A., & Meller, J. (2007). Prediction-based fingerprints of protein-protein interactions. *Proteins, 66*, 630-645.

Pupko, T., Bell, R. E., Mayrose, I., Glaser, F., and Ben-Tal, N. (2002). Rate4Site: An algorithmic tool for the identification of functional regions in proteins by surface mapping of evolutionary determinants within their homologues. *Bioinformatics,* (18 Suppl 1), S71-S77.

Qin, S., & Zhou, H. X. (2007). meta-PPISP: A Meta Web Server for Protein-Protein Interaction Site Prediction. *Bioinformatics.*

Res, I., Mihalek, I., & Lichtarge, O. (2005). An evolution based classifier for prediction of protein interfaces without using protein structures. *Bioinformatics, 21*, 2496-2501.

Shannon, C. E. (2001). A mathematical theory of communication. *ACM SIGMOBILE Mobile Computing and Communications Review.*

Shulman-Peleg, A., Shatsky, M., Nussinov, R., & Wolfson, H. J. (2007). Spatial chemical conservation of hot spot interactions in protein-protein complexes. *BMC Biol, 5*, 43.

Singh, G. P., Ganapathi, M., & Dash, D. (2007). Role of intrinsic disorder in transient interactions of hub proteins. *Proteins, 66*, 761-765.

Tjong, H., Qin, S., & Zhou, H. X. (2007). PI2PE: protein interface/interior prediction engine. *Nucleic Acids Res, 35*, W357-W362.

Uetz, P., Giot, L., Cagney, G., Mansfield, T. A., Judson, R. S., Knight, J. R., Lockshon, D., Narayan, V., Srinivasan, M., Pochart, P., Qureshi-Emili, A., Li, Y., Godwin, B., Conover, D., Kalbfleisch, T., Vijayadamodar, G., Yang, M., Johnston, M., Fields, S., & Rothberg, J. M. (2000). A comprehensive analysis of protein-protein interactions in Saccharomyces cerevisiae. *Nature, 403*, 623-627.

Valdar, W. S. (2002). Scoring residue conservation. *Proteins, 48*, 227-241.

Zhou, H. X., & Qin, S. (2007). Interaction-site prediction for protein complexes: a critical assessment. *Bioinformatics, 23*, 2203-2209.

Zhou, H. X., & Shan, Y. (2001). Prediction of protein interaction sites from sequence profile and residue neighbor list. *Proteins, 44*, 336-343.

Chapter VI
Network Cleansing:
Reliable Interaction Networks

Paolo Marcatili
Sapienza University, Italy

Anna Tramontano
Sapienza University, Istituto Pasteur Fondazione Cenci Bolognetti, Italy

ABSTRACT

This chapter provides an overview of the current computational methods for PPI network cleansing. The authors first present the issue of identifying reliable PPIs from noisy and incomplete experimental data. Next, they address the questions of which are the expected results of the different experimental studies, of what can be defined as true interactions, of which kind of data are to be integrated in assigning reliability levels to PPIs and which gold standard should the authors use in training and testing PPI filtering methods. Finally, Marcatili and Tramontano describe the state of the art in the field, presenting the different classes of algorithms and comparing their results. The aim of the chapter is to guide the reader in the choice of the most convenient methods, experiments and integrative data and to underline the most common biases and errors to obtain a portrait of PINs which is not only reliable but as well able to correctly retrieve the biological information contained in such data.

INTRODUCTION

Interactions play a key role in every cellular activity and function. They are often related to diseases and therefore attract much interest in drug development projects. In the last few years the development of new experimental techniques for protein-protein interaction (PPI) detection together with the wide interest in systemic descriptions of biological processes gave birth to the field of *interactomics* aimed at studying the whole set of molecular interactions in a cell.

As in other -omics sciences, one, and perhaps the most important, issue to be faced concerns the great amount of noisy and unreliable information present in the data. Generally speaking, the data, mostly gathered with high-throughput experiments (Gavin et al., 2006; Gavin et al., 2002; Giot et al., 2003; Ho et al., 2002; Ito et al., 2001; Krogan et al., 2006; Li et al., 2004; Rual et al., 2005; Stelzl et al., 2005; Uetz et al., 2000), contain a sizeable number of false-positives and their reproducibility is not very satisfactory. Several authors have compared the results of the experimental data collected in high-throughput experiments (G. D. Bader & Hogue, 2002; J. S. Bader, Chaudhuri, Rothberg, & Chant, 2004; Gentleman & Huber, 2007; Hart, Ramani, & Marcotte, 2006; Mrowka, Patzak, & Herzel, 2001; Sprinzak, Sattath, & Margalit, 2003; von Mering et al., 2002) and used the overlap between the detected interactions to infer the accuracy and coverage of the complete interactome of various organisms. The results of such comparisons, even if with some differences, seem to depict an extremely problematic situation, with accuracy values ranging from 10% to 40% of true positive interactions and a limited coverage (J. S. Bader, Chaudhuri, Rothberg, & Chant, 2004; Gentleman & Huber, 2007; Sprinzak, Sattath, & Margalit, 2003). In the last few years, thanks to the introduction of more reliable experimental techniques, of new filtering methods and of a somehow deeper understanding of the interactome such pessimistic scenario has been partially revised, but many of the original problems still remain. Despite the many efforts devoted to the analysis of the possible sources of noise in PPI maps and the continuous development of methods aimed at overcoming the problem, a satisfactory solution is still missing. As it has been shown in other -omics contexts, a coordinated effort at both the computational and experimental level is essential for allowing the data to be properly exploited.

Central to the understanding and the analysis of interaction maps is the definition of the interaction itself. Given the continue and fast development of paradigms and methods in interactomics, such definition is not straightforward and, in many cases, strongly dependent on the current scientific expectations and experimental limits.

The binding energy of a physical interaction can vary by orders of magnitude and involve large surface patches of the partners or just small regions; the complex itself can be transient or stable and/or obligatory; the interaction can be essential for the correct three-dimensional folding of both interacting partners and it may occur at specific times and locations in the life of a cell (Nooren & Thornton, 2003). Moreover, the term *interaction* is often used with rather different meaning: for example, we talk of functional or genetic interactions when the behavior of a protein or a gene is dependent upon the effect of another (Boone, Bussey, & Andrews, 2007).

When analyzing PPIs, all these different aspects should in principle be taken into account. This might be very difficult and, in some cases, impossible.

It follows that, in order to understand which information we can reliably extract from a PPI map, we need to define what we mean by interaction. Far from being only a semantic or ontological problem, this is a central issue that affects several of the fundamental questions that we might want to ask, some of which are:

- what do we expect to find as a result of different *experimental studies*;
- what do we define as a *true interaction*;
- which kind of data we need to *integrate* when we want to assign reliability levels to single PPIs or to whole interaction maps;
- how do we assess the results, i.e. which *gold standard* can we use.

This chapter will try and give an overview of the present situation with respect to these issues by first giving a brief definition of protein-protein interaction and next addressing each of the other questions in turn. We will focus on the possible biases, errors and critical choices in PPIs cleansing. Finally, we will describe the state of the art in the field, presenting the different classes of algorithms and comparing their results.

EXPERIMENTAL METHODS FOR PPI DETECTION

Most common experimental methods for PPI detection can be divided on the basis of two main characteristics: low vs high throughput studies and affinity purification-mass spectrometry (AP-MS) (Aebersold & Mann, 2003) versus yeast-two-hybrid (Y2H) technology (Shoemaker & Panchenko, 2007). The pros and cons of each of these methods are well known. For our purposes, it is useful to recall just a few concepts.

Y2H is an *in vivo* technique (even if it makes use of chimeric proteins different from the *wild type* ones) and it does not heavily rely on protein abundance. It is able to detect weak or transient PPIs, but it often fails to detect large complexes or obligate interactions. This method can introduce several artifacts in the data (auto-activators, membrane proteins) (Huang, Jedynak, & Bader, 2007; Vidalain, Boxem, Ge, Li, & Vidal, 2004).

AP-MS permits to identify stable protein complexes with good accuracy. The identification is affected by protein abundance, binding energy and non-proteic contaminants (nucleic acids, small molecules) (G. D. Bader & Hogue, 2002; von Mering et al., 2002).

Both methods can detect physical interactions that might not occur in the cell because they cannot take into account protein localization or timing of expression.

Low-throughput experiments (obtained with different techniques) (Miernyk & Thelen, 2008) are generally considered to have a higher reliability with respect to the high-throughput ones even though, with the advance of experimental techniques, the latter statement might not hold any more. On the other hand, low throughput experiments tend to be biased towards specific classes of PPIs, e.g. interactions between proteins with a high biological or medical relevance (Reguly et al., 2006).

The very low level of intersection between the results of the different methods has been used to infer the false positive and false negative rates in experimental dataset (Gentleman & Huber, 2007; Hart, Ramani, & Marcotte, 2006; Mrowka, Patzak, & Herzel, 2001; Sprinzak, Sattath, & Margalit, 2003; von Mering et al., 2002). However, it is becoming clearer and clearer that all these different techniques can offer a different and complementary portrait of the interactome. Experimental data integration is therefore a hard task, and the paradigm "the more, the better" is not always useful.

Some specific types of interactions are always under-represented in PPI maps, regardless of the experimental method and the quality of the assay (e.g. interaction involving membrane proteins or transient interactions). Although experimental techniques to discover interaction between membrane (Stagljar & Fields, 2002) or extracellular (Sanderson, 2008) proteins are being developed, the problem is unlikely to be completely solved in the near future.

Evidently, datasets should be chosen on the basis of which part of the interaction networks we want to observe. Similarly one has to consider the appropriateness of using the raw experimental data for assessing network reliability (Collins et al., 2007; Gentleman & Huber, 2007; Hart, Ramani, & Marcotte, 2006). Although a large number of PPI datasets can be easily found in the public domain, the corresponding raw data (even for the most used and analyzed experimental PPI maps) are much more difficult to access. This is disappointing as they could provide a large amount of additional useful information with respect to pre-filtered datasets.

"TRUE" INTERACTIONS

As we stated before, the definition of a "true interaction" depends upon the type of events we want to focus on and on the level of detail that we want to achieve. Interaction identification depends on several physical, chemical, cellular, topological, functional and experimental constraints. Such a complex scenario gave birth to a number of different, non-overlapping definitions of interaction.

The simplest situation is when two proteins are physically bound to each other. Such binding is typically mediated by regions of the two proteins which may vary in dimension, amino acidic and structural features (linear or structural motifs, surface patches, disordered regions, domains). Y2H and biochemical low-throughput experiments are the most appropriate methods to detect these interactions.

If several proteins are found to be part of a protein complex we talk about indirect or complex membership interactions. Some of the interactions between proteins in the same complex are direct (physical), but others can be mediated by the other binding partners. AP-MS experiments can detect this kind of interactions, but such experimental data are not readily usable, since, as stated in (Gentleman & Huber, 2007), if a protein D belongs to different complexes ABCD and DEF and it is used as bait in an AP-MS experiment we will detect the prey proteins A,B,C,E,F. If the protein B is used as bait, preys A, C and D will be detected (Figure 1).

One way to define the problem is to distinguish between one-to-many interactions (spoke model) and many-to-many interactions (matrix model). Which of these two models is used for interpreting the interactions can influence the results, as we will discuss.

As we reported in the introduction, the term interaction is often used with a different a meaning: proteins belonging to the same functional or metabolic pathway are often called functional or genetic interactors (Kelley & Ideker, 2005). It is important to underline that such proteins may or may not actually interact in the cell, and

Figure 1. Experimental outcomes for different preys in the same complex. Depending on which prey and which interaction model are used markedly different results are obtained.

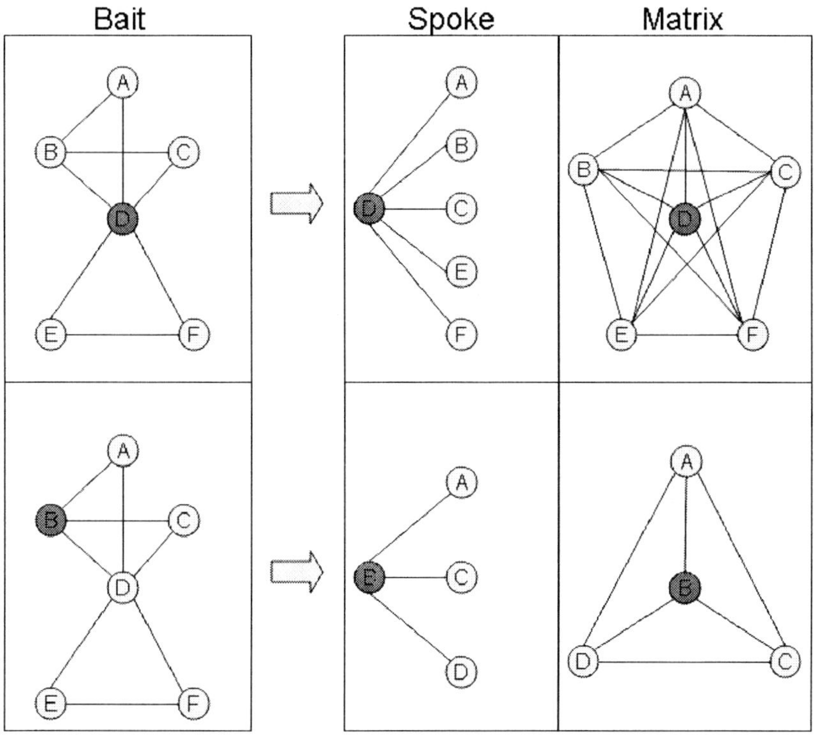

the experimental methods to detect them (synthetic lethality, co-essentiality) are quite different from the other detection method.

Even if we limit ourselves to define only the physical interactions as the true ones, the portrait of the interactome is complex and blurred. Physical interaction may vary in their binding energy, in the stability of their association, in the possibility of the interacting proteins to be correctly folded when unbound, in their spatial and temporal existence and in their specificity (Nooren & Thornton, 2003). Even if stable, energy-favorable and specific interactions are usually easier to detect in every experimental scenario, the substantial differences between the interaction detection techniques can produce a rather different portrait of the interactome. Furthermore, the high-throughput experimental methods able to distinguish whether two proteins are able to interact when co-localized, cannot predict if the interactions do effectively occur in the cell. Auto-activators and sticky proteins are a typical example of the possible artifacts (Vidalain, Boxem, Ge, Li, & Vidal, 2004)

Because of the above, no matter what the coverage and the accuracy of Y2H and AP-MS experiments are, their results in detecting PPIs in an organism are bound to be different and these differences have to be carefully considered in deriving conclusions from their analysis and comparison.

DATA INTEGRATION IN NETWORK CLEANSING

Data integration is the central point in assigning a reliability score to single PPIs or to whole interaction maps. Clearly, the results depend on the underlying model adopted for defining an interaction.

Useful data for network cleansing can be divided into five classes: functional and localization data (GO annotations, text mining), indirect high-throughput data (gene and protein expression, synthetic lethality, co-essentiality), sequence-related data (interologs, domains and motif detection, Rosetta stone, chromosomal position), structural data, network topological features (connectivity, network motifs). Although each of the information sources listed above is unable to produce a complete, reliable and unbiased landscape of the real interactome when considered alone, their integration can be the key to reliable PPI filtering

Functional and localization data are perhaps the most used information in PPI prediction and assessment. Such data has been used in a large number of works and has been shown to be more informative than other types of data (Lin, Wu, Jansen, Gerstein, & Zhao, 2004). Controlled vocabularies have been developed to assign functional and localization annotations, making it possible to compare annotations derived from different experiments, or for different proteins and organisms. Gene Ontology (GO) ("The Gene Ontology (GO) project in 2006", 2006) is the most used among such vocabularies. It has three different categories of annotations, describing function, process and localization of a gene or protein.

The basic rationale behind their usage in our context is that two proteins which share a similar annotation are more likely to interact than proteins with different or no annotation, e.g. the probability that two proteins which share an "rRNA metabolism" annotation (GO:0016072) interact is larger than the probability to find an interaction between proteins which are part of different metabolic pathways. The use of functional annotations has many advantages: a large part of the genomes and proteomes of many organism has associated annotations, the information can be retrieved and used in an automatic manner and several tools for comparison and statistical analysis of the functional terms are publicly available (Aitken, Korf, Webber, & Bard, 2005; Cote, Jones, Apweiler, & Hermjakob, 2006; Mattingly, Rosenstein, Colby, Forrest, & Boyer, 2006; Zhou & Cui, 2004). On the other hand, annotations are often unreliable (e.g. derived from untraceable data or inferred with computational tools), incomplete (the "*shallow annotation*" problem) and biased towards some functional or localization classes. It is possible to partially overcome these problems by only using annotations derived from reliable information sources. In fact, GO annotations provide the user with a label defining the source of information from which they were derived (the GO evidence code), and traceable experimentally derived annotations (e.g. *EXP: Inferred from Experiment* or *IDA: Inferred from Direct Assay*) are usually considered to be the most reliable ones. Text mining methods can now automatically extract a great amount of data from the literature, and they have been used to infer functional linkages or interactions (Ramani, Bunescu, Mooney, & Marcotte, 2005; Reguly et al., 2006) between proteins. These data have a great value in the context of PPI cleansing. They are often generated by low-throughput and specific experiments, and therefore are considered to be more reliable and with less (or at least different) bias than high throughput experiments.

Gene and protein expression data is an extremely informative piece of information in PPI filtering as well. The results for a large number of experiments carried on for the most common organisms and for the largest part of genomes are available. These are mostly gathered using micro-array technologies (Hoheisel, 2006).The similarity of the patterns of expression of different proteins can be measured with several metrics (Pearson correlation, spearman rank correlation among the others) and this has been shown to be a good indicator of real interactions (J. S. Bader, Chaudhuri, Rothberg, & Chant, 2004; Grigoriev, 2001).

Pros and cons in using such data for our purpose are similar to those that we described for functional annotations: there are plenty of experiments and methods already well tested that can be used to assign a reliability score to PPI maps, but the noise level and the biases of expression data is still a concern. Synthetic lethality (Scherens & Goffeau, 2004) and co-essentiality have been used in a number of papers to infer interactions (Lu, Xia, Paccanaro, Yu, & Gerstein, 2005); (Jansen et al., 2003; Lee, Li, & Marcotte, 2007; Singh, Xu, & Berger, 2006). Synthetic lethality is a phenomenon by which two non-lethal mutations yield a lethal phenotype when combined and it might indicate that the two affected proteins share a vital function. Similarly, if deletion of each of two proteins produces the same phenotype it could be assumed that they are both essential for originating the same phenotype. Even if in some cases the use of these data has proven to be useful in revealing true interactions with a high specificity, there are some caveats in their usage to assign a reliability level to large PPI maps, because they are not necessarily indicators of direct physical interaction (Ooi et al., 2006). Protein sequence analysis is an extremely useful source of information for producing reliable interaction maps and several investigations have attempted to develop methods for exploiting the relationships between sequences and interaction, within the same genome or across different ones, but undoubtedly there is still space for great improvements in this direction. The conservation of interactions between homologous pairs of proteins in different organism (often referred as *interologs*) has been considered one of the most reliable indexes of a "true" interaction and is used in several methods. However, several evidences (Mika & Rost, 2006; Saeed & Deane, 2008; Wagner, 2001) have cast shades on the straight-forward application of this paradigm. Interactions seem to be conserved only between very similar proteins and within the same organism rather than between proteins encoded by different genomes. This conclusion depicts an unexpected portrait of the interactome, a large fraction of which seems to be composed by extremely specific and organism-related interactions. A similar conclusion was recently reached by analysing linear motifs (Neduva & Russell, 2006b). Linear motifs play a key role in many cellular activities (interactions between proteins and metabolites, signaling, localization). These are usually short sequences (up to ten residues, just a few of which are indispensable for the correct recognition of the motif) and are less evolutionary constrained than domains. Nevertheless, several papers have reported that linear motifs are useful for predicting or assigning a reliability score to interactions (J. Guo, Wu, Zhang, & Lin, 2008; Marcatili, Bussotti, & Tramontano, 2008; Neduva & Russell, 2006a). The basic idea is that if two proteins share the same interaction motif, they are likely to interact in a similar fashion. The presence of a common motif in the sequences of proteins interacting with the same partner is therefore a strong evidence of the reliability of their interactions. Remarkably, these methods can assess the confidence level of some interactions, but also give additional insights about the specific interaction mode. Not all the physical interactions are mediated by motifs, therefore the motif information alone is not sufficient to assess the reliability of an entire interactome. Protein domains are not only equally useful to assign a reliability score to protein interactions, but are also much more conserved among different species and much easier to detect than linear motifs. Domain interaction conservation (Itzhaki, Akiva, Altuvia, & Margalit, 2006; Jothi, Cherukuri, Tasneem, & Przytycka, 2006) relies on the assumption that, if two domains are often found in interacting protein pairs, two proteins that contain such domains are more likely to interact. Such methods take advantage of domain classification and detection tools and of solved three-dimensional structures. Genomic and phylogenetic considerations have been used to infer and validate interactions. For example, it has been shown that prokaryotic interacting proteins are often encoded by the same operons (Rogozin, Makarova, Wolf, & Koonin, 2004), that interacting proteins often have a strong phylogenetic correlation in different genomes (Fraser, Hirsh, Steinmetz, Scharfe, & Feldman, 2002), being either present or absent together, and that different domains found to be part of a single protein in a genome are likely to interact in other organisms even though they are parts of different proteins (Enright & Ouzounis, 2001). Structural data has been used sparsely in PPI filtering. Docking methods (Aloy et al., 2004; Russell et al., 2004; Szilagyi, Grimm, Arakaki, & Skolnick, 2005; Tovchigrechko & Vakser, 2005) and recognition of structural motifs (Ausiello et al., 2008; Ausiello, Zanzoni, Peluso, Via, & Helmer-Citterich, 2005) can provide information on the interactions, but the poor coverage of structural data and the inherent complexity of the methods has prevented large-scale analysis with these methods, whose applicability is however bound to increase thanks to the flourishing of structural genomics projects and to the steady improvements in protein structure prediction methods.

Another source of information which has not been exploited very thoroughly so far for network cleansing is the analysis of the network topological features, even though the detailed analysis of different biological network from a topological point of view has been exhaustively studied in the last few years (Asthana, King, Gibbons, & Roth, 2004; Chen, Hsu, Lee, & Ng, 2005, , 2006; Chiang, Scholtens, Sarkar, Gentleman, & Huber, 2007; D'Haeseleer & Church, 2004; Gavin et al., 2006; Hakes, Pinney, Robertson, & Lovell, 2008; Huang, Jedynak, & Bader, 2007; Ito et al., 2001; Scott & Barton, 2007). In PPI filtering such features can be used in different ways: they can give an overall index of reliability to a whole map or used for identifying the interactions having a higher degree of confidence. As far as the first scenario is concerned, measures as maps intersection (G. D. Bader & Hogue, 2002), connectivity (J. S. Bader, Chaudhuri, Rothberg, & Chant, 2004; Hakes, Pinney, Robertson, & Lovell, 2008) and essentiality (Hart, Ramani, & Marcotte, 2006; Lin, Wu, Jansen, Gerstein, & Zhao, 2004) have been used to compare the similarity between PPI maps and some gold standards. Even if this approach may help circumventing some of the problems arising from the comparison of different datasets (e.g. different experimental methods), it is highly dependent on the coverage and uniformity of sampling of the current PPI maps with respect to the complete interactome. Some authors have used topological features to infer the reliability of single interactions (Chen, Hsu, Lee, & Ng, 2005, , 2006; Gavin et al., 2006; Hart, Lee, & Marcotte, 2007; Scott & Barton, 2007). It has been reported that biological networks (Barabási, 1999; Barabasi & Oltvai, 2004) and interaction maps in particular (Yook, Oltvai, & Barabasi, 2004) display some specific topologies. These methods have been extremely useful to determine the highly modular and dynamic nature of protein complexes in the cell, that extends the two basic model described above (spoke and matrix models), as we will describe later. As already stated, none of these methods alone is sufficient to gain a complete view of an interactome. Integration (which is often performed using machine-learning, probabilistic or statistical approaches) is a fundamental technique for obtaining reliable and unbiased PPI maps. Several integration methods have been proposed, and the resulting filtered maps share but a small core of "robust" interaction.

GOLD STANDARDS FOR PPI MAPS

Many of the tools developed for PPI cleansing make use of high quality PPI maps. Such "gold standards" are used to assess the results of the method itself or – for machine-learning methods – as training sets. In both cases, the results of the method depend on the effective reliability of the "gold standard" sets. They should be as much as possible unbiased towards specific interaction types. If we use different sets for the training, testing and validation phases of our method we must make sure that such data are independent. For example, if we train an algorithm on experimentally derived complexes and test it using GO annotations, we must be aware that such annotations might have been derived from data that include or are related to the information used in the training stage and therefore the results may be misleading. One of the most used resources is the MIPS complex database (Mewes et al., 2006) which contains information on a large number of co-purified complexes. This database is enriched in indirect (not physical), and stable interactions. A good compromise between reliability and coverage can be achieved by using manually annotated interaction databases, such as dip (Salwinski et al., 2004), the MIPS Mammalian Protein-Protein Interaction Database (Pagel et al., 2005), BIND (Alfarano et al., 2005), HPRD (Mishra et al., 2006) and MINT (Zanzoni et al., 2002), or of interaction maps derived from low-throughput experiments (Reguly et al., 2006). As stated before, the intersection between different experimental dataset is very small and can be biased in stable and specific interactions, nevertheless some authors used the interaction identified in more than one experiment to deduce a putatively reliable interaction map (Hakes, Pinney, Robertson, & Lovell, 2008; von Mering et al., 2002). It is clear that a perfect "gold standard" for PPI does not exist and indeed even the overlap among some of the high-quality datasets reported above is very limited (G.D. Bader & Hogue, 2002; Collins et al., 2007; Hart, Ramani, & Marcotte, 2006). One alternative for inferring a complete, reliable, and unbiased dataset of interactions is to compute general features of the processed dataset and compare them with the same features evaluated in several reliable datasets. This strategy has been used in (Reguly et al., 2006) to compare high-throughput and low-throughput maps, giving interesting results. Although this approach may improve the robustness of the final results, it fails to give information on single PPIs, its biological meaning is unclear, and is still heavily dependent on the reference datasets (Hakes, Pinney, Robertson, & Lovell, 2008). So far we discussed the need for a reliable set of true interactions, but the detection of reliable true negatives is equally important and can be very relevant for evaluating the performance of different methods. Some papers (Jansen et al., 2003) use a simple assumption in order to obtain true negative interactions datasets: proteins that are not co-localized cannot interact. Apart from the errors and the biases of localization annotations themselves, it has been proposed (Ben-Hur & Noble, 2006) that this solution may have another fallback. Machine learning algorithms use the negative training set to discriminate between true and false positives. If this set is "too simple", i.e. if it reports only false interactions between extremely different proteins, the training of the method can be unsatisfactory and the method itself unable to recognize the most "subtle" false positives. A more appropriate choice is to randomize experimental maps. In this case the probability of retaining a true positive interaction is so low that, in most cases, has no effect on the quality of the negative map. Randomized

Figure 2. Different network randomization strategies. A global renaming of nodes (left side) retains every topological feature of the map. A shuffling of each column in the PPI list (right side) keeps the same degree for each protein.

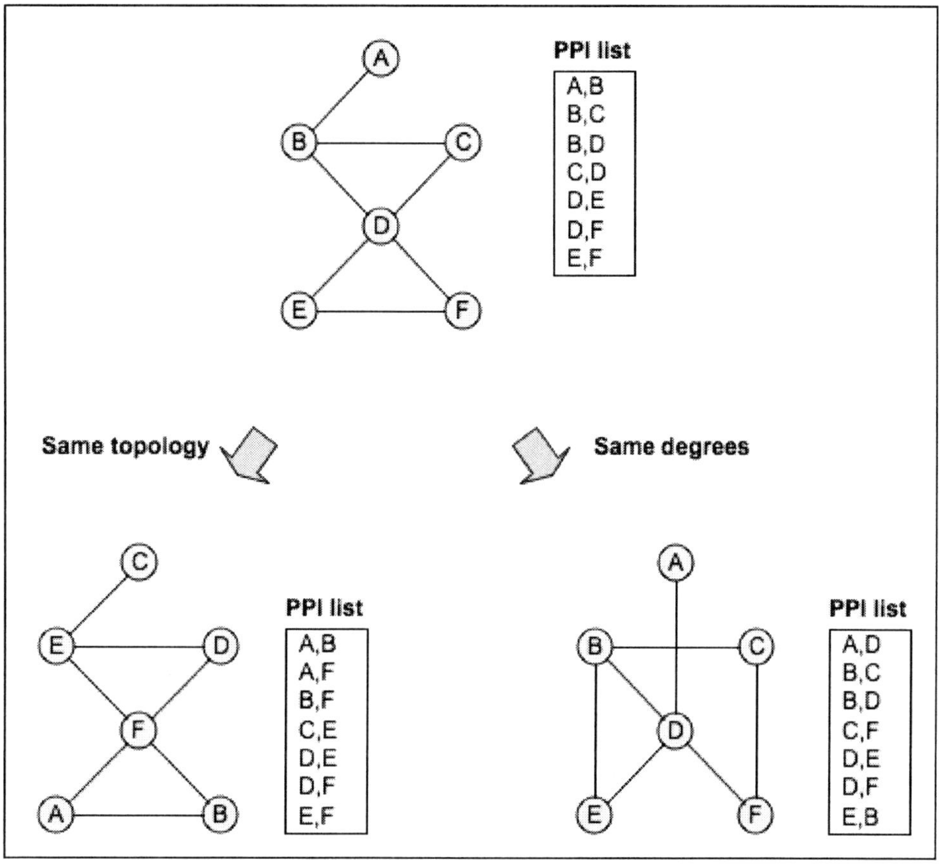

maps can provide a negative interaction datasets, but care should be taken in preserving some of the topological aspects of the respective true interaction network. The most straight-forward randomization method is to randomly rename the proteins in the map (Figure 2, left side). This strategy retains all the topological features of the original map. In (Marcatili, Bussotti, & Tramontano, 2008) a different strategy has been used (figure 2, right side) that takes into account the fact that hub proteins (i.e. proteins that interact with a large number of partners) have been show to possess very specific sequence feature (Ekman, Light, Bjorklund, & Elofsson, 2006). It follows that one needs to preserve this connection between topology and sequences by retaining the original degree of each protein. This can be done by describing the PPI maps as an ordered list of pairs of interacting proteins and by only shuffling the second elements of the pairs.

STATE OF THE ART IN NETWORK CLEANSING

We have described the main issues in network cleansing: integration of extremely heterogeneous data, coherence within a given interaction model, biological and physical meaning, coverage and, of course, accuracy. Several algorithms have been developed in order to produce clean PPI maps, with different emphasis on some of the aspects listed above.

The aim of such algorithms can be roughly divided into three major classes:

- derive an overall index of reliability of a PPI (G. D. Bader & Hogue, 2002; J. S. Bader, Chaudhuri, Rothberg, & Chant, 2004; Deane, Salwinski, Xenarios, & Eisenberg, 2002; Gentleman & Huber, 2007; Grigoriev, 2001; Hakes, Pinney, Robertson, & Lovell, 2008; Hart, Ramani, & Marcotte, 2006; Huang, Jedynak, & Bader, 2007; Mrowka, Patzak, & Herzel, 2001; Qi, Klein-Seetharaman, & Bar-Joseph, 2007; Ramirez, Schlicker, Assenov, Lengauer, & Albrecht, 2007; Sprinzak, Sattath, & Margalit, 2003; Stumpf et al., 2008) and references therein,
- predict reliable PPIs with (Chen et al., 2006; Chen, Hsu, Lee, & Ng, 2005; Collins et al., 2007; Y. Guo, Yu, Wen, & Li, 2008; Hart, Lee, & Marcotte, 2007; Mahdavi & Lin, 2007; Marcatili, Bussotti, & Tramontano, 2008; Patil & Nakamura, 2005; Rhodes et al., 2005; Sardiu et al., 2008; Singh, Xu, & Berger, 2006) or without (Ben-Hur & Noble, 2005; Jansen et al., 2003; Scott & Barton, 2007) the usage of interaction experimental data,
- detect protein complexes from PPI maps(Aloy et al., 2004; Altaf-Ul-Amin, Shinbo, Mihara, Kurokawa, & Kanaya, 2006; Asthana, King, Gibbons, & Roth, 2004; Hart, Lee, & Marcotte, 2007; Krogan et al., 2006; Pang, Krycer, Lek, & Wilkins, 2008; Zhang, Park, Karpinets, & Samatova, 2008).

We will use such classes to present the papers, with the caveat that some in some cases the assignment of a method to one class rather than another is quite arbitrary. Bayesian networks (Jansen et al., 2003; Lin, Wu, Jansen, Gerstein, & Zhao, 2004; Patil & Nakamura, 2005), kernel methods (Ben-Hur & Noble, 2005; Y. Guo, Yu, Wen, & Li, 2008; Shen et al., 2007), random forests (Lin, Wu, Jansen, Gerstein, & Zhao, 2004; Singh, Xu, & Berger, 2006), probabilistic and statistical models (Asthana, King, Gibbons, & Roth, 2004; Collins et al., 2007; Gentleman & Huber, 2007, mixture of feature experts {Qi, Klein-Seetharaman, & Bar-Joseph, 2007; Hart, Lee, & Marcotte, 2007; Jaimovich, Elidan, Margalit, & Friedman, 2006; Kelley & Ideker, 2005; Lee, Date, Adai, & Marcotte, 2004; Lee, Li, & Marcotte, 2007; Rhodes et al., 2005; Sardiu et al., 2008; Scott & Barton, 2007) as well as several ad-hoc methods have been used for these purposes. In the following we describe these methods in the order in which they appeared in the literature as much as possible.

Estimate PINs Reliability

Since the first high-throughput experiments were published, the problem of estimating their reliability has received much attention (Aloy & Russell, 2002; Deane, Salwinski, Xenarios, & Eisenberg, 2002; Mrowka, Patzak, & Herzel, 2001). This has had a strong impact on the interactomics fields, contributing to the general impression that current interaction data are enriched in spurious interactions and they cover but an extremely small fraction of the complete PPI maps.

One of the first papers aiming at estimating the reliability of high-throughput data is by Mrowka and coworkers (2001). A comparison of the expression data of interacting proteins in two Y2H experiments(Ito et al., 2001; Uetz et al., 2000) with that of MIPS complexes was used to estimate the false-positive ratio in the PPIs. The method used was rather simple, nevertheless the results were coherent with current estimates obtained with much more sophisticated techniques.

Next, expression data together with paralogy, functional and localization annotations, were used to estimate the reliability of Y2H and DIP (Salwinski et al., 2004) data and to extract a subset of high-confidence interactions from the DIP database (Deane, Salwinski, Xenarios, & Eisenberg, 2002) .

A very important paper on quality assessment of PPIs was published by von Mering an co-workers (2002). Apart from reporting the estimated coverage and accuracy of several PPI maps experimentally or computationally derived, the authors performed a detailed analysis of the biases and of the most common problems of each method. They analysed, among other features, the tendency of mass-spectrometry methods to preferentially detect abundant proteins, the localization of detected interactions, the ratio of interaction between non-co localized (and so less likely to be interaction partners) proteins for the different datasets and technique.

A similar analysis of mass-spectrometry data was performed by Bader and Hogue (2002) who discussed the internal and mutual coherence of different mass spectrometry datasets. An interesting result presented in the paper is that a "spoke" model that directly pairs bait proteins with associated proteins is roughly threefold more accurate than a "matrix" model that connects all the proteins.

D'haeseleer and Church (2004) used the extent of the overlap between different experiments to infer the fraction of false positives and the coverage of the different datasets. Their method, however, may be too sensitive to biases and hidden variables, and the results only partially agree with other similar estimates.

Hart et al. (2006) reported estimates of the reliability of a number of experiments -including Krogan (2006) and Gavin (2006), the largest available interaction datasets. Their estimates confirm the low quality of interaction data, reporting a somehow unexpected low confidence level for AP-MS methods and estimating the dimension of yeast and human interactome to be of the order of 50.000 and 200.000 interactions respectively. However, recent results (Collins et al., 2007; Hart, Lee, & Marcotte, 2007) seem to indicate that it is possible to obtain highly reliable interaction maps, with accuracy comparable to that of low-throughput experiments, by integrating raw data instead of pre-filtered maps and using simple but effective probabilistic models.

An accurate model of Y2H detection has been proposed by Huang et al. (2007) on the basis of the statistical "capture-recapture" model. They estimated that the false positive rates for yeast (Ito et al., 2001; Uetz et al., 2000), worm (Li et al., 2004) and fly (Giot et al., 2003) interaction datasets range from 25% to 45% with false negative rates of 75% to 90%. These errors are nearly equally due to statistical under-sampling and to systematic errors of the arrays. Another highlight deriving from this work is the tendency of membrane proteins to generate false positive in the Y2H experiments.

Most of the methods for inferring the reliability of PPIs agree on some general aspects of the experimental data quality, biases and coverage, but their detailed results are often different and sometimes even contradictory. In a very interesting report, Gentlemann and Huber (2007) point out that, in many cases, the source of these discrepancies is not to be found in the data itself but rather in the models and the assumptions used in comparing and integrating different information sources.

PPI Filtering

As the work described above indicates, the current representation of the interactome seems more a patchwork of separated and rather fuzzy pieces rather than a clearly readable picture. We obviously miss some of the interactions and the presence of unreliable interactions in the maps makes the picture even less satisfactory.

There are some tools available that attempt the identification of new interactions not present in the experimental data that we will not discuss here. We will instead focus on tools that aim at filtering out the interactions of lower quality mentioning the former only when relevant to the issue of cleansing the PPIs.

As we mentioned, one of the first attempts of cleansing PPIs (Deane, Salwinski, Xenarios, & Eisenberg, 2002) used paralogy information and was able to identify a core of reliable PPIs extracted from DIP. In this case, the gold standards used were composed by two sets of interaction reported in two and three different experimental assays. Deng et al. (Deng, Sun, & Chen, 2003) , using a similar model and the same data, tried to improve the performance of the method using maximum likelihood to infer PPI reliability. Such protocol was tested on physical interactions, complexes and genetic interaction data. This method has been compared with four other methods (J. S. Bader, Chaudhuri, Rothberg, & Chant, 2004); (Qi, Klein-Seetharaman, & Bar-Joseph, 2005; Sharan et al., 2005); (Deane, Salwinski, Xenarios, & Eisenberg, 2002) some of which use more sophisticate models and additional data sources. Interestingly, it turns out that the method of Deng et al., only using a simple model and considering as input only experiment type and expression similarity, performs better than all the other methods.

The results of this comparison are somehow contradicted by the results from Lin et al (2004) who stressed the information contribution of data derived from gene expression, localization, essentiality and functional annotation, indicating the latter (more specifically GO and MIPS functional assignments) as the most informative data source for PPI prediction. The same authors also compared different machine learning approaches in interaction prediction, concluding that random forests algorithm seems to outperform other method such as logistic regression (J. S. Bader, Chaudhuri, Rothberg, & Chant, 2004; Giot et al., 2003) and Bayesian networks.

Ramani et al. (2005) developed a pipeline for the evaluation and the retrieval of human PPI. In particular they generated one of the first PPI datasets derived from the literature using a text mining approach. Such dataset was reported to have a confidence level similar to that of the existing datasets (BIND, HPRD). Human PPIs were also studied by Rhodes et al. (2005) using a probabilistic model for human PPI prediction by integrating interologs, domain, expression and annotation data using a semi naïve Bayes classifiers. Such dataset of predicted PPIs has been used to analyse some cancer-related pathways and some of the predictions reported have been experimentally validated.

Topological features have been used to validate or predict PPIs as well. Interaction generality (Chen et al., 2006), IRAP (Chen, Hsu, Lee, & Ng, 2006), transitivity (Scott & Barton, 2007) and network motifs (Chen et al., 2006) have been proposed as a measure of confidence for PPIs, and the results obtained using them alone or together with other data are quite promising. Topological features have been useful to consolidate or clean experimental data (Gavin et al., 2006; Ito et al., 2001). However, since these features are heavily dependent on the levels of accuracy and sampling of the network (Han, Dupuy, Bertin, Cusick, & Vidal, 2005), such methods can be preferentially applied on large datasets rather than on preliminary maps.

A very interesting result was reported by Reguly and colleagues in 2006. Using a text mining approach refined by a manually curated stage, they were able to populate a dataset of more than 11,000 interactions in *S.cerevisiae* extracted from 3,200 different papers and gathered using several experimental techniques, both *in vivo* (affinity capture followed by identification, co-fractionation, co-localization, co-purification, FRET, Y2H) and *in vitro* (biochemical activity, co-crystal structure, far western, protein-peptide, protein-RNA, reconstituted complex), with an estimated low level of false positive (less than 20%). The comparative analysis between this dataset and high-throughput experimental data revealed a complex scenario, in which different and not completely understood biases affect the two datasets. They also studied the relationships between protein interactions and GO terms, expression, co-regulation, essentiality, connectivity and degree. Even if such features are found to be statistically correlated to PPIs (with a higher significance for the literature-derived datasets), their predictive power seems to be quite poor, resulting useful only for a small fraction of the interactome.

In agreement with this last statement, several works in the last two years have tried to face the problem of high-throughput data noisiness by analysing the raw data rather than by integrating other (noisy as well) data sources (Collins et al., 2007; Hart, Ramani, & Marcotte, 2006; Huang, Jedynak, & Bader, 2007). Collins et al (2007) published one of the most noteworthy paper in this area where they used the AP-MS data from Gavin and Krogan and combined them in a probabilistic framework which models positive and negative evidences of interaction between baits and preys. The resulting dataset is quite different from the two original experiments, and it displays several characteristics of small-scale experiments. Moreover, the authors estimate that their dataset covers approximately 80% of all the interaction in yeast that can be detected using AP-MS technique in conditions similar to the ones of the original experiments.

This last result has set a landmark for the analysis of a large fraction of the *S.cerevisiae* interactome (interactions between highly expressed proteins involved in stable complexes), but several issues are still open: the detection of transient interactions and of PPIs in different organisms is still elusive.

It is worth mentioning some recent papers that attempt to filter and predict PPIs using sequence information alone (J. Guo, Wu, Zhang, & Lin, 2008; Y. Guo, Yu, Wen, & Li, 2008; Marcatili, Bussotti, & Tramontano, 2008; Shen et al., 2007). The rationale behind these methods is that sequences contain most of the information on structure and, therefore, on interactions as well. It is well known that, even if interactions are not completely conserved among different organisms or between similar proteins, the modes of protein interactions are much more conserved (Mika & Rost, 2006; Saeed & Deane, 2008). A recent paper (Marcatili, Bussotti, & Tramontano, 2008) analyzed the presence of conserved linear sequence motifs within the yeast interactome. A large fraction of the proteins which share a common interaction partner display conserved motif, which are likely to act as fingerprints of the interaction. The information content of the motif data is only partially overlapping with that derived from the protein domain presence, and it displays a very low correlation with functional or localization annotation. This would suggest that the low level of evolutionary conservation of motifs may be correlated to the scarce conservation of interactions across species. This concept can be extended: it is known that sequences bring all the information about structure; it is therefore likely that even interactions are "encoded" in the sequence. Shen (2007) and Guo (2008) exploited this idea by computing a number of physico-chemical features of the protein sequences and considering local environments (up to 30 residue in the sequence) to train an SVM with known interactions (HPRD and DIP). Their results confirm that, even using this limited and simplified model, one can partially capture the main modes of protein interactions.

Identification of Protein Complexes

We will dedicate the last part of this survey to the methods devoted at identifying protein complexes with a good reliability. Even if this class of algorithms shares some features with the methods aimed at filtering direct interactions it is useful to dedicate a separate section to their description.

One of the first problems to be faced is the biological interpretation of AP-MS experimental data. As reported before, the paper from Bader and Hogue (2002) reported that the "matrix" model is much more error-prone than the "spoke" one. However, the interactions obtained using the "spoke" model (in which just the interaction between bait and preys are reported) are more likely to be direct interactions. If we are interested in obtaining information on protein complexes, this model could lead to under-sampling of the interactions.

The model proposed by Ashtana et al. (2004) extends the matrix and spoke paradigms using a probabilistic approach. Several other papers have used this approach: (Krogan et al., 2006) uses a pipeline composed by Bayesian networks, decision trees, logistic regression and Markov clustering to deal with the data cleansing and complex identification problem. Also Gavin and co-workers (2006) contributed to the issue by defining a socio-affinity index which is an indicator of the propensity of proteins to interact and is a measure of the log-odds that two proteins are experimentally found to interact with respect to their abundance in the dataset. This index gives an interesting picture of the cellular complexes, populated by modular and dynamic sub-complexes, rather than stable units. Such distinction between "core" and "peripheral" organization of complexes was later confirmed by other authors (Hart, Lee, & Marcotte, 2007; Pang, Krycer, Lek, & Wilkins, 2008), who introduced even more complete and specific models for complex identification.

DISCUSSION AND FUTURE TRENDS

One lesson that we have learned in the last few years is that, in order obtain reliable PPI data, quality, precision (and even linearity) of models should be taken into account more than sheer data quantity. A clear framework for interpreting raw data (based on probabilistic and statistical models) is essential to obtain reliable datasets.

Another point is that data integration is fundamental in the first steps of PPI data filtering. The case of *S.Cerevisiae* is paradigmatic: computational and experimental methods have generated preliminary draft maps, data integration with other sources of information expanded the interactome picture. The availability of new and more reliable experimental data, together with a possibly more realistic model of the yeast interaction network, while reducing the usefulness and the amount of data integration, has made it more accurate and biologically significant. In other words, sometimes more is not always better.

It is becoming more and more evident that the Interactome has a highly complex, dynamic and diversified structure. This must be kept in mind when studying the reliability of PPI data, together with the specific class of biological processes we are analysing.

Several models and paradigms have been proposed to address the complex scenario of protein interactions, often unrelated or in contrast with each other. For example, it is becoming clearer and clearer that data integration (that is the only way to obtain even a partial portrait of PPIs for several organisms) requires reliable, specific and informative models; general and well-studied algorithms (machine learning methods, probabilistic and statistical tools) must be implemented in a problem-specific manner and complemented using ad-hoc methods. The difficulty inherent with the complexity of the PPI maps is mirrored by the problems connected with experimental approaches: Y2H, AP-MS and small scale experiments overlap only partially, both because of the intrinsic noisiness of data and techniques and because they are apt to detect different types of interaction.

Even taking into account all these limitations, giant steps have been taken in the direction of a clearer ad informative vision of the interactome: a large fraction of yeast PPIs has been unraveled, the partial correlation between interaction, expression, regulation and function has been established, the role and the dynamics of protein complexes in the cell have been very much clarified. These advances need to be the starting point for improving our understanding of the interactome. Much work still has to be done: PPI maps of higher organism still have low coverage and reliability, weaker or transient PPI are difficult or impossible to detect with experimental techniques. The role of computational method is even more important in this area than in other biological problems: probabilistic and statistical frameworks and accurate models are essential to manage raw data, machine learning methods can help handling large amount of information, ad-hoc studies can add biological, structural and functional insights to the interaction data.

These approaches can take interactomics to a new more mature phase, and help in the understanding of biological events from a more comprehensive and systemic point of view.

REFERENCES

Aebersold, R., & Mann, M. (2003). Mass spectrometry-based proteomics. *Nature, 422*(6928), 198-207.

Aitken, S., Korf, R., Webber, B., & Bard, J. (2005). COBrA: A bio-ontology editor. *Bioinformatics, 21*(6), 825-826.

Alfarano, C., Andrade, C. E., Anthony, K., Bahroos, N., Bajec, M., Bantoft, K., et al. (2005). The Biomolecular Interaction Network Database and related tools 2005 update. *Nucleic Acids Res, 33*(Database issue), D418-424.

Aloy, P., Bottcher, B., Ceulemans, H., Leutwein, C., Mellwig, C., Fischer, S., et al. (2004). Structure-based assembly of protein complexes in yeast. *Science, 303*(5666), 2026-2029.

Aloy, P., & Russell, R. B. (2002). Potential artefacts in protein-interaction networks. *FEBS Lett, 530*(1-3), 253-254.

Altaf-Ul-Amin, M., Shinbo, Y., Mihara, K., Kurokawa, K., & Kanaya, S. (2006). Development and implementation of an algorithm for detection of protein complexes in large interaction networks. *BMC Bioinformatics, 7*, 207.

Asthana, S., King, O. D., Gibbons, F. D., & Roth, F. P. (2004). Predicting protein complex membership using probabilistic network reliability. *Genome Res, 14*(6), 1170-1175.

Ausiello, G., Gherardini, P. F., Marcatili, P., Tramontano, A., Via, A., & Helmer-Citterich, M. (2008). FunClust: a web server for the identification of structural motifs in a set of non-homologous protein structures. *BMC Bioinformatics, 9 Suppl 2*, S2.

Ausiello, G., Zanzoni, A., Peluso, D., Via, A., & Helmer-Citterich, M. (2005). pdbFun: Mass selection and fast comparison of annotated PDB residues. *Nucleic Acids Res, 33*(Web Server issue), W133-137.

Bader, G. D., & Hogue, C. W. (2002). Analyzing yeast protein-protein interaction data obtained from different sources. *Nat Biotechnol, 20*(10), 991-997.

Bader, J. S., Chaudhuri, A., Rothberg, J. M., & Chant, J. (2004). Gaining confidence in high-throughput protein interaction networks. *Nat Biotechnol, 22*(1), 78-85.

Barabási, A.-L. a. R. A. (1999). Emergence of scaling in random networks. *Science, 286*, 509-512.

Barabasi, A. L., & Oltvai, Z. N. (2004). Network biology: Understanding the cell's functional organization. *Nat Rev Genet, 5*(2), 101-113.

Ben-Hur, A., & Noble, W. S. (2005). Kernel methods for predicting protein-protein interactions. *Bioinformatics, 21, Suppl 1*, i38-46.

Ben-Hur, A., & Noble, W. S. (2006). Choosing negative examples for the prediction of protein-protein interactions. *BMC Bioinformatics, 7 Suppl 1*, S2.

Boone, C., Bussey, H., & Andrews, B. J. (2007). Exploring genetic interactions and networks with yeast. *Nat Rev Genet, 8*(6), 437-449.

Chen, J., Chua, H. N., Hsu, W., Lee, M. L., Ng, S. K., Saito, R., et al. (2006). Increasing confidence of protein-protein interactomes. *Genome Inform, 17*(2), 284-297.

Chen, J., Hsu, W., Lee, M. L., & Ng, S. K. (2005). Discovering reliable protein interactions from high-throughput experimental data using network topology. *Artif Intell Med, 35*(1-2), 37-47.

Chen, J., Hsu, W., Lee, M. L., & Ng, S. K. (2006). Increasing confidence of protein interactomes using network topological metrics. *Bioinformatics, 22*(16), 1998-2004.

Chiang, T., Scholtens, D., Sarkar, D., Gentleman, R., & Huber, W. (2007). Coverage and error models of protein-protein interaction data by directed graph analysis. *Genome Biol, 8*(9), R186.

Collins, S. R., Kemmeren, P., Zhao, X. C., Greenblatt, J. F., Spencer, F., Holstege, F. C., et al. (2007). Toward a comprehensive atlas of the physical interactome of Saccharomyces cerevisiae. *Mol Cell Proteomics, 6*(3), 439-450.

Cote, R. G., Jones, P., Apweiler, R., & Hermjakob, H. (2006). The Ontology Lookup Service, a lightweight cross-platform tool for controlled vocabulary queries. *BMC Bioinformatics, 7*, 97.

D'Haeseleer, P., & Church, G. M. (2004). Estimating and improving protein interaction error rates. *Proc IEEE Comput Syst Bioinform Conf*, (pp. 216-223).

Deane, C. M., Salwinski, L., Xenarios, I., & Eisenberg, D. (2002). Protein interactions: two methods for assessment of the reliability of high throughput observations. *Mol Cell Proteomics, 1*(5), 349-356.

Deng, M., Sun, F., & Chen, T. (2003). Assessment of the reliability of protein-protein interactions and protein function prediction. *Pac Symp Biocomput*, 140-151.

Ekman, D., Light, S., Bjorklund, A. K., & Elofsson, A. (2006). What properties characterize the hub proteins of the protein-protein interaction network of Saccharomyces cerevisiae? *Genome Biol, 7*(6), R45.

Enright, A. J., & Ouzounis, C. A. (2001). Functional associations of proteins in entire genomes by means of exhaustive detection of gene fusions. *Genome Biol, 2*(9), RESEARCH0034.

Fraser, H. B., Hirsh, A. E., Steinmetz, L. M., Scharfe, C., & Feldman, M. W. (2002). Evolutionary rate in the protein interaction network. *Science, 296*(5568), 750-752.

Gavin, A. C., Aloy, P., Grandi, P., Krause, R., Boesche, M., Marzioch, M., et al. (2006). Proteome survey reveals modularity of the yeast cell machinery. *Nature, 440*(7084), 631-636.

Gavin, A. C., Bosche, M., Krause, R., Grandi, P., Marzioch, M., Bauer, A., et al. (2002). Functional organization of the yeast proteome by systematic analysis of protein complexes. *Nature, 415*(6868), 141-147.

The Gene Ontology (GO) project in 2006. (2006). *Nucleic Acids Res, 34*(Database issue), D322-326.

Gentleman, R., & Huber, W. (2007). Making the most of high-throughput protein-interaction data. *Genome Biol, 8*(10), 112.

Giot, L., Bader, J. S., Brouwer, C., Chaudhuri, A., Kuang, B., Li, Y., et al. (2003). A protein interaction map of Drosophila melanogaster. *Science, 302*(5651), 1727-1736.

Grigoriev, A. (2001). A relationship between gene expression and protein interactions on the proteome scale: analysis of the bacteriophage T7 and the yeast Saccharomyces cerevisiae. *Nucleic Acids Res, 29*(17), 3513-3519.

Guo, J., Wu, X., Zhang, D. Y., & Lin, K. (2008). Genome-wide inference of protein interaction sites: lessons from the yeast high-quality negative protein-protein interaction dataset. *Nucleic Acids Res, 36*(6), 2002-2011.

Guo, Y., Yu, L., Wen, Z., & Li, M. (2008). Using support vector machine combined with auto covariance to predict protein-protein interactions from protein sequences. *Nucleic Acids Res, 36*(9), 3025-3030.

Hakes, L., Pinney, J. W., Robertson, D. L., & Lovell, S. C. (2008). Protein-protein interaction networks and biology--what's the connection? *Nat Biotechnol, 26*(1), 69-72.

Han, J. D., Dupuy, D., Bertin, N., Cusick, M. E., & Vidal, M. (2005). Effect of sampling on topology predictions of protein-protein interaction networks. *Nat Biotechnol, 23*(7), 839-844.

Hart, G. T., Lee, I., & Marcotte, E. R. (2007). A high-accuracy consensus map of yeast protein complexes reveals modular nature of gene essentiality. *BMC Bioinformatics, 8*, 236.

Hart, G. T., Ramani, A. K., & Marcotte, E. M. (2006). How complete are current yeast and human protein-interaction networks? *Genome Biol, 7*(11), 120.

Ho, Y., Gruhler, A., Heilbut, A., Bader, G. D., Moore, L., Adams, S. L., et al. (2002). Systematic identification of protein complexes in Saccharomyces cerevisiae by mass spectrometry. *Nature, 415*(6868), 180-183.

Hoheisel, J. D. (2006). Microarray technology: beyond transcript profiling and genotype analysis. *Nat Rev Genet, 7*(3), 200-210.

Huang, H., Jedynak, B. M., & Bader, J. S. (2007). Where have all the interactions gone? Estimating the coverage of two-hybrid protein interaction maps. *PLoS Comput Biol, 3*(11), e214.

Ito, T., Chiba, T., Ozawa, R., Yoshida, M., Hattori, M., & Sakaki, Y. (2001). A comprehensive two-hybrid analysis to explore the yeast protein interactome. *Proc Natl Acad Sci U S A, 98*(8), 4569-4574.

Itzhaki, Z., Akiva, E., Altuvia, Y., & Margalit, H. (2006). Evolutionary conservation of domain-domain interactions. *Genome Biol, 7*(12), R125.

Jaimovich, A., Elidan, G., Margalit, H., & Friedman, N. (2006). Towards an integrated protein-protein interaction network: A relational Markov network approach. *J Comput Biol, 13*(2), 145-164.

Jansen, R., Yu, H., Greenbaum, D., Kluger, Y., Krogan, N. J., Chung, S., et al. (2003). A Bayesian networks approach for predicting protein-protein interactions from genomic data. *Science, 302*(5644), 449-453.

Jothi, R., Cherukuri, P. F., Tasneem, A., & Przytycka, T. M. (2006). Co-evolutionary analysis of domains in interacting proteins reveals insights into domain-domain interactions mediating protein-protein interactions. *J Mol Biol, 362*(4), 861-875.

Kelley, R., & Ideker, T. (2005). Systematic interpretation of genetic interactions using protein networks. *Nat Biotechnol, 23*(5), 561-566.

Krogan, N. J., Cagney, G., Yu, H., Zhong, G., Guo, X., Ignatchenko, A., et al. (2006). Global landscape of protein complexes in the yeast Saccharomyces cerevisiae. *Nature, 440*(7084), 637-643.

Lee, I., Date, S. V., Adai, A. T., & Marcotte, E. M. (2004). A probabilistic functional network of yeast genes. *Science, 306*(5701), 1555-1558.

Lee, I., Li, Z., & Marcotte, E. M. (2007). An improved, bias-reduced probabilistic functional gene network of baker's yeast, Saccharomyces cerevisiae. *PLoS ONE, 2*(10), e988.

Li, S., Armstrong, C. M., Bertin, N., Ge, H., Milstein, S., Boxem, M., et al. (2004). A map of the interactome network of the metazoan C. elegans. *Science, 303*(5657), 540-543.

Lin, N., Wu, B., Jansen, R., Gerstein, M., & Zhao, H. (2004). Information assessment on predicting protein-protein interactions. *BMC Bioinformatics, 5*, 154.

Lu, L. J., Xia, Y., Paccanaro, A., Yu, H., & Gerstein, M. (2005). Assessing the limits of genomic data integration for predicting protein networks. *Genome Res, 15*(7), 945-953.

Mahdavi, M. A., & Lin, Y. H. (2007). False positive reduction in protein-protein interaction predictions using gene ontology annotations. *BMC Bioinformatics, 8*, 262.

Marcatili, P., Bussotti, G., & Tramontano, A. (2008). The MoVIN server for the analysis of protein interaction networks. *BMC Bioinformatics, 9 Suppl 2*, S11.

Mattingly, C. J., Rosenstein, M. C., Colby, G. T., Forrest, J. N., Jr., & Boyer, J. L. (2006). The Comparative Toxicogenomics Database (CTD): a resource for comparative toxicological studies. *J Exp Zoolog A Comp Exp Biol, 305*(9), 689-692.

Mewes, H. W., Frishman, D., Mayer, K. F., Munsterkotter, M., Noubibou, O., Pagel, P., et al. (2006). MIPS: analysis and annotation of proteins from whole genomes in 2005. *Nucleic Acids Res, 34*(Database issue), D169-172.

Miernyk, J. A., & Thelen, J. J. (2008). Biochemical approaches for discovering protein-protein interactions. *Plant J, 53*(4), 597-609.

Mika, S., & Rost, B. (2006). Protein-protein interactions more conserved within species than across species. *PLoS Comput Biol, 2*(7), e79.

Mishra, G. R., Suresh, M., Kumaran, K., Kannabiran, N., Suresh, S., Bala, P., et al. (2006). Human protein reference database--2006 update. *Nucleic Acids Res, 34*(Database issue), D411-414.

Mrowka, R., Patzak, A., & Herzel, H. (2001). Is there a bias in proteome research? *Genome Res, 11*(12), 1971-1973.

Neduva, V., & Russell, R. B. (2006a). DILIMOT: discovery of linear motifs in proteins. *Nucleic Acids Res, 34*(Web Server issue), W350-355.

Neduva, V., & Russell, R. B. (2006b). Peptides mediating interaction networks: new leads at last. *Curr Opin Biotechnol, 17*(5), 465-471.

Nooren, I. M., & Thornton, J. M. (2003). Diversity of protein-protein interactions. *Embo J, 22*(14), 3486-3492.

Ooi, S. L., Pan, X., Peyser, B. D., Ye, P., Meluh, P. B., Yuan, D. S., et al. (2006). Global synthetic-lethality analysis and yeast functional profiling. *Trends Genet, 22*(1), 56-63.

Pagel, P., Kovac, S., Oesterheld, M., Brauner, B., Dunger-Kaltenbach, I., Frishman, G., et al. (2005). The MIPS mammalian protein-protein interaction database. *Bioinformatics, 21*(6), 832-834.

Pang, C. N., Krycer, J. R., Lek, A., & Wilkins, M. R. (2008). Are protein complexes made of cores, modules and attachments? *Proteomics, 8*(3), 425-434.

Patil, A., & Nakamura, H. (2005). Filtering high-throughput protein-protein interaction data using a combination of genomic features. *BMC Bioinformatics, 6*, 100.

Qi, Y., Klein-Seetharaman, J., & Bar-Joseph, Z. (2005). Random forest similarity for protein-protein interaction prediction from multiple sources. *Pac Symp Biocomput*, 531-542.

Qi, Y., Klein-Seetharaman, J., & Bar-Joseph, Z. (2007). A mixture of feature experts approach for protein-protein interaction prediction. *BMC Bioinformatics, 8 Suppl 10*, S6.

Ramani, A. K., Bunescu, R. C., Mooney, R. J., & Marcotte, E. M. (2005). Consolidating the set of known human protein-protein interactions in preparation for large-scale mapping of the human interactome. *Genome Biol, 6*(5), R40.

Ramirez, F., Schlicker, A., Assenov, Y., Lengauer, T., & Albrecht, M. (2007). Computational analysis of human protein interaction networks. *Proteomics, 7*(15), 2541-2552.

Reguly, T., Breitkreutz, A., Boucher, L., Breitkreutz, B. J., Hon, G. C., Myers, C. L., et al. (2006). Comprehensive curation and analysis of global interaction networks in Saccharomyces cerevisiae. *J Biol, 5*(4), 11.

Rhodes, D. R., Tomlins, S. A., Varambally, S., Mahavisno, V., Barrette, T., Kalyana-Sundaram, S., et al. (2005). Probabilistic model of the human protein-protein interaction network. *Nat Biotechnol, 23*(8), 951-959.

Rogozin, I. B., Makarova, K. S., Wolf, Y. I., & Koonin, E. V. (2004). Computational approaches for the analysis of gene neighbourhoods in prokaryotic genomes. *Brief Bioinform, 5*(2), 131-149.

Rual, J. F., Venkatesan, K., Hao, T., Hirozane-Kishikawa, T., Dricot, A., Li, N., et al. (2005). Towards a proteome-scale map of the human protein-protein interaction network. *Nature, 437*(7062), 1173-1178.

Russell, R. B., Alber, F., Aloy, P., Davis, F. P., Korkin, D., Pichaud, M., et al. (2004). A structural perspective on protein-protein interactions. *Curr Opin Struct Biol, 14*(3), 313-324.

Saeed, R., & Deane, C. (2008). An assessment of the uses of homologous interactions. *Bioinformatics, 24*(5), 689-695.

Salwinski, L., Miller, C. S., Smith, A. J., Pettit, F. K., Bowie, J. U., & Eisenberg, D. (2004). The Database of Interacting Proteins: 2004 update. *Nucleic Acids Res, 32*(Database issue), D449-451.

Sanderson, C. M. (2008). A new way to explore the world of extracellular protein interactions. *Genome Res, 18*(4), 517-520.

Sardiu, M. E., Cai, Y., Jin, J., Swanson, S. K., Conaway, R. C., Conaway, J. W., et al. (2008). Probabilistic assembly of human protein interaction networks from label-free quantitative proteomics. *Proc Natl Acad Sci U S A, 105*(5), 1454-1459.

Scherens, B., & Goffeau, A. (2004). The uses of genome-wide yeast mutant collections. *Genome Biol, 5*(7), 229.

Scott, M. S., & Barton, G. J. (2007). Probabilistic prediction and ranking of human protein-protein interactions. *BMC Bioinformatics, 8*, 239.

Sharan, R., Suthram, S., Kelley, R. M., Kuhn, T., McCuine, S., Uetz, P., et al. (2005). Conserved patterns of protein interaction in multiple species. *Proc Natl Acad Sci U S A, 102*(6), 1974-1979.

Shen, J., Zhang, J., Luo, X., Zhu, W., Yu, K., Chen, K., et al. (2007). Predicting protein-protein interactions based only on sequences information. *Proc Natl Acad Sci U S A, 104*(11), 4337-4341.

Shoemaker, B. A., & Panchenko, A. R. (2007). Deciphering protein-protein interactions. Part I. Experimental techniques and databases. *PLoS Comput Biol, 3*(3), e42.

Singh, R., Xu, J., & Berger, B. (2006). Struct2net: integrating structure into protein-protein interaction prediction. *Pac Symp Biocomput*, 403-414.

Sprinzak, E., Sattath, S., & Margalit, H. (2003). How reliable are experimental protein-protein interaction data? *J Mol Biol, 327*(5), 919-923.

Stagljar, I., & Fields, S. (2002). Analysis of membrane protein interactions using yeast-based technologies. *Trends Biochem Sci, 27*(11), 559-563.

Stelzl, U., Worm, U., Lalowski, M., Haenig, C., Brembeck, F. H., Goehler, H., et al. (2005). A human protein-protein interaction network: a resource for annotating the proteome. *Cell, 122*(6), 957-968.

Stumpf, M. P., Thorne, T., de Silva, E., Stewart, R., An, H. J., Lappe, M., et al. (2008). Estimating the size of the human interactome. *Proc Natl Acad Sci U S A, 105*(19), 6959-6964.

Szilagyi, A., Grimm, V., Arakaki, A. K., & Skolnick, J. (2005). Prediction of physical protein-protein interactions. *Phys Biol, 2*(2), S1-16.

Tovchigrechko, A., & Vakser, I. A. (2005). Development and testing of an automated approach to protein docking. *Proteins, 60*(2), 296-301.

Uetz, P., Giot, L., Cagney, G., Mansfield, T. A., Judson, R. S., Knight, J. R., et al. (2000). A comprehensive analysis of protein-protein interactions in Saccharomyces cerevisiae. *Nature, 403*(6770), 623-627.

Vidalain, P. O., Boxem, M., Ge, H., Li, S., & Vidal, M. (2004). Increasing specificity in high-throughput yeast two-hybrid experiments. *Methods, 32*(4), 363-370.

von Mering, C., Krause, R., Snel, B., Cornell, M., Oliver, S. G., Fields, S., et al. (2002). Comparative assessment of large-scale data sets of protein-protein interactions. *Nature, 417*(6887), 399-403.

Wagner, A. (2001). The yeast protein interaction network evolves rapidly and contains few redundant duplicate genes. *Mol Biol Evol, 18*(7), 1283-1292.

Yook, S. H., Oltvai, Z. N., & Barabasi, A. L. (2004). Functional and topological characterization of protein interaction networks. *Proteomics, 4*(4), 928-942.

Zanzoni, A., Montecchi-Palazzi, L., Quondam, M., Ausiello, G., Helmer-Citterich, M., & Cesareni, G. (2002). MINT: a Molecular INTeraction database. *FEBS Lett, 513*(1), 135-140.

Zhang, B., Park, B. H., Karpinets, T., & Samatova, N. F. (2008). From pull-down data to protein interaction networks and complexes with biological relevance. *Bioinformatics, 24*(7), 979-986.

Zhou, M., & Cui, Y. (2004). GeneInfoViz: constructing and visualizing gene relation networks. *In Silico Biol, 4*(3), 323-333.

Section III
Knowledge Discovery from PPI Networks

Section III of the book will focus on knowledge discovery – that is, how do we computationally interrogate the PPI networks to discover new useful biological knowledge such as interaction motifs, network motifs, and protein complexes using sophisticated graph data mining approaches previously inaccessible to the bench biologists. We will also describe the evolutionary analyses of protein interaction networks for gaining insights on molecular evolution and comparative genomics.

Chapter VII
Discovering Interaction Motifs from Protein Interaction Networks

Hugo Willy
National University of Singapore, Singapore

ABSTRACT

Recent breakthroughs in high throughput experiments to determine protein-protein interaction have generated a vast amount of protein interaction data. However, most of the experiments could only answer the question of whether two proteins interact but not the question on the mechanisms by which proteins interact. Such understanding is crucial for understanding the protein interaction of an organism as a whole (the interactome) and even predicting novel protein interactions. Protein interaction usually occurs at some specific sites on the proteins and, given their importance, they are usually well conserved throughout the evolution of the proteins of the same family. Based on this observation, a number of works on finding protein patterns/motifs conserved in interacting proteins have emerged in the last few years. Such motifs are collectively termed as the interaction motifs. This chapter provides a review on the different approaches on finding interaction motifs with a discussion on their implications, potentials and possible areas of improvements in the future.

INTRODUCTION

Protein interaction plays an essential role in a vast number of known biological processes. It is responsible in the formation of multimeric protein complexes, signal transduction, cell regulation and immune response processes. The interaction can be permanent, with relatively high binding affinity, and usually lasts throughout the protein's lifetime. This type of interaction can often be seen in the binding of different subunits of a permanent protein complex (the *obligate* interaction). The second type of interaction is a temporary, mostly of lower affinity (the *transient* interaction). An example of this type of interaction is the protein binding cascade commonly seen in the cellular signal transduction process (Jones & Thornton, 1996; Ofran & Rost, 2003). A good understanding of the mechanism underlying these protein-protein interactions is essential to the understanding of the biological system as a whole and potentially aid research on diseases and finding novel drug targets.

Protein interaction can be identified using quite a number of different biological experiments. Some are of high accuracy but has low throughput – which makes such methods too costly but for a few very important interactions. An example of such experiments would be the methods to determine the structure of protein complexes such as X-Ray crystallography or NMR spectroscopy. Recent years had witnessed the breakthrough of high throughput protein interaction identification technique like the Yeast Two Hybrid (Y2H) techniques and the Tandem Affinity Purification coupled with Mass Spectrometry (TAP-MS). But, as discussed in (Sprinzak *et. al.*, 2003), the higher throughput come with the cost of a significant reduction in accuracy. For yeast-two-hybrid, the expected number of false positive interaction could reach 50% and beyond. Nevertheless, given the wealth of newly generated interaction data, a number of computational methods had been devised and shown to perform reasonably accurate interaction prediction in-silico. Their potential in discovering novel interactions and cleaning up the noisy high-throughput interaction data using relatively inexpensive computational methods has generated a lot of work on this problem. Most of the approaches are machine learning based, combining multiple information source, each of which gives a certain degree of confidence on the protein interaction being investigated. For a broader review on works done on computational approach of protein interaction prediction, readers are referred to the excellent reviews in (Valencia & Pazos, 2002) and a more recent one in (Skrabanek *et. al.*, 2008). However, most of the approaches described above, both biochemical and computational based, are not able to elucidate the mechanism underlying the observed protein interaction. Specifically, most of them are only able to say if two proteins are interacting, but not on how or which parts of the proteins cause them to interact. Such details could only be seen by solving the 3D structure of the interacting protein. To date, the amount of structural data is rapidly increasing but their coverage is still limited in comparison to the existing high-throughput interaction data.

This chapter will discuss several computational approaches emerging over the years on mining the interaction motif. These approaches attempt to mine motifs/signatures within protein sequences which are highly indicative of protein interaction. The interaction motif could be motifs on protein binding sites, enzyme active sites or some structurally important sites which indirectly influence protein interaction. By knowing such motifs, we could possibly predict protein interaction based on the occurrence of interaction motifs in the candidate protein interaction in question. The main underlying assumption on interaction motif mining is that interaction motifs are conserved throughout evolution given its role in protein interaction. Thus one should be able to see such motifs enriched in an adequately large protein interaction data. The motifs can be of two general types; one is the structural, non-linear (e.g. the protein domains) and the other is the linear peptide motif. Since domain interaction is already covered in the other chapter of this book, we would only cover methods which are not solely based on interacting domains (i.e., the domain-linear motifs and linear motifs-linear motifs based methods). It is known that not all protein interactions are mediated by pairs of protein domains. Many of them involve the binding of a protein domain to a short peptide sequence with a defined linear sequence motif, which is the SLiMs (Pawson & Scott, 1997; Sudol, 1998; Neduva & Russell, 2006). The listing of all known SLiMs to date could be found in ELM (Puntervoll *et. al.*, 2003) or MiniMotif (MnM) database (Balla *et. al.*, 2006) .

There are multiple challenges in finding interaction motifs. The most fundamental one would be the limitation of the source interaction data. To be able to confidently say that a motif is essential for interaction, one would expect the motif to be enriched in interacting pair of proteins but not in the non-interacting ones. However, as we mentioned earlier, the available high throughput interaction data is highly noisy. Furthermore, to be able to design a protein interaction prediction system, we must have both interacting and non-interacting protein data. The latter is known to be scarce and most of the time, it is approximated using random pairs or pairs of protein with different cellular localization (Guo *et. al.*, 2008). The second problem is on the nature of linear motifs in particular. Since most of the time linear motifs are degenerate, one could find a huge number of candidate instances of them all over the interaction data. Hence, we would need to design a careful filtering and scoring methodology to fish out the true motif instances to be able to make meaningful inference of an interaction event based on the existence of linear motifs.

The following section will discuss the existing methodologies to find interaction motifs from the protein interaction data. Then we present some discussions on the challenges, issues and possible future work in this area.

EXISTING APPROACHES

Sequence Based Approaches

The methods in this class rely solely on the information derived from protein sequences and the high throughput protein interaction data. Despite the relatively high noise rate, the current amount of such data is sufficiently large for one to perform meaningful large scale data mining. We further classify the methods in this class based on the motif definition that is used.

Methods Finding Interaction Motifs based on Existing Motif Definitions

In this subclass, the methods use the known motifs such as the (longer) protein domains from existing databases like Pfam (Finn *et. al.*, 2006), Prosite (Hulo *et. al.*, 2008), InterPro (Mulder *et. al.*, 2007) and the linear motifs from ELM (Puntervoll *et. al.*, 2003), MnM: Minimotif Miner (Balla *et. al.*, 2006) and Prosite. The advantage of this line of approach would be on the motifs' reliability since most of them had been manually curated (e.g the manually curated subset of Pfam, Pfam-A) or even experimentally verified (e.g ELM, MnM). However, their coverage might be limited since the methods would not be able to discover novel motif candidates.

Sprinzak & Margalit (2001)

One of the first works on interaction motifs is by Sprinzak & Margalit (2001). In this work, they mined pairs of InterPro signatures which occur more than expected by random in known interacting proteins. Specifically, for each interacting protein (P, Q); P contains the InterPro signatures $\{s_{P1}, s_{P2}, s_{P3},.., s_{Pn}\}$ while Q contains $\{s_{Q1}, s_{Q2}, s_{Q3},.., s_{Qm}\}$; increment the counter of the pair (s_{Pi}, s_{Qj}) for $1 \leq i \leq n$ and $1 \leq j \leq m$. Next, given the observed occurrence C_{ij} of each signature pair (s_i, s_j) in the whole interaction data and the individual occurrence C_i and C_j of each signature s_i and s_j, respectively. The significance of the signatures' co-occurrence is computed using the log odd ratio $log_2(C_{ij}/C_iC_j)$.

From the top ranking pairs, the authors managed to find correlations between the *Myb* domain and *Bromodomain* along with several other pairs of signatures implicated in the acetylation of *c-Myb* protein by the *Bromodomain* of human *histone acetyltransferase* p300 protein. The acetylation, however, does not happen on the protein *c-Myb*'s *Myb* domain (ruling out the direct domain-domain interaction) but rather on a region C-terminal to it.

Figure 1. (A) An aspect model, (B) AKM. Image reproduced from Mamitsuka (2005), © [2005] IEEE

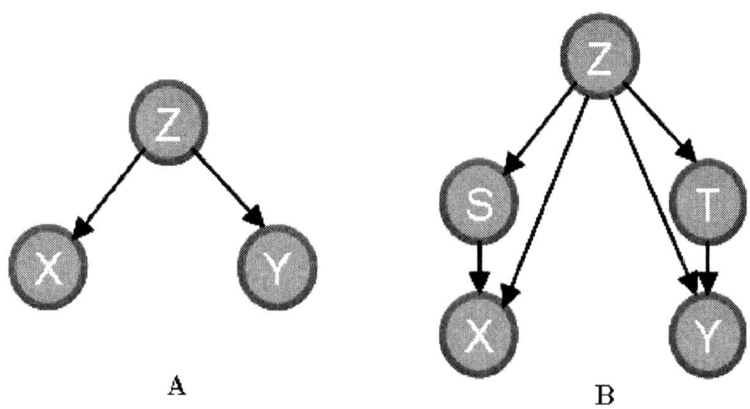

AKM - Mamitsuka (2005).

This work analyzed the factors which cause proteins to interact, termed as the *latent knowledge* of the proteins. The author created an extension of a probabilistic model known as the *aspect model* (Hofmann, 2001) which includes another layer to accommodate the latent knowledge of the protein. The new model is termed the *additive knowledge model* (AKM).

The aspect model, as shown in Fig. 1, is a general probabilistic model for two-mode, co-occurrence data. The model has two observable variables X and Y, and a distinct valued latent variable Z with k different values $z_1, z_2, \ldots z_k$ (in their experiments, the authors empirically set $k=30$). An example using protein interaction would be for X and Y to be proteins and Z to be different factors on which the interaction of X and Y could be facilitated. The probability of observing protein x and y interacting under the a set of parameter θ can be computed by the following formula

$$p(x, y; \theta) = \sum_k p(x \mid z_k; \theta) p(y \mid z_k; \theta) p(z_k; \theta)$$

In Fig. 1(B), the model is extended by two additional latent variables, denoted as S and T, for X and Y, respectively. They represent the signature information derived from the InterPro (Mulder *et~al.*, (2007)) signatures in the proteins. In this model, the relationship between the protein class S and the interacting protein X is modeled by a conditional probability $p(X|S)$. Then, the formula above is extended to handle the new conditional probabilities as follows,

$$p(x, y; \theta) = \sum_k p(z_k; \theta) \sum_l p(x \mid s_l, z_k; \theta) p(s_l \mid z_k; \theta) \sum_m p(y \mid t_m, z_k; \theta) p(t_m \mid z_k; \theta)$$

The model is trained using Expectation Maximization (EM) algorithm (Dempster, Laird & Rubin, 1977) to estimate its optimal parameter θ, which best describes the training protein interaction data D. That is to say, the algorithm would find the optimal θ^{ML} to maximize the likelihood of D where

$$\theta^{ML} = \arg\max_{\theta} \log p(D; \theta)$$

and

$$\log p(D; \theta) = \sum_{(x,y) \in D} \log p(x, y; \theta)$$

The author argued that the training would require linear time and space complexity based on the fact that each protein would only have a roughly a constant number of InterPro signatures. Using 5-fold cross validation, they showed that the model is superior among several other models, including the log-odd ratio based model used in (Sprinzak & Margalit, 2001). In the case of proteins with multiple signatures, the model can estimate the most probable signature in a protein which makes it interacts with some other protein. Furthermore, it can also assess the likelihood of two signatures of being correlated for interaction. They found several pairs of signatures with biological significance such as the signature IPR005828 which is a class of protein transporters found to be correlated with a considerable number of other signatures since they transport a large variety of proteins. Another pair found are IPR001005 (Myb) and IPR002145 (DEAD/DEAH box Helicase) which are implicated in nucleic acid processing.

InSite – Wang et. al. (2007)

InSite offers a probabilistic model to assess the binding sites between two interacting proteins. The main idea is to first train the model on the existing interaction data to come up with a set motif interaction affinity parameters that provides a good explanation of the interaction data. The second step then assesses each hypotheses of the form "motif *M* on protein *A* binds to protein *B*" for each motif *M* on each interacting protein pair *(A,B)*. This is accomplished by (computationally) disallowing the *A-B* binding at the motif *M* by removing any pairing of *M* with any motif in *B* from the model's training data. Then the new set of interactions is re-trained and the model's total score is recomputed. The decrease in the model's overall likelihood score would give an estimate of the importance of the interaction on M, hence the hypotheses' significance. By also checking the converse hypotheses "motif *M'* on protein *B* binds to protein *A*", we would be able to compute the binding motifs of both protein *A* and *B*.

Formally, the authors define the variable $T_{ij}B_{ab}$ to represent whether the pair of protein P_i and P_j interact through motifs *a* in P_i and *b* in P_j, for each pair *(a,b)* occurring in (P_i, P_j). The probability

$$p(T_{ij}B_{ab} = true)$$

Figure 2(A) Pairs of motif (the colored shapes) in different proteins (the rectangles) may mediate the observed interactions among the proteins. The method try to find the minimum number of high affinity motif pair which can best explain the interaction data (preference for simpler explanation). For this example, the interaction can be explained by the motif pair (a,d) which can explain (P1,P3) and (P1,P4); and (b,e) which can explain (P1,P5) and (P2,P5). (B) The steps of InSite method. The model learning involves training a bayesian network on multiple source interaction data. The next step computes the individual motif score based on the protein interactions it occurs in (The images are extracted from Wang et. al. (2007)).

A. B.

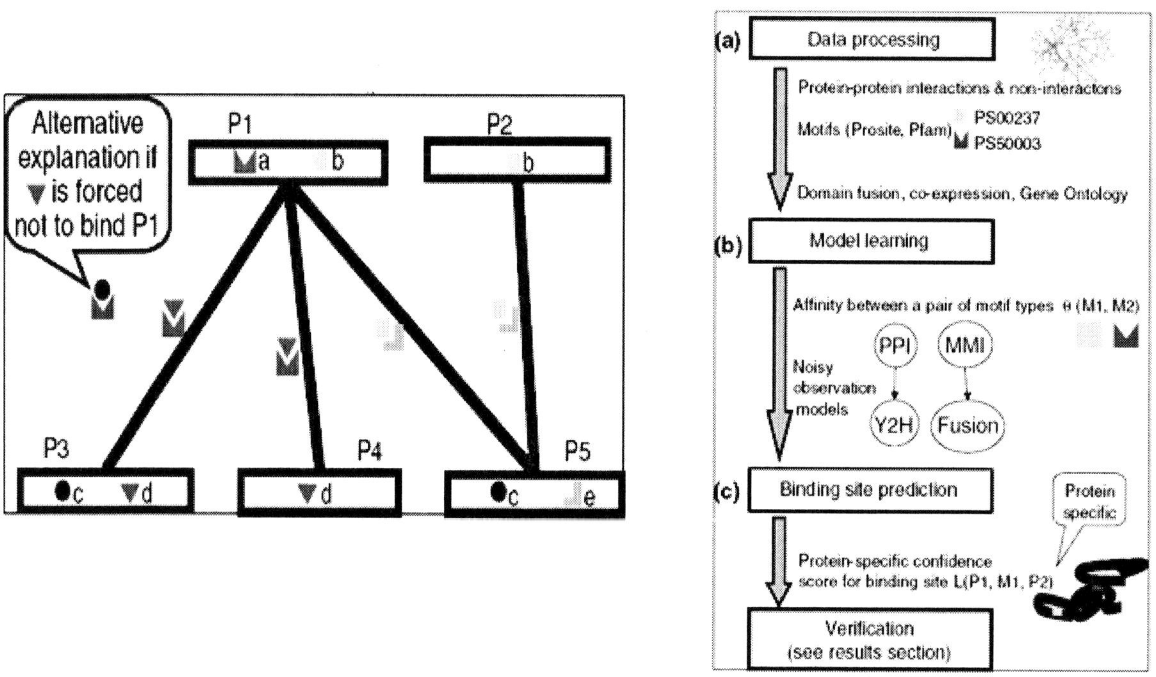

is determined by the affinity of the two motifs termed as θ_{ab}. The authors used the Dirichlet prior distribution over the values of θ_{ab}. To account for the possibility of P_i and P_j interaction occurring at regions outside the motifs, the spurious binding variable $T_{ij}S$ is introduced. The probability of such spurious interaction is computed as

$$p(T_{ij}S = true) = 1 - (1 - \theta_s)^m$$

where θ_s represents the probability of spurious binding at one non-motif position and m is the average length of sequence span not covered by any motif within the two proteins P_i and P_j.

InSite also created a naïve bayes model to account for different prediction accuracy of the interaction data from different experimental assays. It computed the probability of observing interaction of P_i and P_j in an assay of type n, denoted by $T_{ij}O_n$. The author assigned different types of distribution for the probability

$$p(T_{ij}O_n \mid T_{ij}I = true)$$

and

$$p(T_{ij}O_n \mid T_{ij}I = false)$$

based on the best fit to the empirical distribution of each assay. Another naïve bayes model is created to account for the noise in methods to infer motif pair interaction, such as domain fusion and GO category (Ashburner *et. al.*, 2000) overlap. Using the noisy observed interactions from different assays, InSite learned the optimal values of θ_{ab} and θ_s by maximizing the likelihood of the observations using EM algorithm. The importance of a particular motif a in protein P_i for its interaction with P_j is estimated by removing a from P_i and rerunning EM on the model. The difference in the likelihood score of the new model would be the estimates of motif a's strength on explaining (P_i, P_j) interaction.

The set of motifs used are the Pfam domains (Finn *et. al.*, 2006) and Prosite (Hulo *et. al.*, 2008). In the experiments, the authors worked to a great length on validating the correctness of their model. First, the authors noted that both domain definitions are quite limited in term of coverage when checked against the binding site residues in PDB (Berman *et. al.*, 2000). Nevertheless, they showed that the method can be used to predict direct physical interactions usually obscured in TAP-MS or other experimental data of similar bait-prey nature.

Comparison with the results of Gavin *et. al.* (2006) and Krogan *et. al.* (2006) showed that InSite covers more positive interactions. The second validation is from PDB, where InSite's prediction is moderately enriched in actual binding sites (52 – 55% of predicted interaction motifs are actual binding sites in PDB). Since the motif significance is evaluated in the context of two specific proteins e.g. motif M in Protein A binds to protein B, it is possible that M have different score on different interactions. This is different from many existing approaches which would give the same score for each occurrence of an interaction motif.

Methods Based on De-Novo Motif Finding

The methods in this category attempt to find all possible motif pairs found within the input protein sequences and interaction data. The main challenge for the approaches in this line is the combinatorial complexity of the motif space, which is worsened by the quadratic factor of considering pairs of such motifs. The speed-up techniques used in the following works can be categorized into two classes, 1) optimization on the motif model (use of constrained, less general model), 2) Exploitation of the structures in the protein interaction network, such as complete or dense subgraphs, to prune out those motif pairs with low correlation with observed interactions. Usually, the sheer size of motif pair space would hinder most of the methods in this class to use sophisticated statistical function to score and rank all possible motif pairs.

Li et. al. (2006)

This work presented an interesting transformation of the problem on finding interaction motif pairs to the frequent closed itemset mining which is well studied in the data mining field. The method is based on finding two fully interacting protein set termed as the biclique in the protein interaction network formed by the interaction data. They conjectured that such protein sets should share similar motifs which enable them to interact.

The method is basically separated into two steps, namely the biclique mining step and the motif computation step. In the former, they computed all bicliques with size at least five on any side by transforming the problem into the problem of frequent closed pattern mining in the data mining field (Nicolas, P. *et. al.*, 1999). Fig. 3 describes the steps of the transformation. They used an efficient program FPClose* (Grahne & Zhu, 2003) to mine the frequent closed patterns for a selected set of 10000 yeast protein interactions from DIP (Salwinski *et. al.*, 2004). In the next step, they used PROTOMAT (Henikoff & Heinikoff, 1991) to compute singular motifs called *blocks* which are similar to those in the BLOCKS (Pietrokovski *et. al.*, 1996) and PRINTS (Attwood & Beck, 1994) database. An interaction motif pair is then defined by pairing the set of blocks from each protein set on each side of a biclique. Using the software LAMA (Pietrokovski *et. al.*, 1996), they mapped the found blocks to the known blocks in BLOCKS and PRINTS. They also validated their motif pairs using the known interacting Pfam domain in iPfam (Finn *et. al.*, 2005) by mapping the blocks to their corresponding PFAM entries through

Figure 3. The steps of the biclique extraction in the algorithm. The authors transformed the problem of mining complete bipartite graph into the problem of mining frequent closed itemset which is well-studied in the data mining field. The example above uses support n=2 and itemset size m=3.

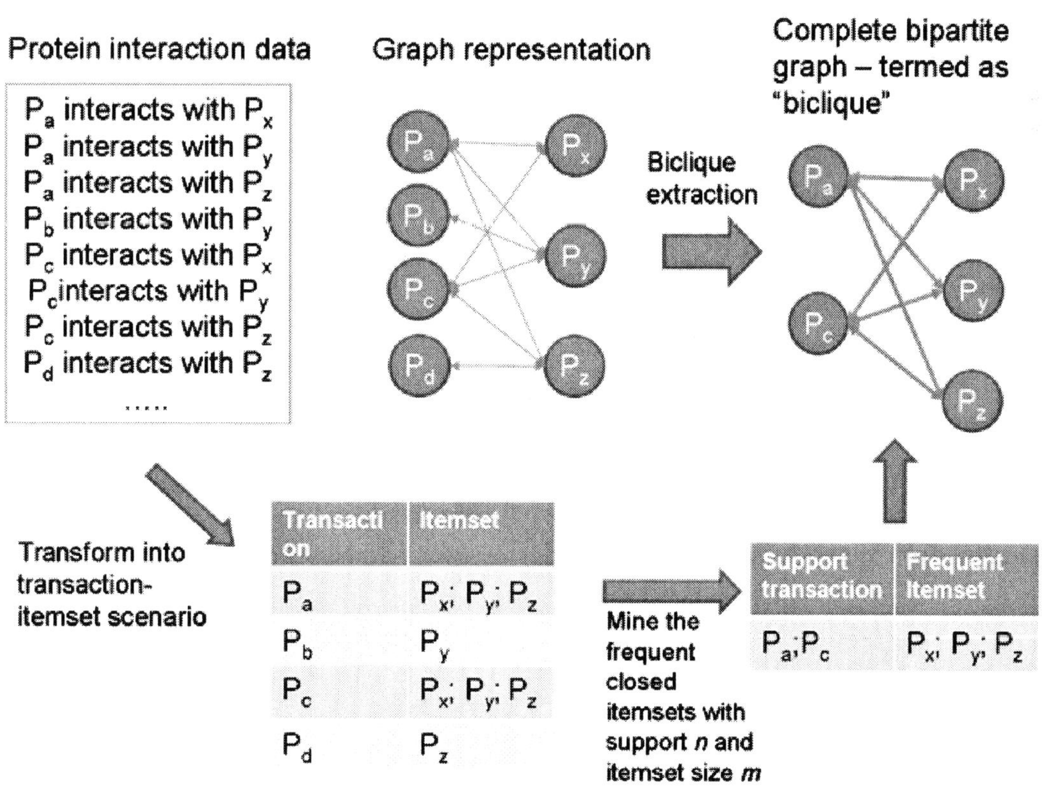

an intermediate mapping to InterPro (Mulder *et. al.*, 2007). Only 18 distinct domain pairs from iPfam's 3045 pairs are covered. The authors asserted that the low coverage might be caused by low coverage of yeast protein complexes in PDB. Validation using another domain interaction database, InterDom (Ng *et. al.*, 2003) managed to cover 203 domain pairs, 94 of which are of high confidence. One biological validation presented in detail involves the domain interaction of LSM domains which is supported by both iPfam and Interdom.

D-STAR - Tan et. al. (2006)

In this paper, Tan *et. al.* improved the generality of the previous work by mining from a densely interacting subgraph of protein interactions, as opposed to the fully connected biclique as in Li *et. al.*, (2006). This, however, imposes a high combinatorial cost as the number of dense bipartite subgraphs is substantially bigger than the fully connected ones. As a result, instead of using the *blocks* motif which is expensive to compute (most of the

Figure 4(A) The concept of correlated motifs. The different shapes represents motifs inside the protein sequence. (B) A pair of motif found by D-STAR validated by the PDB structure 1avz. (C) The detailed instances of the (l,d)-motif pair from the SH3 dataset reveals both class 1 (RxxPxxP) and class 2 (PxxPxR) SH3 binding motifs. The interacting partners are instances from the motif occurring inside the SH3 domain. Interactions are represented by a line connecting two segments from the two protein sets. (The images are extracted from Tan et. al. (2006))

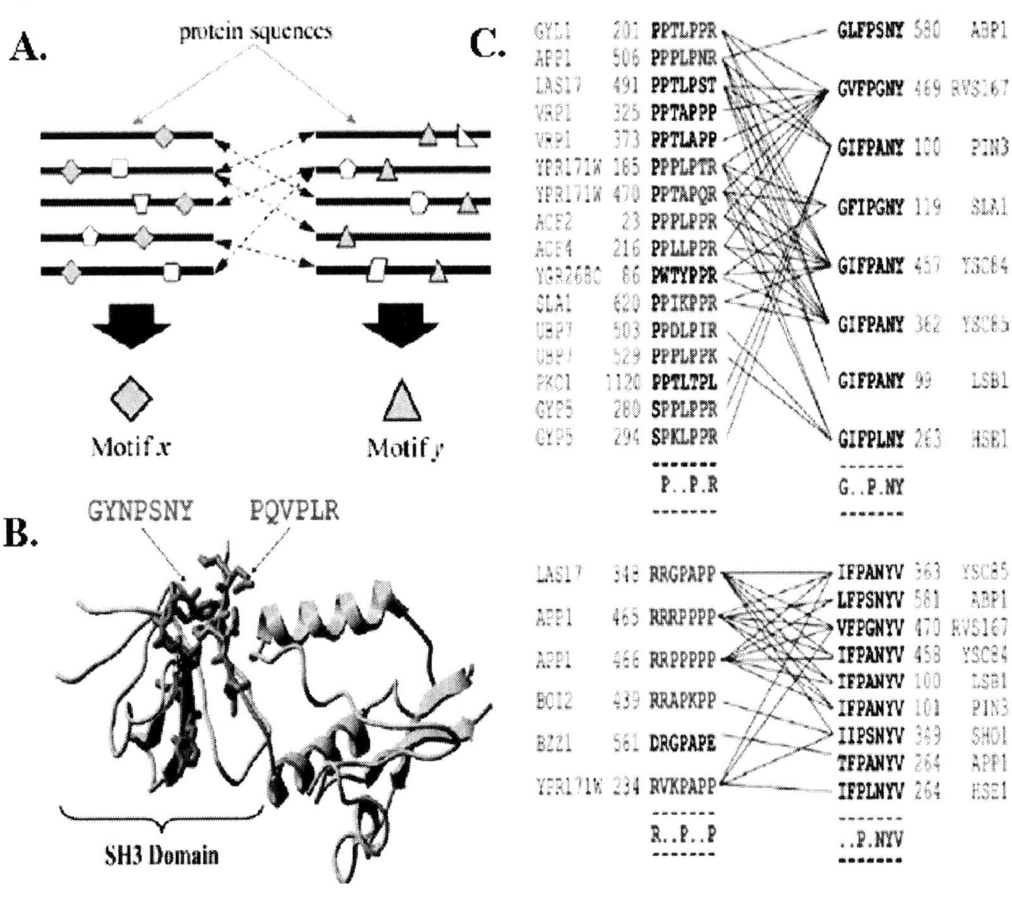

computation time of Li *et. al.*, (2006) is spent on the *blocks*' computation), they employed another model, the *(l,d)-motif*, used in the previous work of WINNOWER (Pevzner & Sze, 2000) and Random Projection (Buhler & Tompa, 2001) on DNA sequences. The motif is basically a string of length *l*, whose instances are strings of similar length have hamming distance at most *d* from the motif. For example, given a *(6,2)-motif* PVPPLR, some of its possible instances would be PSPPLR, PPPPRR, PQRPLR (the underlined residues are the position of the mismatch(es)).

The authors presented one exact algorithm and one approximation algorithms on the *(l,d)-motif pair finding* problem which finds a pair of *(l,d)-motif* occurring in a pair of densely interacting protein sets. The first algorithm, D-MOTIF, tries to find pair of cliques of *(l,d)-motif* instances which has high density interaction between them. Each clique contains instances of an *(l,d)-motif* such that each one of them are at most of distance *d* from one another. D-MOTIF gives the exact solution of the *(l,d)-motif pair finding* problem, but has an exponential running time for the clique finding which is too slow for larger datasets. The algorithm D-STAR is then proposed to give an approximation to the complete solution. Instead of enforcing a clique on the hamming distance of the motif instances, D-STAR relaxed the requirement into a 'star' construct, where one of the instances is set to approximate the *(l,d)-motif* and the other instances are at most of distance *d* from the former. All pairs of instances

Figure 5. The illustration on the (l,d)-motif pair finding problem. D-MOTIF is designed to find the instances of (l,d)-motif pair drawn at the upper model. The motif instances {M_1..M_5} and {M'_1..M'_3} are of distance d from one another, forming a clique. Then the clique pair with significantly high number of interaction between the proteins in which they occur are reported and examined. In the D-STAR model, instead of enforcing a clique, it requires that the motif instances are at most of distance d from a reference motif. In the above example, the reference motifs are M_1 and M'_1 respectively.

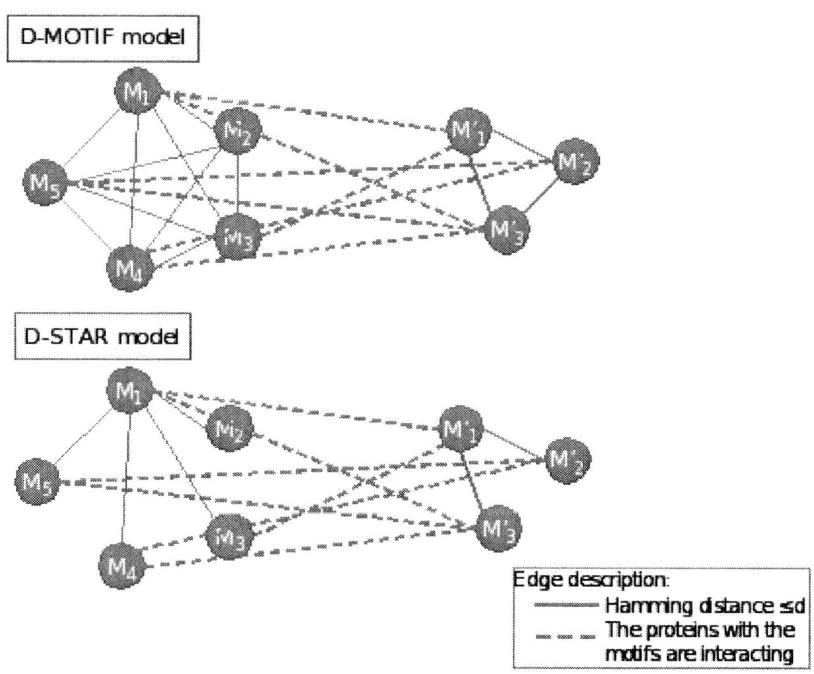

in the star model, however, are still guaranteed to be at most of distance *2d* from one another. The authors showed that D-STAR produces reasonably good approximation results which are validated with both simulation and real biological datasets. The models are depicted in Fig. 5.

Both D-MOTIF and D-STAR employ a *chi-square* scoring function to quantify the significance of each motif pair. The validation on the algorithm is done on the SH3 domain interaction dataset and the TGFβ signaling network dataset. On the first dataset, D-STAR managed to find both class 1 and class 2 binding motifs for the SH3 domain (see Fig. 4C). Further checks in the PDB structural database (Berman *et. al.*, 2000) confirmed that the two motifs found are indeed found at the interaction interface of SH3 and its binding peptide (Fig. 4B). On the TGFβ dataset, D-STAR managed to find motif pairs that is dominated by kinase domain on one side and a significant enrichment of phosphorylation motifs on the other. This is biologically relevant as many signaling pathways are kinase-regulated through protein phosphorylation. However, the main drawback of this method is on its non-scalability since the rather weak *(l,d)-motif* model yields a huge number of motif pair computation.

PIPE - Pitre et. al. (2006)

This work presented a protein interaction prediction engine which is based on the concept of interaction motif. The method is rather straightforward where it collected all pairs of substrings of length 20 from protein interaction data and computed their co-occurrence in the whole dataset. The occurrence of a single segment is determined by local pairwise alignment using PAM120 matrix with a minimum alignment score cutoff set at 35 (empirically determined). The significance of the segment pair is quantified directly by its number of co-occurrence, and the authors set the minimum threshold of 10 for a motif pair to be 'interacting'. As stated by the authors, the works is geared towards proof of concept rather than an algorithmically sophisticated system; the computation of some 15000 interactions involving 6304 yeast proteins collected from DIP took an equivalent of 1000 computational hours.

Nevertheless, the author showed that the system is capable of finding meaningful and some novel protein interactions. The most significant evidence presented is the previously unknown interaction of two yeast proteins, YGL227W and YMR135C. The proteins are implicated in the degradation of *fructose-1,6-bisphosphatase* (FBPase) on which the authors performed TAP experiments to ascertain the interaction. Furthermore, the connectivity within the purified complex was predicted based on the pairwise score produced by PIPE and was verified in genetic experiments using two yeast deletion strains. The strain with a deleted gene, which encodes a protein of a high scoring pair in the complex, was found unable to form the target complex. On the other hand, the strain which has a deleted low scoring protein retained the complex formation.

Leung et. al. (2007)

This paper presents an improvement over the D-STAR algorithm by Tan *et. al.* (2006). The main contribution of this work is a simpler but potentially better motif model which enables an efficient voting-based exact algorithm, FindMotif. The authors also presented a heuristic algorithm, MotifHeuristics, which could mine interaction motif pairs from the whole yeast interaction data. Another contribution is the hypergeometric scoring function which the authors argued to be better in ranking the known motif pairs from their experimental datasets as compared to the chi-square function of D-STAR.

The motif model used in this work is called *wildcard (l,d)-motif*. The motif is a length-*l* string with *d* wildcards, denoted by the character '*x*'. For example, PxxPxR is a *wildcard (6,3)-motif*. An instance of the motif would be a length-*l* string which matches the motif on all of its non-wildcard position. For example, PPLPVR, PQNPLR and PPPPER would all be instances of PxxPxR. This motif is more restrictive than the original *(l,d)-motif* since the mismatches can only take place on the wildcard positions. For a defined *(l,d)*, there are only

$$\binom{l}{d}|\Sigma|^{(l-d)}$$

possible *wildcard (l,d)-motif* string as compared to $|\Sigma|^l$ possible strings of the original *(l,d)-motif* (for amino acids $|\Sigma|=20$). The smaller number of the former enable the exact algorithm, FindMotif, to count all the frequencies of all *wildcard (l,d)-motif* pairs for a reasonably chosen *(l,d)* value. In their experiments, the authors used $l=8$ and $d=5$ which yields

$$\binom{l}{d}^2 |\Sigma|^{2(l-d)} \approx 200\text{G}$$

motif pairs. The running time of such voting procedure would be in $O(mn^2)$ where m is the size of the interaction data and n is the average length of the proteins. When the number of possible motif pairs is too high, they split the voting procedure by fixing one of the motif in the motif pairs at one time which roughly pull down the space complexity to

$$\binom{l}{d}|\Sigma|^{(l-d)}.$$

However, it increases the original $O(mn^2)$ running time by a factor of

$$\binom{l}{d}|\Sigma|^{(l-d)} \text{ too.}$$

The heuristic algorithm, MotifHeuristics, presented an approximation by choosing randomly a number (the program used 200 random motifs) of *wildcard (l,d)-motif* as seeds of the motifs on the left side. They compute the best partner motifs for each of them (those motifs from the same *(l,d)* class which is scored highest with the latter). The partner motifs would in turn be the second set of motif seeds. The experiments are performed on a specialized SH3 domain datasets with 233 interactions of 146 yeast proteins and the set of whole yeast interaction data from both MIPS and DIP.

They showed that both FindMotif and MotifHeuristics are significantly faster and more scalable than D-STAR; especially MotifHeuristics which is capable to handle whole yeast interaction data containing 21225 interactions over 5246 yeast proteins. They find the motif pairs (GxxPxNxV, PxLPxRxx) and (PPxPxRxx, GxxPxNYx) which both express SH3 class 2 binding motif (PxxPxR) with their respective SH3 binding regions. However, both FindMotif and MotifHeuristics failed to find any class 1 binding motifs both in the specialized SH3 datasets and the whole yeast interaction datasets. The reason might be on the hypergeometric scoring function's strong preference on large number of individual occurrences of the motif pairs. This preference prohibits the program to give a high score to SH3 class 1 motif pairs which occur less frequently in yeast interactions (as seen in Fig. 4(C)).

Structure Based Approaches

The main advantage of methods in this class is that one can create a much more accurate model of the interaction sites based on the physical interaction data which might involve finding structural/non-linear motifs. It is known that interactions sites on globular domains usually involves residues that are far apart in the protein sequence but brought together to form the interaction interface by the folding of the protein chain. However, the available structural data is still far less than the known proteins. Furthermore, the available structures are quite biased in the sense that some important proteins (e.g. disease-related ones) have significantly more structural snapshots than others. Another problem is that certain classes of proteins are significantly under-represented because of difficulties in resolving their structures, for example, the trans-membrane and the disordered proteins. That being said, the amount of available data nowadays (around 50000 of solved structures) is quite sufficient for data mining purposes to complement or validate the existing sequence based approaches.

Henschel et. al. (2006)

This paper presented a method to find protein domain binding motifs from the structural dataset in PDB. The authors employed profile Hidden Markov Model (HMM) (Eddy, 1998) to model the protein binding sites where both domain-domain interface and domain-motif interface are considered. In contrast to the contact segment defined by Li & Li (2005) which requires the motif to be consecutive in the sequence space, the algorithm is capable to model discontinuous segments forming the binding sites and combine them in one single HMM profile. As illustrated in Fig. 6(A), the method extracted the binding interface columns from the multiple sequence alignment retrieved from SCOPPI (Winter *et. al.*, 2006). SCOPPI is a database which classifies different interface type of

Figure 6(A) The flowchart of the methodology of extracting the interface descriptors. Using SCOPPI classification, they extracted the binding interface residues as indicated in the accompanying multiple sequence alignments of the two families. For each interface residue columns, a local HMM is built. Whenever an interface is described by more than one HMM, the models are merged by modeling the linker region using insert states. The pairs of HMMs, called the interface descriptors, describing both side of the interface is then searched against Swiss-Prot (B) Validation of the interface descriptor on the ATP binding motif within a set of sequence benchmark from Swiss-Prot and its comparison against Prosite's P-loop motifs . The coverage of both methods are comparable (in the range of 25%). However, the P-loop motif produced significantly more false positives. (C) The ATP motif from one side of the interface descriptor describing the ATP-binding motif (above) compared to the Prosite's P-loop motif (below) (The images are extracted from Henschel et. al. (2006)).

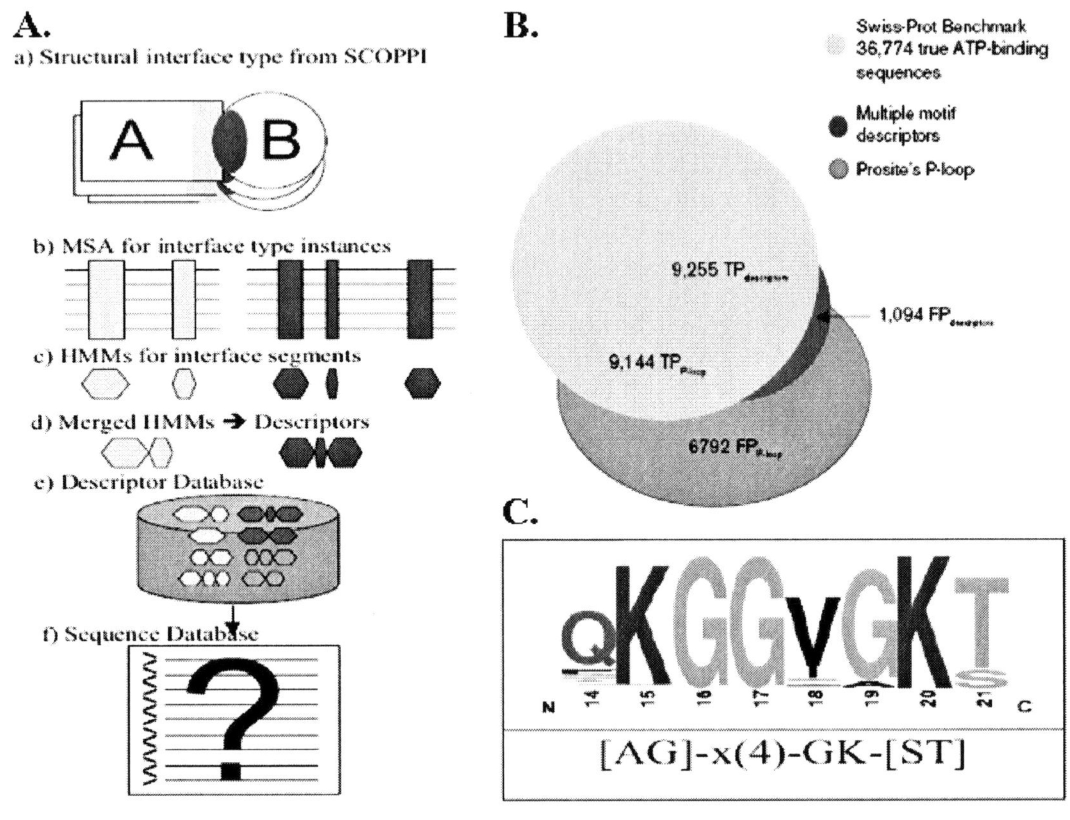

the interaction between SCOP protein families. Alignment columns with more than 50% of its residues observed to be interacting are marked as interface columns. Next, the alignment columns are extended and padded for gaps of length 1 to make the total column width significant enough. For each of such interface columns, a local HMM model is generated to capture its sequence composition. Whenever there are multiple HMMs describing one side of the interaction interface, they are combined into one HMM by adding insert states in between them. The probability for a self loop in the insert states between two linked HMM models is set to $l/(l+1)$ where l is the average length of the region between the two interface columns of the HMMs. The pairs of the combined HMMs describing both side of the interaction interface are called the *interface descriptors*. They are used search against sequence databases to find pairs of potentially interacting proteins using the HMMER packages (Eddy, 1998).

For domain-motif descriptors, the authors build only a descriptor for the domain which binds the ligand. Using a similar approach as SCOPPI, the program starts from the ligand and identify the domain residues that are directly in contact with the ligands, along with a set of conserved neighboring residues. Next, they extended the number of domains sequence in the alignment by including structurally similar domains as defined by HSSP (Sander & Schneider, 1991). With the interface columns from the alignment, they proceed as described above and created one (combined) descriptor for each ligand.

The main validation of the method is on the comparison between the interface descriptors with the well known Prosite P-loop motif to describe the ATP binding motifs. Using 36774 sequences from Swiss-Prot (Boeckmann *et. al.*, 2003) which are annotated as ATP-binding, both the interface descriptors and the P-loop motif achieved similar coverage of 25%. However, the P-loop motif produced five times more false positive hits as compared to the interface descriptor's (see Fig. 6(B)). A comparison between the motif extracted from the interface descriptor's HMM model and the P-loop motif is depicted in Fig. 6(C).

D-MIST - Betel et al. (2007)

The D-MIST algorithm is a specialized method to find domain-motif interaction binding motifs. It uses Gibbs sampling to come up with a Position Specific Scoring Matrix (PSSM) describing the amino acid preference profile of domain binding interfaces extracted from PDB structures. While the method is less general than Henschel *et. al.* (2006) which covered domain-domain and domain-motif interfaces, this paper presented its own wet-lab validations of novel protein interactions predicted by D-MIST.

The algorithm starts by extracting the domain binding motifs from the PDB structures entry based on the mapping in Salama, *et. al.*(2001) for each protein domain. Binding motifs are defined as polypeptide segments of five residues or longer whose the amino acids side chains are 5Å from the interacting domain's side chains on the opposing protein. Two motif residues that are in direct contact with the interacting domain (that is to say, within the 5 Å distance cutoff) could be separated by at most two consecutive non-contacting residues (denoted by the residue 'X' in the resulting motifs). The extracted motifs is then used as priors for Gibbs Sampling motif searching algorithm (Thompson, *et. al.*, 2003) to look for other possible instances of the binding motifs in a non-redundant set of 87894 interactions extracted from 204 species (See Fig. 7). Interactions from structural data are excluded from the latter to avoid self-identification while those from high throughput complex identification or interactions inferred from synthetic lethal studies are excluded as well because they are less reliable in determining direct interaction. The length of the binding motifs from the PDB structures is used to approximate the length of target motifs of the Gibbs sampler.

The resulting PSSM motifs with score $\geq$ 10.0 (as scored by the Gibbs sampler) are used to predict new protein interaction. Two proteins are predicted to interact if one of them contains the domain and the other matches one or more PSSM motifs of that domain. The matching of the PSSMs to a particular protein is defined with a minimum PSSM matching score of 0.2.

The algorithm predicted 18459 interactions between 2313 yeast proteins, 609 of which have supporting literature evidence. A subset of the interactions is further validated by Immunoaffinity Purification – Mass Spectrometry (IP-MS) experiments which confirmed 37 predicted interactions, 23 of which are novel. The authors also performed IP-western experiments reconfirming five interactions from the IP-MS experiments plus an additional of four novel interactions.

Figure 7. The workflow of D-MIST. (A) The binding motifs are extracted from the PDB structures based on the entries in BIND following (Salama et. al., 2001). (B) The motifs are used as priors of the Gibbs Sampling (Thompson et. al., 2003) and the binding motifs are searched against interaction data to generate the Position Specific Scoring Matrix (PSSM) describing the binding motifs. (The images are extracted from Betel et. al. (2007)).

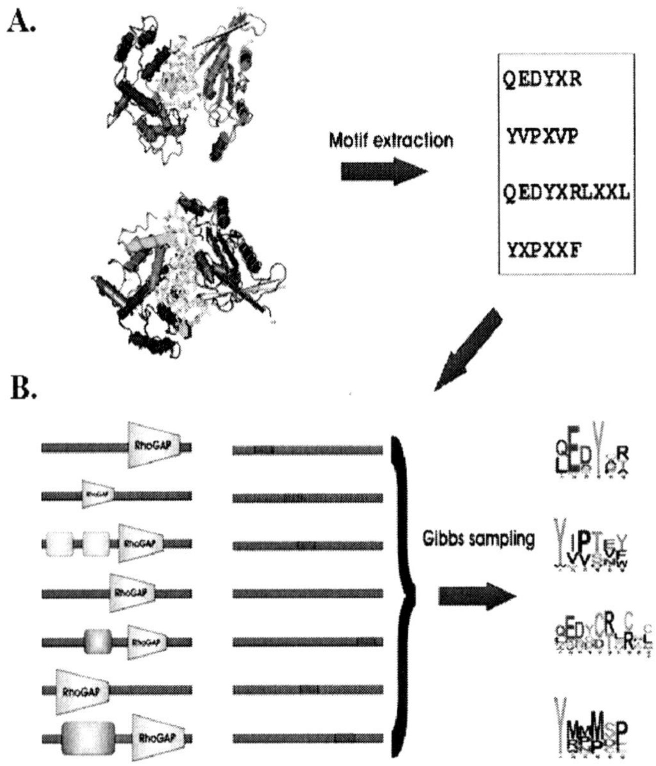

DISCUSSION AND FUTURE WORK

A problem that is prevalent in all of the methods presented is the lack of a standard negative interaction data which are, preferably, manually curated and of high confidence for the methods to train/validate on. For positive datasets, the common practice is to use interactions that are validated by more than one experimental technique. Another alternative is to use high quality datasets, for example the curated dataset from MIPS (Mewes *et. al.*, 2008). On the negative interaction counterpart, most methods either use random pairs not seen in the existing interaction data or pairs of proteins known to localize in different compartments of the cell. The fact that almost every work which considers negative data would construct their own negative data means that there is a certain ambiguity in terms of the comparison of the precision and recall reported among different methods. The Golden Standard Negative (GSN) data collected by Guo *et. al.* (2008) might provide a good starting point for initiating the effort of creating a benchmark negative. One note about the assumption of non-interaction by virtue of cellular localization is that those proteins could be non-interacting not because they do not have the interaction potential, but simply because they never see each other in the cell. This might be caused by some protein localization motif on the protein, or even on the mRNA template of the protein, following evidences that the localization is already done at the mRNA level (Lécuyer *et. al.*, 2007). This fact might present problem on the definition of non-interaction since there might be pairs of proteins seen to be interacting in artificial *in-vitro* experiments while in reality they are localized to different region of the cell.

The second issue is on the non-linearity of the interface. Many *de-novo* approaches assume that existence of a pair of linear motifs could be used as indicators of an interaction, except in Li *et. al.* (2006) where the interaction motif is defined over a (possibly non-contiguous) set of blocks. The fact is most of the known interactions are mediated by two non-linear binding interface like those observed between two globular domains (domain-domain interaction) or by a non-linear interface binding a linear peptide bearing a certain short motif (domain-motif interaction). A way to approximate the modeling of a non-linear interface should be in line with the set of short motif blocks in Li *et. al.*(2006) and the segmented HMMs used in Henschel, A., *et. al.*(2006). These models could be biologically meaningful since there is a recent study suggesting that protein domains are built from the concatenation of ligand binding polypeptide segments (Riechmann & Winter, 2006).

The segmented motif approach could also provide us with the potential ability to explain differences in binding preference among different instances of the same domain. One example of this case would be the binding of phosphotyrosine peptides to SH2 domains. While the binding mechanism of the SH2 to the phosphotyrosine residue is almost universal, it is known that different SH2 domains have different preference on the residues C-terminal of the phosphotyrosine on the peptides (and in some less frequent cases, the residues N-terminal of the phosphotyrosine as well). This difference can be traced to the difference on the residues forming the binding interface where, for example, more hydrophobic interface would prefer hydrophobic residues C-terminal to the phosphotyrosine (e.g the SH2 domain of the P85α, PLCγ, and SHP2); while the polar ones would prefer more polar peptides (e.g. the SH2 domains of Src, Hck, Crk) (Songyang *et. al.*, 1993; Huang *et. al.*, 2007). Such difference in binding preferences among different proteins containing the same domain is quite ubiquitous and thought to be involved in the fine tuning of the multitude of protein interactions regulating the living cell. Hence, methods which treat domains as a single entity with an underlying assumption of similar binding properties might not be able to accurately reflect the true binding mechanisms.

The last observation is that the interaction interfaces seems more conserved in terms of physico-chemical properties rather than amino acid residues. As can be seen in the results of the structure based approaches, the methods' effectiveness is dependent on allowing some mismatch on the extracted binding motifs either by local alignment (Li & Li, 2005) or by using a probabilistic approach (Henschel *et. al.*, 2006; Betel *et. al.*, 2007). This means that we should design the motif models to approximate such flexibility. A simple mismatch model, where at the mismatch position, any residues can be replaced by any other residues, might not be good enough. The reason is, unlike nucleotides which has only four types and a well defined pairing preferences (with some exceptions), the amino acids consist of twenty different residues with highly varied physico-chemical properties and the binding of preference between any pair is not well-defined. One possible approach is to collect the residue's physico-chemical properties and mine motifs of the amino acid groups instead (like the algorithm by Li *et. al.* (2004)). Nevertheless, we believe that the increasing amount of structural data in PDB, albeit still far from complete, would provide us with a better chance to come up with a better and more realistic motif model and improve the accuracy of interaction motif mining and computational protein interaction prediction in general.

CONCLUSION

Considerable work had been done on finding motifs which are responsible of mediating the protein interactions, which are collectively called the interaction motifs. Existing approaches can be divided in general into structural and sequence based approach. Structural approaches yield more specific results with a higher expected accuracy but suffer from limitation on the coverage of the known physical interactions in the structural databases. Sequence based approaches, on the other hand, have much more interaction data to mine from but need to deal with a considerable amount of experimental noise.

One subclass of the sequence based approach uses the existing known domains and motifs to infer the correlations between such entities within the available protein interaction data. By using existing validated domains and motifs, these methods offers a relatively higher precision. However, the coverage of the interaction data explainable by the existing domains and motif definitions might be limited. The *de-novo* motif approaches offer, potentially, more complete motif coverage but suffer from the huge complexity of the motif space. Furthermore, the approaches in this line should be able to correctly model non-linear binding interface, perhaps by using a set of short linear motifs since most interactions involves at least one non-linear interaction interface (in the case of domain-motif interaction). We also pointed out the importance of a standardized negative interaction data as a benchmark for the research community in this field to effectively construct reliable interaction models and objectively compare their effectiveness.

REFERENCES

Ashburner, M. et. al. (2000). Gene ontology: tool for the unification of biology. The Gene Ontology Consortium. *Nature Genetics, 25*(1), 25-29.

Balla, S., et. al. (2006). Minimotif Miner: a tool for investigating protein function. *Nature Methods, 3*(3), 175-177.

Berman, H. M., et. al. (2000). The Protein Data Bank, *Nucleic Acids Research, 28*(1), 235-242.

Betel, D., et. al. (2007). Structure-templated predictions of novel protein interactions from sequence information. *PloS Computational Biology, 3*(9), 1783-1789.

Boeckmann, B., et. al. (2003). The SWISS-PROT protein knowledgebase and its supplement TrEMBL in 2003. *Nucleic Acids Research, Database issue 31*, D365-D370.

Buhler, J., & Tompa, M. (2002). Finding motifs using random projections. *Journal of Computational Biology, 9*(2), 225-242.

Ceol, A., et. al. (2007). DOMINO: A database of domain peptide interactions. *Nucleic Acids Research, Database issue 35,* D557-D560.

Dempster, A., Laird, N., & Rubin, D. (1977). Maximum Likelihood from Incomplete Data via the EM Algorithm. *Journal of the Royal Statistical Society, 39*(B), 1-38.

Doray, B., & Kornfiel, S. (2001). Gamma subunit of the AP-1 adaptor complex binds clathrin: implications for cooperative binding in coated vesicle assembly. *Molecular Biology of the Cell, 12*(7), 1925-1935.

Eddy, S. R. (1998). Profile Hidden Markov Models. *Bioinformatics, 14*, 755-763.

Finn, R. D., et. al. (2005). iPfam: visualization of protein-protein interactions in PDB at domain and amino acid resolutions. *Bioinformatics, 21*(3), 410-412.

Finn, R. D., et. al. (2006). Pfam: clans, web tools and services. *Nucleic Acids Research, Database issue 34*, D247-D251.

Gavin, A.L., et. al. (2006). Proteome survey reveals modularity of the yeast cell machinery. *Nature, 440*, 631-636.

Grahne, G., & Zhu, J. (2003). Efficiently using prefix-trees in mining frequent itemsets. *Proceedings of the Workshop on Frequent Itemset Mining Implementations 2003.*

Greenman, C. *et. al.* (2007). Patterns of somatic mutation in human cancer genome. *Nature, 446*, 153-158.

Guo, J., et. al. (2008). Genome-wide inference of protein interaction sites: lessons from the yeast high-quality negative protein-protein interaction dataset. *Nucleic Acids Research, 36*(6), 2002-2011.

Hamosh, A., et. al. (2005). Online Mendelian Inheritance in Man (OMIM), a knowledge base of human genes and genetic disorders. *Nucleic Acids Research. Database issue 33*, D514-517.

Henschel, A., et. al. (2006). Using structural motif descriptors for sequence-based binding site prediction. *BMC Bioinformatics, 8*(Suppl 4), S5.

Hofmann, T. (2001). Unsupervised learning by probabilistic latent semantic analysis. *Machine Learning, 42*, 177-196.

Huang, H., et. al. (2007). Defining the specificity space of the human Src-homology 2 domain. *Molecular & Cellular Proteomics, 7*(4), 768-784.

Hulo, N., et. al. (2008). *The 20 years of PROSITE. Nucleic Acids Research, Database issue 36*, D245-D249.

Jones, S., & Thornton, J. M., (1996). Principles of protein-protein interactions. *Proceedings of the National Academy of Sciences U S A., 93*(1), 13-20.

Krogan, N.J., et. al. (2006). Global landscape of protein complexes in the yeast Saccharomyces cerevisiae. *Nature, 440*, 637-643.

Leung, H.C., et. al. (2007). Finding linear motif pairs from protein interaction networks: a probabilistic approach. *Computational Systems Bioinformatics Conference, 6*, 111-119.

Li, H. et. al. (2004). Discovery of binding motif pairs from protein complex structural data and protein interaction sequence data. *Pacific Symposium on Biocomputing*, (pp. 312-323).

Li, H. et. al. (2006). Discovering motif pairs at interaction sites from protein sequences on a proteome-wide scale. *Bioinformatics, 22*(8), 989-996.

Li, H., & Li, J. (2005). Discovery of stable and significant binding motif pairs from PDB complexes and protein interaction datasets. *Bioinformatics, 21*(3), 314-324.

Lécuyer, E. et. al. (2007). Global analysis of mRNA localization reveals a prominent role in organizing cellular architecture and function. *Cell, 131*(1), 174-187.

Mamitsuka, H. (2005). Essential latent knowledge for protein interactions: Analysis by an unsupervised learning approach. *IEEE/ACM Transactions on Computational Biology and Bioinformatics, 2*(2), 119-130.

Mewes, H. W., et. al. (2008). MIPS: analysis and annotation of genome information in 2007. *Nucleic Acids Research, Database issue 36*, D196-D201.

Mulder, N. J., et. al. (2007). New developments in the InterPro database. *Nucleic Acids Research, Database issue 35*, D224-D228.

Neduva, V., & Russell, R. B., (2006). Peptides mediating interaction networks: new leads at last. *Current Opinion in Biotechnology, 17*(5), 465-471.

Ng, S. K., et. Al. (2003). Integrative approach for computationally inferring protein domain interactions. *Bioinformatics, 19*, 923-929.

Nicolas, P. et. al. (1999). Discovering frequent closed itemsets for association rules. *Proceedings of the 7th International Conference on Database Theory*, Israel, (pp. 398–416).

Ofran, Y., & Rost, B. (2003). Analysing six types of protein-protein interfaces. *Journal of Molecular Biology, 325*(2), 377-387.

Pawson, T., & Scott, J. D. (1997). Signalling through scaffold, anchoring, and adaptor proteins. *Science, 278*, 2075-2080.

Pevzner, P. A., & Sze, S-.H. (2000). Combinatorial Approaches to Finding Subtle Signals in DNA Sequences. *Proceedings of the International Conference on Intelligent Systems for Molecular Biology, 8,* 269-278.

Pitre, S., et. al. (2006). PIPE: a protein-protein interaction prediction engine based on the re-occurring short polypeptide sequences between known interacting protein pairs. *BMC Bioinformatics, 7,* 365.

Puntervoll, P., et. al. (2003). ELM server: a new resource for investigating short functional sites in modular eukaryotic proteins. *Nucleic Acids Research, 31,* 3625-3630.

Riechmann, L., & Winter, G. (2006). Early protein evolution: Building domains from ligand binding polypeptide segments. *Journal of Molecular Biology, 363,* 460-468.

Salama, J. J., et. al. (2001). Automatic annotation of BIND molecular interactions from three-dimensional structures. *Biopolymers, 61,* 111-120.

Salwinski, *et. al.* (2004). The database of interacting proteins: 2004 update. *Nucleic Acids Research, Database issue 32,* D449-D451.

Sander, C., & Schneider, R., (1991). Database of homology-derived protein structures and the structural meaning of sequence alignment. *Proteins, 9,* 56-68.

Skrabanek, L., et. al. (2008). Computational prediction of protein-protein interactions. *Molecular Biotechnology, 38*(1), 1-17.

Songyang, Z., et. al. (1993). SH2 domains recognize specific phosphopeptide sequences. *Cell, 72*(5), 767-778.

Sprinzak, E., & Margalit, H. (2001). Correlated sequence-signatures as markers of protein-protein interaction. *Journal of Molecular Biology, 311*(4), 681-692.

Sprinzak, E, et. al. (2003). How reliable are experimental protein-protein interaction data? *Journal of Molecular Biology, 327*(5), 919-923.

Sudol, M. (1998). From Src Homology domains to other signaling modules: Proposal of the 'protein recognition code'. *Oncogene, 17,* 1469-1474.

Tan, S.-H., et. al. (2004). Discovering Novel Interacting Motif Pairs from Large Protein-Protein Interaction Datasets. *4th IEEE International Symposium on BioInformatics and BioEngineering (BIBE), (pp.* 568-575).

Tan, S.-H., et. al. (2006). A correlated motif approach for finding short linear motifs from protein interaction networks. *BMC Bioinformatics, 7,* 502.

Thompson, W. et. al. (2003). Gibbs recursive sampler: Finding transcription factor binding sites. *Nucleic Acids Research, 31,* 3580-3585.

Valencia, A., & Pazos, F., (2002). Computational methods for the prediction of protein interactions. *Current Opinion in Structural Biology, 12,* 368-373.

Wang, H., et. al. (2007). InSite: a computational method for identifying protein-protein interaction binding sites on a proteome-wide scale. *Genome Biology, 7,* R192.

Winter, C., et. al. (2006). SCOPPI: A structural classification of protein protein interfaces. *Nucleic Acids Research, Database issue 34,* D310-D314.

Wu, T. D. et. al. (1996). Discovering Empirically Conserved Amino Acid Substitution Groups in Databases of Protein Families. *Proceedings of the International Conference on Intelligent Systems for Molecular Biology, 4,* 230-240.

Wu, X., & Brutlag, D. L. (2006). Prediction of yeast protein-protein interaction network: insights from the gene ontology and annotations. *Nucleic Acids Research, 34,* 2137-2150.

Chapter VIII
Discovering Network Motifs in Protein Interaction Networks

Raymond Wan
Kyoto University, Japan

Hiroshi Mamitsuka
Kyoto University, Japan

ABSTRACT

This chapter examines some of the available techniques for analyzing a protein interaction network (PIN) when depicted as an undirected graph. Within this graph, algorithms have been developed which identify "notable" smaller building blocks called network motifs. The authors examine these algorithms by dividing them into two broad categories based on two definitions of "notable": (a) statistically-based methods and (b) frequency-based methods. They describe how these two classes of algorithms differ not only in terms of efficiency, but also in terms of the type of results that they report. Some publicly-available programs are demonstrated as part of their comparison. While most of the techniques are generic and were originally proposed for other types of networks, the focus of this chapter is on the application of these methods and software tools to PINs.

INTRODUCTION

Data from high-throughput experimental methods are currently being used to construct complex biological networks which include transcription regulatory networks, protein interaction networks (PINs), and metabolic networks. As more and more of these networks become available, an increasing level of attention is being paid to characterize these large, complex structures.

One technique originally proposed by Milo et al. (2002) is to break down these networks into smaller components called *network motifs*. These network motifs can be seen as the networks' basic building blocks. Determining the number and structure of these network motifs allows researchers to provide a general overview of the network and even assign each to a so-called "superfamily" (Milo, Itzkovitz, Kashtan, Levitt, Shen-Orr, et al., 2004). While the initial work with biological network motifs focused on transcription regulatory networks, Milo, Itzkovitz, Kashtan, Levitt, Shen-Orr, et al. (2004) and others since have extended this work to protein interaction networks.

The focus of this chapter is to examine the algorithms and software tools that are available to users for network motif discovery. Since Milo et al. (2002) highlighted the importance of network motifs, various researchers have sought various methods of finding them. The underlying theme in this chapter is to examine two definitions of "network motifs" that stem from their seminal work. Our aim is to compare and contrast the methods within the context of their corresponding algorithms and software systems. A recent review paper on network motifs for PINs provides a broader range of methods (Ciriello & Guerra, 2008); this chapter focusses on several selected algorithm within the context of this theme. Readers are encouraged to consult this review to gain another perspective on the motif discovery problem.

MOTIVATION

The analysis of protein interaction networks (hereafter abbreviated as PIN) is based on the work on many types of networks, including other biological networks and even the structure between web pages on the World Wide Web. Because of this, we explore the motivation for analyzing a PIN by occasionally digressing to other types of networks.

A PIN can be represented as an undirected graph G such that each node represents a protein and unweighted edges are placed between two nodes if an interaction exists between them. By depicting proteins in this manner, we can examine the patterns of interacting proteins.

Starting at the node-level, one such analysis examines the number of immediate neighbors of each node, which is defined as the *degree*. Using tools such as the power law, it has been shown that PINs and other biological networks are scale-free based on the distribution of the node degrees (Barabási & Albert, 1999). That is, many nodes have few edges and few nodes are *hub nodes* and have many edges. In the context of PINs, this means that a random loss of a protein would have a small effect on the overall biological system. However, a focussed attack on a specific protein could have severe consequences if it was a hub node.

Motifs represent the next level in biological network analysis. According to one definition, they are small subgraphs which are repeated and conserved and used to form the building blocks of larger modules (Wolf & Arkin, 2003). However, as suggested by Wolf and Arkin, the separation between motifs and modules is unclear and remains a topic for future work. In their work and the subsequent review by Alon (2007), functionality was the focus of the definition of a motif, and in both cases, the discussion was centered on transcription regulation networks.

An alternate view of motifs is to focus on the topology of the network by giving priority to the configuration of each motif. If motifs are found through their structure, they can then be used to describe the structure and functionality of the PIN. Wuchty, Oltvai, and Barabási (2003) showed that certain topological motifs are important since they are conserved through evolution. To arrive at this conclusion, protein interaction data of yeast (*S. cerevisiae*) was shown to have certain motifs of two to five nodes which also existed in 5 higher eukaryotes. In a more recent study, the evolution of the yeast PIN was modeled from a "preduplication network" by inferring the duplication and divergence of genes using motifs from a PIN (Presser, Elowitz, Kellis, & Kishony, 2008).

Thus, once motifs of a PIN have been found, additional insights can be obtained from the data. In this chapter, our focus is on discovering such motifs using the topology of the PIN. Following the work of Milo et al. (2002), these motifs are subgraphs which occur with frequencies that are higher than one would expect by chance. This definition resembles the usage of "motifs" in other areas of theoretical biology. DNA sequence motifs are short DNA sequences which occur more frequently than some background distribution of nucleotide frequencies. For the networks examined by Milo et al. (2002), the motifs were dubbed *network motifs* to differentiate them from the more general usage of "motifs" above. An alternative definition of network motifs is based on the frequency of the subgraph. We explore both definitions in this chapter. As a matter of convenience, since we concentrate on this topological definition for PINs for the remainder of this chapter, we use "motifs" and "network motifs" interchangeably.

BACKGROUND

We begin with some terminology and notation from graph theory before we formally provide the two definitions of network motif.

Definitions

We denote the undirected graph representation of a PIN as G(V, E) where V is the set of vertices (or nodes) and E is the set of edges between interacting proteins. An edge $e_{ij} \in E$ between vertices v_i and $v_j \in V$ is said to be *incident* on those vertices, or those vertices are incident to that edge. The *degree* of a node indicates the number of edges connected to it. In this chapter, we exclude the possibility of *self-loops*, or edges whose both ends are incident on the same node.

A k-subgraph s of G(V, E) is a subset of k nodes from V such that for any two vertices v_i and v_j in s, if an edge is incident between them in G, then they are also incident between v_i and v_j in s. The term "subgraphs" does not indicate the size of the graph, but merely that it is extracted from a much larger graph G. A special type of k-subgraph is a clique where every node is connected to every other node in the subgraph, excluding self-loops.

Other terms have also come into use in the network motif literature. A k-subgraph is also a *graphlet* (Pržulj, Corneil, & Jurisica, 2006). Due to the interest in small subgraphs, 3-subgraphs and 4-subgraphs are also called triads and tetrads, respectively (Milo, Itzkovitz, Kashtan, Levitt, Shen-Orr, et al., 2004).

Graph Representation

Software implementations of graph-based algorithms represent graphs in one of two ways: as an adjacency list or as an adjacency matrix. The choice between the two depends on the trade-offs between time and memory space. Generally, a graph with few edges are represented using an adjacency list, while a dense graph is coded using an adjacency matrix. A thorough comparison of these two representations can be found in any textbook on graph theory (for example, see Cormen, Leiserson, Rivest, and Stein (2001) or Sedgewick (2002)).

While PINs are generally sparse (see the last column in Table 5), we focus on adjacency matrices since they are also relevant to some implementations of algorithms for finding network motifs. An adjacency matrix A for an undirected graph G(V, E) is a matrix of size $|V| \times |V|$. If an edge exists between vi and vj, then matrix position Aij has a 1, otherwise a 0. Since these graphs are undirected, the matrix is symmetric across the diagonal and $A_{ji} = A_{ij}$.

Figure 1. Comparison of identifiers for a connected graph of three nodes and two edges which differ in their assigned node labels

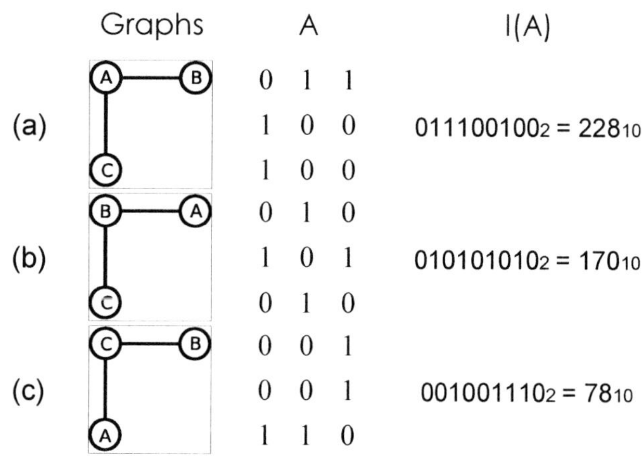

Figure 2. All possible configurations of undirected graphs of size k =3 and k =4

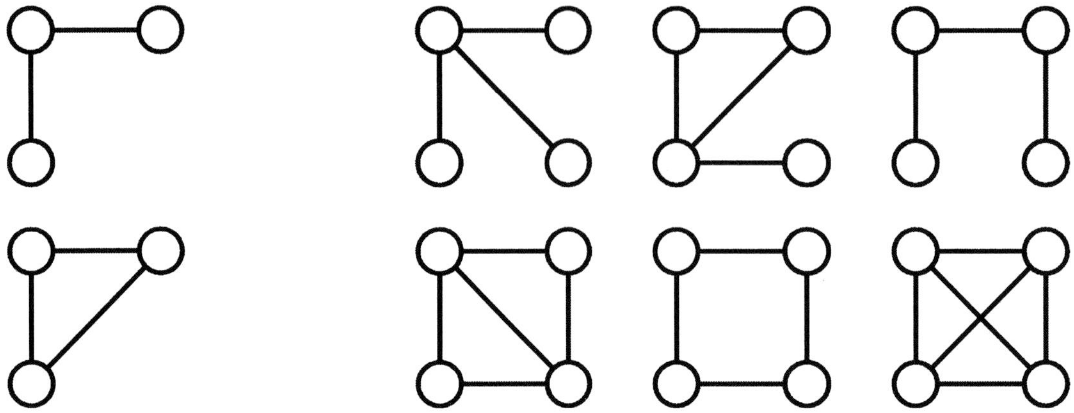

(a) k =3 (b) k =4

If every row of this adjacency matrix is concatenated, a bit string of length $|V|^2$ acts as a unique identifier I(A). We illustrate this with an example in Figure 1. The example consists of the same connected graph of 3 nodes and two edges, but with the nodes relabeled so that the node with degree 2 changes each time. Next to each graph is the adjacency matrix representation with the rows and columns always assigned to A, B, and C, in that order. Also shown is the result from concatenating the rows of the matrix to form a binary bit string and the decimal equivalent (subscripts indicate the base, which we drop from now on since we are interested in the decimal equivalent).

Since the matrix is symmetric, the bit string can also be derived from either the upper or lower triangular matrices. This discussion is of course equally relevant to subgraphs, and in fact, the decimal representation is how some programs report the shape of a subgraph to users.

Graph Isomorphism

The three graphs in Figure 1 are isomorphic to each other. Two graphs s and t are *isomorphic* to each other if, given a mapping function f and for every e_{ij} in s, $e_{f(i)f(j)}$ is also in t, and vice versa. Alternatively, if every vertex in s can be mapped to a vertex in t such that the two graphs overlap with each other, then these two graphs are isomorphic to each other. We consider two graphs which have the same number of vertices but which are not isomorphic to each other to be of different *configurations* (others use the term *classes*). For example, undirected subgraphs of k =3 and k =4 nodes have 2 and 6 different configurations, respectively, as shown in Figure 2.

Isomorphism is important to network motif discovery because after a subgraph has been identified, it needs to be categorized into the proper configuration. The set of all subgraphs with the same configuration c and its size (number of subgraphs with that configuration) are denoted as S_c and $|S_c|$, respectively, where c is a subscript starting from c =1. So, if the graphs of Figure 2(b) each indicated a configuration, then they would be labeled S_1 to S_6. Note that subgraph configurations have no pre-defined order and the purpose of different subscripts is only to indicate that two subgraphs from two configurations are not isomorphic to each other. The bit string identifier may be used to impose an order, but this is purely for convenience.

In general, the problem of determining if two graphs are isomorphic to each other is in the class NP (Fortin, 1996). The more specific problem of graph isomorphism of trees is in the class P and can be solved in linear time (Aho, Hopcroft, & Ullman, 1974, pg. 84–86), a trait that is used by one of the algorithms which we discuss later. Fortin (1996) summarizes the problem and gives an overview of some of the solutions available, including a backtracking algorithm and the Nauty algorithm of McKay (1981). Invariant tests that can be used to quickly rule out whether or not two graphs are isomorphic to each other are also often used. One common example is a comparison of the degrees of the vertices (for example, see Gross and Yellen (2005) for this and other invariant tests).

Another idea is to use the *automorphisms* of s coupled with a *canonical labeling* scheme. The automorphism of a subgraph is essentially a re-labeling of s, just as we have done in Figure 1. To determine the canonical labeling, the identifier of the adjacency matrix (I(A)) of every automorphism of the subgraph is calculated. The canonical label of s is the bit string with the smallest decimal equivalent. As illustrated by the example of Figure 1, a connected subgraph of 3 nodes and 2 edges would have the canonical label of 78. We should emphasize that this technique alone does not "solve" the graph isomorphism problem since every automorphism still has to be generated (Fortin, 1996).

In our case, subgraphs are "small" in comparison to the entire PIN. However, graph isomorphism needs to be tested many times since the subgraphs need to be categorized into their corresponding configurations. Because of this, complex implementations to solve the graph isomorphism problem may be unnecessary. Moreover, canonical labels alone may enable faster grouping of subgraphs. Nevertheless, since several programming languages already provide graph libraries that can be used with little additional effort, it may be advantageous to use them. Two example libraries with graph isomorphism tests include the Graph module for Perl[1] and the Boost graph library for C++ (G.Siek, Lee, & Lumsdaine, 2002).

Network Motifs

Network motifs are *connected* subgraphs which satisfy one of the following two definitions:

Definition 1 *A network motif is a subgraph whose frequency in a network is statistically higher than what one would expect in a set of* N *random graphs (Statistically-based).*

Definition 2 *A network motif is a subgraph whose frequency in a network is higher than some frequency threshold* ft *(Frequency-based).*

We denote these two definitions as statistically-based and frequency-based, respectively. The underlying theme of both definitions is that a network motif is a subgraph of a much larger network whose structure occurs "often". That is, in both cases, the topology of the network is important and not the type of protein each node represents. The difference between the two views lies in what "often" means.

The first definition of network motifs was first proposed for biological networks by Milo et al. (2002). In that seminal paper, Milo et al. defined network motifs as "recurring, significant patterns of interconnections". A subsequent paper by others extended this definition by suggesting that network motifs have "a small number of nodes (typically three to five)" (Artzy-Randrup, Fleishman, Ben-Tal, & Stone, 2004).

In order to determine whether a subgraph is a network motif under Definition 1, several steps are required. First, for a given subgraph of size k, the frequency of all k-subgraphs is obtained from the PIN. This counting needs to take into account isomorphic subgraphs such that the subgraphs are sorted into their respective configurations. These two steps are then repeated for each of a set of N randomly generated networks (where Milo et al. (2002), for example, had chosen N =1,000). Finally, for each subgraph configuration, the number of appearances in the PIN is compared against the randomly generated networks. Subgraphs which occur above some threshold are *overrepresented subgraphs* and are networks motifs. Those that do not occur often enough are *underrepresented subgraphs*.

Later, researchers from computer science, proposed Definition 2 as an alternative. Definition 2 stems from the field of *frequent subgraph mining*. As the name implies, a frequency threshold parameter f_t is required to define what "frequent" means. The use of an explicit threshold means that applying the method to a set of N random networks is now optional, since the motifs of interest are frequent, instead of overrepresented.

These two definitions of network motifs are summarized in Figure 3 by illustrating the steps required for each definition. Note that this description is merely a guide and variations on these definitions abound. In the next two sections, we focus on some of the earlier algorithms for motif detection, starting with statistically and then frequency-based ones. A comparison of the two definitions is deferred until Table 9 at the final section of this chapter.

Figure 3. Steps in network motif detection using the two definitions for network motifs: statistically and frequency-based

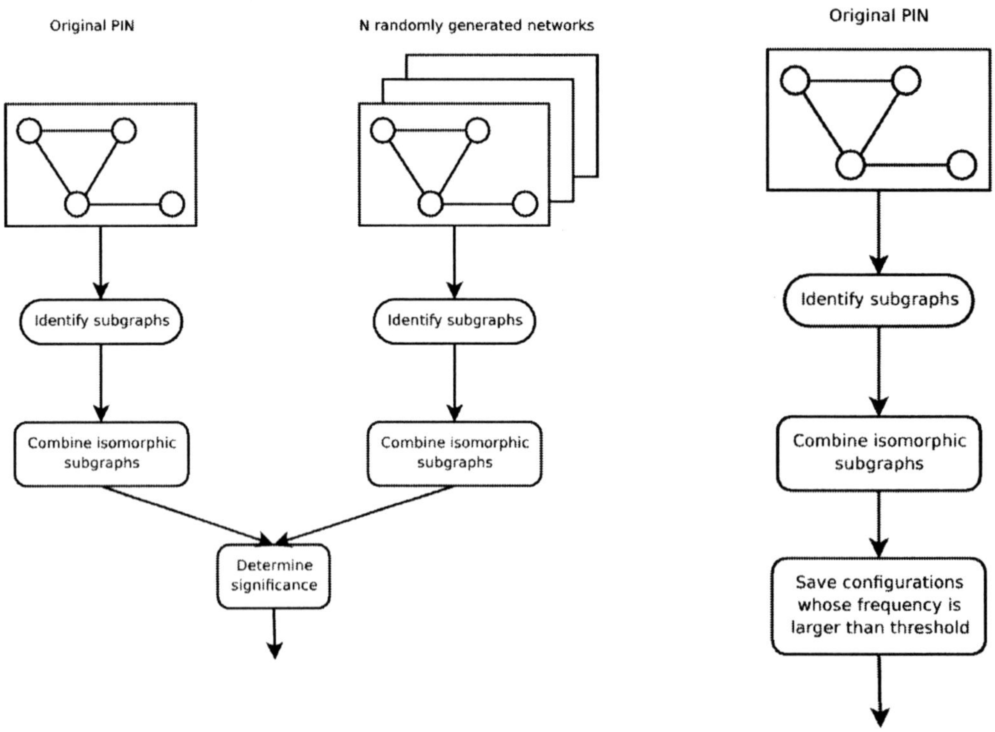

(a) Statistically-based *(b) Frequency-based*

STATISTICALLY-BASED MOTIF DISCOVERY

Statistically-based network motif detection usually relies on some form of the random null hypothesis and consists primarily of two steps:

1. Identify and count all subgraphs of k nodes in the PIN, taking into account isomorphic subgraphs.
2. Compare the frequency of each subgraph configuration against a set of N random networks by repeating the above step for *each* random network.

While others have considered subgraph isomorphism to be a separate third step to emphasize its importance (Wernicke, 2006), we combined it with the first step as the problem has already been discussed earlier.

Counting k-Subgraphs

Algorithms that count the number of subgraphs in a network G are provided with a single parameter k indicating the size of the subgraphs to search for. This is a common trait for the algorithms which we examine in this chapter and is needed to limit the search space, which grows significantly as k increases. This is in contrast to other problems in motif discovery, such as with DNA sequences, where the length of the DNA sequence motif of interest is not always known beforehand.

Table 1 lists algorithms that we examine in this section. As the table shows, there are two main characteristics of statistically-based methods: Exact and Parallelization.

Exact Subgraph Counting

As the name implies, exact subgraph counting gives the exact number of k-subgraphs for each of the possible subgraph configurations. A logical starting point is to simply find all of them in a brute-force, exhaustive manner. This is the idea behind the work on network motifs by Milo et al. (2002). Algorithmic complexity was a lower priority and, in fact, the aim of their work was the application of their technique on a wide range of networks such as transcription regulatory networks and electronic circuits. This method, which we call EXHAUSTIVE, was computationally intensive and the maximum motif size that could be handled was $k = 4$. (Their method and all of the methods described in this section were applied to *directed* graphs. One would expect that the computational requirements for *undirected* graphs should be less due to the number of isomorphic subgraphs. We return to this topic at the end of this chapter). As expected, the downside of this approach is that computation time increases with the size of the PIN as well as the number of hub nodes.

An improved version of EXHAUSTIVE is ESU (for Enumerate Subgraphs) which builds an explicit search tree T of height k to avoid duplicates (Wernicke, 2006). At first, integral labels are assigned to every node in G. The neighboring nodes of each node are explored by the algorithm up to a distance k, but under the constraint that the explored nodes must be numerically larger. Thus, any subgraph can only be found when the search starts from the node with the smallest label. A minor consequence of this constraint is that T is imbalanced. Searches starting from the node with the smallest label will record every subgraph found; searches starting from the highest label will not report any subgraph at all. The advantage of this method over EXHAUSTIVE is that every possible subgraph is discovered once and each one will appear as a leaf in the search tree. The construction of a search tree by ESU also allows it to better estimate its search progress.

Subgraph Counting Based on Sampling

Randomly sampling from the PIN is an obvious alternative to the previous exact methods. Sampling from a large graph to obtain a smaller, representative version has been explored by others (Leskovec & Faloutsos, 2006). Motif finding applies sampling in a different way since a random sample of subgraphs of a fixed size k is taken to obtain approximate counts for every possible k-subgraph configuration. The accuracy of the sampling improves with the size of the sample, at the cost of increased execution time. Thus, in addition to specifying the subgraph size k, a sampling probability p needs to be chosen.

This possibility was first tackled by an algorithm which we refer simply as SAMPLING (Kashtan et al., 2004). The sampling technique is conceptually easy. First, the algorithm selects an edge at random from G. Then, from the set of neighboring edges, a subsequent edge is chosen. The next step is to select a third edge that is adjacent to either of the two previously selected edges. Thus, this process is repeated until a subgraph of k nodes is formed. This is of course not the same as selecting k edges due to cycles in G. Each subgraph is sampled *independently*. However, the probability of selecting a subgraph depends on whether the section of the graph being sampled is sparsely or densely filled with edges. In order to rectify this problem, the count of each sampled subgraph is corrected by the probability with which the subgraph could have been found. While necessary, this step adds additional computational time that further complicates what would have been a simple method. Nevertheless,

Table 1. List of some statistically-based network motif detection algorithms

Algorithm	Exact	Parallel	Citation
EXHAUSTIVE	Yes	No	Milo et al. (2002)
SAMPLING	No	No	Kashtan, Itzkovitz, Milo, and Alon (2004)
ESU	Yes	No	Wernicke (2006)
RANDESU	No	No	Wernicke (2006)
PARALLELMOTIF	Yes	Yes	Wang, Touchman, Zhang, Suh, and Xue (2005)

their experiments demonstrate that in a directed network of hyperlinks on a subset of the Web with 287 million 3-subgraphs, sampling as little as 5,000 is enough to give reasonable approximations.

The developers of ESU extended their method to RANDESU (for Random Enumerate Subgraphs) as a way of fixing the problems of SAMPLING (Wernicke, 2006). Wernicke lists some of the problems with SAMPLING, including the computation time for calculating the correction probability and its choice of allowing sampling with replacement. A problem related to this last point is that since there is no way of keeping track of what has been sampled, how much of the network has been sampled, and when we have sampled more than what the network has to offer.

The search tree T of ESU has k levels. Instead of requiring the user to provide a single probability, a set of k probabilities p_1 to p_k is needed (which of course can all have the same value). At each node during the tree traversal, the neighboring nodes which have not yet been visited are examined with a probability determined by the level in the search tree. As a result, instead of randomly sampling the PIN, RANDESU randomly samples the search tree at each branching point.

Subgraph Counting Based on Parallelization

Based on the discussion so far and how desktop computers can now be easily arranged into clusters, a natural next step is to parallelize the algorithms above across multiple CPUs.

Wang et al. (2005) devised an algorithm that performs exact counting using parallelization which we call PARALLELMOTIF. As with other parallelization algorithms, there is a single master CPU and multiple slaves. Bookkeeping is done by the master, including a unique sorting and labeling of the nodes (similar to what ESU does). PARALLELMOTIF places emphasis in labeling nodes according to the sorted degrees of the nodes. This allows it to better distribute the workload to each slave CPU.

The network is partitioned into neighborhoods. While each CPU has the entire network in-memory, a slave CPU only explores the nodes that have been assigned to it by the master. Since the assignment is done using the node degrees, balancing the load across the slave CPUs can be achieved in a round-robin fashion. A CPU which first obtains a node with a small degree, will next obtain a node with a large degree. PARALLELMOTIF also eliminates subgraph redundancy using the same method as ESU– by comparing the numerical labels on the nodes. However, subgraph isomorphism must be handled by the master CPU as part of its bookkeeping. Of course, each slave CPU can perform part of this work so that a shorter list of subgraphs is returned to the master.

They applied their method on a cluster of 32 machines, each with two 2.4 GHz processors. The additional cost of communicating among the 32 processors seems to pay off. Their results indicate that their method was faster than EXHAUSTIVE and more accurate than SAMPLING since it finds all motifs instead of relying on a sample. One of the most challenging tasks in parallelization is to properly distribute each part of the problem to each slave CPU to minimize the total amount of idle CPU time. While the amount of idle time was not explicitly reported by Wang et al. (2005), this challenge was the focus of their method.

Random Networks

The second part of statistically-based methods is the construction of a set of N random graphs to compare against the number of each subgraph configuration from G. The aim is to determine whether the counts |Sc | are large enough that it could not have happened by chance. Several problems are evident with this approach:

1. What definition of "random graph" is suitable?
2. How should randomization be done and at which point do we know we have done enough?
3. How many random graphs should be generated?

A proper definition of "random graph" requires detail which is beyond the scope of this chapter (for example, see Bollobás (2001) for a more complete coverage). The problem lies in not just defining what "happening by chance" means but also keeping the definition relevant to PINs.

Returning to the original definition of network motifs proposed by Milo et al. (2002), they suggested several criteria for random graph generation. Obviously, the random graphs should have the same number of nodes as the original graph $G(V, E)$. This implies that the original network should be used as the starting point. More importantly, the degrees of the nodes should remain the same, guaranteeing that the number of edges in the random graphs is equal to $|E|$. Milo et al. (2002) specified an additional constraint that ensures that a search for subgraphs of size k has random graphs whose structure ensures that all counts of subgraphs of size $k - 1$ are retained. In subsequent work by the same research group (Kashtan et al., 2004), this latter restriction has been removed and it would appear that other researchers have since followed suit.

In the end, they devised two methods for random graph generation. One method is based on a Markov chain which guides the swapping of edges between two pairs of nodes; the other method removes all of the edges in the graph but keeps track of the number of the degree of each node. Nodes are matched up in a random fashion. (Since the algorithm was discussed in the context of a directed graph, the method also has to keep track of the incoming and outgoing edges of each node.) In other work, these methods are called the "switching algorithm" and the "matching algorithm", respectively (Milo, Kashtan, Itzkovitz, Newman, & Alon, 2003). A third method based on the Monte Carlo procedure was also proposed which they called "Go with the Winners". It was found to be more time consuming and yielded little difference to the other two methods.

As for the number of random graphs to build, past works have chosen values ranging from $N = 100$ (Wang et al., 2005) to $N = 1{,}000$ (Milo et al., 2002). Selecting a value for N represents the trade-off between execution time and statistical accuracy. Experiments later in this chapter show that even 1,000 can be insufficient (see the discussion corresponding to Figure 5(a)).

Random graph generation this way has garnered some criticism (Artzy-Randrup et al., 2004). Artzy-Randrup et al. noted that such a random null hypothesis for networks is inadequate and to prove this, conceived a "toy experiment" which generated an artificial network G using a Gaussian distribution such that the closer two nodes are, the higher chance they would be connected. This toy experiment showed that network motifs that were found by Milo et al. in a real network could also be found using this "toy experiment". They also applied the "rule of preferential attachment" of Barabási and Albert (1999) in two ways, and found that a graph structure identified by Milo et al. using a similar method for graph generation was underrepresented using one method, but overrepresented using the other.

In response, Milo, Itzkovitz, Kashtan, Levitt, and Alon noted that the toy experiment identified "network motifs" that were not found in statistically significant levels in the real networks (Milo, Itzkovitz, Kashtan, Levitt, & Alon, 2004). Thus, which subgraphs were not network motifs is as important as noting which ones were. However, they conceded that employing a null random hypothesis is merely a "first step" in understanding a network's structure and the techniques they describe does not preclude the use of wet lab experiments for validation. It would seem that the use of random networks will always attract attention due to the uncertainty inherent in random graphs.

FREQUENCY-BASED MOTIF DISCOVERY

Algorithms for frequency-based network motif discovery require at least two noticeable changes compared to statistically-based methods. While the parameter k is still necessary to indicate the subgraph size of interest, the user must also provide a frequency threshold f_t. If the number of subgraphs in class Sc is greater than or equal to f_t, then the corresponding subgraph is a network motif. Due to this frequency parameter, this method resembles the frequent subgraph mining problem. Also, as the frequency threshold is used to determine when a subgraph is a network motif, the set of N random graphs is optional.

In this section, we begin with a brief discussion about algorithms related to frequent subgraph mining. While they are not directly related to motif discovery, the ideas used for motif discovery borrow ideas from this area. Then, we examine two frequency-based algorithms: FPF and NEMOFINDER, as shown in Table 2. Their definitions of what a motif is differ from each other. How they use the frequency threshold and NEMOFINDER's use of random graphs sets each method apart. We also briefly discuss the parallel implementation of FPF (Schreiber & Schwöbbermeyer, 2005a).

Table 2. Some frequency-based network motif detection algorithms

Algorithm	Overlaps	Random graphs	Citation
FPF	Yes	No	Schreiber and Schwöbbermeyer (2005a)
NEMOFINDER	No	Yes	Chen, Hsu, Lee, and Ng (2006)

Frequent Subgraph Mining

In frequent subgraph mining, we seek to determine the number of networks in a *set* of networks G with at least one occurrence of the subgraph S_c configuration of interest. This number is the support of S_c, or *support*(S_c). The configuration is frequent if its support is greater than or equal to some user-provided threshold ft.

There are several differences between frequent subgraph mining and network motif discovery. Frequent subgraph mining expects smaller networks than PINs. One algorithm that performs frequent subgraph mining is GSPAN (Yan & Han, 2002). In their experiments, they applied their method to a data set of 340 chemical compounds (networks) with around 27 atoms (vertices) each. A PIN can have thousands of nodes and edges, as exemplified by the data sets in Table 5. Moreover, frequent subgraph mining identifies the existence of a subgraph configuration in each network independently. Motif discovery aims to find every occurrence of the configuration in the network.

Also, since frequent subgraph mining disallows overlapping subgraphs, the *downward closure* property coupled with the frequency threshold can be combined to reduce the overall search space. The downward closure property is stated below:

Definition 3 *The frequency of a subgraph configuration in a network monotonically decreases with increasing size (downward closure property).*

Alternatively, the property states that if a subgraph configuration of k nodes has a frequency of $|S_c|$, then extending it by a vertex would yield a lower (or same) frequency. So, once a configuration yields a frequency that is less than f_t, then further extensions are not required.

However, in the context of motif discovery, there is one disadvantage to disallowing overlaps. Depending on the order in which subgraphs are chosen, the overall count of the subgraph configuration will change. This can be easily seen in the graphs of Figure 4. If the configuration being sought is the square-shaped subgraph of four

Figure 4. Example demonstrating that disallowing overlapping subgraphs can yield different counts depending on the order in which they are chosen. The configuration being sought is the 4-node, square-shaped subgraph. The identified subgraphs are shown in solid lines; omitted ones in dashed lines

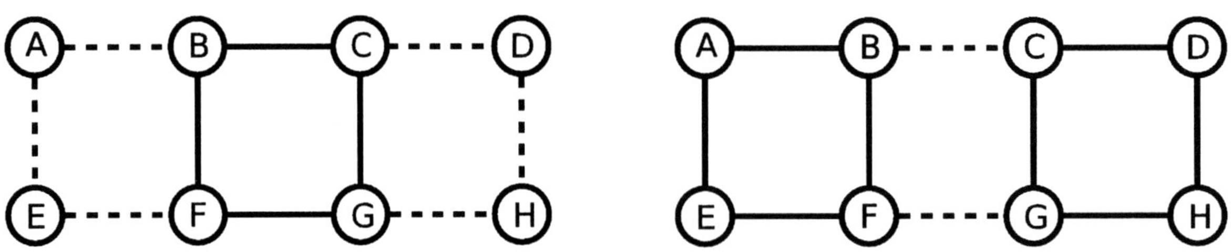

(a) One subgraph found (b) Two subgraphs found

nodes and edges, then after BCFG is found in Figure 4(a), no other one can be found. In contrast, if either ABEF or CDGH is identified in Figure 4(b), then the other one will also be located, giving 2 subgraphs instead of 1.

The Flexible Pattern Finder

The Flexible Pattern Finder (FPF) by Schreiber and Schwöbbermeyer (2005a) reports the frequency of subgraphs in a network using two parameters: the subgraph size k and one of three *concepts* ($\mathcal{F}_x$) for judging if a subgraph is valid. Essentially, the chosen concept determines whether or not subgraphs can overlap and if they do, then how. Schreiber and Schwöbbermeyer (2005a) labels the three concepts as $\mathcal{F}_1$, $\mathcal{F}_2$, and $\mathcal{F}_3$ and defines them as follows:

Definition 4 *The* k*-subgraphs of* G *are the set of all subgraphs with* k *nodes ($\mathcal{F}_1$).*

Definition 5 *The* k*-subgraphs of* G *are the set of all subgraphs with* k *nodes such that no two subgraphs share the same edge ($\mathcal{F}_2$).*

Definition 6 *The* k*-subgraphs of* G *are the set of all subgraphs with* k *nodes such that no two subgraphs share the same edge or node ($\mathcal{F}_3$).*

The least restrictive definition is $\mathcal{F}_1$ since overlapping subgraphs are permitted. This means it is also the definition used by all statistically-based methods described in the previous section. The remaining two methods are more restrictive and, as outlined above, have the potential to (a) give different frequencies depending on the order in which subgraphs are chosen and (b) understate the counts compared to the statistically-based methods. Note that the combination where edges can overlap but not their corresponding vertices is obviously not possible. The advantage of $\mathcal{F}_2$ and $\mathcal{F}_3$ is the application of the downward closure property for improved efficiency.

All three concepts are handled by FPF. The basis of their method is to construct a *pattern tree* that resembles the search tree of ESU except that "extending" a subgraph refers to adding an edge instead of adding a node. In order to bring their method in line with other algorithms, they continue growing branches of the pattern tree until k vertices have been found. There are other more critical differences. Since this is a frequency-based algorithm, the parameter f_t can be used to stop growing parts of the pattern tree if the frequency drops below f_t. Also, every k-subgraph has a single *generating parent* of size $k - 1$. While extending subgraphs, if a subgraph is created from a subgraph which is not its generating parent, then the extension is abandoned. This limits the number of subgraphs generated.

Table 3. The three concepts available to FPF, as defined by (Schreiber & Schwöbbermeyer, 2005a). The last combination where edges can overlap but not vertices is impossible.

Concept ($\mathcal{F}_x$)	Overlaps: Vertices	Overlaps: Edges
$\mathcal{F}_1$	Yes	Yes
$\mathcal{F}_2$	Yes	No
$\mathcal{F}_3$	No	No
Not possible	No	Yes

The Network Motif Finder

The Network Motif Finder (or NEMOFINDER) of Chen et al. (2006) combines ideas with statistically-based algorithms by making use of random graphs. Nevertheless, its underlying scheme is still a frequency-based method.

NEMOFINDER extends earlier subgraph mining algorithms such as FFSM and SPIN (Huan, Wang, & Prins, 2003; Huan, Wang, Prins, & Yang, 2004). NEMOFINDER is based on the formation of trees from G of size 2 to k and then using these trees to build more general graphs. Efficiency is achieved through an adjacency matrix representation that allows both ease in decomposing subgraphs into smaller components and more efficient comparison of isomorphic subgraphs. Their technique builds on the work of SPIN since they apply their method to a large PIN with the possibility of overlapping subgraphs ($\mathcal{F}_1$).

Another feature of NEMOFINDER is to declare subgraphs as motifs if they are *both* frequent and unique. In this case, the definition of frequent is the same as other frequent subgraph mining techniques. As for *unique*, an additional parameter U is required. A unique subgraph is a subgraph which occurs more then f_t times in a proportion of U of the N random graphs. Thus, this approach bears some resemblance to statistically-based methods. However, there is one major difference. Labeling a subgraph as being unique is a different and arguably weaker statement than saying its appearance in G is overrepresented when compared to the distribution of the subgraph configuration across the random networks. Nevertheless, their approach demonstrates an alternative use of the random graphs.

Parallelization of FPF

As with statistically-based methods, an opportunity to parallelize the algorithms based on subgraph mining also exists. Schreiber and Schwöbbermeyer (2005a) describes their extension to the FPF algorithm which segments the pattern tree and gives each part to a different CPU. In their experiments, they noted that CPUs are sometimes left idle. This is particularly true if concepts F_2 and F_3 are used because the pattern tree is pruned in a way that the load among CPUs cannot be properly balanced in advance.

REPORTING RESULTS

While the discovery of network motifs is the most difficult and time consuming task, proper presentation of the results is just as important. In papers that describe novel algorithms for motif discovery, running time and space complexity are obvious factors to compare. However, if the application of the algorithm to a real network is the primary concern, then the results from applying the method to G needs to be provided.

The easiest way of presenting motif results is to explicitly illustrate them, as we have done in Figure 2. However, listing them in this manner does not indicate to the reader the confidence we have of each motif. In this section, we summarize how others have chosen to report their results.

Since there are at least two definitions of motifs, it is not surprising that there should be at least two ways of presenting network motif results. Unfortunately, the situation is more complicated since each definition has multiple techniques as well. This also means that results between two papers on network motifs may not be readily comparable.

Results from Statistically-Based Methods

We commence with the evaluation measures proposed by Milo et al. (2002): concentration and standardized scores. The *concentration* for a configuration c and a subgraph size k is the proportion of subgraphs of k nodes that have this configuration. We denote this value as C_c:

$$C_c = \frac{|S_c|}{\sum_i |S_i|} \quad (1)$$

Standardized scores, or variants of it, have also been used extensively to measure the importance of a configuration in G relative to its existence in a set of random graphs (Milo et al., 2002; Milo, Itzkovitz, Kashtan, Levitt, Shen-Orr, et al., 2004). Standardize scores, or Z-scores, are unitless values which are typically used to explain how far an observation is from the population using the population's average and standard deviation. To be precise, it gives the number of standard deviations the observation is from the mean. Within the context of network motifs, the set of random graphs is considered to be the "population". For each subgraph configuration, the average and standard deviation across all N random graphs is calculated, which we denote as μ_c and σ_c, respectively. With the corresponding value for the PIN being $|S_c|$, the Z-score (Z_c) is expressed as:

$$Z_c = \frac{|S_c| - \mu_c}{\sigma_c} \quad (2)$$

For example, a Z-score of 2.0 means that the subgraph configuration occurs at a frequency which is two times the standard deviation of its appearance across all N random graphs. A suitable cut-off for the Z-score needs to be chosen such that all Z-scores above this value correspond to overrepresented subgraph configurations, and hence, network motifs.

However, this information is still insufficient since knowing the significance of underrepresented subgraphs helps in understanding how overrepresented are the identified subgraphs. Thus, showing all subgraph configurations is needed. The solution is to build a *significance profile* across all of the subgraphs of the same size. The significance profile (SP) is a vector of Z-scores with a dimension equal to the number of possible configurations for that subgraph size. While no order exists for the configurations, a possible option is to order by increasing canonical labels. The vector is normalized to length 1 as follows:

$$SP_c = \frac{Z_c}{\sqrt{\sum_i Z_i^2}} \quad (3)$$

As a further complication, it was observed that 4-subgraphs of undirected graphs showed a "significant dependence on the network size" (Milo, Itzkovitz, Kashtan, Levitt, Shen-Orr, et al., 2004, pg. 1541). Because of this, they proposed an alternative for four-node subgraphs called the abundance (Δ_c) of a subgraph configuration c:

$$\Delta_c = \frac{|S_c| - \mu_c}{|S_c| + \mu_c + \varepsilon} \quad (4)$$

where ε is a constant which ensures that the abundance is "not misleadingly large when the subgraph appears very few times in both the real and random networks" (Milo, Itzkovitz, Kashtan, Levitt, Shen-Orr, et al., 2004, pg. 1541). In their work, they selected $\varepsilon=4$ without further justification. Like Z-scores, the abundance is normalized as follows to obtain a *subgraph ratio profile*:

$$SRP_c = \frac{\Delta_c}{\sqrt{\sum_i \Delta_i^2}} \quad (5)$$

Other than Z-scores and abundance, no other measure was proposed for subgraphs larger than $k = 4$ by Milo, Itzkovitz, Kashtan, Levitt, Shen-Orr, et al. (2004). Both significance and subgraph ratio profile are vectors and several comparison methods are available. The first is as a graph where the normalized Z-score or relative subgraph ratio is along the vertical axis and each configuration is on the horizontal axis. Lines can be drawn for illustrative purposes even in the absence of any fixed order. The idea is that networks that yield similar line graphs may be from the same "superfamily".

A second method is to apply a technique commonly used by researchers using microarray data. The correlation using, say, Pearson correlation coefficient, can be calculated between every pair of vectors and a heat map drawn to indicate which networks have correlated profiles using color or various shades of gray.

In summary, concentrations indicate what proportion of the total number of subgraphs *within* a given real network have a configuration c. Standardized scores and abundance gives an indication of how frequent is a subgraph compared to its presence in the random graphs. Profiles extend these measurements to normalized vectors, where the dimension of the vector is equal to the number of possible configuration for the k-subgraphs. Profiles of networks can be compared as a means of grouping networks into superfamilies using tools that apply to vectors such as correlation coefficient and heat maps. We depict some of these methods later in this chapter.

Results from Frequency-Based Methods

In comparison, results derived from frequency-based methods are easier to report. The most common way is as a frequency for each corresponding configuration. Of course, the frequencies reported will be greater than f_t. In the case of NEMOFINDER, the identified motifs need to satisfy a second criteria related to the uniqueness threshold U.

SOFTWARE TOOLS

In this section, we explore some of the software that is available for public download and use the methods described in the previous section for presenting the results. Software for network motifs can generally be divided into two purposes: motif discovery and visualization. Occasionally, a program may perform both functions.

Data Sets

We consider the PINs of several species for this evaluation, whose data sets were obtained from the Database of Interacting Proteins (Salwinski et al., 2004). The data are stored in an XML format developed by the Proteomics Standards Initiative of the Human Proteome Organization (HUPO) for describing molecular interactions (PSI-MI) (Hermjakob et al., 2004). The data files are formatted according to PSI-MI v2.5. The species-specific data used were downloaded from DIP at http://dip.doe-mbi.ucla.edu/dip/Download.cgi. The version of our data files are dated July 8, 2008.

Table 4 summarizes the six data sets obtained. In the first column, we list the common name which we refer to the data sets later, and for four species, their scientific names. The other two species are the bacteria *Helicobacter pylori* and *Escherichia coli.* Since there may be several species with similar names, we indicate their unique NCBI Taxonomy Identifications in the second column. These numerical identifications are included in the PSI-MI data files and can be queried at http://www.ncbi.nlm.nih.gov/Taxonomy/ as one of the available NCBI services (Wheeler et al., 2000). The number of proteins and interactions are provided next.

In order to use this data for our methods, the following pre-processing was performed. An interaction from the data files contains one or more proteins. While most of the proteins are from the species of interest, occasionally, information on protein interactions between the species of interest and other species is also shown. So, in order to remove self-loops, protein interactions involving a single protein are removed. Then, protein interactions that involve species other than the one specific to the data file are dropped. The result from this pre-processing is reported as the last column of Table 4.

Since interactions in our processed data sets have two or more proteins each, further processing is required so that G can be constructed. When there are more than two proteins, we represent this interaction as a clique such that an edge exists between every two proteins that contribute to this interaction. While this is not an entirely correct representation, this assumption is suitable for our purposes. The final network may be disconnected though, since we are using the entire network and not just the single, largest connected component (as others have done, such as Chen et al. (2006)).

This transformation of interactions to edges results in the statistics described in Table 5. Note that we have changed the terminology from Table 4. After the processing, it is more correct to call the proteins and interactions as nodes and edges, respectively. As expected from the creation of cliques from interactions, the number of edges increases, except when the network is small and the number of removed self-loops outweigh the gain in edges. The fourth column in the table indicates the percentage of nodes which have only a single edge incident on it. The next two columns indicate the average and maximum node degrees. Together, these three columns demonstrate the scale-free nature of PINs. As an extreme example, the network of *E. coli* has a node with 716 edges while 13.5% of the nodes have only a single edge.

In the last column of this table, we show the *density* $\mathcal{D}$ of each network as the percentage of edges in the network compared to a complete undirected graph where every node is connected to every other node, except with itself:

$$\mathcal{D}(G(V,E)) = \frac{|E|}{(|V|^2 - |V|)/2} * 100 \tag{6}$$

As the statistics show, PINs are very sparse since they have no more than 3% of the total number of possible edges.

Available Software

The publicly available programs that were considered are shown in Table 6. In addition to the version of the software, we also indicate whether the source code is available and under what platforms can the source or binary be used (without significant changes to the source). This information is valid as of June 2008. The four platforms listed are: (W)indows, (M)ac OS, (L)inux, and (J)ava. FPF is the only system that runs under Java, or more specifically, Java Web Start[2]. If Java Web Start and all of its prerequisites are properly installed, then FPF runs under any operating system. All of them can support both directed and undirected networks.

Table 4. The names of several species whose PIN data sets are used in this chapter. Also shown is their respective NCBI Taxonomy Identifications and the number of proteins and protein-protein interactions before and after pre-processing. The number of proteins after pre-processing is the same as the number of nodes in the graph and is given in Table 5.

Species	Taxonomy ID (NCBI)	Original		Filtered
		Proteins	Interactions	Interactions
mouse (*M. musculus*)	10,090	529	445	205
H. pylori	85,962	712	1421	1415
human (*H. sapiens*)	9,606	1523	1855	1429
E. coli	562	1872	7428	7413
yeast (*S. cerevisiae*)	4,932	4974	18286	18243
fly (*D. melanogaster*)	7,227	7497	22853	22819

Table 5. The sizes of the PIN graphs corresponding to the six species. Also shown are the percentage of nodes with a single edge, average and maximum node degrees, and the edge density $\mathcal{D}$ as a percentage of the maximum number of edges possible.

Species	Graph size		Degree 1	Node Degree		$\mathcal{D}$ (%)
	Nodes	Edges	Nodes (%)	Average	Maximum	
mouse	224	183	65.2	1.63	8	0.733
H. pylori	700	1354	38.7	3.87	54	0.553
human	1105	1362	51.8	2.47	29	0.223
E. coli	1533	31790	13.5	41.47	716	2.707
yeast	4934	56639	20.3	22.96	428	0.465
fly	7455	22645	31.5	6.08	178	0.082

MFINDER The MFINDER software embodies the original brute-force and subsequent sampling algorithms of EXHAUSTIVE and SAMPLING (Milo et al. (2002); Kashtan et al. (2004)). Random graph generation and comparison against random graphs is included in its functionality. Three methods for random graph generation which are described elsewhere by Milo et al. (2003) are also provided. According to the documentation, the software is suitable for subgraph sizes of up to size $k = 5$ due to an intermittent bug for sized 6 subgraphs. Availability: http://www.weizmann.ac.il/mcb/UriAlon/groupNetworkMotifSW.html.

FANMOD and **RANDESU-CL** The FANMOD and RANDESU-CL programs are two of several implementations of the ESU and RANDESU algorithms that are available for download (Wernicke & Rasche, 2006). The FANMOD system encompasses the ESU and RANDESU algorithms within a graphical user interface and supports subgraphs of up to size 8. Results can be shown in both text and HTML format. RANDESU-CL is an older version of the software which is a command line tool distributed as C++ source code. Other ports such as a Microsoft Visual C++ version made by other developers can also be found on the same web site. Availability: http://www.minet.uni-jena.de/~wernicke/motifs/.

MDRAW The MDRAW visualization tool has no direct relationship with motif discovery, but instead, complements MFINDER. Its purpose is for visualizing and laying out graphs by regulating the attractive and repulsive forces between connected graph components. If the MFINDER executable is available, MDRAW can call it and present its results graphically. It can be downloaded from the same site as MFINDER.

MAVISTO The MAVISTO visualization program employs the FPF algorithm and the three concepts to return frequency-based network motifs (Schreiber & Schwöbbermeyer, 2005b). The features of MAVISTO appear more advance than MDRAW as it also includes the ability to draw networks as well as laying out existing ones in a manner similar to MDRAW. Unlike the FPF algorithm on which MAVISTO is based, the ability to make use of random graphs to determine statistical significance is also included. Statistics for all three concepts are displayed, in addition to coloring of the nodes and edges of the network based on the subgraphs they contribute to. Availability: http://mavisto.ipk-gatersleben.de/.

The above is a brief description of each system; further information in the form of a user manual or a web-based tutorial exists at their respective web pages. Below, we explore a few systems to provide an idea about what each program offers with the main aim of demonstrating the reporting techniques outlined in the previous section.

In our analysis, we do not provide comparisons in execution time due to the differences in the various implementations. Generally, statistically-based methods take more time than frequency-based methods. As a guideline, the difference for subgraphs of 4 nodes range from more than a week to hours, respectively, on a 3.0 GHz Dual-Core AMD Opteron(tm) Processor 2222 SE with 18 GB RAM for the largest networks. In practice, we advise that an estimate of the running time be obtained first with a smaller data set and a conservative value for k to gauge the amount of time required.

Table 6. List of some publicly-available programs for network motif discovery and visualiza¬tion. The abbreviations for the supported platforms are: (W)indows, (M)ac OS, (L)inux, and (J)ava.

Software	Version	Algorithms	Source	Platform			
				W	M	L	J
Network Motif Discovery							
MFINDER	1.2	EXHAUSTIVE, SAMPLING	Yes	Y	N	Y	N
FANMOD	2006/12/23	ESU, RANDESU	Yes	N	N	Y	N
RANDESU-CL	"WABI paper"	ESU, RANDESU	Yes	Y	Y	Y	N
Network Motif Visualization							
MDRAW	1.0	N/A	No	Y	N	N	N
MAVISTO	2.7.0	FPF	No	N	N	N	Y

Determining Network Motifs

We applied the command line version of the ESU algorithm (RANDESU-CL) by selecting full enumeration with N =1,000 random graphs. Random graphs were made using the default option of keeping the number of edges globally constant. The RANDESU-CL system reports the frequency and mean across the N random graphs as a percentage of the total number of subgraphs found. We converted these values back to raw ones and in Table 7, report our results for the two possible types of triads.

For each PIN, Table 7 reports the size of each configuration ($|S_c|$), the average frequency and standard deviation across all N random graphs (μ_c and σ_c). The Z-score (Equation (2)) and concentration (Equation (1)) are also calculated. As the results show, the top subgraph of two edges always appear in frequencies higher in the random graphs. Thus, this subgraph is underrepresented in the original PIN, even though the concentration is noticeably high (at least 80% of the total 3-subgraphs). In contrast, the clique of 3 nodes is overrepresented in all 6 PINs, despite the lower concentrations. Note that the Z-scores of one configuration is the same as the other, but opposite in sign. One reason for this phenomenon is that there are only two configurations of 3-subgraphs and their distributions are tied by the method of network permutation which keeps the number of edges globally constant.

As there are only two configurations of 3-subgraphs, a graphical representation of Table 7 is unnecessary. There are however six configurations of 4-subgraphs, which we depict as graphs in Figure 5, similar to Milo, Itzkovitz, Kashtan, Levitt, Shen-Orr, et al. (2004). From top-to-bottom, we show the significance profile ($\mathcal{SP}$), the subgraph ratio profile ($\mathcal{SRF}$), and the concentration ($\mathcal{C}$). Evenly spaced along the horizontal axes in all of the graphs is the 6 possible configurations, which we have ordered according to the bit string identifier I(A). In base 10, from left-to-right, they are 4,382, 4,958, 8,598, 13,278, 27,030, and 31,710.

In a few instances, the number of subgraphs for a particular configuration was not found across 1,000 random networks and no Z-score could be obtained. In Figure 5(a), this occurs twice for mouse and once in human. Usually, the solution is to increase the number of random networks; however, we have chosen to simply assign them a Z-score of 0 in order to illustrate this problem. Recall that Milo, Itzkovitz, Kashtan, Levitt, Shen-Orr, et al. (2004) stated that the significance profiles of tetrads depends on the size of the networks. This appears true

Table 7. Statistics for the two types of 3-subgraphs on the six PINs

		$\lvert S_c \rvert$	μ_c	±	σ_c	Z_c	C_c
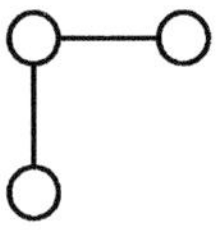	mouse	192	267.00	±	0.09	-790.57	0.88
	H. pylori	13915	14133.00	±	17.94	-12.15	0.99
	human	5482	6633.60	±	1.76	-653.09	0.93
	E. coli	4169800	6001400.00	±	107730.00	-17.00	0.86
	yeast	3705300	5386200.00	±	13194.00	-127.40	0.87
	fly	470090	475210.00	±	104.00	-49.23	1.00
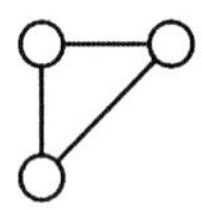	mouse	25	0.00	±	0.03	790.53	0.12
	H. pylori	75	2.29	±	5.98	12.16	0.01
	human	384	0.12	±	0.59	653.12	0.07
	E. coli	658530	48013.00	±	35909.00	17.00	0.14
	yeast	561430	1136.50	±	4397.90	127.40	0.13
	fly	1716	8.95	±	34.67	49.24	0.00

according to Table 4 since the fly's network has the most nodes and we can see that the shape of its line differs greatly since it always remain close to the horizontal axis.

Their solution was to plot a graph of subgraph ratio profiles using the normalized abundance defined in Equation (5) with $\varepsilon = 4$. This is shown in Figure 5(b), which now gives similar lines for all six species. The purpose of these graphs is to illustrate that profiles form "superfamilies". Some of the superfamilies demonstrated by Milo, Itzkovitz, Kashtan, Levitt, Shen-Orr, et al. (2004) included the connectivity of a power grid, protein structures, and Internet connectivity. Thus, it would be expected that all six PINs are part of the same superfamily and that their lines should be similar. Of course, some differences between the lines do exist and a better way is to calculate the all-pairs Pearson correlation coefficient between the six vectors and produce a heat map. This is shown in Figure 6. As the heat map shows, the highest correlations (lightest shade of gray, excluding the diagonal) are between human and mouse and between yeast and fly.

As for the concentrations of each configuration, the last graph shows that mouse and *E. coli* have graphs that differ from the other species. Comparing Figure 5(b) to Figure 5(c), note how concentration does not have a correspondence with a high significance profile. This observation coincides with the table of triads in Table 7.

Visualizing Network Motifs

Instead of simply finding network motifs using the command-line, others have adopted a more visual approach. One example of this is FANMOD, whose opening screen is shown in Figure 7 (Wernicke & Rasche, 2006). FANMOD incorporates both full subgraph enumeration (ESU) as well as the sampling variant (RANDESU) for finding subgraphs of size 3 to 8 nodes (point A in the figure). The ability to set options related to random graph generation is also available, including the number of graphs N and the number of exchanges allowed (point B). As noted earlier, one advantage of ESU over EXHAUSTIVE is its ability to build a search tree to ensure that each subgraph is found exactly once and, as a result, the total number of subgraphs can be estimated. This allows the progress of the system to be indicated using a progress bar (point C). After processing has completed, the results are saved into text format which can then be exported to HTML.

The second system that we consider provides not only a graphical interface to motif finding, but a way of visualizing motifs directly on the network. The MAVISTO software is originally based on the FPF algorithm. However, the implementation moves beyond the algorithm and incorporates ideas from statistically-based methods, such as the use of N random graphs and Z-scores instead of simply reporting frequency values.

Figure 5. Graphs comparing the profiles and the concentrations of the PIN data sets, as calculated using the ESU algorithm of the RANDESU-CL system with N =1,000 random graphs.

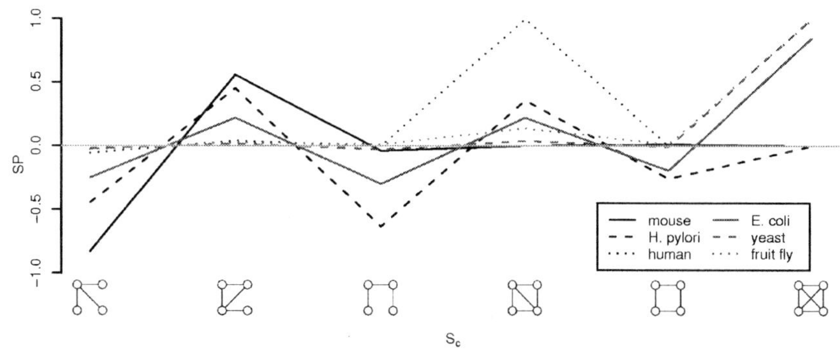

(a) Subgraph profile ($\mathcal{SP}$)

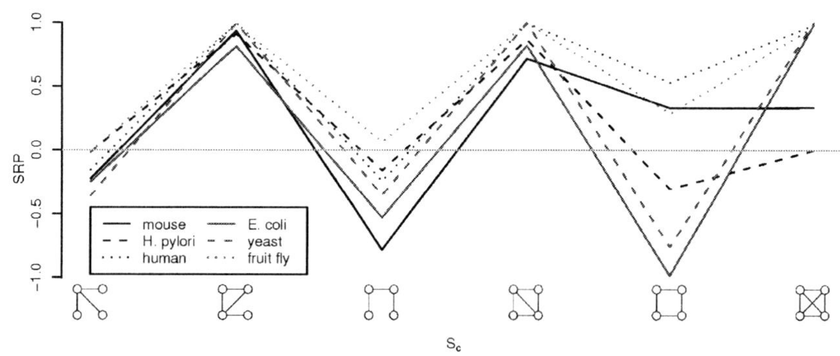

(b) Subgraph ratio profile ($\mathcal{SRP}$)

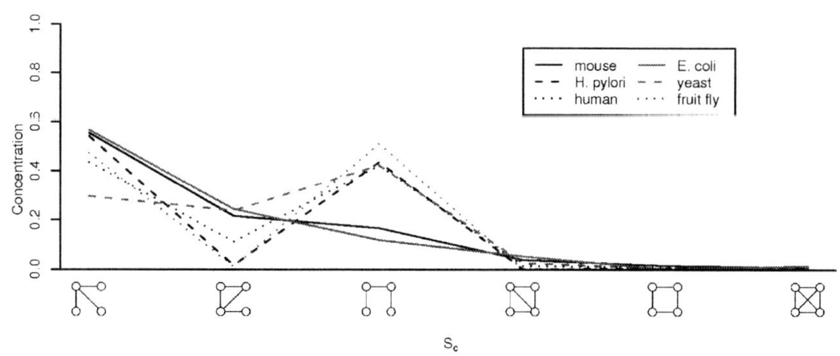

(c) Concentration ($\mathcal{C}$)

A sample application of the human PIN is shown in Figure 8. On the right at point A, some choices and statistics are made available to the user. We enlarge this figure later as Figure 9. We have chosen frequency concept $\mathcal{F}_1$ (overlapping nodes and edges) and the 4-subgraph shown below at point B. An undirected subgraph of 4 nodes has 6 possible configurations. The distribution of each configuration is given in the lower right corner of the screen as a bar graph (point C).

In the main part of the screen, a portion of the entire PIN is shown (point D). All 4-subgraphs with the configuration selected is highlighted using various colors, with the corresponding legend for the colors on the

Figure 6. Heat map for the subgraph ratio profiles (SRP) of the six species

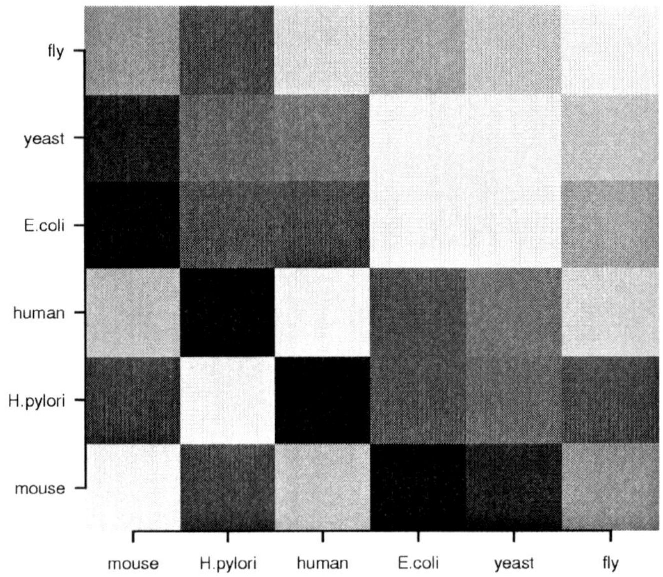

Figure 7. Main screen of the FANMOD software

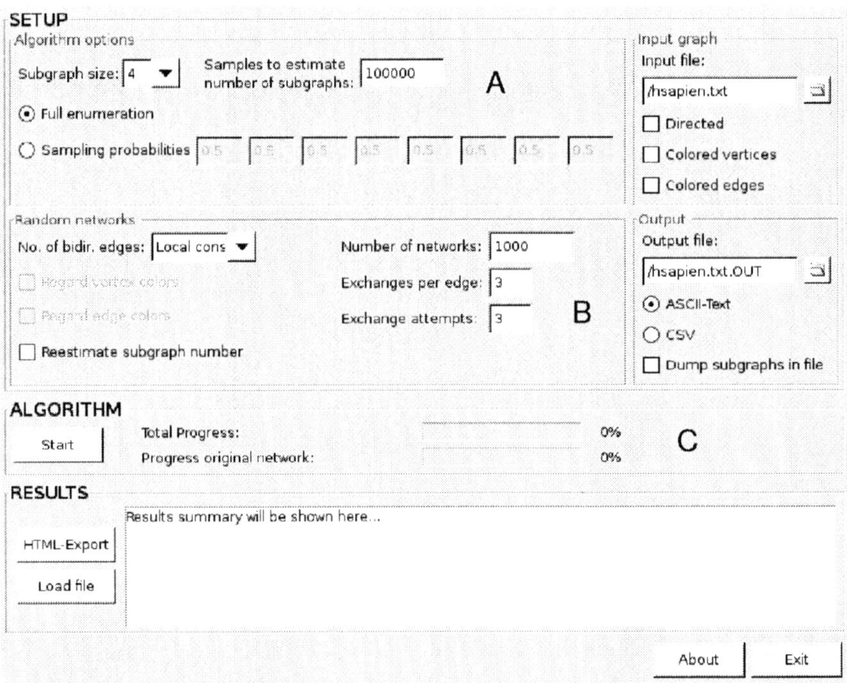

far left (point E). Since $\mathcal{F}_1$ permits overlapping nodes and edges, the coloring scheme indicates the number of subgraphs that use a particular node or edge.

An enlarged version of point A of Figure 8 is presented as Figure 9. In order to produce Figure 8, subgraphs with 4 vertices are chosen, together with 100 randomized graphs. The 6 configurations are listed, each with a unique label, and the corresponding vertex and edge sizes. The frequency counts under all three frequency concepts are also shown. At the right-hand side of the table are values related to the randomized graphs: p-values, Z-scores, means, and standard deviations.

BEYOND PINS

This chapter has focused exclusively on PINs, which are represented as undirected graphs whose nodes and edges are unlabeled. Before concluding this chapter, we examine other problem settings and remark on the differences when the methods outlined in this chapter are applied to other types of networks.

In PINs, every node can potentially be labeled with the name of the protein. However, labeling nodes in the context of network motifs typically means that each node is assigned to one out of a set of categories and these categories are used by the motif search algorithm. Algorithms that make use of labeled nodes generalize their method descriptions by coloring nodes. As far as we are aware, few researchers have demonstrated the utility of such an approach for PINs.

In the context of other networks, Lacroix, Fernandes, and Sagot (2006) presented an algorithm for finding subgraphs in metabolic pathways that have certain, pre-specified labels and a set number of nodes whose labels are unimportant ("gaps"). That is, a motif in this case is not limited by the size in nodes, but by whether all of the labels are included within the allowable search area, as defined by the gap parameter. In separate work, Parida (2007) showed a more general algorithm, but for a different problem: maximal motifs. The labels are again pre-

Figure 8. A screen shot of MAVISTO for the human PIN data set. Frequency concept $\mathcal{F}_1$ has been chosen as well as the 4-subgraph at point B in the image.

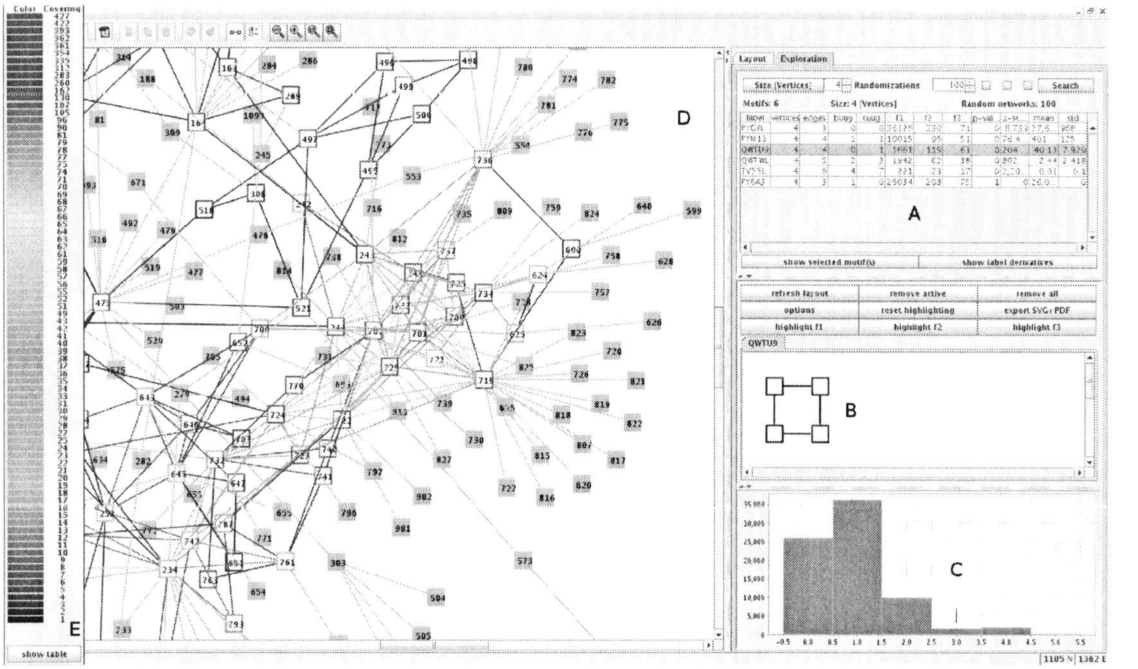

Figure 9. The table of statistics from point A of Figure 8 has been enlarged. A specific subgraph configuration has been chosen by the user, as well as 100 randomized graphs. The label of the graph, assigned by the system, is "QWTU9".

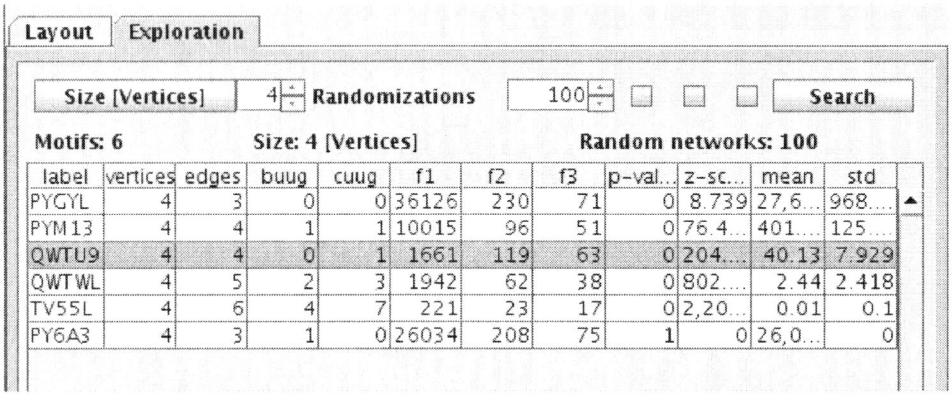

label	vertices	edges	buug	cuug	f1	f2	f3	p-val...	z-sc...	mean	std
PYGYL	4	3	0	0	36126	230	71	0	8.739	27,6...	968....
PYM13	4	4	1	1	10015	96	51	0	76.4...	401....	125....
QWTU9	4	4	0	1	1661	119	63	0	204....	40.13	7.929
QWTWL	4	5	2	3	1942	62	38	0	802....	2.44	2.418
TV55L	4	6	4	7	221	23	17	0	2,20...	0.01	0.1
PY6A3	4	3	1	0	26034	208	75	1	0	26,0...	0

specified so that only a set number of neighbors need to be considered. By combining this problem setting with a compact notation, a more efficient search algorithm is realized.

Many of the algorithms surveyed in this chapter were originally conceived in the context of directed networks. Some networks represented as directed graphs which have had network motif algorithms applied to them include metabolic networks (as indicated above), transcription regulatory networks, food webs which indicate predator/prey relationships, electronic circuits, and hyperlinks on the World Wide Web. In these networks, edges have direction and between two nodes v_i and v_j, there can now be two edges – one from v_i to v_j and another in the reverse direction. Transcription regulatory networks further complicate the situation since there are two types of edges: activation and repression.

Generally, algorithms for directed graphs can be applied to undirected ones, and many of the algorithms described in this chapter fall into this category. The only exception from this chapter is NEMOFINDER since it and other frequent subgraph mining algorithms are based on the existence or absence of a connection between two nodes. It is therefore unclear how they can be modified in order to support directed ones. Therefore, some motif finding algorithms may not be easily generalized from undirected to directed graphs since they sometimes rely on certain assumptions.

In addition to how motifs are found, another notable difference is how isomorphic subgraphs are handled. Given a k-subgraph s and the n known subgraph configurations so far, an implementation has to compare s to every possible configuration. If a match is found, then the count for that configuration is incremented. If a new one is found, then the new configuration S_n+1 is added with a count of 1. Changing the problem from undirected to directed graphs does not change the definition of isomorphism, but it greatly increases the number of possible configurations to check. In Table 8, we list the differences as the maximum number of configurations for subgraphs of sizes up to k =8. This table shows the number of *connected* directed and undirected graphs, whose differences increase rather dramatically with subgraph size (Harary & Palmer, 1973, pgs. 241-242).

CONCLUDING REMARKS

Information about protein interaction networks (PINs) continue to grow as better techniques are developed and more experiments are conducted to validate protein-protein interactions. Discovering network motifs represents one way of characterizing a PIN by identifying the small components from which it is constructed.

A search in the US National Library of Medicine's Pubmed or the digital library of the Association of Computing Machinery (ACM) for the phrase "network motif" will yield many results that have appeared since Milo et

Table 8. Number of isomorphisms for connected directed and undirected graphs of size k =2 to k =8 vertices. Values obtained from Harary and Palmer (1973, pgs. 241-242).

Number of nodes	Number of combinations	
	Directed	Undirected
2	2	1
3	13	2
4	199	6
5	9,364	21
6	1,530,843	112
7	880,471,142	843
8	1,792,473,955,306	11,117

al. (2002) initially published their work on this topic. Their methods have since been applied to various networks (including PINs) and others are investigating the problem of finding faster and more efficient algorithms, a few of which we have described in this chapter.

The aim of this chapter was to describe these methods by highlighting the two definitions that have evolved since that 2002 work. One method is statistically-based and the other comes from the field of frequent subgraph mining. Since systems such as MAVISTO incorporates methods which combine both approaches, the separation between the two approaches is not always clear in practice. However, we have chosen to emphasize their differences in order to aid the reader in deciding which approach is more suitable. We summarize the advantages and disadvantages of these two types of methods in Table 9.

Frequency-based methods require at least an additional parameter in the form of an explicit threshold f_t. The problem with this approach is that users need to select a value that represents their notion of what is "frequent enough". Statistically-based methods are not much better since they have many less obvious parameters which are all necessary for indicating how many random networks are required and how each one should be generated. So, while both methods rely on parameters whose purpose is to properly model the type of motifs we would like to find, what the best settings should be in either case is not obvious. In the end, regardless of the method chosen, some trial and error may be required.

Table 9. A general comparison of the two classes of algorithms for network motif discovery. Individual algorithms may vary. Optional inputs are inputs that are required by some, but not all systems within that class.

	Statistically-based	Frequency-based
Input	- Subgraph size (k) - Number of random graphs (N) - Parameters for generating random graphs - Cut-off for statistical significance	- Subgraph size (k) - Frequency threshold (f_t)
Optional Inputs	- Sampling probability (p)	- A frequency concept ($\mathcal{F}_x$) - Uniqueness (U)
Output	- Overrepresented subgraphs	- Frequent subgraphs
Advantages	- Motifs are statistically significant	- Able to find larger subgraphs efficiently - Usually faster
Disadvantages	- Random null hypothesis may be inappropriate - Usually slower	- If node or edge overlaps are disallowed, the order subgraphs are selected affects the configuration frequency

In terms of execution time, frequency-based methods are generally faster and more able to find larger subgraphs than their statistically-based counterparts. However, this is a side-effect of re-interpreting the definition of "network motif" as originally proposed by Milo et al. (2002).

While its been stated that functionally important subgraphs may exist which are not statistically significant, the reverse is also true since a statistically significant subgraph may not be important to the formation of the network (Milo et al., 2002; Schreiber & Schwöbbermeyer, 2005a). However, this same statement can also be said of frequent subgraphs. In the absence of an unambiguous definition of what "network motifs" are, we are left with selecting one of the two heuristics with no clear answer as to which is the "better" definition.

A user who wants to find network motifs needs to first decide which definition of network motifs is more appropriate to their work. Then, the parameters of the respective method needs to be chosen, perhaps in an iterative manner, until the motifs found are reasonable. As noted by Milo, Itzkovitz, Kashtan, Levitt, and Alon (2004), in the end, network motif detection needs to be coupled with validation using experimental techniques as these computational tools are simply guides.

Making Use of Motifs

Earlier in this chapter, we gave examples that showed how discovered motifs can be used to infer the evolution of PINs. We close this chapter with two other uses that are of particular interest since they are directly related to one of the algorithms covered in this chapter, NEMOFINDER.

Chen et al. (2006) notes that data in PINs have a high number of false positives (interactions which are actually not correct). Their solution to this problem involves using NEMOFINDER to score each interaction in the network based on the *strength* of the motif in which they are a part of. A motif's strength is calculated from combining its frequency and uniqueness. They then checked to see if high-scoring interactions matches information from other data sources, including annotated protein functions and gene expression correlation in microarray data. Compared with two other metrics, they demonstrated the usefulness of NEMOFINDER in this regard.

The same research group subsequently developed a method which they dubbed LAMOFINDER for Labeled Motif Finder (Chen, Hsu, Lee, & Ng, 2007). LAMOFINDER is essentially a post-processor to a network motif finding system for protein interaction networks, such as NEMOFINDER. Their system makes extensive use of the Gene Ontology (GO) (The Gene Ontology Consortium, 2000). The Gene Ontology is a controlled vocabulary for genes, arranged in a hierarchical order as three separate trees (biological processes, cellular components, and molecular functions). As each gene can have multiple GO terms assigned to them, the task at hand is to label the nodes of each motif in a way that ensures that it is frequent in the now *labeled* PIN. As the labeling permits more general terms from being used, the problem is the selection of the most appropriate combination of labels. Note that their work differs from the algorithms described in the previous section by Lacroix et al. (2006) and Parida (2007), since the labeling is a post-processing step and not a characteristic of the network that is used to affect what motifs are found. From these two developments alone, it would seem that motifs from PINs can be used to not only summarize the structure of the PIN, but also to check whether the subunits found match protein functions which have been annotated in other databases.

In the future, we expect that new algorithms for PINs may be needed as more information is incorporated into the PINs. For example, the edges may be labeled with weights indicating the reliability of the interactions, or the nodes may be assigned to specific categories. Thus, algorithms and data need to be developed in parallel so that the algorithms meet the needs of the researchers analyzing the data. At the moment, one immediate need is the increase in the size of PINs. This requires more efficient network motif finding algorithms to be devised so that larger subgraphs can be found.

ACKNOWLEDGMENT

Our interest in network motifs stem from an earlier collaboration with Nelson Hayes (Kyoto University). RW was supported by a postdoctoral fellowship from the Japan Society for the Promotion of Science. This work has been supported in part by BIRD of the Japan Science and Technology Agency (JST).

REFERENCES

Aho, A. V., Hopcroft, J. E., & Ullman, J. D. (1974). *The design and analysis of computer algorithms*. Reading, USA: Addison-Wesley Publishing Co.

Alon, U. (2007, June). Network motifs: theory and experimental approaches. *Nature Reviews Genetics*, *8*(6), 450–461.

Artzy-Randrup, Y., Fleishman, S. J., Ben-Tal, N., & Stone, L. (2004, August). Comment on 'Network motifs: Simple building blocks of complex networks' and 'Superfamilies of evolved and designed networks'. *Science*, *305*(5687), 1107c.

Barabási, A.-L., & Albert, R. (1999, October). Emergence of scaling in random networks. *Science*, *286*(5439), 509–512.

Bollobás, B. (2001). *Random graphs* (Second ed., Vol. 73). Cambridge, England: Cambridge University Press.

Chen, J., Hsu, W., Lee, M. L., & Ng, S.-K. (2006). NeMoFinder: dissecting genome-wide protein-protein interactions with meso-scale network motifs. In *Proc. 12th ACM International Conference on Knowledge Discovery and Data Mining (SIGKDD)* (pp. 106–115). New York, USA: ACM Press.

Chen, J., Hsu, W., Lee, M. L., & Ng, S.-K. (2007, April). Labeling network motifs in protein interactomes for protein function prediction. In *Proc. IEEE 23rd International Conference on Data Engineering* (pp. 546–555). Washington, DC, USA: IEEE Computer Society Press.

Ciriello, G., & Guerra, C. (2008). A review on models and algorithms for motif discovery in protein-protein interaction networks. *Briefings in Functional Genomics and Proteomics*.

Cormen, T. H., Leiserson, C. E., Rivest, R. L., & Stein, C. (2001). *Introduction to algorithms* (Second ed.). Cambridge, USA: The MIT Press.

Fortin, S. (1996). *The graph isomorphism problem* (Tech. Rep. No. 96-20). University of Alberta: Dept. of Computer Science.

Gross, J. L., & Yellen, J. (2005). *Graph theory and its applications (Discrete mathematics and its applications)* (Second ed.). Boca Raton, USA: Chapman & Hall/CRC.

G Siek, J., Lee, L.-Q., & Lumsdaine, A. (2002). *The Boost graph library: User guide and reference manual*. New York, USA: Addison-Wesley Publishing Co.

Harary, F., & Palmer, E. M. (1973). *Graphical enumeration*. New York, USA: Academic Press.

Hermjakob, H., et al. (2004). The HUPO PSI's Molecular Interaction format – A community standard for the representation of protein interaction data. *Nature Biotechnology*, *22*, 177–183. (URL: http://www.psidev.info/index.php?q=node/60)

Huan, J., Wang, W., & Prins, J. (2003). Efficient mining of frequent subgraphs in the presence of isomorphism. In *Proc. 3rd IEEE international conference on data mining* (pp. 549–552). Washington, DC, USA: IEEE Computer Society Press.

Huan, J., Wang, W., Prins, J., & Yang, J. (2004). SPIN: Mining maximal frequent subgraphs from graph databases. In *Proc. 10th ACM international conference on knowledge discovery and data mining (SIGKDD)* (pp. 581–586). New York, USA: ACM Press.

Kashtan, N., Itzkovitz, S., Milo, R., & Alon, U. (2004). Efficient sampling algorithm for estimating subgraph concentrations and detecting network motifs. *Bioinformatics*, *20*(11), 1746–1758.

Lacroix, V., Fernandes, C. G., & Sagot, M.-F. (2006). Motif search in graphs: Application to metabolic networks. *IEEE/ACM Transactions on Computational Biology and Bioinformatics*, *3*(4), 360–368.

Leskovec, J., & Faloutsos, C. (2006). Sampling from large graphs. In *Proc. 12th ACM International Conference on Knowledge Discovery and Data Mining (SIGKDD)* (pp. 631–636). New York, USA: ACM Press.

McKay, B. D. (1981). Practical graph isomorphism. *Congressus Numerantium*, (pp. 45–87).

Milo, R., Itzkovitz, S., Kashtan, N., Levitt, R., & Alon, U. (2004, August). Response to comment on 'Network motifs: Simple building blocks of complex networks' and 'Superfamilies of evolved and designed networks'. *Science*, *305*(5687), 1107d.

Milo, R., Itzkovitz, S., Kashtan, N., Levitt, R., Shen-Orr, S., Ayzenshtat, I., et al. (2004,March). Superfamilies of evolved and designed networks. *Science*, *303*(5663), 1538–1542.

Milo, R., Kashtan, N., Itzkovitz, S., Newman, M. E. J., & Alon, U. (2003). On the uniform generation of random graphs with prescribed degree sequences. *cond-mat*, *0312028*. Available from http://arxiv.org/abs/cond-mat/0312028

Milo, R., Shen-Orr, S., Itzkovitz, S., Kashtan, N., Chklovskii, D., & Alon, U. (2002, October). Network motifs: Simple building blocks of complex networks. *Science*, *298*(5594), 824–827. (Software: http://www.weizmann.ac.il/mcb/UriAlon/groupNetworkMotifSW.html)

Parida, L. (2007). Discovering topological motifs using a compact notation. *Journal of Computational Biology*, *14*(3), 300–323. (US Patent: 20070027662)

Presser, A., Elowitz, M. B., Kellis, M., & Kishony, R. (2008, January). The evolutionary dynamics of the saccharomyces cerevisiae protein interaction network after duplication. *Proc. National Academy of Sciences USA*, *105*(3), 950–954.

Pržulj, N., Corneil, D. G., & Jurisica, I. (2006). Efficient estimation of graphlet frequency distributions in protein-protein interaction networks. *Bioinformatics*, *22*(8), 974–980.

Salwinski, L., Miller, C. S., Smith, A. J., Pettit, F. K., Bowie, J. U., & Eisenberg, D. (2004, January). The database of interacting proteins: 2004 update. *Nucleic Acids Research*, *32*, D449–D451. (URL: http://dip.doe-mbi.ucla.edu/)

Schreiber, F., & Schwöbbermeyer, H. (2005a). Frequency concepts and pattern detection for the analysis of motifs in networks. *Transactions on Computational Systems Biology III*, *3737*, 89–104.

Schreiber, F., & Schwöbbermeyer, H. (2005b). MAVisto: A tool for the exploration of network motifs. *Bioinformatics*, *21*(17), 3572–3574. (Software: http://mavisto.ipk-gatersleben.de/)

Sedgewick, R. (2002). *Algorithms in C, Part 5: Graph algorithms* (Third ed.). New York, USA: Addison-Wesley Publishing Co.

The Gene Ontology Consortium. (2000). Gene Ontology: tool for the unification of biology. *Nature Genetics*, *25*(1), 25–29. (URL: http://www.geneontology.org/)

Wang, T., Touchman, J. W., Zhang, W., Suh, E. B., & Xue, G. (2005, October). A parallel algorithm for extracting transcriptional regulatory network motifs. In *Proc. 5th IEEE Symposium on Bioinformatics and Bioengineering* (pp. 193–200). Washington, DC, USA: IEEE Computer Society Press.

Wernicke, S. (2006). Efficient detection of network motifs. *IEEE/ACM Transactions on Computational Biology and Bioinformatics*, *3*(4), 347–359.

Wernicke, S., & Rasche, F. (2006). FANMOD: A tool for fast network motif detection. *Bioinformatics*, *22*(9), 1152–1153. (Software: http://www.minet.uni-jena.de/~wernicke/motifs/)

Wheeler, D. L., Chappey, C., Lash, A. E., Leipe, D. D., Madden, T. L., Schuler, G. D., et al. (2000, January). Database resources of the National Center for Biotechnology Information. *Nucleic Acids Research*, *28*(1), 10–14.

Wolf, D. M., & Arkin, A. P. (2003, April). Motifs, modules and games in bacteria. *Current Opinion in Microbiology*, *6*(2), 125–134.

Wuchty, S., Oltvai, Z. N., & Barabási, A.-L. (2003, October). Evolutionary conservation of motif constituents in the yeast protein interaction network. *Nature Genetics*, *35*(2), 176–179.

Yan, X., & Han, J. (2002). gSpan: Graph-based substructure pattern mining. In *Proc. 2002 IEEE International Conference on Data Mining* (pp. 721–724). Washington, DC, USA: IEEE Computer Society Press.

ENDNOTES

1 URL: http://search.cpan.org/~jhi/Graph-0.84/

2 URL: http://java.sun.com/products/javawebstart/

Chapter IX
Discovering Protein Complexes in Protein Interaction Networks

Clara Pizzuti
ICAR, Consiglio Nazionale delle Ricerche, Italy

Simona Ester Rombo
Università della Calabria, Italy

ABSTRACT

In this chapter a survey on the main graph-based clustering techniques proposed in the literature to mine protein-protein interaction networks (PINs) is presented. The detection of putative protein complexes is an important research problem in systems biology. In fact it may help in understanding the mechanisms regulating cell life, in deriving conservations across species, in predicting the biological functions of uncharacterized proteins, and, more importantly, for therapeutic purposes. Different kind of approaches are described and classified. Furthermore, some validation techniques commonly exploited in this context are illustrated. The goal of the chapter is to provide a useful guide and reference for both computer scientists and biologists. Computer scientists may have a complete vision of what has already been made and which are the new challenges about PINs clustering, taking them as a starting point for further researches and new proposals; on the other hand, biologists may find in the chapter the necessary material to select the most appropriate methods to apply for their specific purposes.

INTRODUCTION

In the last few years the development of new high-throughput technologies (von Mering, 2002) to determine protein interactions has made available large volumes of experimental data that reflect the interplay between proteins in complex cellular networks. An important aspect of the protein-protein interaction networks (*PINs*) analysis concerns the discovery of *modules*, or *complexes*, made of proteins characterized by dense sets of interactions. The interest in this kind of analysis is motivated by the observation that PINs have modular structure (Hartwell, 1999; Pereira-Leal, 2004), being organized into different putative protein complexes each performing specific tasks in the cell. It is worth to point out that "protein complexes" and "function modules" have different biological meanings. A protein complex is a molecular machine that consists of several proteins that bind each other at the same place and time. On the contrary, a functional module consists of a few proteins that control or perform a particular cellular function through interactions between themselves (these proteins do not necessarily interact at the same time and place). However, in many cases it is hard to distinguish these two concepts because the analyzed pair-wise protein interactions are not associated with temporal and spatial information. Thus, in the following we will refer indifferently to protein complexes or functional modules.

The detection of putative protein complexes may help in understanding the mechanisms regulating cell life, in deriving conservations across species, in predicting the biological functions of uncharacterized proteins, and, more importantly, for therapeutic purposes. Indeed, it has been recognized by biologists that proteins interacting with each other often participate in the same biological processes. Furthermore, protein modules may be often associated with specific biological functions and proteins belonging to a specific module are more related each other than w.r.t. the members of other modules (Tornow, 2003). Thus, discovering putative protein complexes may be useful also to detect processes in which uncharacterized proteins are involved. Unfortunately, isolating protein complexes through the PINs exploration is per se a difficult problem, due to the intricateness of interaction relationships characterizing them (see Figure 1, where only a small part of *S. cerevisiae* protein interactions are shown).

Recently, clustering techniques (Jain, 1988) proved to be successful for the identification of putative protein complexes in complex networks. Clustering consists in grouping data objects into groups (clusters) such that the objects in the same cluster are more similar each other than with objects in the other clusters. In PINs, clustering means grouping together proteins which share a larger number of interactions, thus clusters may be assimilated to functional modules. Possible uncharacterized proteins in a cluster may be assigned to the biological function recognized for that module. A number of clustering approaches have been proposed to extract relevant modules from PINs. Some of them rely on distance-based hierarchical clustering methods (Blatt, 1996; Arnau, 2004; Pei, 2005). In these approaches suitable similarity metrics have to be defined. Others consider the topology of the PIN and apply different strategies such as searching for sub-graphs having maximum density (Bader, 2003; Derenyi, 2005; Palla, 2005; Adamcsek, 2006; Altaf, 2006; Lubovac, 2006; Chua, 2007; Li, 2007; Pizzuti, 2007; Pizzuti, 2008); optimization of a cost function (King, 2004; Spirin, 2003); flow simulation (Enright, 2002; Pereira, 2004; Cho, 2006; Hwang, 2006; Cho, 2007); statistical approaches (Samantha, 2003; Farutin, 2006). Some technical and biological distinction among such approaches may be driven by considering if they possibly admit clustering overlapping, i.e., they allow proteins in the same network to participate in several clusters. In fact, as will be underlined in the following sections, there exist proteins which are involved in several biological processes, thus, methods that output overlapping clusters are biologically more relevant.

Figure 1. Some of the S. Cerevisiae protein interactions

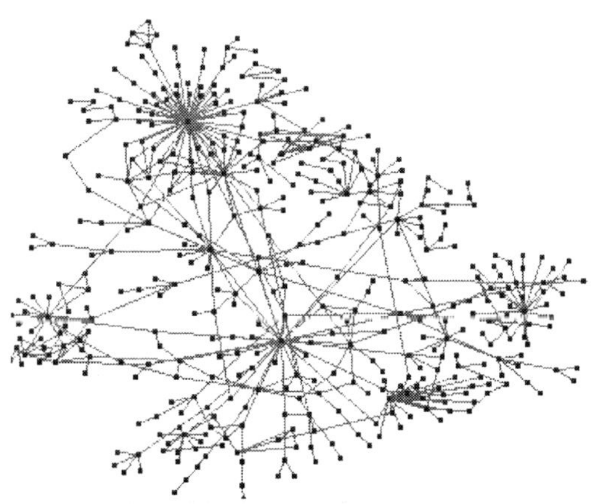

It is important to point out that the application of traditional clustering techniques "as they are" is often not appropriate to solve the particular problem under consideration. Indeed, PINs have specific characteristics which have to be taken into account and specific clustering algorithms are necessary for this purpose. In the clustering approaches described in this chapter, a PIN is modelled as an undirected graph where nodes represent proteins and edges represent interactions.

In order to validate the clusters obtained by applying the different approaches on PINs, specific metrics have been introduced such as the *clustering coefficient* (Watt, 1999), that takes into account only the nodes of a network and how they are linked together, and the *p-value*, representing the probability that a given cluster of proteins occurs by chance. Furthermore, automatic tools such as the Gene Ontology Term Finder tool (Boyle, 2004), may also be exploited to understand the biological mining of the obtained clusters, as will be explained later in this work.

In this chapter, we propose a detailed analysis of the main graph-based clustering approaches that have been presented in the literature to mine PINs, comparing some of them and individuating some open issues for the problem. The goal of this chapter is that of providing a useful guide and reference for both computer scientists and biologists. In fact, computer scientists may have a complete vision of what has already been made and which are the new challenges about PINs clustering, taking them as a starting point for further researches and new proposals; on the other hand, biologists may find in the chapter the necessary material to select the most appropriate methods to apply for their specific purposes.

The rest of the chapter is organized as follows. Section *Methods* illustrates some preliminary notions, the main metrics adopted to validate the PINs clusters and some public tools which may be exploited to understand the biological meaning of the obtained clusters. In Section *Classification* the techniques proposed by different groups are discussed and compared, and also a full example of clustering/validation is provided. In Section *Discussions and future trends* we summarize the main features of the described methods, their possible limitations and the open problems. In Section *Concluding remarks* we draw our conclusion and, finally, in Section *Resources* we list and briefly describe some online tools and database resources relevant to PINs clustering.

METHODS

In this section some preliminary notions necessary for the full understanding of the chapter are first presented. Then the main methods to validate PIN clusters are illustrated.

Preliminaries

A PIN N can be modelled as an undirected graph $G_N=(V,E)$ where V is a set of nodes, each corresponding to a specific protein, and E is the set of undirected edges corresponding to the pairwise interactions. Two nodes of a graph that are connected by an edge are said *adjacent*, and two edges that share a node are said *adjacent* as well. The edge connecting the nodes u and v is represented by (u,v). The *neighbourhood* of a node v is $N(v) = \{u \mid (u, v) \in E\}$. Nodes belonging to the neighbourhood of v are said its *neighbours*. A (partial) *sub-graph H* of a graph $G=(V_G,E_G)$ is a graph $H=(V_H,E_H)$ with $V_H \subset V_G$ and $E_H \subset E_G$.

If a network is constituted by n proteins, the associated graph can be represented with its $n \times n$ adjacency matrix M, where the entry at position (i, j) is 1 if there is an edge from node i to node j, 0 otherwise. Since the associated graph G_N is undirected, the adjacency matrix is symmetric.

The problem of clustering PINs may be interpreted as that of finding dense regions in a PIN N, that is, finding dense sub-graphs of the graph G_N associated with N.

Cluster Validation

A relevant aspect of the clustering based analysis of PINs is the validation of clustering results, in order to understand how much they are biologically significant. As will be explained in the following, such a validation may be achieved by both exploiting suitable metrics and considering on-line available tools, such as the Gene Ontology Term Finder provided by the GO Consortium (Asburner, 2000). The following definitions are useful to illustrate the main metrics adopted in the literature for both technical and biological validation.

Technical Validation

One of the most important validation metric to asses the quality of the PINs clustering results is the *clustering coefficient*. The concept of clustering coefficient has been defined in (Watt, 1999), by taking into account the nodes of a network and how they are linked together.

Definition 1 *(Clustering coefficient of a node)* Given a node i, let n_i be the number of links connecting the k_i neighbours of i to each other. The clustering coefficient of i is defined as:

$$C_i = 2n_i / k_i(k_i - 1)$$

Note that n_i may be also interpreted as the number of triangles passing through i, while $k_i(k_i - 1)/2$ represents the number of possible triangles that could pass through the node i.

Example 1 Consider the network illustrated in Figure 2. We show how the clustering coefficient of the node *3* is computed. The neighbours of node *3* are the four nodes *1*, *2*, *4* and *7*. The number of links connecting such neighbours to each other is $n_3=4$. Then:

$$C_3 = 2 \cdot 4/4(4-1) = 0.667$$

Definition 2 *(Clustering coefficient)* The clustering coefficient C_{Bj} of a cluster B_j, is the average of the clustering coefficients of all the proteins belonging to B_j. The clustering coefficient C_B of a clustering $B = \{B_1, \dots, B_h\}$ is defined as:

$$C_B = \frac{\sum_{j=1}^{h} C_{Bj}}{h}$$

Figure 2. Example for computing the clustering coefficient

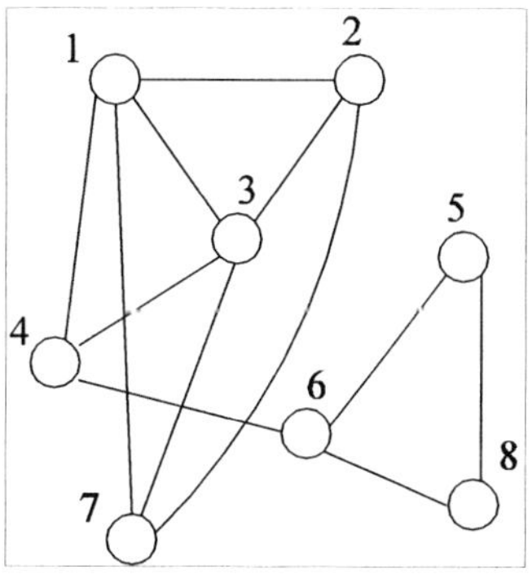

Example 2 Assume that, applying a clustering technique to the graph shown in Figure 2, the following clusters are obtained:

- B_1={1,2,3,4,7}
- B_2={5,6,8}

The clustering coefficients of each single node are: C_1=0.667, C_2=1, C_3=0.667, C_4=0.333, C_5=1, C_6=0.333, C_7=1 and C_8=1. Then, C_{B1}=0.667 and C_{B2}=0.778, from which C_B=0.723.

Biological Validation

In PINs analysis it is important to verify if the obtained clusters correspond to a function meaningful from a biological point of view. This validation can be done by using the known biological associations from the Gene Ontology Consortium Online DataBase (Asburner, 2000). The Gene Ontology database provides three vocabularies of known associations:

- *molecular function*, describing the tasks performed by individual gene products (e.g., *transcription factor* and *DNA binding*);
- *cellular component*, encompassing subcellular structures, locations, and macromolecular complexes (e.g., *nucleus*, *telomere*, and *origin recognition complex*);
- *biological process*, describing broad biological goals, such as *mitosis* or *purine metabolism*.

An important metric exploited for clusters validation is the *p-value*, that is, a measure of the functional homogeneity of a cluster. In particular, a protein cluster may be associated with a list of genes, each corresponding to a protein in the cluster. To determine the statistical significance of the association of a particular GO term with a group of genes in the list, it is possible to calculate the probability that a given set of proteins occurs by chance, as better illustrated by the next definition.

Definition 3 *(p-value)* Given a cluster of size *n* with *m* proteins sharing a particular biological annotation, then the probability of observing *m* or more proteins that are annotated with the same GO term out of those *n* proteins, according to the Hypergeometric Distribution, is:

$$p - value = \sum_{i=m}^{n} \frac{\binom{M}{i}\binom{N-M}{n-i}}{\binom{N}{n}}$$

where *N* is the number of proteins in the database with *M* of them known to have that same annotation.

Thus, the closer the p-value to zero, the more significant the associated GO term. The biological significance of a group is settled by using a cut-off value to distinguish significant from insignificant groups. If the p-value of a cluster is below the cut-off then it is considered insignificant. It is worth pointing out that the p-value tests the significance of the clusters discovered against random clusters only. Further biological validation (possibly supported, for example, by expert biologists) should be driven to confirm that the candidate clusters are really protein complexes or parts of biological pathways. When datasets manually curated by biologists are available, also other metrics may be introduced, such as *precision* and *recall* (Li, 2007).

As observed in (Ucar, 2006), it could be useful to have a global measure of an obtained clustering, instead of the p-value of a single group.

Definition 4 *(clustering score)* The clustering score CS_B of a clustering *B* is defined as:

$$CS_B = 1 - \frac{\sum_i^{n_S} \min(p_i) + (n_I \times cutoff)}{(n_I + n_S) \times cutoff}$$

where $\min(p_i)$ is the smallest p-value of the partition i, n_S is the number of significant partitions, that is, partitions scoring a p-value < 0.01, and n_I is the number of insignificant partitions.

As already pointed out, besides validation metrics, another way to understand if the obtained clusters are biologically significant is referring to the GO annotations recalled above. In particular, the Gene Ontology Term Finder Tool (Boyle, 2004), available at http://db.yeastgenome.org/cgi-bin/GO/goTermFinder.pl, comprises a set of object-oriented Perl modules for accessing Gene Ontology (GO) information and evaluating and visualizing the collective annotation of a list of genes to GO terms. Figure 3 shows the main page of the Gene Ontology Term Finder, where a query on the nine proteins PFS2, PTI1, MPE1, REF2, YTH1, FIP1, CFT1, CFT2 and PTA1 is considered. In particular, the process vocabulary is selected for the query, and the system returns results both in graphic and in tabular form, as shown in Figure 4 and in Figure 5, respectively. In the graphic form, p-values in different ranges are indicated by different colours, and genes are associated with the GO term(s) to which they are directly annotated. To keep the graph size reasonable, only significant hits with a p-value < 0.01 and terms descended from them are illustrated. In Figure 5 the top 10 entries of the tabular results are reported. In particular, the table returned by the Gene Ontology Term Finder contains a number of entries corresponding to different processes in which all or some proteins in the query participate. Such entries are sorted according to the associated p-value. The first column of the table shows the GO term corresponding to the process. In the second column the cluster frequency, i.e. the number of proteins in the cluster, out of the total number, that participate in that process is reported. The third and fourth columns report the background frequency and the p-value associated with the process and the proteins of the correspondent entry, respectively; the p-value is simply calculated using the hypergeometric distribution and it is not corrected for multiple comparisons. The last column contains the

Figure 3. The Gene Ontology Term Finder tool (Boyle, 2004)

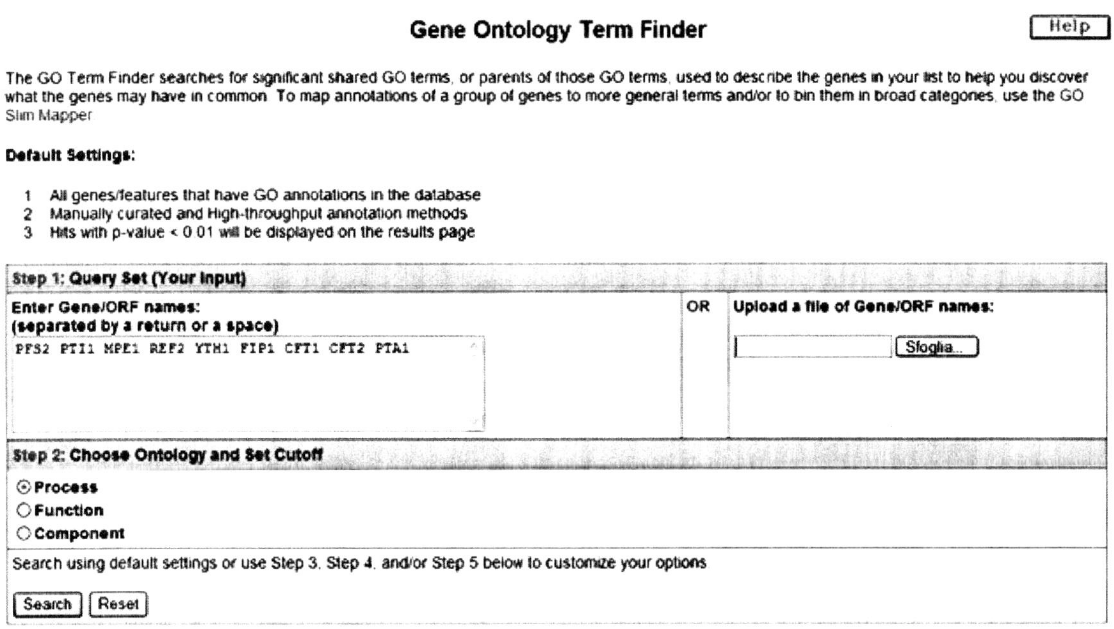

names of genes annotated to the term in the first column. By choosing the function or the component ontology in the query process the GO Term Finder returns results in a similar format.

Note that also a stand-alone version of the GO Term Finder tool is downloadable from http://search.cpan.org/dist/GO-TermFinder/. In particular, such a version allows a client program to provide a list of genes, from which it will determine p-values for the annotation associated with that list of genes, by using the hypergeometric

Figure 4. Graphic results for proteins: PFS2, PTI1, MPE1, REF2, YTH1, FIP1, CFT1, CFT2, PTA1

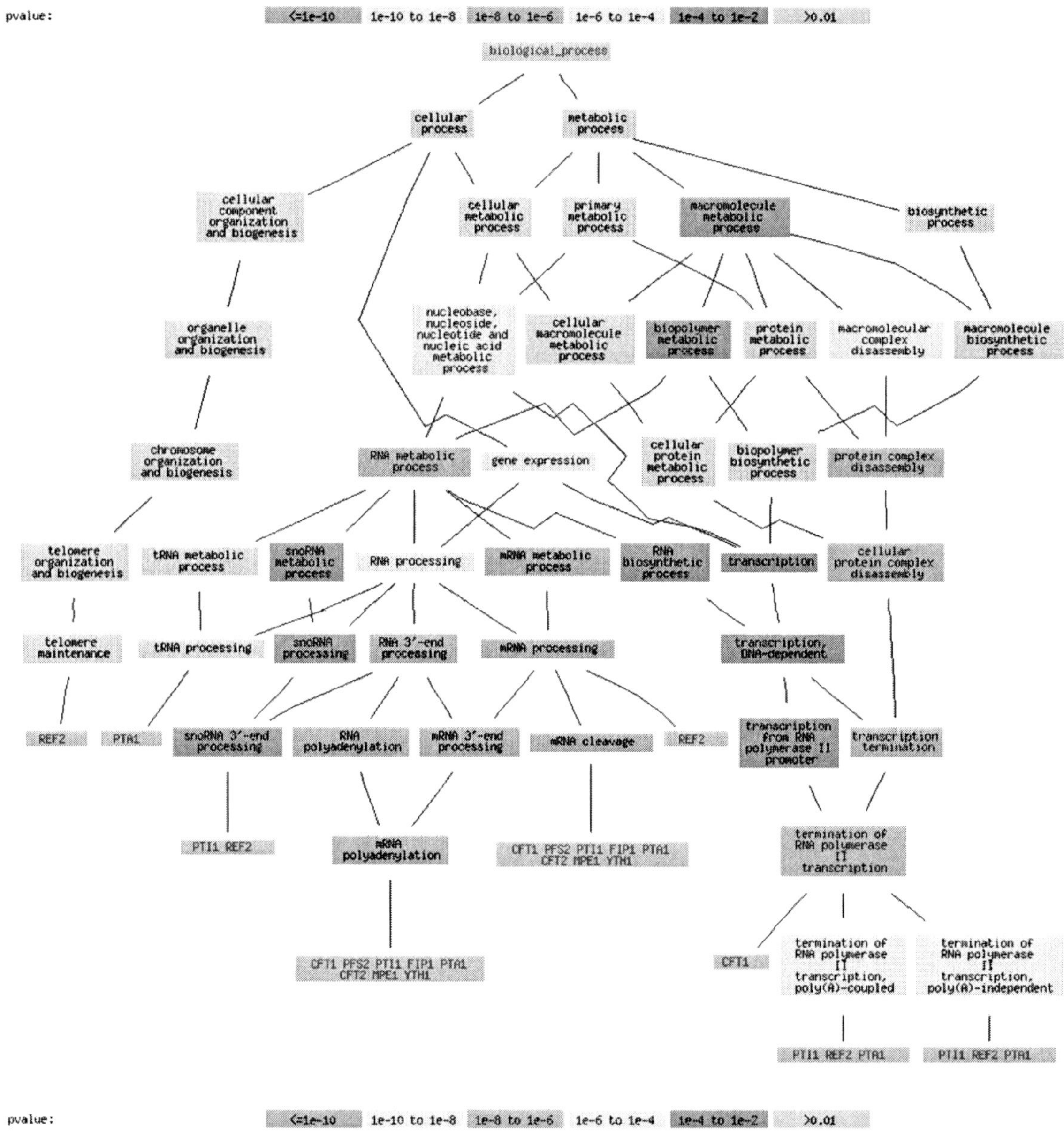

Figure 5. Top 10 tabular results for proteins: PFS2, PTI1, MPE1, REF2, YTH1, FIP1, CFT1, CFT2, PTA1

Terms from the Process Ontology				
Gene Ontology term	**Cluster frequency**	**Background frequency**	**P-value**	**Genes annotated to the term**
RNA 3'-end processing | AmiGO	9 out of 9 genes, 100.0%	38 out of 7159 background genes, 0.5%	4.45e-20	PTA1, REF2, CFT1, PTI1, FIP1, MPE1, CFT2, PFS2, YTH1
mRNA polyadenylation | AmiGO	8 out of 9 genes, 88.9%	18 out of 7159 background genes, 0.3%	8.53e-20	PTA1, CFT1, PTI1, FIP1, MPE1, CFT2, PFS2, YTH1
RNA polyadenylation | AmiGO	8 out of 9 genes, 88.9%	24 out of 7159 background genes, 0.3%	1.43e-18	PTA1, CFT1, PTI1, FIP1, MPE1, CFT2, PFS2, YTH1
mRNA 3'-end processing | AmiGO	8 out of 9 genes, 88.9%	25 out of 7159 background genes, 0.3%	2.10e-18	PTA1, CFT1, PTI1, FIP1, MPE1, CFT2, PFS2, YTH1
mRNA cleavage | AmiGO	8 out of 9 genes, 88.9%	26 out of 7159 background genes, 0.4%	3.04e-18	PTA1, CFT1, PTI1, FIP1, MPE1, CFT2, PFS2, YTH1
mRNA processing | AmiGO	9 out of 9 genes, 100.0%	150 out of 7159 background genes, 2.1%	2.26e-14	PTA1, REF2, CFT1, PTI1, FIP1, MPE1, CFT2, PFS2, YTH1
mRNA metabolic process | AmiGO	9 out of 9 genes, 100.0%	212 out of 7159 background genes, 3.0%	5.48e-13	PTA1, REF2, CFT1, PTI1, FIP1, MPE1, CFT2, PFS2, YTH1
RNA processing | AmiGO	9 out of 9 genes, 100.0%	484 out of 7159 background genes, 6.8%	1.01e-09	PTA1, REF2, CFT1, PTI1, FIP1, MPE1, CFT2, PFS2, YTH1
RNA metabolic process | AmiGO	9 out of 9 genes, 100.0%	768 out of 7159 background genes, 10.7%	6.67e-08	PTA1, REF2, CFT1, PTI1, FIP1, MPE1, CFT2, PFS2, YTH1
termination of RNA polymerase II transcription | AmiGO	4 out of 9 genes, 44.4%	18 out of 7159 background genes, 0.3%	1.29e-07	PTA1, REF2, CFT1, PTI1

distribution. In addition, a corrected p-value is calculated, to correct for multiple hypothesis testing. The correction factor used is the total number of nodes to which the provided list of genes are annotated, excepting any nodes which have only a single annotation in the background. This version of the GO Term Finder may be used also for organisms different from the yeast, by providing the correspondent ontologies.

CLASSIFICATION

Clustering approaches to PPI networks can be broadly categorized as *distance-based* and *graph-based* ones (Lin, 2006). Distance-based clustering approaches apply traditional clustering techniques, such as hierarchical clustering, by employing the concept of distance between two proteins (Blatt, 1996; Rives, 2003; Arnau, 2004; Pei, 2005). Graph-based clustering techniques consider the topology of the PIN. Then they find the clusters by applying different strategies. One of the most popular approach searches for sub-graphs having maximum density (Bader, 2003; Derenyi, 2005; Palla, 2005; Adamcsek, 2006; Altaf, 2006; Lubovac, 2006; Chua, 2007; Li, 2007; Pizzuti, 2007; Pizzuti, 2008), by using different notions of when a sub-graph is considered dense. Another approach partitions the graph by optimizing a cost function (King, 2004; Spirin, 2003). The concept of flow simulation, though applied in different ways, is exploited in (Enright, 2002; Pereira, 2004; Cho, 2006; Hwang, 2006; Cho, 2007). A statistical approach to protein clustering is taken instead in (Samantha, 2003; Farutin, 2006). Many other clustering algorithms not cited here have been proposed (Hartuv, 2000; Tornow, 2003; Brun, 2004; Dunn, 2005; Ucar, 2005; Xiong, 2005; Ucar, 2006; Chen, 2006; Demirel, 2006; Zotenko, 2006; Hwang, 2007; Asur, 2007). A complete list of all the proposals is beyond the aim of this chapter. Since other surveys already described and compared a number of methods presented in the literature (Przulj, 2005; Brohèe, 2006; Lin, 2006; Aittokallio, 2006; Sharan, 2007), in the following a description of the most known and recent graph based approaches will be given in greater details, by trying to point out the advantages and drawbacks of each method. We classified the considered approaches in four main categories, that are:

1. local neighbourhood density search (LNDS);
2. local search cost based (LSCB);
3. flow simulation (FS);
4. statistical measures based (SMB);

For each of the categories listed above, we describe the most popular and up-to-date techniques presented in the literature.

Local Neighbourhood Density Search

The approaches classified as local neighbourhood density search have the objective of finding dense sub-graphs within the network and of maximizing the density of each sub-graph.

The first method we consider is the *Molecular COmplex DEtection* (*MCODE*), presented in (Bader, 2003). MCODE detects dense and connected regions by weighting nodes on the basis of their local neighbourhood density. To this end, the *k-core* concept is applied. A *k-core* is a sub-graph in which each vertex has degree at least *k*. The highest *k-core* of a graph is the most densely connected sub-graph. The core-clustering coefficient of a node *v* is the density of the highest *k-core* of the vertices directly connected to it, included *v* itself. The weight of a node is then defined as the product of the node core-clustering coefficient and the highest *k-core* of its neighbourhood. The authors note that the core-clustering coefficient amplifies the weighting of highly interconnected regions and that the presence of singly linked nodes to a highly connected node in a dense region of a graph does not reduce the core-clustering coefficient value.

MCODE performs three steps:

1. vertex weighting,
2. complex prediction,
3. optional post-processing to add or remove proteins.

In the first step, nodes are weighted as explained above. In the second step, the vertex with the highest weight is selected as seed cluster, and neighbourhood's nodes are included in the cluster if their weight is above a fixed threshold, called vertex weight percentage (VWP). If a vertex is included in the cluster, its neighbours are recursively checked to verify if they can be part of the complex. When no more nodes can be added to the cluster, the process stops and it is repeated for the next-highest unexamined node. In such a way the densest regions of the graph are identified. Post-processing is finally optionally executed to filter proteins according to certain connectivity criteria. Clusters that do not contain at least a *2-core* are excluded. The algorithm may use a "fluff" option which increases the size of the modules to find and allows for overlapping among the complexes obtained since the nodes added to a cluster are not marked as already used. MCODE, tested on the *S. Cerevisiae* network, effectively find densely connected regions, however, because of the density threshold to set, it produces a high percentage of unassigned proteins.

Spirin and Mirny (Spirin, 2003) suggested an exhaustive enumeration of the complete subgraphs, i. e., cliques, starting from the smallest statistically significant size, which is 4, and increasing the size until all cliques are found. Actually the authors, in the same paper, proposed another method that will be described later, thus this one is included for completeness.

Another technique based on LNDS is that proposed in (Adamcsek, 2006), where *CFinder*, a program for detecting and visualizing densely interconnected and overlapped groups of nodes, is described. CFinder uses the *Clique Percolation Method* (Derenyi, 2005; Palla, 2005) to find *k-clique percolation* clusters, i.e., groups of nodes that can be reached via chains of k-cliques and the link in these cliques. The idea behind this approach is that a cluster can be interpreted as the union of small fully connected sub-graphs that share nodes. A k-clique is a complete sub-graph constituted by *k* nodes such that there is an edge for each pair of nodes. Two k-cliques are said adjacent if they have *k-1* common nodes. A *k-clique-community* is then defined as the union of all the k-cliques that can be reached through adjacent k-cliques.

The algorithm implemented in CFinder extracts all the *maximal complete sub-graphs*, i.e., the maximal cliques. Then a clique-clique overlap matrix is built in which each entry contains the number of common nodes between the two corresponding cliques, and each diagonal entry is the clique size. The k-cliques-communities can be found by deleting every entry off the diagonal having a value less than *k-1*, and every diagonal entry less than *k*. The remaining separate components will be the k-cliques-communities. The parameter *k* has to be provided in input. Increasing *k* shrinks community size because nodes must belong to at least a clique of size *k*. One of the characteristics of CFinder is that it allows overlap between communities, thus each node can be member of several different modules at the same time.

DPClus (Altaf, 2006) discovers protein complexes in large interaction graphs by using the concepts of density and neighbourhood. The authors introduce the concepts of cluster property and node weight that take into account the common neighbours of nodes belonging to the same cluster. The density d_k of any cluster k is the ratio of the number of edges present in the cluster and the maximum number of possible edges in the cluster. The *cluster property* of a node u with respect to a cluster k of density d_k and size $|N_k|$ is defined as

$$cp_{uk} = \frac{|E_{uk}|}{d_k \times |N_k|},$$

where E_{uk} is the total number of edges between the node u and all the nodes of cluster k. The weight w_u of a node u is the sum of the weights of the edges connected to a node, where the weight of an edge (u,v) is the number of common neighbours of u and v. DPClus needs two input parameters: a minimum density value d_{in} and a minimum value of cluster property cp_{in} to decide the insertion of a new neighbouring node in a cluster. These two parameters influences the kind and number of obtained clusters. The algorithm starts by ranking all the nodes with respect to their weights and choosing as seed a node having the highest weight. Thus, the cluster containing the seed node is expanded by adding neighbouring nodes, provided that the density and the cluster property of the expanded cluster do not diminish below the fixed initial input values.

Once a cluster is generated, its nodes are removed from the graph and the next cluster is built on the remaining nodes until all the nodes have been assigned to a module. However the algorithm allows to maintain the nodes already members of a cluster, thus permitting the generation of overlapping clusters. An experimental study for different input values shows that if high values for either cp_{in} or d_{in} are used, the clusters obtained have high density but are smaller in size and more in number. The authors argue that a reasonable balanced value for cp_{in} is 0.5, higher values tend to produce complexes having more spherical structure.

The algorithm SWEMODE (Semantic WEights for MODule Elucidation), presented in (Lubovac, 2006), identifies dense sub-graphs by introducing two network measures that combine functional information with topological properties of the networks. These measures, *weighted cluster coefficient* and *weighted nearest-neighbours degree*, compute the strengths of interactions between the proteins by using their semantic similarity based on the Gene Ontology terms of the proteins. The method ranks all the nodes of the network by assigning a weight according to their weighted neighbourhood cohesiveness. The highest ranked nodes are considered as seeds for candidate modules. The algorithm then searches in the neighbourhood of each seed protein to detect densely connected proteins with high functional similarity. Proteins satisfying a parameter named *Node Weight Percentage (NWP)* are included in the current module. This module prediction is repeated for the second highest weighted node and so on until all the nodes have been examined. The parameter NWP determines the number of obtained clusters. High values of NWP result in modules containing proteins weakly functionally correlated to the seed node. On the contrary, low values of NWP produce many modules with few proteins, the majority being constituted by two or three proteins. The authors found that a value of 0.4 is a good compromise to obtain meaningful modules.

DECAFF (Dense-neighbourhood Extraction using Connectivity and conFidence Features), proposed by (Li, 2007), is an algorithm to mine protein complexes in PIN that tries to address two major limitations plaguing protein interaction data, namely incompleteness and noise. The method consists of three main steps: detection of local dense neighbourhoods of each protein, merging of the local sub-graphs on the base of the similarity degree between neighbourhoods, filtering away possible false complexes detected. In order to find the local dense neighbourhood of a node, first its local cliques are obtained by adapting the method LCMA (Li, 2005) of the same authors. Local cliques are found by iteratively removing nodes with lowest degree. However, dense non-clique sub-graphs for each node could be parts of complexes.

Thus, a Hub Removal algorithm is applied that removes hub proteins (i.e., those having the highest degree) and their edges for the graph, and then repeating this process on its connected components, until a dense group of proteins is obtained. Hub proteins are finally inserted back into the cluster. As regards merging, the authors introduce the concept of neighbourhood affinity between two overlapping neighbourhoods. The neighbourhood affinity between two sub-graphs A and B is defined as:

$$NA(A,B) = \frac{|A \cap B|^2}{|A| * |B|}.$$

Neighbourhoods affinity measures the similarity between two neighbourhoods. Thus, maximal dense neighbourhoods are obtained by merging local dense neighbourhoods having an affinity value above a given threshold. The last step deletes dense sub-graphs having low reliability scores. The protein interaction network is modelled as a weighted graph, where the weights correspond to the reliability of the interactions. Dense sub-graphs containing edges with low reliability are deleted. Experiments on three protein-protein interaction data sets for yeast showed that DECAFF produces better results with respect to MCODE (Bader, 2003) and RNSC (King, 2004).

In (Chua, 2007) a method for the identification of protein complexes that exploits the shared interaction partners of proteins, i.e., the level-2 neighbours, instead of the direct connections, is proposed. The protein-protein interaction network is preprocessed by expanding it with the indirect interactions, and then computing a topological weight, named FS-Weight, for both direct and indirect edges. The FS-Weight estimates the reliability of the interactions, i.e. the likelihood that two proteins share functions. Direct interactions having a weight below a fixed threshold are removed, while level-2 interactions having a weight above the threshold are added into the original network. The modified network can then be used to discover dense sub-graphs by any existing algorithm. However the authors propose a new method, *Protein Complex Prediction (PCP)*, that first finds all the maximal cliques of the modifies network, and then merges them by exploiting the concept of *Inter-Cluster density (ICD)*. The ICD measures the interconnectedness between two sub-graphs by computing the FS-Weight density of the inter-cluster interactions between the proteins not belonging to both the sub-graphs. Given two clusters, a high value of their ICD means that the two clusters are highly connected. The merge procedure considers an initial graph constituted by *partial cliques*, i.e., strongly connected components composed by cliques, and adds an edge between two partial cliques if their ICD value is above a fixed threshold.This is repeated until no further merge is possible. Experiments show that PCP performs better than MCODE (Bader, 2003), RNSC (King, 2004) and MCL (Enright, 2002) methods.

The last method we consider in this section is the algorithm *MF-PINCoC, Multi-Functional PPI network Co-Clustering*, (Pizzuti, 2008) an extension of the method PINCoC (Pizzuti, 2007) that search for overlapping functional modules in protein-protein interaction networks. MF-PINCoC is based on the concept of co-clustering. Co-clustering methods (Madeira, 2004), differently from clustering approaches, aim at simultaneously grouping both the dimensions of a data set.

MF-PINCoC considers the binary adjacency matrix A of the interaction graph, where rows and columns correspond to proteins and a 1 entry at the position (i, j) means that the proteins i and j interact. The algorithm searches for, eventually overlapping, dense sub-matrices containing the maximum number of ones by using a greedy local search technique. The notion of *quality* of a sub-matrix, that reflects the concept of maximal dense sub-matrix, is introduced. High quality sub-matrices should correspond to modules of the interaction graph having a significant biological function. The algorithm starts with an initial random solution constituted by a single protein and searches for a locally optimal solution by adding/removing connected proteins that best contribute to improve the quality function. In order to escape poor local maxima, with a fixed probability, the protein causing the minimal decrease of the quality function is removed. To enable participation of a protein to more groups, its degree, i.e. the number of other proteins with which it is connected, is computed. A protein can be added to the current cluster if the number of clusters to which it has already been assigned is less than its degree. When the algorithm cannot improve any more the solution found so far, the computed cluster is returned. At this point a new random protein is chosen, and the process is repeated until all the proteins are assigned to a group. Experiments on the *S. Cerevisiae* interaction network showed that the algorithm is able to efficiently isolate those groups of proteins corresponding to the most compact and biologically meaningful sets of interactions.

The approaches described in this subsection has the objective of finding dense subgraphs within the network and to maximize the density of each subgraph. MCODE, DPClus, and SWEMODE have a similar strategy. They define the weight of each node, the node with highest weight is chosen as seed cluster, neighbouring nodes are added to the current cluster if threshold parameters are satisfied. The main difference lies in the definition of weight. To this end, MCODE and DPClus use the network topology, while SWEMODE combines also semantic information coming from the Gene Ontology database. Thus the latter approach should give increased confidence in the predicted function. However, SWEMODE does not allow the participation of a protein to more than one group. DECAFF and PCP follow a similar approach too. Both first find cliques and then merge them to obtain larger dense sub-graphs. The merging procedures of the two methods use different criteria. Furthermore PCP performs a pre-processing on the interaction graph. The other two approaches, CFinder and MF-PINCoC, generate overlapping clusters, and use the concepts of k-clique and co-cluster to find dense sub-graphs. All the methods need some parameters that biases the number and kind of clusters obtained.

Local Search Cost Based

The local search cost based methods extract modules from the interaction graph by partitioning the graph into connected subgraphs. The partitioning is obtained by deleting edges deemed not relevant. In this category two methods are examined.

The first LSCB based algorithm we consider is the *Restricted Neighbourhood Search Clustering* (*RNSC*), proposed by King et al in (King, 2004). RNSC is a cost-based local search algorithm that explores the solution space of all the possible clusterings to minimize a cost function that reflects the number of inter-cluster and intra-cluster edges. RNSC uses two cost functions. The first, called the *naive cost function*, for each node *v*, computes the number of bad connections incident with *v*, i.e., one that exists between *v* and a node not belonging to the same cluster of *v*, or one that does not exist between *v* and another node in the same cluster as *v*. The second one, called the scaled cost function, measures the size of the area that *v* effects in the clustering. The algorithm begins with a random clustering, and attempts to find a clustering with low naive cost by moving a vertex from a cluster to another one. Then it tries to improve the solution by searching for a clustering with low scaled cost. Since the approach is greedy, the problem of getting stack into poor local minima is dealt by making diversification moves that mix up the clustering by scattering the clusters at random. Furthermore, RNSC maintains a list of tabular moves that forbid to cycle back to previously examined solutions. In order to output clusters likely to correspond to true protein complexes, thresholds for minimum cluster size, minimum density, and functional homogeneity must be set. Only clusters satisfying these criteria are given as the final result. This obviously implies that many proteins are not assigned to any cluster.

Spirin and Mirny (Spirin, 2003) proposed a Monte Carlo (MC) simulation that finds clusters of a prefixed number of nodes *m*. The method begins by picking a random set of *m* nodes and by computing for each pair of nodes their shortest path. The sum of all the shortest paths is denoted by l_0. The next step selects a node at random from the set, and a node at random outside the set. Then the new sum l_1 of shortest paths is computed by considering the set from which the former node is deleted and the latter is added. If $l_1 < l_0$ then the replacement is accepted, otherwise the replacement takes place with a probability depending on a temperature parameter. The process is repeated until the original set converges to a complete sub-graph or the predetermined number of steps is over. In this case, the tightest sub-graph obtained is recorded. The found clusters are finally tested for statistical meaning and then merged if they overlap. The authors compared MC with the Super-paramagnetic Clustering (SPC) method of Blatt et al. {Blatt,1996), a hierarchical clustering method inspired from the physical properties of a ferromagnetic model subject to fluctuations at nonzero temperature.

They found that MC performs better on dense graphs, while SPC is able to find clusters having few links to the rest of the graph. The main drawback of the MC approach is that the size of the clusters to determine must be given a priori.

Flow Simulation

The flow simulation approaches mimic the spread of information on a network. We describe five approaches, however two of them (Pereira, 2004; Cho, 2007) are specializations of the others. In particular, the first representative of this class, MCL (Enright, 2002), simulates the behaviour of many walkers starting from the same point, that move within a graph in a random way. The approach of Cho et al. in (Cho, 2006), instead, tries to imitate the scattering of information inside the network from some informative nodes to determine the proteins influenced from the starting nodes. The main difference between the two approaches is that the former exploits only the network topology, while the latter uses semantic information. The latter approach (Hwang, 2006) models the dynamic relations between proteins using a signal transduction model.

The Markov Clustering algorithm (MCL) presented in (Enright, 2002) is based on the concept of random walk on a graph to retrieve cluster structure and uses algebraic operations on a distance matrix associated with the graph. It was originally designed for graph clustering. In a random walk the direction to be followed at each node is given by chance. MCL simulates many random walks (or flows) within a graph by strengthening flow where it is strong, and weakening it where it is weak. By repeating this process, a number of regions with strong internal flow (the clusters), separated by boundary with no flow will appear. The flow is simulated by algebraic operations on a stochastic Markov matrix. The edges of the graph are weighted according to a sequence similarity score obtained from an algorithm like BLAST. The weight of an edge represents the degree of similarity between two proteins. A Markov matrix is then built in which each entry is the similarity between two proteins. The algorithm use an operator of flow expansion by computing the square power of the Markov matrix, and an inflation operator which raises each entry of the matrix to a given power, and then rescales the matrix so that the column sum equals 1. By repeating a number of times squaring, inflating, and scaling the matrix tends to an equilibrium state that shows the cluster structure. The inflation parameter influences the number of clusters.

Another technique based on flow simulation is that presented by Pereira et al. in (Pereira, 2004), where the interaction graph is transformed into the corresponding line graph, in which edges represent nodes and nodes represent edges.

This approach is often used in graph theory and has a number of advantages. First, it can recover the original network, thus all the information content is maintained. Second, it takes into account not only the direct neighbours of a node. Third, it is more highly structured of the original graph, i.e., the number of connections between the nodes increases. In fact, it has been verified that the line graph has a higher clustering coefficient than the original graph.

Furthermore, the line graph clustering approach produces an overlapping graph partitioning of the original protein-protein interaction graph, thus allowing proteins to be present in multiple functional modules. The authors apply the graph clustering algorithm MCL of (Enright, 2002) to group the interaction network corresponding to the line graph.

Cho et al. in (Cho, 2006) proposed an information flow-based approach to identify overlapping functional modules. The algorithm integrates topological and biological knowledge to select a number of *informative* proteins and simulates the information flow through the network from each informative protein. An informative node, through the information flow simulation, determines the nodes it influences that should belong to the same module. To this end, in order to define the information that flows from a node v_i to a connected node v_j, the concepts of *weighted degree* d_i of a node v_i and the *flow* f_s of the information of a node v_s from a node v_i to a node v_j are introduced. The weighted degree d_i of a node v_i is defined as the sum of the weights of the edges between v_i and its neighbours, where the weight w_{ij} of an edge (v_i, v_j) is computed using the correlation of the expression profiles between the two genes that encode the proteins v_i and v_j. Thus the weighted degree gives a semantic information. The information rate of v_s on v_i, denoted $inf_s(v_i)$, is the amount of information of v_s that is flowing on v_i. Finally, the flow f_s is defined as the product of the weight w_{ij} and the information rate normalized by the number of neighbours.

It is worth to note that the information flow diminishes each time it passes through an edge. This means that an initial informative node v_s, during a path, influences a set of proteins. The most strongly influenced ones will be considered belonging to the same module of v_s.

The algorithm performs three steps:

1. selection of the informative nodes,
2. detection of the preliminary modules by information flow simulation for each informative node,
3. merging of the preliminary modules.

The informative nodes are selected by taking the nodes having the highest value of the weighted degree. The number of informative nodes will determine the number of preliminary modules detected by the algorithm. In the second step the flow of an informative node v_s is delivered into all its neighbours, visiting all the nodes through every possible walk. This means that around the densely connected area of v_s more information is conveyed. All the nodes for which the flow is greater than a threshold θ_{inf} are included in the same module. The third step, finally, performs a post-processing merge of the preliminary modules to produce the final clusters.

In order to merge two modules, a similarity measure must be introduced. The similarity between two clusters *Ms* and *Mt* is defined as the ratio of the sum of the edge weights between all the pairs of nodes $v_i \in Ms$ and $v_j \in Mt$ for which the edge (v_i, v_j) exists and the size of the smaller module. Two modules with the strongest interconnections are than iteratively merged, provided that their similarity is greater than a user-defined threshold.

Thus, the algorithm needs three parameters: the number of informative nodes to use in the first step; the threshold θ_{inf} to decide when a flow is considered significant for a protein, the similarity threshold value to determine when two preliminary modules can be merged. These parameters affect the number, the size and the shape of the obtained clusters.

In the experiments the authors used different values of these parameters in order to choose the proper values leading to the better results. Based on this study, they found that a value of 0.1 for θ_{inf}, a similarity threshold of 6.0, and preliminary module numbers of 5%, 7% and 10% the number of proteins, produces results that outperform other methods in terms of accuracy.

A variant of the approach is presented by the same authors in (Cho, 2007) to compute the weight of an edge and, consequently, to select the informative proteins. Two metrics based on the Gene Ontology annotations are introduced: the *semantic similarity* and the *semantic interactivity*. The semantic similarity of two proteins is computed by considering the common GO terms on which the two proteins are annotated. The semantic interactivity of two proteins *x* and *y*, instead, is computed by taking into account the probability that *x* interacts with the annotated proteins on the GO terms, whose annotation includes *y*.

A comparison with other approaches shows that this method finds more modules than the others considered, the average size of the found modules is larger and the p-value is lower.

The last approach we describe is the algorithm STM (Signal Transduction Model), presented in (Hwang, 2006), that finds clusters of arbitrary shape by modelling the dynamic relationships between proteins of a PPI network as a signal transduction system. The overall signal transduction behaviour between two proteins *v* and *w* of the network is defined in order to evaluate the perturbation of protein *v* on the other protein *w* both biologically and topologically. The signal transduction behaviour is modelled using the Erlag distribution:

$$F(c) = 1 - e^{-\frac{x}{b}} \sum_{k=0}^{c-1} \frac{(\frac{x}{b})^k}{k!}$$

where $c>0$ is the shape parameter, $b>0$ is the scale parameter, $x \geq 0$ is the independent variable, usually time. The authors used $x/b = 1$ and *c* equals to the number of edges between a source protein node *v* and the target protein node *w*. In this case the perturbation at the target protein *w* is assessed when the perturbation at the nearest neighbour of the source protein *v* reaches 1/e of the initial value.

The signal transduced from node *v* to *w* is then defined as:

$$S(v \to w) = \frac{d(v)}{\prod_{i \in P(v,w)} d(i)} F(c)$$

where *d(i)* is the degree of node *i*, *P(v,w)* is the set of all the nodes visited on the shortest path from *v* to *w*, and *F(c)* is the signal transduction behaviour function. The term on the denominator represents the dissipation on each node along the shortest path from *v* to *w*.

The algorithm starts by computing for all the protein pairs the signal transduced. Then, the cluster representatives are selected for each node. The cluster or module representatives are the most influential nodes, where influential means having the highest scores of the signal transduced on each node of the module. From these nodes, preliminary clusters are formed by aggregating each node *w* to the module having as representative the node *v* for which $S(v \rightarrow w)$ is the highest. Finally, these clusters are merged if there exists a substantial number of interconnections.

A peculiarity of STM is that if a tie score appears on a node, then that node will be assigned to all the cluster representatives corresponding to that score. This allows for overlaps among clusters. However, this aspect is not discussed in the experimental section, thus the effective degree of overlapping among clusters is not known.

By the comparison with other approaches the authors show that STM outperforms all the methods by obtaining clusters of arbitrary shape, higher size, better p-values, low density, and very low percentage of unassigned proteins.

Statistical Measures Based

The two following approaches relies on statistical concepts to cluster proteins. The former, (Samantha, 2003), deems important the number of shared neighbours between two proteins. The latter, (Farutin, 2006), counts on the notion of preferential attachment of the members of a module to the others of the same module.

Samantha and Liang in (Samantha, 2003) propose a clustering approach based on the idea that if two proteins share a number of common interaction partners larger than what is expected for a random network, then they should be clustered together. The method thus rank the statistical significance of forming shared partnership for all the pairs of proteins by using the concept of p-value of a pair of proteins. The p-value of a protein pair computes the probability that the two proteins, with n_1 and n_2 neighbours respectively, have m shared neighbours. Once the p-values for all possible proteins pairs have been computed and stored in a similarity matrix, the protein pair with the lowest p-value is chosen to form the first group and the corresponding rows and columns are merged in a new row and column. The new p-value of the merged row/column is the geometric mean of the separate p-values of the corresponding elements. This process is repeated to add new proteins to the original cluster until a threshold is reached. Then the protein pair with the remaining lowest p-value is selected to generate the next cluster. The process is repeated until all the proteins have been clustered. Experiments on the *Saccharomyces cerevisiae* network revealed many reliable associations. From the derived modules the authors were able to predict functions for 81 unannotated proteins with high certainty.

Farutin et al. in (Farutin, 2006) presented a statistical approach for the identification of protein clusters, called also *communities*, based on the concept of preferential interaction among the members of a module. The authors use a novel metric to measure the community strength. The community strength is gauged by the preferential attachment of each member of a module to the other of the same module. This concept of preferential attachment was quantified by how unlikely it is observed in a random graph.

Since it is necessary to count the number of edges in the graph, the author assume a random graph as the null model where the edge is the random variable. This measure of community strength is local, since it is a function of the sub-graph induced by a set of proteins and their degrees. Given a list L of nodes and a node $v \in L$, let x be the number of edges between v and the members of L. Then the attachment of v to the others nodes of L is defined as the probability of observing x or more edges between v and L. The community score of L is thus defined as the average of the attachments of all its nodes. To identify the clusters, a greedy approach that searches for a set of nodes in the network with small values of community strength is adopted. At the beginning a list of two adjacent nodes is considered. The list is then grown by adding the node that leads to the largest decrease of the community score until no such node exists. This is repeated for each connected node pair, thus the obtained clusters can partially overlap. The algorithm has been applied to three protein networks, two coming from the yeast proteins and one from human interacting proteins. The p-values obtained on functional categories of the Gene ontology were very small, thus indicating that the found modules are significant.

A Small Example

After that an extensive overview of the main techniques proposed in the literature for PINs clustering has been provided, we show in this section an example where four of the discussed methods are queried on a small *S. cerevisiae* (yeast) sub-network, and both the returned results and their validation are illustrated. The aim of this example is that of making the reader more familiar with the usage of clustering and validation tools, and also of facilitating the full understanding of the topics discussed in the chapter.

Consider the 13 proteins of *S. cerevisiae* whose correspondent gene names are listed below, together with a brief description of their function taken from the SWISSPROT database (Bairoch, 1996):

1. BNR1, may organize microtubules by mediating spindle positioning and movement in the budding process. It is a potential target of the RHO family members.
2. PFY1, binding to actin and affecting the structure of the cytoskeleton.
3. TRM5, specifically methylates guanosine-37 in various tRNAs.
4. BNI1, required for the assembly of F-actin structures, such as actin cables and stress fibers. It may organize microtubules by mediating spindle positioning and movement in the budding process. It is a potential target of the RHO family members.
5. AIP1, involved in the depolymerization of actin filaments. It enhances the filament disassembly activity of cofilin and restricts cofilin localization to cortical actin patches.
6. CDC42, involved in development of cell polarity during the cell division cycle, and essential for bud emergence.
7. SVL3, may have a vacuolar function.
8. CRN1, may have a vacuolar function.
9. COF1, controls reversibly actin polymerization and depolymerization in a pH-sensitive manner. It has the ability to bind G- and F-actin in a 1:1 ratio of cofilin to actin. Such a protein complex is important for the polarized growth of yeast cells.
10. SRV2, whose N-terminal domain binds to adenylyl cyclase, thereby enabling adenylyl cyclase to be activated by upstream regulatory signals, such as Ras. The C-terminal domain is required for normal cellular morphology and growth control.
11. ACT1, belonging to actins. Actins are highly conserved proteins that are involved in various types of cell motility and are ubiquitously expressed in all eukaryotic cells.

Figure 6. The network corresponding to the 13 yeast proteins considered in the small example

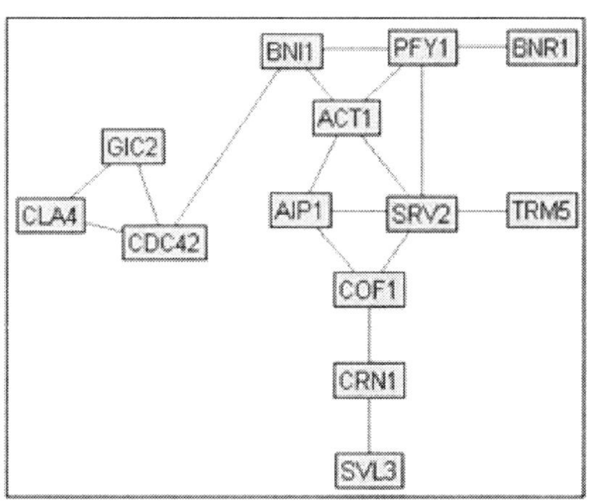

Table 1. Clustering coefficients of the nodes in the network represented in Figure 6

Proteins	**Clustering coefficient of the node**
GIC2, CLA4	1
AIP1	0.667
ACT1	0.5
CDC42, BNI1, PFY1, COF1	0.333
SRV2	0.3
BNR1, TRM5, CRN1, SVL3	0

12. CLA4, involved in budding and cytokinesis.
13. GIC2, required for cell size and shape control, bud site selection, bud emergence, actin cytoskeleton organization, mitotic spindle orientation/positioning, and mating projection formation in response to mating pheromone.

As it is evident by the list above, all the 13 considered yeast proteins are well characterized, and this makes easier the validation process. By downloading interactions by databases such as DIP (Xenarios, 2000), it is possible to build the network connecting such 13 proteins, that is a sub-network of the yeast PIN. This sub-network is shown in Figure 6, drawn by using PIVOT (Orlev, 2004), and the clustering coefficients of its nodes are reported in Table 1.

The first tool we queried on the 13 considered yeast proteins is MF-PINCoC (Pizzuti, 2008). In particular, to query such tool it is necessary to provide a file containing the matrix of interactions associated with the PIN. MF-PINCoC returned six different clusters, that are:

(a) {PFY1, BNI1, SRV2, ACT1};
(b) {CDC42, CLA4, GIC2};
(c) {SVL3, CRN1};
(d) {AIP1, COF1, SRV2, ACT1};
(e) {BNR1, PFY1};
(f) {TRM5, SRV2}.

Note that proteins SRV2 and ACT1 participate to more than one cluster. After validating the six clusters by the GO Term Finder, it results that the most probable process associated with cluster (a) is *Cytoskeleton organization and biogenesis*, involving all the four proteins, with p-value $7.19e^{-05}$. Proteins of cluster (b) are all involved in *Rho protein signal transduction*, with p-value $5.34e^{-07}$, whereas proteins of cluster (d) are found to participate in *Cytoskeleton organization and biogenesis*, with p-value $2.24e^{-05}$. No significant terms have been found for clusters (c), (e) and (f), each made of only two proteins; by observing Figure 6, we can note that such three clusters actually correspond to chains or leafs. The clustering coefficient of the clustering returned by MF-PINCoC is equal to 0.319.

The second tool we considered is MCODE (Bader, 2003), which returned the two clusters:

(a) {AIP1, COF1, SRV2, ACT1}
(b) {CDC42, CLA4, GIC2}

Note that the cluster (a) of MCODE is the same of the MF-PINCoC cluster (d) and that the cluster (b) of MCODE is the same of the MF-PINCoC cluster (b). Differently from the first considered tool, MCODE clustering does not take into consideration all the proteins in the network. The clustering coefficient for the MCODE clustering is equal to 0.614.

We queried the tool CFinder (Adamcsek, 2006) on the same small yeast sub-network, obtaining two "communities" (according to the term used in the paper):

(a) {ACT1, AIP1, BNI1, COF1, PFY1, SRV2}
(b) {CDC42, CLA4, GIC2}

Again, cluster (b) is the same of clusters (b) returned by the other two methods considered above. Cluster (a), instead, is the union of the clusters (a) and (d) obtained by MF-PINCoC. The first process in the list returned by the GO Term Finder is *cytoskeleton organization and biogenesis*, involving all the proteins in that community, and the associated p-value is equal to $8.12e^{-08}$. The clustering coefficient is equal to 0.6.

Finally, the last tool we considered in this example is RNSC (King, 2004), which returned the following groups:

(a) {TRM5}
(b) {GIC2, CLA4, CDC42}
(c) {ACT1, BNI1, PFY1}
(d) {BNR1}
(e) {CNR1, SVL3}
(f) {AIP1, SRV2, COF1}

Note that only clusters (b), (c), (e) and (f) may be validated by the GO Term Finder, since (b) and (d) contain singletons. In particular, cluster (b) is the same of clusters (b) obtained also by MF-PINCoC, MCODE and CFinder; proteins BNI1 and PFY1 of cluster (c) are found to participate in *actin nucleation* with p-value $2.45e^{-05}$; no significant GO terms may be associated with cluster (e), whereas only proteins COF1 and AIP1 of cluster (f) are involved in *actin filament depolymerization* with p-value $1.87e^{-06}$. The clustering coefficient of the RNSC clustering is equal to 0.267.

Summarizing, we can observe that all the considered methods agree on the cluster {CDC42, CLA4, GIC2}. The validity of such cluster is also supported by both technical and biological validations, as highlighted by the associated clustering coefficient, that is equal to 0.778, and by the low p-value. For the other groups, the variety of the results returned by the different methods might be associated with the variety of the processes involving the proteins under consideration. This is also confirmed by the fact that, applying a method that allows for multiple-clusters participations, there exist proteins belonging to different clusters.

DISCUSSIONS AND FUTURE TRENDS

After surveying on the main PPIs clustering graph-based approaches, we summarize in this section the main features of the described techniques, and also discuss the possible limitations of the presented methods, in order to give the starting point for further solutions. Finally, we point out some open issues about the topic discussed in this chapter.

We tried to synthesize the characteristics of each method in Table 2. The features chosen to emphasize have been the structure of the found clusters, the kind of approach, on the base of the classification of the previous section, whether the clusters are found simultaneously or one at a time, the capability of allowing the participation of a protein to many modules, the requirement about the number of clusters to find, and, finally, if the method assigns each protein to a cluster. Table 2 reports all the algorithms in chronological order. It is worth to point out that none of the presented methods, except the Monte Carlo method of Spirin and Mirny, need the number of clusters to be given a priori. However, all the algorithms have some input parameters that influence the number

Table 2. Summarizing the main methods characteristics. In the first column the methods, in chronological order, are listed. The second column highlights the structure of clusters each method searches for, while the approach each method is based on is reported in the third one. The fourth column specifies if the approaches find the clusters simultaneously and the fifth column if they admit multiple-clusters participation. Finally, the last two columns point out if the number of clusters is required at the beginning and if the methods return some unassigned proteins.

Method	Structure	Approach	Simultaneity	Overlapping	N. of requested clusters	Unassigned proteins
MCL (Enright, 2002)	arbitrary	FS	yes	no	no	no
(Spirin, 2003), MC	dense sub-graphs	LSCB	no	no	yes	no
(Spirin, 2003), clique enumeration	cliques	LNDS	no	no	no	yes
(Samantha, 2003)	arbitrary	SMB	no	no	no	no
MCODE (Bader, 2003)	dense sub-graphs	LNDS	no	yes	no	yes
RNSC (King, 2004)	dense sub-graphs	LSCB	yes	no	no	yes
(Pereira, 2004)	dense sub-graphs	FS	yes	yes	no	no
STM (Hwang, 2006)	arbitrary	FS	yes	yes	no	yes
SWECODE (Lubovac, 2006)	dense sub-graphs	LNDS	no	no	no	yes
DPClus (Altaf-Ul-Amin, 2006)	dense sub-graphs	LNDS	no	yes	no	yes
(Cho, 2006; Cho, 2007)	arbitrary	FS	no	yes	no	yes
(Farutin, 2006)	arbitrary	SMB	no	yes	no	no
CFinder (Adamcsek, 2006)	dense sub-graphs	LNDS	yes	yes	no	yes
PCP (Chua, 2007)	dense sub-graphs	LNDS	no	yes	no	yes
DECAFF (Li, 2007)	dense sub-graphs	LNDS	no	yes	no	yes
MF-PINCoC (Pizzuti, 2008)	arbitrary	LNDS	no	yes	no	no

of the clusters produced, the size, the density, and the structure. The LSDS methods, except CFinder, obtain the modules one at a time because they select a seed node and expand it until a condition, generally related to cluster density, is satisfied. Thus they can be considered bottom-up approaches: individual nodes are grouped together until all the graph has been examined. Approaches that simultaneously find the clusters have instead a top-down approach. They consider the whole graph and try to partition it in connected components by cutting edges. Because of the threshold constrains many methods require to be satisfied, in order to decide when a group of connected nodes is a cluster, nodes with few interactions often are discarded. In fact, the presence of these proteins would reduce the value of the function to optimize. The elimination of sparsely connected node has two drawbacks. First it prevents the possibility of having topological shapes different from maximally dense sub-graphs. Second, important information on the network structure could be lost. In (Hwang, 2006; Cho, 2006) the authors observe that bottom-up approaches discard a high percentage of nodes, though the clusters obtained have a relatively more accurate p-value. On the other hand, top down approaches have a lower number of unassigned proteins, yield modules with larger size, but the average p-value is lower.

All the presented methods showed the presence, in protein interaction networks, of groups of highly connected proteins involved in common cellular processes. The modular nature of biological networks hypothesized by Hartwell et al. (Hartwell, 1999) can thus be deemed effective. The computational methods described can be considered a valuable help in identifying network components and the connections among such components to do a step forward the understanding of cell biology.

A Look At The Future

An important aspect of PINs analysis is that, due to the nature of high-throughput experimental techniques (Ito, 2001; Krogan, 2006) and computational methods (Miller, 2005; von Mering, 2002) often exploited to discover new protein interactions, many of the stored interactions are not completely reliable (Sprinzak, 2003), as also testified by several studies (Bader, 2004; Deng, 2003). The potentially limited reliability of such data may affect any attempt to extract useful information, also in the case of detecting protein complexes. Some techniques recently presented in the literature (e.g., Cho, 2007) propose the introduction of weights in PINs representation in order to limit this problem. As already illustrated in the previous sections, Cho et al. use Gene Ontology (GO) annotations to measure the reliability of protein-protein interactions. An interesting direction of analysis might be that of enriching such weights by other kind of reliability information; for example, there exist protein interaction databases, such as MINT (Chatraryamontri, 2006), which provide useful information about the reliability of stored interactions in terms of numeric coefficients associated with interactions. The exploitation of numeric coefficients directly available on public interaction databases could make easier the process of taking into account reliability in PINs analysis.

Another issue to look at for the future research concerns the idea, already appeared in (Cho, 2007; Lubovac, 2006), of exploiting semantic and functional information related to protein nodes. In our opinion, a further direction of investigation might be the consideration of node-labels in PINs representation, related to the similarities between pairs of proteins. Such a similarity could be computed, for example, by protein sequence alignment, protein domain comparison or protein neighbourhoods exploration. Node-labels might also be associated, e.g., with expression profiles of the annotated genes.

The observations reported above may also be enforced by the fact that both node- and edge-labels consideration showed already successful in the context of other problems involving PINs, such as motif extraction (Chen, 2007), network alignment (Fionda, 2007) and network querying (Fionda, 2008), as also clarified in the chapter of this book dealing with "Protein Interaction Network Querying Techniques".

CONCLUDING REMARKS

In this chapter, we dealt with the problem of mining complexes from PINs. In particular, one of the main goal of this chapter has been that of introducing the reader to the problem of clustering protein interaction networks. With this aim in mind, we first described some preliminary notions and the main clusters validation techniques, and then provided an overview of the main and most recent graph-based clustering approaches that have been presented in the literature. In particular, we classified them in four categories, and discussed their main features, advantages and limitations. We also showed a small example in order to make the reader familiar to the process of querying clustering tools and validating their results.

As already point out in the previous section, interesting challenges about the topics discussed in this chapter are still open, and this represent a strong motivation for further researches in this field.

RESOURCES

In this section, we provide a list of the available tools and other resources useful for the discovery of protein complexes in PINs, with the corresponding links to web sites, whenever available.

1. **CFinder** - http://www.cfinder.org/
   CFinder is a free software for finding overlapping dense groups of nodes in networks, based on the Clique Percolation Method (CPM) of (Palla, 2005). CFinder offers a fast and efficient method for clustering data represented by large graphs, such as genetic or social networks and microarray data. CFinder is also very efficient for locating the cliques of large sparse graphs. For more information, see (Adamcsek, 2006).
2. **DIP** - http://dip.doe-mbi.ucla.edu/
   The Database of Interacting Proteins stores experimentally determined interactions between proteins. It combines information from a variety of sources to create a single, consistent set of protein-protein interactions. The data stored within the DIP database were curated, both, manually by expert curators and also automatically using computational approaches that utilize the the knowledge about the protein-protein interaction networks extracted from the most reliable, core subset of the DIP data. For more information, see (Xenarios, 2000).
3. **DPClus** - http://kanaya.naist.jp/DPClus/
   A graph clustering software, based on an algorithm which tends to isolate densely connected regions of a graph as clusters. Users have freedom to choose two input parameters within reasonable range and thus to affect the outcome of the clustering up to certain extent. Though this software can be used for graph clustering in general but it mainly focuses on detection of protein complexes in interaction networks. The software makes it possible to detect and visualize clusters of proteins in interaction networks which mostly represent molecular biological functional units. For more information, see (Altaf-Ul-Amin, 2006).
4. **GO Term Finder Tool** - http://db.yeastgenome.org/cgi-bin/GO/goTermFinder.pl
   A tool comprising a set of object-oriented Perl modules for accessing Gene Ontology (GO) information and evaluating and visualizing the collective annotation of a list of genes to GO terms. This tool may be exploited in order to validate the biological significance of PINs clusters, since each protein in a cluster corresponds to a specific gene annotated by the GO. For more information, see (Boyle, 2004).
5. **MCL** - http://micans.org/mcl/
   The MCL algorithm is short for the Markov Cluster Algorithm, a fast and scalable unsupervised cluster algorithm for graphs based on simulation of (stochastic) flow in graphs. The MCL algorithm simulates flow using (alternating) two simple algebraic operations on matrices. The basic interface to the MCL algorithm is very simple. For more information, see (Enright, 2002).
6. **MCODE** - ftp://ftp.mshri.on.ca/pub/BIND/Tools/MCODE
   A tool for detecting densely connected regions in large protein-protein interaction networks that may represent molecular complexes. The method is based on vertex weighting by local neighbourhood density and outward traversal from a locally dense seed protein to isolate the dense regions according to given parameters. For more information, see (Bader, 2003).
7. **MF-PINCoC**- *available upon request from the authors*
   The tool is an extension of PINCoC to generate overlapping clusters of the interaction graph in input. Due to this peculiarity, multi-facets proteins are allowed to belong to different groups corresponding to different biological processes. For more information, see (Pizzuti, 2008).

8. **MINT**- http://mint.bio.uniroma2.it/mint/Welcome.do
   The Molecular INTeraction database focuses on experimentally verified protein-protein interactions mined from the scientific literature by expert curators. The curated data can be analyzed in the context of the high throughput data and viewed graphically with the "MINT Viewer". For more information, see (Chatraryamontri, 2006).
9. **PIVOT** - http://www.cs.tau.ac.il/~rshamir/pivot
   The Protein Interaction VisualizatiOn Tool (PIVOT) is a visualization tool for protein-protein interactions. It allows the user to create personal data sets of interactions by combining information from private and public data sources. The user can also employ PIVOT to predict unknown interactions among proteins, based on interactions among their homologous proteins in other species. For more information, see (Orlev, 2004).
10. **RNSC** - *available upon request from the authors*
   A tool to partition networks into clusters using a cost function. The algorithm it is based on explores the solution space of all the possible clusterings to minimize a cost function that reflects the number of inter-cluster and intra-cluster edges. For more information, see (King, 2004).

REFERENCES

Adamcsek, B., Palla, G., Farkas, I. J., Dernyi, I., & Vicsek, T. (2006). Cfinder: locating cliques and overlapping modules in biological networks. *Bioinformatics, 22*(8), 1021-1023.

Altaf-Ul-Amin, M., Shinbo, Y., Mihara, K., Kurokawa, K., & Kanaya, S. (2006). Development and implementation of an algorithm for detection of protein complexes in large interaction networks, *BMC Bioinformatics, 7*, 207.

Aittokallio, T., & Schwikowski, B. (2006). Graph-based methods for analysing networks in cell biology. *Briefing in Bionformatics, 7*(3), 243-255.

Arnau,V., Mars, S., & Marìn, I. (2004). Iterative cluster analysis of protein interaction data. *Bioinformatics, 21*(3), 364–378.

Asur, S., Ucar, D., & Parthasarathy, S. (2007). An ensemble framework for clustering protein-protein interaction networks. *Bioinformatics, 23*, i29-i40.

Asburner, S. et al. (2000). Gene ontology: Tool for the unification of biology. The gene ontology consortium. *Nature Genetics, 25*, 25-29.

Bader, J. S., Chaudhuri, A., Rothberg, J. M., & Chant, J. (2004). Gaining confidence in high-throughput protein interaction networks. *Nature Biotechnology, 22*, 78-85.

Bader, G., & Hogue, H. (2003). An automated method for finding molecular complexes in large protein-protein interaction networks. *BMC Bioinformatics, 5*(2).

Bairoch, A., & Apweiler, R. (1996) The swiss-prot protein sequence data bank and its new supplement trembl. *Nucleic Acids Research, 24*(1), 21–25.

Blatt, M., Wiseman, S., & Domany, E. (1996). Superparamagnetic clustering of data. *Phisical Review Letters, 76*(18), 3251-3254.

Boyle, E. I., Weng, S., Gollub, J., Jin, H., Botstein, D., Cherry, J. M., & Sherlock, G. (2004). GO TermFinder--open source software for accessing Gene Ontology information and finding significantly enriched Gene Ontology terms associated with a list of genes. *Bioinformatics 20*(18), 3710-5.

Brohèe, S., & van Helden, J. (2006). Evaluation of clustering algorithms for protein-protein interaction networks. *BMC Bioinformatics, 7*.

Brun, C., Herrmann, C., & Guenoche, A. (2004) Clustering proteins from interaction networks for the prediction of cellular functions. *BMC Bioinformatics, 5.*

Chatraryamontri, A., Ceol, A., Montecchi Palazzi, L., Nardelli, G., Schneider, M. V., Castagnoli, L., & Cesarei, G. (2006). Mint: the molecular interaction database. *Nucleic Acid Research, 35*(Database issue), D572-D574.

Chen, J., & Yuan, B. (2006). Detecting Functional Modules in the yeast protein-protein interaction network. *Bioinformatics, 22*(18), 2283-2290.

Chen, J., Hsu, W., Lee, M. L. & Ng, S.-K. (2007). Labeling network motifs in protein interactomes for protein function prediction. In *IEEE 23rd International Conference on Data Engineering* (pp. 546-555).

Cho, Y.-R., Hwang, W., & Zhang, A. (2006). Identification of Overlapping Functional Modules in Protein interaction networks: Information Flow-based Approach, in Proceedings of the *Sixth International Conference on Data Mining-Workshops (ICDMW'06).*

Cho, Y.-R., Hwang, W., Ramanathan, M., & Zhang, A. (2007). Semantic integration to identify overlapping functional modules in protein interaction networks. *BMC Bioinformatics*, 8:265.

Chua, H. N., Ning, K., Sung, W. K., Leong, H. W., & Wong, L. (2007). Using indirect protein-protein interactions for protein complex prediction. In Proceedings of *IEEE Computational Systems Bioinformatics Conference (CSB07)* (pp. 97-109).

Demirel, M. C., & Cherny, D. (2005). Clustering and diversity of fluctuations for proteins. *Nanomedicine: Nanotechnology, Biology, and Medicine, 1*, 41-46.

Deng, M., Sun, F., & Chen, T. (2003) Assessment of the reliability of protein protein interactions and protein function prediction. In *Pac. Symp. Biocomput. (PSB2003)* (pp. 140-151).

Derenyi, I., Palla, G., & Vicsek, T. (2005). Clique percolation in random networks. *Physical Review Letters, 94*, 160202.

Drees, B. L., Sundin, B., & et al. (2001) A protein interaction map for cell polarity development. *Journal of Cellular Biology, 154*, 549-571.

Dunn, R., Dudbridge, F., & Sanderson, C. M. (2005). The use of edge-betweenness Clustering to Investigate Biological Function in Protein Interaction Networks. *BMC Bioinformatics, 6*, 39.

Enright, A. J., Dongen, S. V., & Ouzounis, C.A. (2002). An efficient algorithm for large-scale detection of protein families. *Nucleic Acids Res, 30*(7), 1575-84.

Farutin, V., Robinson, K., et al. (2006). Edge-Count Probabilities for the Identification of Local Protein Communities and their Organization. *Proteins: Structure, Function, and Bioinformatics, 62*, 800-818.

Fionda, V., Panni, S., Palopoli, S., & Rombo, S. E. (2007). Bi-GRAPPIN: Bipartite graph based protein-protein interaction networks similarity search. In *the IEEE International Conference on Bioinformatics and Biomedicine (BIBM)* (pp. 355-61).

Fionda, V., Palopoli, L., Panni, S., & Rombo, S. E. (2008). Protein-protein interaction network querying by a "focus and zoom" approach. In Proceedings of *BioInformatics Research and Development (BIRD2008)*, To appear.

Hartuv, E., & Shamir, R. A. (2000). Clustering Algorithm based Graph Connectivity. *Information Processing Letters, 76*, 175-181.

Hartwell, L. H., Hopfield, J. J., Leibler, S., & Murray, A. W. (1999). From molecular to modular cell biology. *Nature, 402*(Suppl), C47-C52.

Hwang, W., Cho, Y.-R., Zhang, A., & Ramanathan, M. (2006). A novel functional module detection algorithm for protein-protein interaction networks. *Algorithms for Molecular Biology, 1*(24).

Hwang, W., Kim, T., Cho, Y.-R., Zhang, A., & Ramanathan, M. (2007). SIGN: reliable protein interaction identification by integrating the Similarity In GO and the similarity in protein interaction Networks. In Proceedings of the *7th IEEE International Conference on Bioinformatics and Bioengineering (BIBE'2007)*, (pp. 1384-1388).

King, A. D., Przulj, N., & Jurisica, I. (2004). Protein complex prediction via cost-based clustering. *Bioinformatics, 20*(17), 3013-3020.

Krogan, N. J., Cagney, G. et al. (2006) Global landscape of protein complexes in the yeast saccharomyces cerevisiae. *Nature, 440*(7084), 637–643.

Ito, T., Chiba, T., Ozawa, R., Yoshida, M., Hattori, M., & Sakaki, Y. (2001). A comprehensive twohybrid analysis to explore the yeast protein interactome. *Proceedings of the National Academy of Sciences of the USA, 98*(8), 4569-4574.

Li, X.-L, Foo, C. S., & Ng, S.-K. (2007) Discovering protein complexes in dense reliable neighbourhoods of protein interaction networks. In *Proceedings of IEEE Computational Systems Bioinformatics Conference (CSB07)* (pp. 157-168).

Li, X.-L., Tan, S.-H., & Ng, S.-K. (2005). Interaction Graph Mining for Protein Complexes Using Local Clique Merging. *Genome Informatics,* (16), 260.

Lin, C., Cho, Y., Hwang, W., Pei, P., & Zhang, A. (2006) Clustering Methods in Protein-Protein Interaction Network. *Knowledge Discovery in Bioinformatics: Techniques, Methods and Application.* John Wiley & Sons, Inc.

Lubovac, Z., Gamalielsson, J., & Olsson, B. (2006). Combining Functional and Topological Properties to identify Core Modules in Protein Interaction Networks. *Proteins: Structure, Function, and Bioinformatics, 64*, 948-959.

Hartwell L. H., Hopfield J. J., Leibler S., & Murray A. W. (1999). From molecular to modular cell biology. *Nature, 402*, C47-C52.

King, A. D., Przulj, N., & Jurisica, I. (2004). Protein complex prediction via cost-based clustering. *Bioinformatics, 20*(17), 3013–3020.

Jain., R. D. A. (1988). *Algorithms for Clustering Data.* Prentice Hall.

Madeira, S. C., & Oliveira, A. L. (2004). Biclustering algorithms for biological data analysis: A survey. *IEEE Transactions on Computational Biology and Bioinformatics, 1*(1), 24-45.

Miller, J. P., Lo, R. S., Ben-Hur, A., Desmarais, C., Stagljar, I., Stafford Noble, W., & Fields, S. (2005) Large-scale identification of yeast integral membrane protein interactions. *Proceedings of the National Academy of Sciences of the USA, 102*(34), 12123-12128.

Orlev, N., Shamir, R., & Shiloh Y. (2004). Pivot: Protein interaction visualization tool. *Bioinformatics, 20*(3), 424–425.

Palla, G., Derényi, I., Farkas, I., & Vicsek, T. (2005). Uncovering the overlapping community structure of complex networks in nature and society. *Nature, 435*(7043), 814-818.

Pei, P., & Zhang , A. (2005). A two-step approach for clustering proteins based on protein interaction profiles. In Proceedings of the *IEEE Int. Symposium on Bioinformatics and Bioengeneering (BIBE'2005)*, (pp. 201-209).

Pereira-Leal, J. B., Enright, A. J., & Ouzounis, C. A. (2004). Detection of functional modules from protein interaction networks. *Proteins, 54*, 49-57

Pereira, J. B., Enright, A. J., & Ouzounis, C. A. (2004). Detection of functional modules from protein interaction networks. *Proteins: Structure, Fuctions, and Bioinformatics*, (20), 49-57.

Pizzuti, C., & Rombo, S. E. (2007). PINCoC: a Co-Clustering based Method to Analyze Protein-Protein Interaction Networks. In *the 8th International Conference on Intelligent Data Engineering and Automated Learning (IDEAL'07)* (pp. 821-830).

Pizzuti, C., & Rombo, S. E. (2008). Multi-functional Protein Clustering in PPI Networks. In *Proceedings of the 2nd International Conference on BioInformatics Research and Development (BIRD'08)*, Vienna, To appear.

Przulj, N., Wigle, D. A., & Jurisica, I. (2004). Functional topology in a network of protein interactions. *Bioinformatics*, *20*(3), 340–348.

Rives, A. W., & Galitski, T. (2003). Modular organization of cellular. *Proceedings of the National Academy of Sciences of the USA*, *100*(3), 1128-1133.

Samantha, M. P., & Liang, S. (2003). Redundancies in large-scale protein interaction networks. *Proceedings of the National Academy of Science of the USA*, (pp. 12579-12583).

Sharan, R., Ulitisky, I., & Shamir, R. (2007). Network-based prediction of protein function. *Molecular Systems Biology*, *3*(88).

Spirin, V., & Mirny, L. A. (2003). Protein Complexes and functional modules in molecular networks. *Proceedings of the National Academy of Sciences of the USA*, *100*, 12123-12128.

Sprinzak, E., Sattath, S., & Margalit, H. (2003). How reliable are experimental protein-protein interaction data? *Journal of Molecular Biology*, *327*(5), 919-923.

Tornow, S., & Mewes, H. (2003). Functional modules by relating protein interaction networks and gene expression. *Nucleic Acid Research*, *31*(21), 6283-6289.

Ucar, D., Asur, S. Catalyurek, U. V., & Parthasarathy, S. (2006). Improving functional modularity in protein-protein interactions graphs using hub-induced subgraphs. In *10th European Conference on Principles and Practice of Knowledge Discovery in Databases (PKDD)* (pp. 371-382).

Ucar, D., Parthasarathy, S., Asur, S., & Wang, C. (2005). Effective pre-processing strategies for functional clustering of a protein-protein interaction network. In *IEEE Int. Symposium on Bioinformatics and Bioengeneering (BIBE)*, (pp. 129-136).

von Mering, C., & et al. (2002). Comparative assessment of a large-scale data sets of protein-protein interactions. *Nature*, *417*(6887), 399-403.

Watt, D. J. (1999) *Small worlds*. Princeton University Press.

Xenarios, I., Rice, D. W., Salwinski, L., Baron, M. K., Marcotte, E. M., & Eisenberg, D. (2000) DIP: The Database of Interacting Proteins. *NAR*, *28*, 289-91.

Xiong, H., He, X., Ding, C., Zhang, Y., Kumar, V., & Holbrook, S. R. (2005) Identification of Functional Modules in Protein Complexes via Hyperclique Pattern Discovery. *Pacific Symposium on Biocomputing*, *10* 221-232.

Zotenko, E., Guimarães, K. S., Jothi, R., & Przytycka, T. M. (n/d). Decomposition of Overlapping Protein Complexes: A Graph Theoretical Method for Analyzing Static and Dynamic Protein Associations. *Algorithms for Molecular Biology*, *1*(7).

Chapter X
Evolutionary Analyses of Protein Interaction Networks

Takashi Makino
University of Dublin, Trinity College, Ireland

Aoife McLysaght
University of Dublin, Trinity College, Ireland

ABSTRACT

This chapter introduces evolutionary analyses of protein interaction networks and of proteins as components of the networks. The authors show relationships between proteins in the networks and their evolutionary rates. For understanding protein-protein interaction (PPI) divergence, duplicated genes are often compared because they are derived from a common ancestral gene. In order to reveal evolutionary mechanisms acting on the interactome it is necessary to compare PPIs across species. Investigation of co-localization of interacting genes in a genome shows that PPIs have an important role in the maintenance of a physical link between neighboring genes. The purpose of this chapter is to introduce methodologies for analyzing PPI data and to describe molecular evolution and comparative genomics insights gained from such studies.

INTRODUCTION

Protein-protein interactions (PPIs) are one of the most important components of biological networks. An understanding of the evolution of PPIs is crucial to elucidating how the evolution of biological networks has contributed to diversification of extant organisms. The amount of information about PPIs has grown rapidly due to the development of a high-throughput two-hybrid system, mass spectrometry of co-immunoprecipitated protein complexes, and bioinformatics approaches such as text mining from the many individual studies reported in the literature. The extensive data allow us to analyze the protein interaction networks from an evolutionary aspect.

Evolutionary studies of protein interaction networks can be classified into at least five topics based on their focus (Fig. 1). The most studied topic is the relationship between the number of PPIs of a protein (**connectivity**) and protein evolution (Fig. 1A). It has been shown that protein **connectivity** correlates with evolutionary rates (Fraser *et al.*, 2002; Fraser *et al.*, 2003). Protein **connectivity** can be directly calculated from the protein interaction network. To estimate evolutionary rates of proteins, we must first make multiple sequence alignments by using an application such as CLUSTAL W (Thompson *et al.*, 1994). The number of amino acid substitutions is estimated from the number of differences between the aligned sequences with a biologically realistic statistical model, e.g. Kimura's method. Many of the most widely used methods are implemented in the PHYLIP software package (http://evolution.genetics.washington.edu/phylip.html). The evolutionary distances can be translated into relative evolutionary rates by comparison with an older homologous sequence (an outgroup). When evolutionary rates of proteins are slow, the proteins are conserved even in distantly related species. In general, **orthologous** proteins between two species are identified as a pair of sequences that show reciprocal best hits in the sequence similarity search using all protein sequences from both species. In other words, a reciprocal best hit is evidence that the protein pair is related through a speciation event. BLAST (http://www.ncbi.nlm.nih.gov/blast/Blast.cgi) is the most commonly used sequence similarity search software. The very fact of easily detectable **orthologs** between distantly related species implies strong functional constraint and slow evolutionary rates without explicit rate estimation. The ability to detect **orthologs** in distantly-related genomes is one measure of conservation; the more distantly related the species where an **ortholog** is detectable the greater the conservation of the protein.

Not only the **connectivity** but also the structure of the protein interaction networks influence functional constraint on proteins (Fig. 1B). There are several indicators to describe **network structure** such as motif constituents, clustering coefficients, betweenness, centrality and so on. Wuchty and his colleagues perform the first survey of the PPI patterns between groups of two, three, four and five proteins in the yeast protein interaction network (defined as motif constituents) and examined their evolution (Wuchty *et al.*, 2003). When we focus on two proteins in a protein interaction network there is only one possible interaction. When there are three proteins, there are two possible interaction patterns. The most cohesive motif in the definition consists of five proteins and 10 PPIs (i.e. all proteins interacting with all others). Alternatively, clustering of proteins in the interaction network may be simplified into a classification of dense parts and sparse parts of the networks (Makino & Gojobori, 2006). Sparse and dense parts are defined using the clustering coefficients (Watts & Strogatz, 1998). Note that the clustering coefficients differ from motif constituents. The more PPI partners a protein has, the greater the chance the protein has a cohesive motif. On the other hand, the more PPI partners a protein has, the smaller the chance the protein has a high clustering coefficient because the number of possible interactions between those PPI partners increases.

Gene pairs with highly similar sequences in an organism's genome must be derived from a gene duplication event. Duplicated pairs are defined as pairs of sequences that 'hit' each other in a similarity search (e.g. BLAST as before) using protein sequences from a single organism. To understand the evolution of protein interaction networks it is useful to study the PPIs of **duplicated genes**. This is because immediately after duplication these genes will produce identical proteins, which, by definition, will behave identically in the protein interaction network (Fig. 1C). In fact, **duplicated gene** pairs often share PPI partners because they have a common ancestor (Deane *et al.*, 2002; Wagner, 2001). We can classify duplicated pairs into two groups depending on the absence or presence of common PPI partners as diverged or non-diverged pairs, respectively.

Most reports on the evolution of protein interaction networks deal with yeast protein interaction networks because yeast PPI data have been most abundant (Gavin *et al.*, 2006; Gavin *et al.*, 2002; Ho *et al.*, 2002; Ito *et al.*, 2000; Krogan *et al.*, 2006; Tarassov *et al.*, 2008; Uetz *et al.*, 2000). However recent global studies on PPIs have been performed in the prokaryotes *Helicobacter pylori* (Rain *et al.*, 2001) and *Escherichia coli* (Butland *et al.*, 2005), and also eukaryotes *Plasmodium falciparum* (LaCount *et al.*, 2005), *Caenorhabditis elegans* (Li *et al.*, 2004), *Drosophila melanogaster* (Formstecher *et al.*, 2005; Giot *et al.*, 2003) and Human (Rual *et al.*, 2005; Stelzl *et al.*, 2005). These comprehensive PPI data of multi species allow us to compare PPIs between species; evolutionary conserved interactions are known as **interologs** (Walhout *et al.*, 2000; Fig. 1D).

A functional **gene cluster** is defined as a gene pair having both functional and physical links on a genome (Fig. 1E). The type of functional link that is considered differs between studies and includes PPIs, interactions between subunits in a protein complex, and gene expression profiles (Hurst *et al.*, 2004; Makino & McLysaght, 2008; Poyatos & Hurst, 2006; Teichmann & Veitia, 2004). When functional links are mapped onto a genome one can observe that some genes have functional links to neighboring (i.e., genetically linked) genes in the genome.

Before identifying the **gene clusters**, we have to exclude tandem **duplicated genes** from the analysis, because **duplicated gene** pairs are often arrayed tandemly on a genome due to the mechanism of duplication (replication

Figure 1. Evolutionary analyses of protein interaction networks. Circles and black lines indicate proteins and protein-protein interactions (PPIs), respectively. Proteins of interest are black or gray and their PPI partners are white. Proteins may differ in their connectivity (A) or clustering (B) in the PPI network. The PPI partners of paralogs (C) and orthologs (D) are often conserved, and the relative physical location of genes whose protein products interact is often conserved over evolutionary time (E). (A) The black protein is said to have a greater connectivity because the number of PPIs is larger than that of gray one. (B) Black proteins are clustered more tightly than gray ones. (C) Black proteins are derived from a gene duplication. Duplicated pairs (paralogs) often have the same PPI partners, although there may be loss or gain of PPI partners during evolution. If a gene has a self-interaction before gene duplication, the duplicated pair may interact with each other after gene duplication. (D) Broken lines indicate orthology between human and yeast genes. The black and gray proteins interact in human and in yeast. (E) Rectangles and gray lines indicate genes and genomes, respectively. Broken lines indicate orthology between human and dog. The black and gray genes are located close together in the human genome and their protein products interact. Interacting gene clusters such as this in the human genome are likely to be conserved in other vertebrate genomes.

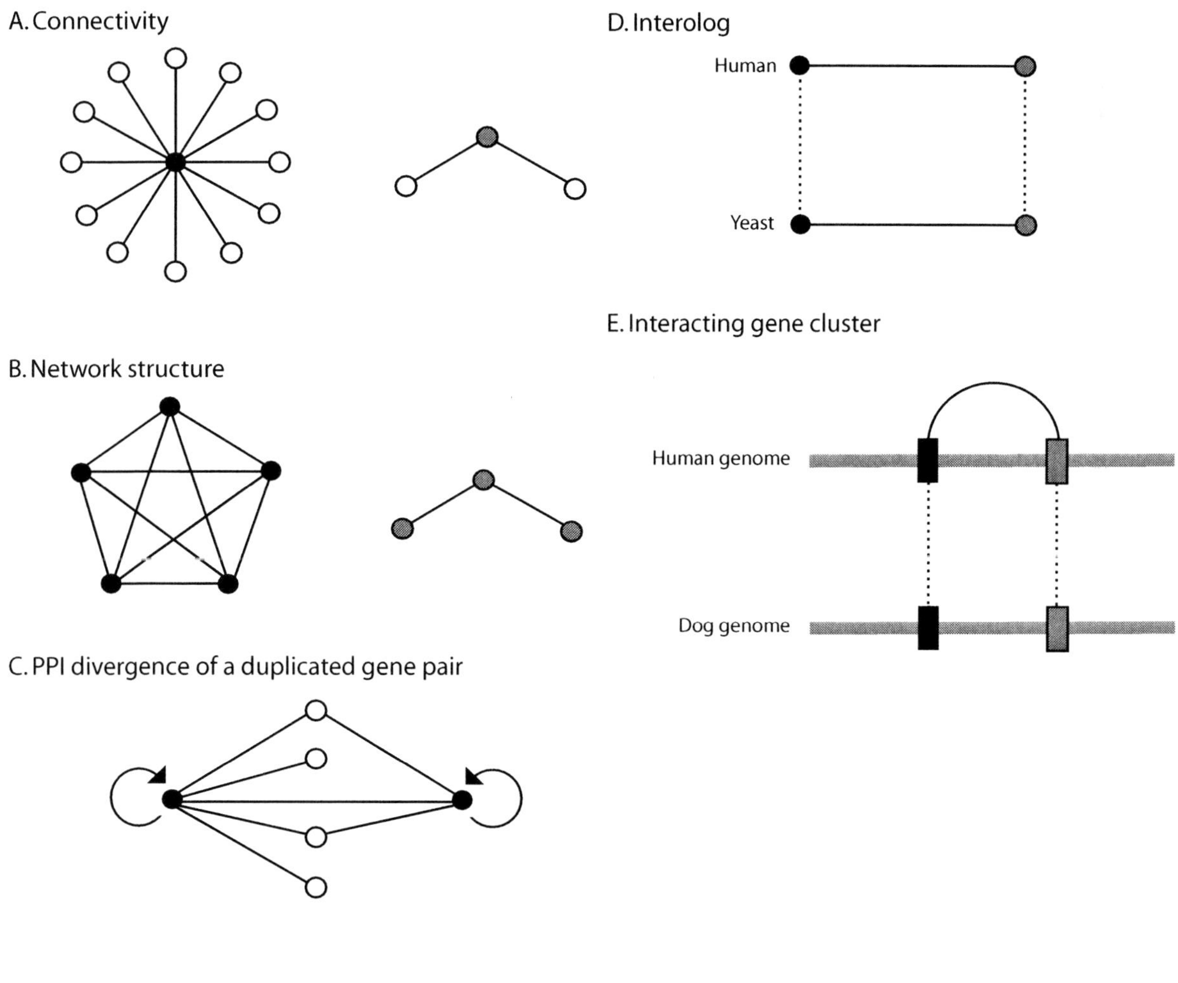

error) rather than any biological significance of clustering. Makino and McLysaght suggest that the interference from the tandem duplication effect on **gene cluster** identification when using PPIs is weaker than the effect when using expression data (Makino & McLysaght, 2008). However, the effect is not completely absent because, if a self-interacting gene is duplicated in tandem, the gene would possibly interact with its duplicated copy (Pereira-Leal *et al.*, 2007; Presser *et al.*, 2008; Wagner, 2001; Fig. 1C) and form a **gene cluster**. So it is necessary to exclude tandemly **duplicated gene** pairs when identifying interacting **gene clusters**. Note that randomly selected gene pairs may make **gene clusters** on a genome by chance. Therefore we need to test whether the number of **gene clusters** is larger than expected estimated by simulation.

Comparative genomics is a framework to understand evolutionary relationships among different organisms by comparison of their genome sequences. It has been shown that interacting **gene clusters** tend to have been conserved on vertebrate genomes during evolution (Makino & McLysaght, 2008).

We describe the details and problems for each topic in the following sections.

RELATIONSHIP BETWEEN THE NUMBER OF PROTEIN-PROTEIN INTERACTIONS AND PROTEIN EVOLUTION

The evolutionary rate of a protein is useful for characterising protein evolution. It is usually defined as amino acid substitutions per site per year. This rate reflects the functional constraints acting on proteins or protein domains, with stronger constraint manifesting in fewer sequence substitutions. A dramatic, classic example of this is the Histone 4 protein which has only two amino acid differences between the human and wheat gene despite the long divergence time (Graur & Li, 2000). Several determinants of functional constraint on proteins have been reported such as expression levels, protein dispensability, and so on (Pal *et al.*, 2006). Protein-protein interactions may also determine functional constraint on protein sequences. Fraser and colleagues showed that the number of PPIs negatively correlated with evolutionary rates for proteins in yeast protein interaction networks (Fraser et al., 2002; Fraser *et al.*, 2003), although there are some arguments that the **connectivity** does not correlate with evolutionary rates for proteins (Agrafioti *et al.*, 2005; Batada *et al.*, 2006; Bloom & Adami, 2003; Hahn *et al.*, 2004; Jordan *et al.*, 2003). The number of PPIs for each protein is counted in the yeast PPI dataset, and the evolutionary rates of proteins are estimated by comparison of protein sequences between yeast and worm. Following the publication of these seminal studies, PPIs have received major attention as a determinant of functional constraints.

Proteins with many interacting partners in the protein interaction network are defined as hubs. As mentioned above, hubs evolve slowly (Fraser et al., 2002; Fraser *et al.*, 2003); hubs are often encoded by essential genes (He & Zhang, 2006; Jeong *et al.*, 2001; Yu *et al.*, 2004); and are less prone to loss during evolution than are those with fewer interacting partners (Krylov *et al.*, 2003; Wuchty, 2004; Wuchty *et al.*, 2006). By viewing this observation from a different angle, the older a gene is, the more PPIs a protein encoded by the gene has (Eisenberg & Levanon, 2003; Kunin *et al.*, 2004). He and Zhang reported that hubs tend to be essential but that is not caused by their **network structure** (He & Zhang, 2006). This is explained by the simple fact that hubs have many interacting partners, namely they are related to many biological processes. Interestingly, duplicate genes encoding hubs tend to have been conserved in the genome during evolution (Kafri *et al.*, 2008). Producing genetic redundancy of hubs by gene duplication seems to be important in protein interaction networks.

RELATIONSHIP BETWEEN THE STRUCTURE OF PROTEIN INTERACTION NETWORKS AND PROTEIN EVOLUTION

Protein interaction networks may be represented as a two-dimensional graph with each protein represented by a node and edges (lines) connecting interacting proteins, resulting in a very complex structure of spider-web-like networks. It has been reported that proteins tightly clustered in these networks may share biological properties (Spirin & Mirny, 2003). Proteins sharing a particular functional class tend to appear in the same part of a protein interaction network in a so-called "functional module" (Yook *et al.*, 2004). Proteins in functional module tend to be clustered in the protein interaction networks. It has been shown that these structures of yeast protein interaction networks are related to protein evolution (Wuchty *et al.*, 2003). Motif constituents in protein interaction networks are defined as interaction patterns consisting of five or fewer proteins. The greater the number of interactions among proteins in a motif, the more cohesive is the motif (Fig. 1B). Proteins in cohesive motifs are well conserved in human, mouse, fly, worm and plant. The results are robust to the definition of a cluster in the protein interaction network and the authors report similar results using hypergeometric clustering coefficients instead of motif constituents (Wuchty et al., 2006). In particular, proteins in cohesive motifs tend to be subunits in a protein complex (Wuchty *et al.*, 2003). These results indicate that conserved proteins are not distributed randomly in the protein interaction networks, and the structure of protein interaction networks influences protein evolution.

Interactions between subunits in a protein complex and other protein-protein interactions may be defined as stable and transient interactions respectively. It has been observed that subunits in protein complexes evolve slower than the other proteins in the yeast protein interaction network (Teichmann, 2002). When proteins interact with each other in a protein complex, the proteins must have large areas on their surfaces for simultaneous interactions with their partners in the complex. In this case, functional constraint on the proteins would be strong and broad. In fact, by comparison of protein interaction regions and protein evolution, it has been shown that amino acids in the region of the subunit surface in a protein complex evolve slowly (Mintseris & Weng, 2005). In addition, the interacting regions co-evolve with those in interacting partners. In other words, when an amino acid in an interacting region of a protein is substituted, the interacting amino acid of its interacting partner is likely to be substituted, probably to preserve the interaction. On the other hand, amino acids on the surface of the other proteins in protein interaction networks are not likely to be conserved, and did not co-evolve with their interacting partners.

Subunits in protein complexes are under strong functional constraint as described above (Teichmann, 2002; Wuchty et al., 2003). Following the convention of Spirin and Mirny (Spirin & Mirny, 2003), the protein complex is a good example for a functional module. On the other hand, there are proteins interacting with other proteins having different functional classes in protein interaction networks, though, in general, these are not clustered in protein interaction networks (Makino & Gojobori, 2006). Makino and Gojobori define the region in the networks by contrast with a functional module as a framework module (Makino & Gojobori, 2006). Note that the determinants of the modules are derived from not only the protein itself but also its interacting partners. In comparison with evolutionary rates of proteins in functional modules, those in framework modules evolve slower, although subunits in protein complexes are likely to be in functional modules (Makino & Gojobori, 2006). Loss of function by amino acid substitutions in proteins in framework modules would influence multiple different biological processes. The results indicate that features of the interacting partners influence the evolutionary rate of a protein, and propose that not only proteins clustered in protein interaction networks but also other proteins in the networks are important for organisms.

DIVERGENCE OF PROTEIN-PROTEIN INTERACTIONS OF DUPLICATED GENES

After gene duplication, functional constraint on one or both copies of the gene may become weak because of functional redundancy, and their amino acid sequences are likely to be mutable. It has been shown that **duplicated genes** usually become pseudogenes by accumulation of mutations, however some **duplicated genes** are conserved in a genome as functional genes. One possible evolutionary outcome is that one of the duplicate copies acquires a novel function through the accumulation of mutations.

Duplicated gene pairs provide us with a unique opportunity of making fair comparisons of the genes under the same initial conditions. Because a **duplicated gene** pair is derived from a common ancestral gene, the pair often shares interacting partners (Deane *et al.*, 2002; Wagner, 2001). Immediately after gene duplication, the **duplicated gene** pair must have exactly the same interacting partners. Over evolutionary time interacting partners may be gained or lost and some duplicated pairs might lose all common interacting partners. It has been shown that the number of shared PPI partners between **duplicated gene** pairs has been decreasing as time advances (Wagner, 2001). Furthermore, PPIs have diverged rapidly (Beltrao & Serrano, 2007; Wagner, 2001). In addition, gain and loss of interactions through mutations in existing proteins is shown to be much faster than gene duplication (Berg *et al.*, 2004), and PPI divergence has occurred asymmetrically during evolution (Makino et al., 2006; Wagner, 2002). Psator-Satorras and his colleagues propose an evolutionary model of protein interaction networks by gene duplication and PPI re-wiring (Pastor-Satorras *et al.*, 2003). For a given interacting protein, PPI divergence must relate to functional divergence. Baudot and his colleagues report divergence or conservation of function of proteins from their PPI divergence (Baudot *et al.*, 2004). He and Zhang propose a new model of subneofunctionalization of PPIs after gene duplication (He & Zhang, 2005); PPI divergence of **duplicated genes** is likely to involve subfunctionalization and neofunctionalization simultaneously.

In the case of yeast and some other organisms, a subset of the **duplicated genes** is derived from a whole genome duplication (WGD). WGD occurred in yeast lineage about 100 million years ago (Kellis *et al.*, 2004; Wolfe & Shields, 1997). The merit of using the duplicates from genome duplication is that their divergence times, that are exactly the same, are known. Therefore many studies focus on the evolution of PPIs of WGD gene pairs. To exclude the influence of time passage and focus on relationship between amino acid substitutions and losses of PPI partners for duplicated pairs, Makino and his colleagues use WGD gene pairs and examine whether loss of common PPI partners for duplicated pairs influence protein evolution (Makino et al., 2006). The study shows that duplicated pairs not sharing PPI partners have evolved more asymmetrically than those sharing PPI partners. The result indicates that gene duplication leads to the functional differentiation of the **duplicated gene** pairs through the losses and/or gains of the PPI partners, resulting in a change in their evolutionary rates. In addition, recent studies show that an enrichment of synergy between WGD gene pairs sharing protein–protein interaction partners (DeLuna *et al.*, 2008; Musso *et al.*, 2008). WGD gene pairs are more likely to share function and PPI partners than other **duplicated gene** pairs (Guan *et al.*, 2007). Furthermore, proteins encoded by WGD gene pairs are likely to be on the same protein complex (Musso *et al.*, 2007). The evolutionary comparison of the PPI partners of one copy in a duplicated pair with those of the other copy give an important clue for understanding their functional differentiations through PPI network divergence.

Proteins often have an interaction with themselves. Some studies focus on the evolution of proteins having self-interactions. It has been reported that there are more self-interacting proteins then expected in human, fly, worm and yeast, and these proteins are likely to have many more PPI partners than others (Ispolatov *et al.*, 2005). One early study concerned the evolution of self-interacting proteins after gene duplication, but the available dataset was too small to refer to their evolutionary history (Wagner, 2001). A recent study showed that duplication of genes encoding self-interacting proteins tends to give rise to protein complexes formed by paralogous proteins (Pereira-Leal et al., 2007). In addition, proteins encoded by WGD gene pairs tend to interact with each other (Presser et al., 2008). The results indicate that genes encoding self-interacting proteins are preserved during evolution despite the extensive gene loss that has occurred after WGD. Self-interaction of proteins has not been studied well, however it must be one of central factors affecting the evolution of protein interaction networks.

COMPARISON OF PROTEIN-PROTEIN INTERACTIONS AMONG DIFFERENT SPECIES

It is usually assumed that **orthologous** gene pairs have similar function. Therefore **orthologous** genes may have the same protein interaction partners in each species. Conserved PPIs among different species are defined as **interologs** (Fig. 1D), and some **interologs** between yeast and other organisms are detected (Matthews *et al.*, 2001; Walhout et al., 2000). In addition, extensive PPIs between gene pairs having high similarity to pairs in other species are conserved in fly, worm and yeast (Sharan *et al.*, 2005).

Human **interolog** networks are reconstructed using conserved PPIs (Huang *et al.*, 2007), and the properties of **interologs** have been investigated in the following reports. Dutkowski and Tiuryn reconstruct ancestral protein interaction networks using PPIs in yeast, worm and fly, and show that functional modules consist of conserved ancestral PPIs (Dutkowski & Tiuryn, 2007). Liang and his colleagues refer to not only conserved **interologs** but also divergence of them using PPIs in seven different species (Liang *et al.*, 2006). Another characteristic of **interologs** was discovered when it was shown that **evolutionary conservation** of human interactions in the yeast protein interaction networks correlates with the number of PPIs (Brown & Jurisica, 2007). The result is consistent with past reports in yeast (Fraser et al., 2002; Fraser et al., 2003). Although **interologs** were not explicitly studied, Hahn and Kem showed that protein interaction networks in fly, yeast and worm have similar structure (Hahn & Kern, 2005).

Only 16 **interologs** are detected as conserved PPIs among human, yeast, worm and fly, when only one-to-one **orthologous** gene pairs are used for detecting **interologs** (Gandhi *et al.*, 2006). The authors propose that extensive divergence of **interologs** has occurred even when protein sequences have been conserved during evolution. That is, PPI divergence of **orthologous** pairs could rapidly induce functional but not sequence divergence. This idea is supported by recent study showing that null mutations in human and mouse **orthologs** often result in different phenotypes (Liao & Zhang, 2008). PPI divergence of proteins might contribute to their functional divergence and change phenotype even in **orthologous** pairs. However, Mika and Rost warn that homology based inferences of PPIs lack reliability (Mika & Rost, 2006). The prediction of PPI partners through sequence similarity is more reliable for protein pairs within the same species than for pairs between species. The authors show that discrepancies between different high throughput PPI data are large. We need further discussion about the issue. The inconsistent observations about the evolution of **interologs** seem to be derived from accuracy and incompleteness of protein-protein interaction data (Cesareni *et al.*, 2005). In yeast, the whole PPI network is estimated to consist of 30,000 PPIs (von Mering *et al.*, 2002), and in human the whole PPI network is estimated to consist of 650,000 PPIs (Stumpf *et al.*, 2008), but currently available datasets are significantly smaller. We see merely the tip of the iceberg.

CO-LOCALIZATION OF INTERACTING GENE PAIRS ON A GENOME

Many recent studies have reported non random gene order in eukaryotic genomes using expression data (Hurst *et al.*, 2004). Makino and McLysaght suggest that PPIs constitute a preferable biological relationship to study non random gene order, because identification of **gene clusters** in the genome based on PPIs avoids the bias caused by chromatin structure that is inherent to the analysis of co-expressed genes (Makino & McLysaght, 2008). Genes encoding proteins in the same protein complex tend to be co-localized on a yeast genome (Teichmann & Veitia, 2004). Genes with other interactions are also likely to be co-localized (Poyatos & Hurst, 2006). Note that we have to control the influence of tandem duplicate genes to identify **gene clusters** on a genome as mentioned above (Hurst et al., 2004).

It is more difficult to identify interacting **gene clusters** in large genomes such as those of vertebrates than in the more compact yeast genomes because of the increased size of the proteome and the relative deficit of experimental data. Makino and McLysaght searched for evidence of the co-localization of interacting genes within the human genome using human PPI data (Makino & McLysaght, 2008). The authors found 83 interacting gene pairs that were located within 1 Mbp on a human genome. The number of interacting **gene clusters** is significantly more than expected by chance, and is not the result of tandem duplications. The importance of interacting **gene clusters** might be derived from co-regulation by bidirectional promoters (Trinklein *et al.*, 2004), shared enhancers (West *et al.*, 2002) and other unknown factors.

If **gene clusters** are under some functional constraint, the physical links on a genome must have been conserved during evolution. Interestingly, co-expressed gene pairs in *S. cerevisiae* are conserved in *Candida albicans* (Huynen *et al.*, 2001; Lercher *et al.*, 2002). In addition, there are fewer breakpoints within co-expressed **gene clusters** in both human and mouse than expected (Singer *et al.*, 2005). These results indicate that co-expressed **gene clusters** have been conserved during evolution. Makino and McLysaght examined whether interacting **gene clusters** on a human genome are conserved in other vertebrate genomes by **comparative genomics** (Makino & McLysaght, 2008). They investigated both **orthologs** of genes in each **gene cluster** and their genomic locations on 13 chordate genomes derived from the Ensembl database. The interacting **gene clusters** are significantly more conserved across vertebrate genomes than other pairs of genes located within 1 Mbp in the human genome (Makino & McLysaght, 2008). The interacting **gene clusters** would have been under some functional constraints as mentioned above during vertebrate geome evolution.

CONCLUDING REMARKS

We clearly point out the advantage of utilizing a vast amount of information about PPIs in the **molecular evolution** and **comparative genomics** studies of biological networks in this chapter. In particular, we show that the evolution of proteins as the components of protein interaction networks can be understood through the number of PPIs or the structure of protein interaction network. **Duplicated genes** are a uniquely fertile ground to investigate and understand PPI divergence. In addition, comparison of PPIs among different species is hot topic in evolution although it is very difficult to find consistent conclusions because of differing qualities of PPI data (Kiemer & Cesareni, 2007). We also show that PPIs influence genomic locations of genes encoding proteins in protein interaction networks during evolution. We feel that this line of studies will yield important insights into and understanding of the evolutionary processes influencing protein interaction networks.

Evolutionary analyses of protein interaction networks will become increasingly popular with the advance of accurate and quantitative PPIs (Kiemer & Cesareni, 2007). Note that PPI is not a static biological phenomenon. The PPIs among proteins must be dramatically changed in a cell over time and development. Evolutionary analysis of dynamic PPI patterns with the changes in the gene expression is also interesting. In the near future, these dynamic PPI data will be produced and available for evolutionary analysis of PPIs (Levy & Pereira-Leal, 2008). These approaches will contribute to the understanding of the phenotype from genotype relationship and the development of systems biology (Benfey & Mitchell-Olds, 2008).

RESOURCES

DIP (Database of Interacting Proteins): One of the biggest protein interaction databases. Almost all known PPIs for yeast, fly and worm are registered in the database. The selected yeast CORE PPI dataset is often used for protein interaction analyses (http://dip.doe-mbi.ucla.edu/).

HPRD (Human Protein Reference Database): The most abundant PPI database for human. There are not only PPIs but also other annotations for human proteins in the database (http://www.hprd.org/).

MIPS (Munich Information center for Protein Sequences): Protein interaction data and functional classes for yeast proteins are available (http://mips.gsf.de/)

Ensembl: Vertebrate genome browser. There are many kinds of annotations for vertebrate genomes (including several invertebrate genomes). The data may be accessed through a web browser or through the Ensembl API which makes automation easy to manage (http://www.ensembl.org)

BLAST and PSIBLAST: Algorithm for finding DNA or protein sequence similarity. PSIBLAST allows us to execute more sensitive homology searches (http://www.ncbi.nlm.nih.gov/blast/Blast.cgi).

CLUSTALW: DNA or protein sequence alignment application. (http://www.ebi.ac.uk/Tools/clustalw2/index.html)

PHYLIP: Software package for inferring phylogenetic trees. We can also estimate evolutionary distances between aligned DNA or protein sequences (http://evolution.genetics.washington.edu/phylip.html)

ACKNOWLEDGMENT

This work is supported by Science Foundation Ireland.

REFERENCES

Agrafioti, I., Swire, J., Abbott, J., Huntley, D., Butcher, S., & Stumpf, M. P. (2005). Comparative analysis of the saccharomyces cerevisiae and caenorhabditis elegans protein interaction networks. *BMC Evol Biol, 5*(1), 23.

Batada, N. N., Hurst, L. D., & Tyers, M. (2006). Evolutionary and physiological importance of hub proteins. *PLoS Comput Biol, 2*(7), e88.

Baudot, A., Jacq, B., & Brun, C. (2004). A scale of functional divergence for yeast duplicated genes revealed from analysis of the protein-protein interaction network. *Genome Biol, 5*(10), R76.

Beltrao, P., & Serrano, L. (2007). Specificity and evolvability in eukaryotic protein interaction networks. *PLoS Comput Biol, 3*(2), e25.

Benfey, P. N., & Mitchell-Olds, T. (2008). From genotype to phenotype: Systems biology meets natural variation. *Science, 320*(5875), 495-497.

Berg, J., Lassig, M., & Wagner, A. (2004). Structure and evolution of protein interaction networks: A statistical model for link dynamics and gene duplications. *BMC Evol Biol, 4*(1), 51.

Bloom, J. D., & Adami, C. (2003). Apparent dependence of protein evolutionary rate on number of interactions is linked to biases in protein-protein interactions data sets. *BMC Evol Biol, 3*, 21.

Brown, K. R., & Jurisica, I. (2007). Unequal evolutionary conservation of human protein interactions in interologous networks. *Genome Biol, 8*(5), R95.

Butland, G., Peregrin-Alvarez, J. M., Li, J., Yang, W., Yang, X., Canadien, V., et al. (2005). Interaction network containing conserved and essential protein complexes in escherichia coli. *Nature, 433*(7025), 531-537.

Cesareni, G., Ceol, A., Gavrila, C., Palazzi, L. M., Persico, M., & Schneider, M. V. (2005). Comparative interactomics. *FEBS Lett, 579*(8), 1828-1833.

Deane, C. M., Salwinski, L., Xenarios, I., & Eisenberg, D. (2002). Protein interactions: Two methods for assessment of the reliability of high throughput observations. *Mol Cell Proteomics, 1*(5), 349-356.

DeLuna, A., Vetsigian, K., Shoresh, N., Hegreness, M., Colon-Gonzalez, M., Chao, S., et al. (2008). Exposing the fitness contribution of duplicated genes. *Nat Genet, 40*(5), 676-681.

Dutkowski, J., & Tiuryn, J. (2007). Identification of functional modules from conserved ancestral protein-protein interactions. *Bioinformatics, 23*(13), i149-158.

Eisenberg, E., & Levanon, E. Y. (2003). Preferential attachment in the protein network evolution. *Phys Rev Lett, 91*(13), 138701.

Formstecher, E., Aresta, S., Collura, V., Hamburger, A., Meil, A., Trehin, A., et al. (2005). Protein interaction mapping: A drosophila case study. *Genome Res, 15*(3), 376-384.

Fraser, H. B., Hirsh, A. E., Steinmetz, L. M., Scharfe, C., & Feldman, M. W. (2002). Evolutionary rate in the protein interaction network. *Science, 296*(5568), 750-752.

Fraser, H. B., Wall, D. P., & Hirsh, A. E. (2003). A simple dependence between protein evolution rate and the number of protein-protein interactions. *BMC Evol Biol, 3*(1), 11.

Gandhi, T. K., Zhong, J., Mathivanan, S., Karthick, L., Chandrika, K. N., Mohan, S. S., et al. (2006). Analysis of the human protein interactome and comparison with yeast, worm and fly interaction datasets. *Nat Genet, 38*(3), 285-293.

Gavin, A. C., Aloy, P., Grandi, P., Krause, R., Boesche, M., Marzioch, M., et al. (2006). Proteome survey reveals modularity of the yeast cell machinery. *Nature, 440*(7084), 631-636.

Gavin, A. C., Bosche, M., Krause, R., Grandi, P., Marzioch, M., Bauer, A., et al. (2002). Functional organization of the yeast proteome by systematic analysis of protein complexes. *Nature, 415*(6868), 141-147.

Giot, L., Bader, J. S., Brouwer, C., Chaudhuri, A., Kuang, B., Li, Y., et al. (2003). A protein interaction map of drosophila melanogaster. *Science, 302*(5651), 1727-1736.

Graur, D., & Li, W. H. (2000). Fundamentals of molecular evolution (second edition). *Sunderland, Massachusetts: Sinauer.*

Guan, Y., Dunham, M. J., & Troyanskaya, O. G. (2007). Functional analysis of gene duplications in saccharomyces cerevisiae. *Genetics, 175*(2), 933-943.

Hahn, M. W., Conant, G. C., & Wagner, A. (2004). Molecular evolution in large genetic networks: Does connectivity equal constraint? *J Mol Evol, 58*(2), 203-211.

Hahn, M. W., & Kern, A. D. (2005). Comparative genomics of centrality and essentiality in three eukaryotic protein-interaction networks. *Mol Biol Evol, 22*(4), 803-806.

He, X., & Zhang, J. (2005). Rapid subfunctionalization accompanied by prolonged and substantial neofunctionalization in duplicate gene evolution. *Genetics, 169*(2), 1157-1164.

He, X., & Zhang, J. (2006). Why do hubs tend to be essential in protein networks? *PLoS Genet, 2*(6), e88.

Ho, Y., Gruhler, A., Heilbut, A., Bader, G. D., Moore, L., Adams, S. L., et al. (2002). Systematic identification of protein complexes in saccharomyces cerevisiae by mass spectrometry. *Nature, 415*(6868), 180-183.

Huang, T. W., Lin, C. Y., & Kao, C. Y. (2007). Reconstruction of human protein interolog network using evolutionary conserved network. *BMC Bioinformatics, 8*, 152.

Hurst, L. D., Pal, C., & Lercher, M. J. (2004). The evolutionary dynamics of eukaryotic gene order. *Nat Rev Genet, 5*(4), 299-310.

Huynen, M. A., Snel, B., & Bork, P. (2001). Inversions and the dynamics of eukaryotic gene order. *Trends Genet, 17*(6), 304-306.

Ispolatov, I., Yuryev, A., Mazo, I., & Maslov, S. (2005). Binding properties and evolution of homodimers in protein-protein interaction networks. *Nucleic Acids Res, 33*(11), 3629-3635.

Ito, T., Tashiro, K., Muta, S., Ozawa, R., Chiba, T., Nishizawa, M., et al. (2000). Toward a protein-protein interaction map of the budding yeast: A comprehensive system to examine two-hybrid interactions in all possible combinations between the yeast proteins. *Proc Natl Acad Sci U S A, 97*(3), 1143-1147.

Jeong, H., Mason, S. P., Barabasi, A. L., & Oltvai, Z. N. (2001). Lethality and centrality in protein networks. *Nature, 411*(6833), 41-42.

Jordan, I. K., Wolf, Y. I., & Koonin, E. V. (2003). No simple dependence between protein evolution rate and the number of protein-protein interactions: Only the most prolific interactors tend to evolve slowly. *BMC Evol Biol, 3*, 1.

Kafri, R., Dahan, O., Levy, J., & Pilpel, Y. (2008). Preferential protection of protein interaction network hubs in yeast: Evolved functionality of genetic redundancy. *Proc Natl Acad Sci U S A, 105*(4), 1243-1248.

Kellis, M., Birren, B. W., & Lander, E. S. (2004). Proof and evolutionary analysis of ancient genome duplication in the yeast saccharomyces cerevisiae. *Nature, 428*(6983), 617-624.

Kiemer, L., & Cesareni, G. (2007). Comparative interactomics: Comparing apples and pears? *Trends Biotechnol, 25*(10), 448-454.

Krogan, N. J., Cagney, G., Yu, H., Zhong, G., Guo, X., Ignatchenko, A., et al. (2006). Global landscape of protein complexes in the yeast saccharomyces cerevisiae. *Nature, 440*(7084), 637-643.

Krylov, D. M., Wolf, Y. I., Rogozin, I. B., & Koonin, E. V. (2003). Gene loss, protein sequence divergence, gene dispensability, expression level, and interactivity are correlated in eukaryotic evolution. *Genome Res, 13*(10), 2229-2235.

Kunin, V., Pereira-Leal, J. B., & Ouzounis, C. A. (2004). Functional evolution of the yeast protein interaction network. *Mol Biol Evol, 21*(7), 1171-1176.

LaCount, D. J., Vignali, M., Chettier, R., Phansalkar, A., Bell, R., Hesselberth, J. R., et al. (2005). A protein interaction network of the malaria parasite plasmodium falciparum. *Nature, 438*(7064), 103-107.

Lercher, M. J., Urrutia, A. O., & Hurst, L. D. (2002). Clustering of housekeeping genes provides a unified model of gene order in the human genome. *Nat Genet, 31*(2), 180-183.

Levy, E. D., & Pereira-Leal, J. B. (2008). Evolution and dynamics of protein interactions and networks. *Curr Opin Struct Biol, 18*(3), 349-357.

Li, S., Armstrong, C. M., Bertin, N., Ge, H., Milstein, S., Boxem, M., et al. (2004). A map of the interactome network of the metazoan c. Elegans. *Science, 303*(5657), 540-543.

Liang, Z., Xu, M., Teng, M., & Niu, L. (2006). Comparison of protein interaction networks reveals species conservation and divergence. *BMC Bioinformatics, 7*, 457.

Liao, B. Y., & Zhang, J. (2008). Null mutations in human and mouse orthologs frequently result in different phenotypes. *Proc Natl Acad Sci U S A, 105*(19), 6987-6992.

Makino, T., & Gojobori, T. (2006). The evolutionary rate of a protein is influenced by features of the interacting partners. *Mol Biol Evol, 23*(4), 784-789.

Makino, T., & McLysaght, A. (2008). Interacting gene clusters and the evolution of the vertebrate immune system. *Mol Biol Evol, 25*(9), 1855-1862.

Makino, T., Suzuki, Y., & Gojobori, T. (2006). Differential evolutionary rates of duplicated genes in protein interaction network. *Gene, 385*, 57-63.

Matthews, L. R., Vaglio, P., Reboul, J., Ge, H., Davis, B. P., Garrels, J., et al. (2001). Identification of potential interaction networks using sequence-based searches for conserved protein-protein interactions or "interologs". *Genome Res, 11*(12), 2120-2126.

Mika, S., & Rost, B. (2006). Protein-protein interactions more conserved within species than across species. *PLoS Comput Biol, 2*(7), e79.

Mintseris, J., & Weng, Z. (2005). Structure, function, and evolution of transient and obligate protein-protein interactions. *Proc Natl Acad Sci U S A, 102*(31), 10930-10935.

Musso, G., Costanzo, M., Huangfu, M., Smith, A. M., Paw, J., San Luis, B. J., et al. (2008). The extensive and condition-dependent nature of epistasis among whole-genome duplicates in yeast. *Genome Res, 18*(7), 1092-1099.

Musso, G., Zhang, Z., & Emili, A. (2007). Retention of protein complex membership by ancient duplicated gene products in budding yeast. *Trends Genet, 23*(6), 266-269.

Pal, C., Papp, B., & Lercher, M. J. (2006). An integrated view of protein evolution. *Nat Rev Genet, 7*(5), 337-348.

Pastor-Satorras, R., Smith, E., & Sole, R. V. (2003). Evolving protein interaction networks through gene duplication. *J Theor Biol, 222*(2), 199-210.

Pereira-Leal, J. B., Levy, E. D., Kamp, C., & Teichmann, S. A. (2007). Evolution of protein complexes by duplication of homomeric interactions. *Genome Biol, 8*(4), R51.

Poyatos, J. F., & Hurst, L. D. (2006). Is optimal gene order impossible? *Trends Genet, 22*(8), 420-423.

Presser, A., Elowitz, M. B., Kellis, M., & Kishony, R. (2008). The evolutionary dynamics of the saccharomyces cerevisiae protein interaction network after duplication. *Proc Natl Acad Sci U S A, 105*(3), 950-954.

Rain, J. C., Selig, L., De Reuse, H., Battaglia, V., Reverdy, C., Simon, S., et al. (2001). The protein-protein interaction map of helicobacter pylori. *Nature, 409*(6817), 211-215.

Rual, J. F., Venkatesan, K., Hao, T., Hirozane-Kishikawa, T., Dricot, A., Li, N., et al. (2005). Towards a proteome-scale map of the human protein-protein interaction network. *Nature, 437*(7062), 1173-1178.

Sharan, R., Suthram, S., Kelley, R. M., Kuhn, T., McCuine, S., Uetz, P., et al. (2005). Conserved patterns of protein interaction in multiple species. *Proc Natl Acad Sci U S A, 102*(6), 1974-1979.

Singer, G. A., Lloyd, A. T., Huminiecki, L. B., & Wolfe, K. H. (2005). Clusters of co-expressed genes in mammalian genomes are conserved by natural selection. *Mol Biol Evol, 22*(3), 767-775.

Spirin, V., & Mirny, L. A. (2003). Protein complexes and functional modules in molecular networks. *Proc Natl Acad Sci U S A, 100*(21), 12123-12128.

Stelzl, U., Worm, U., Lalowski, M., Haenig, C., Brembeck, F. H., Goehler, H., et al. (2005). A human protein-protein interaction network: A resource for annotating the proteome. *Cell, 122*(6), 957-968.

Stumpf, M. P., Thorne, T., de Silva, E., Stewart, R., An, H. J., Lappe, M., et al. (2008). Estimating the size of the human interactome. *Proc Natl Acad Sci U S A, 105*(19), 6959-6964.

Tarassov, K., Messier, V., Landry, C. R., Radinovic, S., Molina, M. M., Shames, I., et al. (2008). An in vivo map of the yeast protein interactome. *Science, 320*(5882), 1465-1470.

Teichmann, S. A. (2002). The constraints protein-protein interactions place on sequence divergence. *J Mol Biol, 324*(3), 399-407.

Teichmann, S. A., & Veitia, R. A. (2004). Genes encoding subunits of stable complexes are clustered on the yeast chromosomes: An interpretation from a dosage balance perspective. *Genetics, 167*(4), 2121-2125.

Thompson, J. D., Higgins, D. G., & Gibson, T. J. (1994). Clustal w: Improving the sensitivity of progressive multiple sequence alignment through sequence weighting, position-specific gap penalties and weight matrix choice. *Nucleic Acids Res, 22*(22), 4673-4680.

Trinklein, N. D., Aldred, S. F., Hartman, S. J., Schroeder, D. I., Otillar, R. P., & Myers, R. M. (2004). An abundance of bidirectional promoters in the human genome. *Genome Res, 14*(1), 62-66.

Uetz, P., Giot, L., Cagney, G., Mansfield, T. A., Judson, R. S., Knight, J. R., et al. (2000). A comprehensive analysis of protein-protein interactions in saccharomyces cerevisiae. *Nature, 403*(6770), 623-627.

von Mering, C., Krause, R., Snel, B., Cornell, M., Oliver, S. G., Fields, S., et al. (2002). Comparative assessment of large-scale data sets of protein-protein interactions. *Nature, 417*(6887), 399-403.

Wagner, A. (2001). The yeast protein interaction network evolves rapidly and contains few redundant duplicate genes. *Mol Biol Evol, 18*(7), 1283-1292.

Wagner, A. (2002). Asymmetric functional divergence of duplicate genes in yeast. *Mol Biol Evol, 19*(10), 1760-1768.

Walhout, A. J., Sordella, R., Lu, X., Hartley, J. L., Temple, G. F., Brasch, M. A., et al. (2000). Protein interaction mapping in c. Elegans using proteins involved in vulval development. *Science, 287*(5450), 116-122.

Watts, D. J., & Strogatz, S. H. (1998). Collective dynamics of 'small-world' networks. *Nature, 393*(6684), 440-442.

West, A. G., Gaszner, M., & Felsenfeld, G. (2002). Insulators: Many functions, many mechanisms. *Genes Dev, 16*(3), 271-288.

Wolfe, K. H., & Shields, D. C. (1997). Molecular evidence for an ancient duplication of the entire yeast genome. *Nature, 387*(6634), 708-713.

Wuchty, S. (2004). Evolution and topology in the yeast protein interaction network. *Genome Res, 14*(7), 1310-1314.

Wuchty, S., Barabasi, A. L., & Ferdig, M. T. (2006). Stable evolutionary signal in a yeast protein interaction network. *BMC Evol Biol, 6*, 8.

Wuchty, S., Oltvai, Z. N., & Barabasi, A. L. (2003). Evolutionary conservation of motif constituents in the yeast protein interaction network. *Nat Genet, 35*(2), 176-179.

Yook, S. H., Oltvai, Z. N., & Barabasi, A. L. (2004). Functional and topological characterization of protein interaction networks. *Proteomics, 4*(4), 928-942.

Yu, H., Greenbaum, D., Xin Lu, H., Zhu, X., & Gerstein, M. (2004). Genomic analysis of essentiality within protein networks. *Trends Genet, 20*(6), 227-231.

Section IV
Biological Applications Using PPI Analysis

Section IV of the book will present the various potential biological applications of protein interaction networks, e.g. the identification of lethal proteins, the functional annotation of unknown proteins, and the enrichment of functional genomics analyses. We will discuss how to use the protein interaction network and other biological resources for drug discovery, how to uncover the underlying disease pathways, and how to exploit the PPI networks to understand the complex dynamics of cellular processes.

Chapter XI
Discovering Lethal Proteins in Protein Interaction Networks

Kar Leong Tew
Institute for Infocomm Research, Singapore

Xiao-Li Li
Institute for Infocomm Research, Singapore

ABSTRACT

This chapter introduces state-of-the-art computational methods which discover lethal proteins from Protein Interaction Networks (PINs). Lethal proteins are an interesting subject in understanding the minimal condition for cellular development and survival. A dysfunctional research subject or absence of a lethal protein would result in fatality of the cell. Biological experiments have been conducted to systematically detect such proteins. However, such processes are time consuming and requires huge amount of effort to conduct. The researchers have developed a series of computational methods which take advantage of the network properties of individual proteins to detect lethal proteins in PINs. In this chapter, each computational method is studied in depth with an analysis on its pros and cons. Finally, a discussion on the possible further research directions will conclude the chapter.

INTRODUCTION

Proteins play a key role in the operation and survival of a cell. They rarely function alone, but perform biological tasks through interactions with other biological entities such as DNA, RNA and other proteins. Interactions between proteins within an organism collectively form a **P**rotein **I**nteraction **N**etwork (**PIN**). The most common representation of a PIN is a graph where proteins are depicted by nodes and an edge between two nodes indicates a detected interaction between the two proteins. A lethal (or essential) protein is one that renders the cell unviable upon its removal. Lethal proteins play an intricate role in the development and survival of the cell and their characteristics and detection is an interesting research topic in proteomics.

A protein's lethality has been used in various biological and medical researches in recent years. In the study of protein evolution and conservation, it was predicted that proteins differing in their importance (lethality) are subjected to different evolution rate (Kimura *et al.* 1974; Wilson *et al.* 1977). Their prediction has been put to test with the availability of large gene-knockout data by various research groups (Hurst *et al.* 1999; Liao *et al.* 2006;

Pal *et al.* 2003; Yang *et al.* 2003; Zhang *et al.* 2005). Due to its biological significance, even with a substantial number of genes with unknown lethality profiles (Jeong *et al.* 2003), lethal proteins are used as the focus of their work. The results of their investigations do agree that lethal proteins evolve at a different rate from normal proteins. However, there is a dispute over whether does lethal proteins evolve *faster* (to better adapt to changing environment) or *slower* (to avoid drastic changes).

Cross species studies have also utilized information pertaining to protein lethality. A study on the PIN of *Saccharomyces cerevisiae*, *Caenorhabditis elegans*, and *Drosophila melanogaster* indicates an association between protein evolution, centrality, and protein lethality (Hahn *et al.* 2005). Conservation of lethal proteins has also been detected between *Saccharomyces cerevisiae* and *Saccharomyces mikatae* (Seringhaus *et al.* 2006). Such work demonstrates the usefulness of lethal proteins in aiding biological research while investigating new species.

Associations between lethal proteins, disease, and human gene morbidity have also been established by various research groups. Steinmetz *et al.* (2002) used the *Saccharomyces cerevisiae* deletion mutants to identify 256 new human mitochondrial proteins with a fivefold greater selection than gene expression analysis. Kondrashov *et al.* (2004) first stated the close relationship between morbidity and protein lethality and further found that morbid genes are more similar to lethal proteins of *Drosophila melanogaster*. Furney *et al.* (2006) made a finer division of disease genes into dominant or recessive mutations and lethal proteins were found to have a higher correlation with dominant genes.

All the above mentioned research work reflects the biological insights and discovery related to lethal proteins. Currently lethal proteins are found based on gene knockout experiments which can be time-consuming and cost-prohibitive when the biological experiments are done on a large-scale basis. Because of this reason, a large number of proteins are still without known lethality information, making it desirable to formulate alternative detection approaches, e.g. computational methods, which can harness the knowledge of social network analysis, graph mining and biological information. Such computational approaches could help generating a ranked list of potential lethal proteins and the cost to detect lethal proteins is thus largely reduced.

The impact of removing a lethal protein from a PIN is similar to that of a "*Key Player*" in a social network where the node is strategically located within the network (Bonacich 1987; Everett *et al.* 1999; Stephen 2006). Their absence or dysfunctional would create an adverse disruption to the topological stability of the network which in the case of PIN – biological lethality. This laid the foundation where computation methods based on topological features are developed to better detect lethal proteins (Estrada 2006; Jeong *et al.* 2001; Joy *et al.* 2005; Schmith *et al.* 2005; Yu *et al.* 2004). Recently, there is also a new trend where biological information is integrated into their models (Saha *et al.* 2006; Tew *et al.* 2007; Ulitsky *et al.* 2007). The first category of computational methods can be broadly classified under "*Topological*" and the second category as "*Integrative*" approaches.

This chapter will provide an overview of the present state-of-the-art methods to detect the lethal proteins from protein interaction networks. The basic concepts for each individual technique will be first covered in the Methods section. This is followed by Definitions and Formulas Section which provides the detailed computational description for the different techniques and compares their strengths and weakness of each method. Finally, a discussion on the possible further research directions will conclude the chapter.

METHODS

There has been a series of work in developing computational methods to detect the lethal proteins from PIN. Each of this work is based on their own assumptions, i.e. because of some proteins have some important network properties or functionalities, their removal will thus disrupt the topological stability or impair the biological functions of the cell (they are thus essential to the cell). These network properties are also known as the **centrality** of the node. It accounts for the different node characteristics that permit them to be ranked in order of importance in the network.

Two broad categories can be used to classify the available computational prediction methods, a) "*Topological approaches*" which is solely based on the network characteristic, without using other biological information and b) "*Integrative approaches*" which make use of various biological information and properties. The details of each method would be discussed in this section while the comparisons will be done later.

Topological Approaches

This category of computational method is the first set of computational methods developed to provide novel properties of each node in a network. In this category, the inputs are just the protein interaction data. The list of computational methods under this group is as follows:

- Connectivity/Degree – Jeong *et al.* 2001
- Clustering Coefficient – Yu *et al.* 2004
- Betweenness – Joy *et al.* 2005
- Damage Score – Schmith *et al.* 2005
- Subgraph Centrality – Estrada 2006

Connectivity

Jeong *et al.* (2001) first reported that the lethality of a protein is positively correlated to its connectivity (or degree) in the protein interaction network. They first analyzed the network which are highly inhomogeneous scale-free in which a few highly connected proteins play a central role in mediating interactions among numerous, less connected proteins. Inhomogeneous structure has the property of being highly resilient against random errors coupled with fragility against removal of the most connected nodes. In the *Saccharomyces cerevisiae* PIN, they found that, random removal of selected yeast proteins do not affect the overall topology. In contrast, when the most connected proteins are computationally eliminated, the network diameter increases rapidly, which means the proteins need longer paths to communicate with each other.

Based on their study, they formed a hypothesis that on average highly connected proteins should prove more lethal than less connected ones, i.e. connectivity measures can be used to detect lethal proteins. A clear correlation was found between the proteins lethality and the number of interactions the proteins have. Statistical studies have shown that only ~21% of proteins with five or less links are lethal. On the other hand for "hub" proteins with 15 edges or more, ~62% of them cause fatality when removed. Highly-connected proteins (hubs) in a central position in the PIN are therefore three times more likely to be lethal than proteins with lower number of interactions.

Connectivity was the first work to establish an association between a topological feature and lethality of a protein. This is followed by other discoveries of various network properties which can also be used to detect lethal proteins.

Clustering Coefficient

Yu *et al.* (2004) followed on the work where they studied various network properties of proteins and their correlation with lethality. On top of confirming the effectiveness of connectivity as an indicator of lethality, they found that lethal proteins tend to be more tightly connected than random. This was supported by their experiments that lethal proteins have a higher clustering coefficient value (or more cliquish behavior) than non-lethal proteins.

Both connectivity and clustering coefficient are capturing the local network architecture surround each individual protein. Their works are based on the fact that lethal proteins have a higher level of activity due to the "*key role*" in the network.

Betweenness

Joy *et al.* (2005) has designed a more macro approach in analyzing the overall position of the protein as opposed to the micro view of connectivity and clustering coefficient. They viewed the edges between nodes as passageway for information transfer between proteins. This stems from the traditional assumption of the existence of distinct signaling modules that can be associated with particular cellular functions. They sought to seek out nodes having the most influence over the "information transfer" in the PIN.

To fulfill this target, they applied the betweenness measure. Betweenness value is the number of times a node is a member of the set of shortest paths connecting all pairs of nodes in the network. This measure indicates the relative importance of the node in the context of the entire network. According to their work, on average, the betweenness value was only $6.6 * 10^{-4}$ for all proteins but $1.2 * 10^{-3}$ for lethal proteins, representing a significant increase of 82%. The performance of using betweenness to detect lethal proteins was to be at the same level as with connectivity.

Betweenness was an illustration on the wide range of possibilities of topological features that could be associated with lethal proteins. There is a stronger emphasis in the work related to the importance of proteins in terms of information transmission in the PIN.

Damage Score

Schmith *et al.* (2005) took a completely different approach in the identification of lethal proteins within PIN. All the previous approaches were based on analyzing the existing network. In contrast, what they proposed was an approach named damage Score (damage for short) that measures the consequences on the topological properties resulting from the deletion of a protein.

This concept is based on the fact that perturbations on hubs would more likely lead to lethality as the impact on the network is more significant. It is an extension of the topological measure of the importance of a node in the network. Damage takes into account the influence of a node not only on its immediate neighbors, but also on the integrity and strength of the entire PIN. It is a quantitative criterion for the deleterious effect of the removal of a node from the network. Not surprisingly, there is a correlation between connectivity and damage.

Subgraph Centrality

Estrada (2006) introduced a mathematical model that considers the participation of the node in the network. This is done via computing the number of closed walks (subgraphs) that starts and end at the node. Strength of each closed walk decreases as the order of the walk (path length) increases. This measure computes the number of different subgraphs that a node participates in, with smaller subgraph having higher importance.

Subgraph centrality takes into account not only the immediate effects of the closest nodes but also the long range effects "transmitted" through the participation of a node in all subgraphs existing in the network, making it a more realistic model on the potential impact of a node in the network.

Other than being tested on PINs, Estrada *et al.* (2005) had also tested it on six other real-world networks. The nodes discovered have been shown to be largely novel due to the fact that subgraph centrality does not show a strong correlation with other tested measures.

Integrative Approaches

All the previous works have a common characteristic that they require only the network information, i.e. protein interaction data, as input for their lethal protein prediction. Although the network is highly enriched with topological information about the proteins and their interactions, no other biological information are actually used. As noted by the researchers, there actually exists a correlation between their high ranked proteins and proteins with certain network properties, such as high connectivity. This is unsurprising as proteins with more interactions have a higher sense of "presence" in the network.

With the rapid development of biology in the recent years, the biological information related to proteins has increased in their availability for research. It provides semantic information to the nodes, i.e. proteins, in protein interaction networks. Examples of such information include protein sequences, biological functions such as transcription and translation, cellular components, gene expression data and the degree of paralogy of proteins etc. Recently, some computational methods have been proposed to detect the lethal proteins which exploit the rich biological knowledge. In this chapter, these methods that make use of both biological and topological information are categorized as "*Integrative approaches*". Through integrating biological with topological information, these methods have improved the existing work and uncovered new insights. The list of integrative methods is as follows:

- Pivot Proteins – Ulitsky *et al.* 2007
- Lethal Protein Classifier – Saha *et al.* 2006
- Neighborhood Functional Centrality – Tew *et al.* 2007

Pivot Proteins

Following the success of the topological approaches, it is natural that the inclusion of additional biological information could improve lethal protein prediction. Ulitsky *et al.* (2007) began with harnessing the knowledge that cellular organization is largely modular based in nature. Such modules are often termed *pathways* represent physically interacting proteins involved in carrying out a particular function. Depending on the detection methods, pathways may represent molecular complexes or signaling cascades.

In their work, they defined that a pathway is simply a connected subnetwork in the PIN. Based on their definition, they developed a between-pathway-model (BPM) that is able to find and provide explanations while maintaining the significant functional content of each pathway model. Each BPM is formed by multiple proteins and their corresponding interactions. Their model also allows more direct interpretation of specific buffering cases. The pathways found from their model were then established to represent coherent functional units.

From each pathway, they seek out proteins which are densely connected to other pathways and named them pivot proteins. They discovered a total of 124 distinct pivots in 40 models. A systematic analysis on the biological activities that the group of pivots engages in was performed for protein complex participation, multiple pathway involvement, and multicomplexed genes. The pivots were found to be enriched in all three sets of biological activities, and further tested for correlation with lethality. They found that out of the 124 pivots found, 72 of them were found to be essential. Although they observed a general strong correlation between connectivity and lethality, the high prevalence of lethal proteins among pivots is far beyond what can be explained by their connectivity alone.

In this integration model, biological knowledge (modularity of cellular organization and prevalence of between-pathways genetic interaction) is used to design and verify the results of their model (functional enrichment of the models and the essentiality and evolutionary retention of pivot proteins). They have also shown that there exists an abundance of new features waiting to be uncovered from PINs.

Lethal Protein Classifier

Saha *et al.* (2006) took a more traditional machine learning approach of creating a classifier for lethal proteins. Their goal was to complement the existing experimental methods with an *in silico* prediction of the phenotypic effects of protein absence in the PIN. This approach was based on the knowledge that lethal proteins share distinctive features, such as they are more likely to have homologs in other organisms and they have fewer paralogs, than non-essential proteins.

They gathered a set of biological information and transformed them into feature values associated with each individual protein. The list of features used is length of protein sequence, degree of paralogy, connectivity, and lastly phylogenetic conservation score against other organisms. The set of organism used are bacteria, fungi (excluding *Ascomycota*), *Ascomycota* (excluding *Saccharomyces cerevisiae*), plants and mammals.

After determining the set of features to be used, they then developed an automatic classification scheme using an ensemble of weighted k-nearest neighbor (KNN) and support vector machines (SVM) for identifying lethal proteins. They used simulated annealing which is a global optimization algorithm to adjust the weights of the features (Kirkpatrick *et al.* 1983). The weight of each individual feature was adjusted using a greedy approach.

Their evaluation showed that through an ensemble method, they are able to improve the prediction of lethal proteins over using a single classifier system (either KNN or SVM). This is in line with the knowledge that in general, multiple classifier system could illuminate the biases of individual classifier and would thus be able to perform better than a single classifier system.

Neighborhood Functional Centrality

Tew *et al.* (2007) makes use of biological information different from the above mentioned methods. They combined the ideologies of the topological approaches with the notion of functional modules which are groups of interconnected proteins performing discrete functions in the PIN. Both protein complexes and biological pathways are instance of functional modules. The reason is that lethal proteins are key coordinators within functional modules. As such, their removal will maximally disrupt the operations of the modules which impact cell fitness.

They hypothesize that these key proteins should also be centrally positioned within functional modules in order to carry out their roles effectively. Removal of centrally-located proteins in the functional modules will cripple the modules more easily than the removal of proteins lying at the modules' peripheral. They thus expect that lethal proteins would be surrounded by proteins with similar functions in the underlying PIN.

They introduced the concept of neighborhood functional centrality which quantifies the extent of whether a protein is surrounded by functionally consistent neighboring proteins in the PIN. In order for such a model to work, on top of the knowledge of biological functional modules, they incorporated biological function information from the Gene Ontology (Ashburner *et al.* 2000). This work follows the trend of topological approaches where they generate a ranked list of proteins on the probability of them causing lethality when removed.

In their algorithm, for each protein, a neighborhood is gathered based on their distance from the protein. Each protein in the neighborhood is subsequently compared for functional similarity with the source protein. This provides the functional consistency score of the source protein. This score is then normalized globally based on the size of the neighborhood.

In their evaluation, neighborhood functional centrality has been able to outperform the topological based approaches. In addition, they were the first to provide favorable ranking for low-connectivity proteins (connectivity ≤ 5). Statistically, low-connectivity proteins form the bulk of proteins within a PIN resulting in most lethal proteins falling under this category. Biologically, low-connectivity proteins are interesting as they are mostly loosely connected to the PIN — topologically their absence in PIN would have minimal disruption, but biologically they lead to lethality.

DEFINITIONS AND FORMULAS

Common Definitions

Protein interaction data used in all the above mentioned methods can be modeled as a large undirected graph $G_{PIN} = (V_{PIN}, E_{PIN})$, where V_{PIN} represents the set of the interacting proteins and E_{PIN} denotes all the detected pair-wise interactions between two proteins from V_{PIN}. An edge $e_{ij} \in E_{PIN}$ connects node *i* with node *j*.

For simplicity, all edges have a uniformed weight of 1 and self interactions are typically removed. Multiple interactions may exist between proteins but are mapped out as a single interaction. All proteins within the network would have at least one interaction with another protein in order for any decent computation.

Connectivity

Connectivity is one of the most commonly used measures in network. It is the fundamental value describing a node within any given network. It represents the number of interacting partners that a node has within the network. The value type of connectivity is an integer and begins from 1 for any connected graph.

$$Connectivity(u) = |\{v \mid e_{uv} \in E_{PIN}\}| \quad (1)$$

Figure 1. Shows an example of the connectivity, connectivity (1)=5, connectivity (2)=3, connectivity (3)=4, connectivity (4)=4, connectivity (5)=2, connectivity (6)=2.

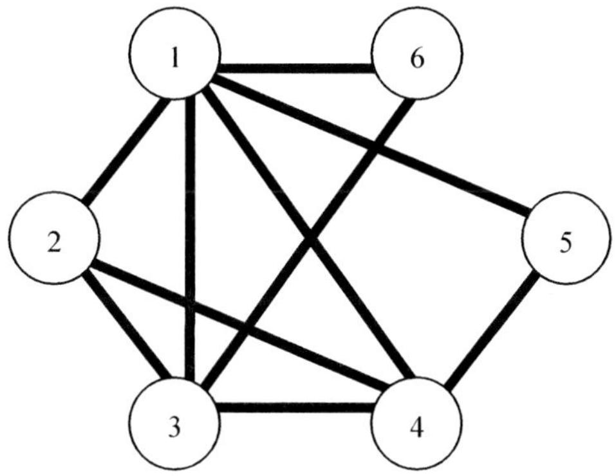

Strengths

Connectivity is the fastest, easiest, and most intuitive to understand due to its simplistic nature. For such a simple method, it has yielded one of the highest and best performances in detecting lethal proteins. This phenomena is evident in the works by (Saha *et al.* 2006) where connectivity was found to be one of the most important feature, and (Tew *et al.* 2007) where it was the next best performing method compared to a few existing methods (neighborhood functional centrality method is the best).

Weaknesses

One of the key weaknesses is that the values are integer values. In a PIN, there exist a large number of proteins and many of them can share the same number of interacting partners. In such a situation, the connectivity values would be exactly the same and unless a secondary feature is used. In other words, there is no means of differentiating the proteins without using an additional measure. This is further aggravated by the scale-free nature of PIN where approximately 75% of proteins have low-connectivity (connectivity ≤ 5), making it impossible to detect low-connectivity lethal proteins.

Clustering Coefficient

The clustering coefficient of a vertex in a graph quantifies how close the vertex and its neighbours are from being a clique (Watts *et al.* 1998).

Given a vertex v_i, its neighbourhood graph N_i formed by its immediate neighbors (excluding v_i) is defined as follows:

$$N_i = \{v_j \mid e_{ij} \in E_{PIN}\} \quad (2)$$

In nature, clustering coefficient measures the level of connectivity surrounding a particular protein.

Let $k_i = Connectivity(v_i)$, and the clustering coefficient (CC) value for node u is defined as follows:

Figure 2. Illustrations of a clique CC

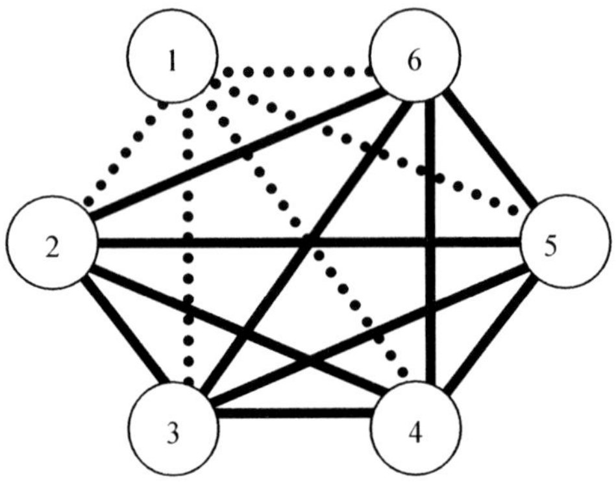

Figure 3. Illustrations of non-cliques CC

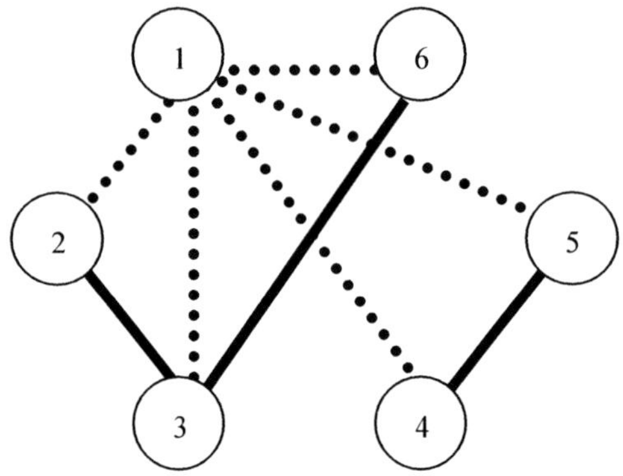

$$CC(v_i) = \frac{2|\{e_{jk}\}|}{k_i(k_i - 1)} : v_j, v_k \in N_i, e_{jk} \in E_{PIN} \quad (3)$$

The value of clustering coefficient is a float that ranges from 0 to 1. Given a vertex v_i, when its neighbors are connected with each other (the solid links, note that the dashed links will not be counted in equation 3), its clustering coefficient value is 1, presented as a clique.

For example, Figure 2 shows vertex1's neighborhood graph, according to equation (3), $cc(1) = 2*10/(5*4) = 1$. Figure 3 shows another example of vertex 1's neighborhood graph where $cc(1) = 2*3/(5*4) = 0.3$.

Clustering coefficient was modeled to detect proteins embedded in highly connected set of proteins. In our example, vertex 1 in Figure 2 is located in highly connected regions and thus has a bigger cluster coefficient. Also, due to the nature of PIN, a vast number of proteins actually do have only one interacting partners. In such situations, clustering coefficient would be regarded as 0, instead of 1.

Strengths

Clustering coefficient is a measure that one would first use for detecting any underlying structures in a network. It is simple to compute, and provides good connectivity information about the surrounding neighborhood for a given node.

Weaknesses

Although the values range from [0, 1], there still exists a large number of proteins with the same score on both end of the spectrum. In fact, the most common values are 0, 0.25, 0.5, 0.75, and 1. This is largely due to the higher probability of smaller neighborhoods having similar number of links, thus resulting in the same score. As the object of each model is to generate a ranked list of proteins, equally scored proteins would not be differentiated from each other. There exist a fair number of lethal proteins which has only one interacting partner and as they would be assigned a score of 0, they would be unfavorably ranked.

Betweenness

Betweenness characterizes how influential a node is in the communication process between all pairs of nodes in the PIN. It is computed based on the number of times that the node is a member of the set of shortest paths that connect all pairs of node in the network.

Let σ_{st} be the number of shortest paths from s to t, and $\sigma_{st}(u)$ be the number of shortest paths from s to t that passes through node u. The betweenness score of node u is as follows:

$$Betweenness(u) = \sum_{s \neq u \neq t \in V_{PIN}} \frac{\sigma_{st}(u)}{\sigma_{st}} \tag{4}$$

Strengths

It covers a global phenomenon of the network in consideration that all pairs of proteins would need to communicate with each other. There also exists a variety of shortest distance algorithm available which eases the implementation. There is also a low likelihood that the scores would be an exact same which suits the objective of ranking the proteins.

Weaknesses

The number of shortest paths between two nodes within a PIN can be extremely huge. This is a result of both a scale-free network and having a uniformed weight (PINs are commonly represented as an undirected graph and each edge has a weight of 1) for all edges. A uniformed weight would result in multiple shortest alternative paths between two nodes. Hence, betweenness is a computationally expensive method when the size of the network is relatively dense.

Damage Score

Damage score measures the difference of the network when a node is removed. The damage score of a node is computed in two phases. In the first phase, the largest cluster C_{PIN} of network G_{PIN} is found (the largest cluster in PIN typically has majority of nodes). Let n represent the number of nodes in cluster C_{PIN}. For the last phase, removes a target node from the network, gathers the largest possible cluster C'_{PIN} computes n', the number of nodes in cluster C'_{PIN}, and subtracts n from n'.

Let n represent the number of nodes in cluster C_{PIN}, and n' be the number of nodes in cluster C'_{PIN}, damage score of node u is as follows:

$$Damage(u) = n - n' \tag{5}$$

Strengths

This is the sole method that calculates the topological change in the network due to the absence of a node. The computation is fairly simple, and for a well-connected network, the results are favorable.

Weakness

The main question looms over how the main cluster C_{PIN} should be obtained. There exists several different clustering algorithms and there are subtle differences in the clusters they obtain. For a loosely connected network where there exist partitions and bridges, damage will not be suitable as those nodes will have no impact on the main cluster (Figure 4).

Figure 4. There will be no effect on the "main" cluster of the graph as it is in a separate "partition".

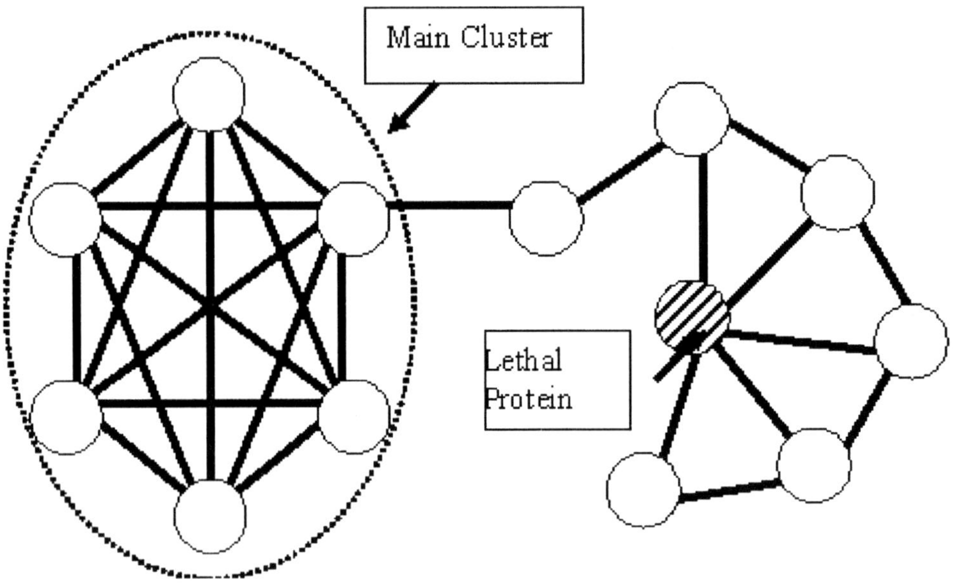

Subgraph Centrality

Subgraph centrality measures the number of closed walks in the graph starting and ending at the same node. Each closed walk is appropriately weighted such that their influence on the score decreases as the length of the walk increases. A closed walk is in conjunction a connected subgraph, meaning that this measure counts the times that nodes takes part in different connect subgraphs of the network.

The subgraph centrality (SC) is defined as:

$$SC(i) = \sum_{l=0}^{\infty} \frac{\mu_l(i)}{l!} = \sum_{j=1}^{N} [v_j(i)]^2 e^{\lambda_j} \quad (6)$$

where $\mu_l(i)$ is the number of walks started and ending at node *i*.

Strengths

This measure has the most complete interpretation of the global PIN and has a weighting that decreases as each subgraph grows. This is the first case at which a dynamic parameter and distance (length of closed walk) is taken into consideration. Although the formula might seem complicated, the eigenvectors and eigenvalues can easily be computed using a software package like MATLAB.

Weaknesses

As higher weighting will be allocated to smaller subgraphs (closed walks with shorter distances), this results in a slight biasness towards nodes with higher connectivity. This is due to the fact that nodes with higher connectivity are more likely to be involved in a larger number of small sized subgraphs compared to a node with a low-connectivity. The experimental results show that there are large numbers of high connectivity proteins in the highly ranked proteins.

Pivot Proteins

In this model, lethal proteins are found to be the pivot proteins connecting two pathways in the model found by the Betweenness-Pathway-Model (BPM). The additional information required in this model is Genetic Interaction (GI) network. The GI network is formed by synthetic lethality and synthetic fitness interaction data of the same species as the PIN.

Let $G^G = (V, E^G)$ be the GI network and let $G^P = (V, E^P)$ be the PIN. A BPM is a pair of disjointed sets V_1, V_2, such that (a) $|V_1|, |V_2| \geq 2$, (b) each V_i induces a connected subgraph in G^P and (c) there are unusually many GIs between V_1 and V_2. Each V_i is called a pathway. Given a model (V_1, V_2), nodes that are densely connected to both pathways in the physical network are identified as pivot nodes.

Let $N_i(v)$ denote the nodes in V_i that are adjacent to v in G^P. Node v is regarded as a pivot if, for $i = 1, 2$, $|N_i(v)| \geq l_{min}$ and $|N_i(v)|$ is significant, given the degree of v in G^P ($P < P_{max}$, using hypergeometric test).

Strengths

In this model, it uses two networks (Genetic Interaction and Protein Interaction) which differ in the representation of the edges. The inclusion of the GI network provided another paradigm of interpretation and detection of important nodes in the network. The success of this method has also provided another insight into the purpose and inner workings of lethal proteins within a PIN.

Weaknesses

The main weakness of this method is that it requires the additional GI network, which may not be readily available for some species. When there is no information about the GI network, the BPM is unable to produce any model. This model was designed mainly to elucidate the pathway models, and pivot proteins belong to a subset of proteins within each model.

Lethal Protein Classifier

This classifier used a combination of weighted *k*-nearest neighbor (Cover *et al.* 1967) and support vector machine (Vapnik 1999). Individual features have their weights adjusted through simulated annealing. The set of features used in the training and testing of the classifiers are listed below:

- Protein sequence length
- Degree of paralogy
- Protein-protein interactions (connectivity)
- Phylogenetic Conservation Score against
    - Viruses
    - Archaea
    - Bacteria
    - Protista
    - Fungi
    - Ascomycota
    - Plants
    - Invertebrates
    - Vertebrates
    - Mammals

The sequence comparisons for the degree of paralogy and phylogenetic conservation scores were done using Blastp, with the BLOSUM62 matrix and an e-value threshold of 10^{-6}. The paralogy of a protein was obtained by dividing the score of the best Blastp hit found in the organism (excluding self-comparison) by the Blastp score of the hit against itself. The phylogenetic conservation score for each taxonomic group is obtained by dividing the Blastp score of the closest homologue in the taxonomy by the Blastp score of the protein against itself.

Strengths

The key advantage of this classifier is that it comprises of well-established methods of KNN and SVM. Therefore, with the features generated and annotated to each protein, a proper training of each classifier will likely ensure a decent performance by each classifier.

Weaknesses

In order to have a proper classifier, a rich set of features is required. As such, this method requires the most amount of additional biological information. The computational time required to obtain the model is highly dependent on the size of the data and training parameters used.

Neighborhood Functional Centrality

Neighborhood functional centrality is one of the more complicated techniques and requires a substantial amount of additional information in order for it to work. Additional information required is from the Gene Ontology (GO) where for each protein their associated biological function can be obtained. A parameter θ is required in order to determine the radius of the neighborhood for each individual protein.

Computation of the value is divided into three main steps. In the first step, a neighborhood is grouped together based on θ, a user-defined parameter to control the radius (or size) of the neighborhood graph of the node. Following that, an initial neighborhood functional centrality (nfc) score will be computed for each protein. For the last step, the significance of each protein is assessed by computing its corresponding Z-Score Z_{nfc}. The distribution of the nfc scores of similar-sized neighborhoods will be used as the normalizing factor. The three steps are described in sequence here.

1. Construct the neighborhood graph

For each vertex $u \in V_{PIN}$, its neighborhood graph is defined as $G_u = (V_u, E_u)$,

$$V_u = \{u\} \cup \{v \mid v \in V_{PIN} \wedge dist(u,v) \leq \theta\} \tag{7}$$

$$E_u = \{(v_j, v_k) \mid (v_j, v_k) \in E_{PIN} \wedge v_j, v_k \in V_u\} \tag{8}$$

Where function *dist(u, v)* returns the shortest distance between *u* and *v*.

2. Compute an initial neighborhood functional centrality (nfc) score

The nfc score quantifies the level of functional consistency between protein *u* and all other proteins in its neighborhood graph G_u. It is thus necessary to define the functional similarities between the various functions.

As the functions used to annotate proteins are in an ontology structure which follows a "true path rule" (The Gene Ontology Consortium 2001), a model is required to provide a similarity score between functions. Relative Specificity Similarity (RSS) is the model adopted to obtain a similarity score for two functions (Wu *et al.* 2006).

Relative Specificity Similarity for two GO terms is defined as,

$$RSS(term_i, term_j) = \frac{Depth^{GO}}{Depth^{GO} + \gamma} \cdot \frac{\alpha}{\alpha + \beta} \tag{9}$$

Where the parameters $Depth^{GO}$, α, β, γ represents the maximal depth of the GO, maximum number of common ancestors between the terms, extent of generality of the terms, and the local distance between the terms respectively. For details, please refer to Wu *et al.* 2006.

RSS only provides a score for two functions, but a protein could be biologically involved in different process and therefore associated with multiple GO annotations. The next definition defines the extent of similarity between two proteins.

The protein functional similarity between two proteins *u* and *v* is defined as

$$protein\_funsim(u,v) = \frac{\sum_{i=1}^{|F_u|} \sum_{j=1}^{|F_v|} RSS(F_{(u,i)}, F_{(v,j)})}{(|F_u| * |F_v|) * dist(u,v)} \tag{10}$$

Where $F_{(u,i)}$ and $F_{(v,j)}$ denotes protein *u*'s *i-th* and protein *v*'s *j-th* function respectively.

The denominator *dist*(*u*, *v*) has been included to provide higher weighting for protein pairs that are closer together in the PIN. This quantifies the magnitude of functional similarity between two proteins which may have multiple functions. Following which, an initial neighborhood functional centrality score will be calculated based on the protein's neighborhood functional centrality score.

The neighborhood functional centrality (nfc) score of a protein is defined as

$$nfc(u) = \sum_{u \in V_u, v \neq u} protein\_funsim(u,v) \tag{11}$$

The score quantifies the level of functional consistency between protein *u* and all other proteins in its neighborhood graph. A high value of *nfc*(*u*) indicates that protein *u* is functionally central in G_u. As the score is obtained through a summation of the number of proteins in G_u, there exist biasness towards larger-sized graphs.

3. Normalize the initial nfc score into Z-score

The last step in this computation tackles this issue by using the Z-score or standard score. The statistical distributions of *nfc*(*u*) may be different when located in neighborhood of different sizes. The significance of a particular *nfc*(*u*) value is therefore dependent on the underlying distribution with respect to the size of the local neighborhood chosen for *u*.

The Z-score of each protein is defined as

$$Z_{nfc}(u,s) = \frac{nfc(u) - \mu_s}{\sigma_s} \tag{12}$$

where *s* indicates the size of protein *u*'s neighborhood, μ_s and σ_s are the mean and standard deviation of the distribution of nfc values computed.

Strengths

The process of measuring the significance of the score allows low-connectivity lethal proteins to be highly ranked. This is particularly interesting as proteins of this category are having considerably lower levels of activity in the PIN but still results in lethality when absent. It encompasses both the notion of biological modules and their relative importance in the module. The algorithm has been proven to sustain the performance level with only 50% of the proteins having functional annotations.

Weaknesses

Although proven that a good level of performance is achieved with incomplete (50% unknown) functional annotation, it is still necessary for them to be available for the algorithm to work. Compared to the other methods, this is one of the most sophisticated with multiple intermediate states required in order to obtain the final results.

RESOURCES

The main organism used in all the above methods belong to the *Saccharomyces cerevisiae* (*baker's yeast*) which has the most intensively studied PIN and lethal protein gold standard. Unless stated otherwise, all resources listed here are specific to *Saccharomyces cerevisiae.*

Lethal Proteins Gold Standard

Biological experiments have been developed to determine the phenotypic effects of removing a protein from the PIN. Several initiatives and research work have been done to biologically experiment and determine the list of lethality causing proteins in the network.

Saccharomyces Cerevisiae

Giaever *et al.* (2002) derived a comprehensive set of lethal proteins using PCR-based gene deletion strategy (Baudin *et al.* 1993; Wach *et al.* 1994). They generated a total 1106 list of proteins known to result in fatality when removed from the PIN. This is by far the most comprehensive and accurate set of lethal protein gold standard for *Saccharomyces cerevisiae.*

Escherichia Coli

Smalley *et al.* (2003) summarized the series of efforts in profiling the phenotypic effects of different genes. In their work, they introduced a brand new concept of "mutually exclusive" and "mutually inclusive" genes which adds a new dimension to the study of the minimal genome.

"Mutually exclusive" describes genes that can be deleted independently of one another, but when both genes are deleted, the result is a lethal phenotype. This is also commonly known as synthetic lethal proteins where there exists a backup in the genome.

"Mutually inclusive" describes cases where a single gene deletion results in fatality but when coupled with a secondary deletion, the initial deletion is tolerated. An example is the deletion of the antitoxin gene would result in the build-up of toxin and consequent cell death (Gotfredsen *et al.* 1998). Deletion of both the antitoxin and toxin genes negates the lethality.

Bacillus Subtilis

Kobayashi *et al.* (2003) worked out a list *Bacillus subtilis* genes which are essential to sustaining the life of the cell. The approach used to determine the lethality profile of a gene is based on a systematic gene inactivation method developed by Vagner *et al.* (1998). In total, they identified 271 genes which results in fatality when inactivated (nonfunctional) in the genome.

Web Applications

Tew *et al.* (2007) has made the results of their work available in a web-based application (http://nfc.i2r.a-star.edu.sg/). In this application, they provided a basic query and visualization applet which allows simple interaction to occur between the selected neighborhoods of a selected protein. Screenshots of the application are shown in Figure 5 and Figure 6.

In this example, protein **YJR022W** is used as the query input for the dataset **FYI** used in their work (Tew *et al.* 2007). The results are shown in three mini-windows, *Protein List*, *Function List*, and *Network Layout.*

Protein List shows the listing of proteins that are visualized in the applet. The query protein will always be the first and highlighted in yellow. Description on each column is provided at Table 1.

Function List provides a listing on all the GO terms that are annotated to the proteins listed in *Protein List. Name* is the unique identifier of the GO term, *Freq* is the number of occurrence the function has in this set of proteins, and *Source* indicates if the source protein (**YJR022W** in this example) has been annotated with the function.

Network Layout visualizes the first and second degree neighborhood of the search protein (**YJR022W**). The numbers in the box indicates the distance from the source ('1' means direct interaction neighbors of YJR022W and '2' means the neighbors' neighbors). **RED** denotes lethal proteins while **BLACK** denotes non-lethal proteins. A close-up of is shown in Figure 6.

Figure 5. Screenshot of the web-application where both the first degree neighborhood and second degree neighborhood of protein YJR022W is shown.

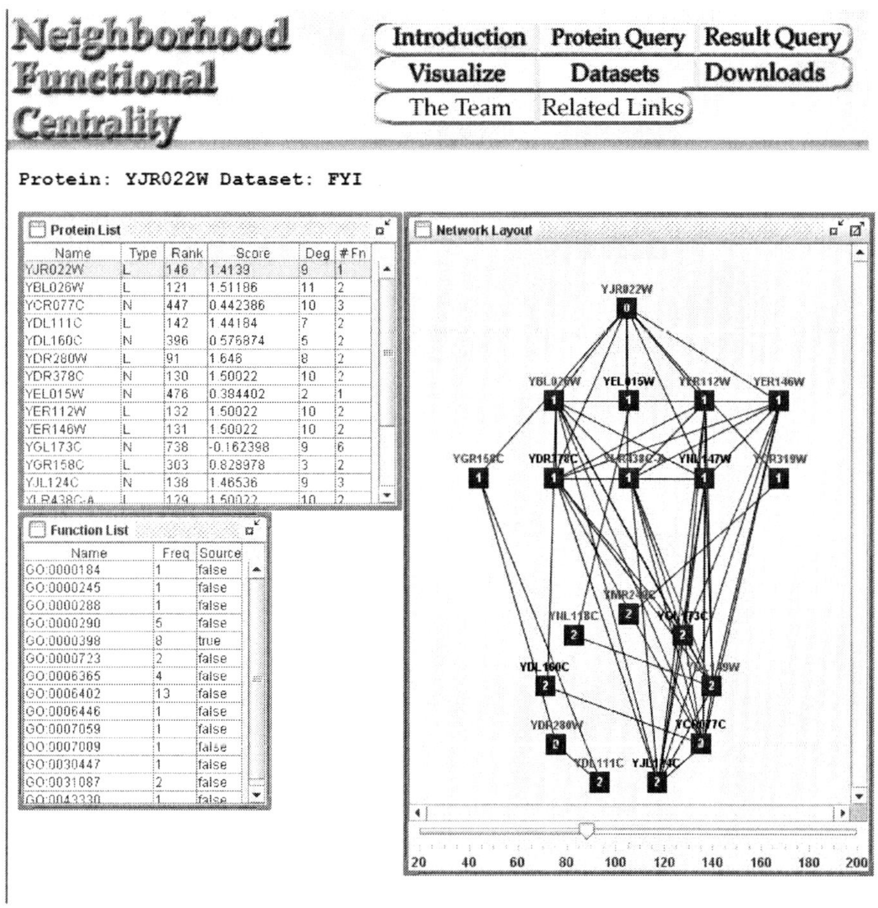

Table 1. Description of the list of proteins surrounded with the query protein

Name	Unique identifier of the protein
Type	Lethal protein (**L**) or Non-lethal protein (**N**)
Rank	The rank achieved by the protein in the dataset
Score	NFC score computed for the protein
Deg	The number of neighbours the protein has
# Fn	The number of associated GO functions.

Figure 6. Screenshot of a close-up of the network of protein YJR022W. The box indicates the details (type, rank, score, etc) of the selected protein.

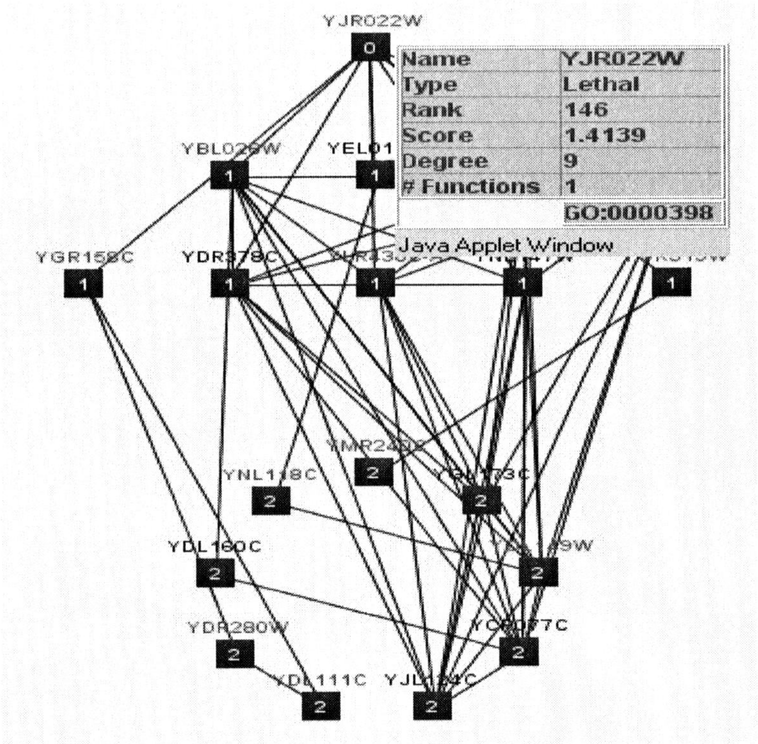

In Figure 5, by clicking on any proteins, the following information will be showed in a popup box:

1. The type of protein (Lethal/Non-lethal)
2. The rank of the protein by NFC method
3. The protein's nfc scores
4. The protein's degree/connectivity
5. The protein's GO function

A protein name that is **RED** indicates it is a lethal protein, while **BLACK** indicates a non-lethal protein. A black line represents the presence of biological interaction between the two proteins.

DISCUSSIONS AND FUTURE TRENDS

Detection of lethal proteins is a very crucial research topic in both theoretical and practical biology study. A clearer understanding on the mechanisms of lethal proteins would be another step in understanding the minimal requirements for cellular survival and development. It will also aid in the following studies, rate of protein evolution (Hurst *et al.* 1999; Liao *et al.* 2006; Pal *et al.* 2003; Yang *et al.* 2003; Zhang *et al.* 2005), cross species study (Hahn *et al.* 2005; Seringhaus *et al.* 2006), and the association between lethal proteins and diseases (Furney *et al.* 2006; Kondrashov *et al.* 2004; Steinmetz *et al.* 2002). It will also complement the relatively costly experimental approaches such as PCR-based gene deletion strategy.

Computational approaches have been proposed recently to predict lethal proteins and they may provide an explanation on the properties of lethal proteins. Beginning from the topological approaches, lethal proteins have been found to operate like a hub (connectivity), member of a cohesive unit (clustering coefficient), vital passageway for information transfer (betweenness), fundamental component in the entire PIN (damage score), and lastly deeply embedded in the network (subgraph centrality).

Following the success of topological approaches, a series of approaches have been developed harnessing additional biological knowledge and information. Integrative approaches have made use of the additional biological information and knowledge, for example, identifying the common proteins between various biological pathways (pivot proteins), inclusion of paralogs and conservation score against other taxonomy as features for a machine learning techniques (lethal protein classifier), and lastly the utilization of biological functional information and the concept of functional modules in PIN (neighborhood functional centrality).

All the various methods have been applied onto at least one protein interaction network which proves their suitability for their objective. As mentioned earlier, all the networks have a uniformed weight. This representation has been the most common interpretation of interaction data which vastly simplifies the network. In reality, the interactions occur in different time of the cell's development and might be initiated by certain proteins. Such information provides an opportunity to provide a more accurate and realistic representation of the network where the edges are weighted. How to detect lethal proteins from the weighed protein interaction networks is still an open problem.

Lastly, with the understanding that certain interactions only occur during certain stages of the cell development, some proteins might only cause lethality when removed at a stage that it is required. Therefore, a better categorization and interpretation of the lethality profiles of the proteins would shed more light on the minimal condition for cellular survival and development.

CONCLUDING REMARKS

Advancement in biological research and bioinformatics has resulted in a better understanding of the intricacies of cellular development. There exists a category of essential proteins that bring about lethality when they are removed from the cell. Such lethal proteins are interesting because they are important for understanding cellular survival and development, cross species studies, and disease studies. Computational approaches have been developed to help generate potential drug targets for biological experiments. The shift from topological approaches to integrative approaches signifies the need for additional biological information in order for further developments. Coupled with new developments in protein interaction detection and biological knowledge, a more realistic representation of the protein interaction network will be extremely useful in the development novel techniques to detect lethal proteins for various species.

REFERENCES

Ashburner, M., Ball, C. A., Blake, J. A., Botstein, D., Butler, H., Cherry, J. M., Davis, A. P., Dolinski, K., Dwight, S. S., Eppig, J. T., Harris, M. A., Hill, D. P., Issel-Tarver, L., Kasarskis, A., Lewis, S., Matese, J. C., Richardson, J. E., Ringwald, M., Rubin, G. M., & Sherlock, G. (2000). Gene ontology: tool for the unification of biology. The Gene Ontology Consortium. *Nat Genet*, *25*(1), 25-9.

Baudin, A., Ozier-Kalogeropoulos, O., Denouel, A., Lacroute, F., & Cullin, C. (1993). A simple and efficient method for direct gene deletion in *Saccharomyces cerevisiae*. *Nucleic Acids Res, 21*(14), 3329-30.

Bonacich, P. (1987). Power and Centrality: A Family of Measures. *The American Journal of Sociology, 92*(5), 1170-1182.

Cover, T., & Hart, P. (1967). Nearest neighbor pattern classification. *IEEE Transactions on Information Theory, 13*(1), 21-27.

Estrada, E. (2006). Virtual identification of essential proteins within the protein interaction network of yeast. *Proteomics, 6*(1), 35-40.

Estrada, E., & Rodríguez-Velázquez, J. A. (2005). Subgraph centrality in complex networks. *Phys Rev E Stat Nonlin Soft Matter Phys, 71*(5 Pt 2).

Everett, M. G., & Borgatti, S. P. (1999). The centrality of groups and classes. *Journal of Mathematical Sociology, 23*(3), 181-201.

Furney, S. J., Alba, M. M., & Lopez-Bigas, N. (2006). Differences in the evolutionary history of disease genes affected by dominant or recessive mutations. *BMC Genomics, 7*, 165.

Giaever, G., Chu, A. M., Ni, L., Connelly, C., Riles, L., Veronneau, S., et al. (2002). Functional profiling of the *Saccharomyces cerevisiae* genome. *Nature, 418*(6896), 387-91.

Gotfredsen, M., & Gerdes, K. (1998). The *Escherichia coli* relBE genes belong to a new toxin-antitoxin gene family. *Mol Microbiol*, *29*(4), 1065-76.

Hahn, M. W., & Kern, A. D. (2005). Comparative genomics of centrality and essentiality in three eukaryotic protein-interaction networks. *Mol Biol Evol, 22*(4), 803-6.

Hurst, L. D., & Smith, N. G. (1999). Do essential genes evolve slowly? *Curr Biol 9*(14), 747-50.

Jeong, H., Mason, S. P., Barabasi, A. L., & Oltvai, Z. N. (2001). Lethality and centrality in protein networks. *Nature*, *411*(6833), 41-2.

Jeong, H., Oltvai, Z. N., & Barabási, A. L. (2003). Prediction of Protein Essentiality Based on Genomic Data. *Complexus*, *1*(1), 19-28.

Joy, M. P., Brock, A., Ingber, D. E., & Huang, S. (2005). High-betweenness proteins in the yeast protein interaction network. *J Biomed Biotechnol, 2005*(2), 96-103.

Kimura, M., & Ota, T. (1974). On some principles governing molecular evolution. *Proc Natl Acad Sci U S A.*, *71*, 2848-52.

Kirkpatrick, S., Gelatt, C. D., & Vecchi, M. P. (1983). Optimization by Simulated Annealing. *Science, 220*(4598), 671-680.

Kobayashi, K., Ehrlich, S. D., Albertini, A., Amati, G., Andersen, K. K., Arnaud, M., et al. (2003). Essential *Bacillus subtilis* genes. *Proc Natl Acad Sci U S A, 100*(8), 4678-83.

Kondrashov, F. A., Ogurtsov, A. Y., & Kondrashov, A. S. (2004). Bioinformatical assay of human gene morbidity. Nucleic Acids Res, *32*(5), 1731-7.

Liao, B. Y., Scott, N. M., & Zhang, J. (2006). Impacts of gene essentiality, expression pattern, and gene compactness on the evolutionary rate of mammalian proteins. *Mol Biol Evol*, *23*(11), 2072-80.

Pal, C., Papp, B., & Hurst, L. D. (2003). Genomic function: Rate of evolution and gene dispensability. *Nature*, *421*(6922), 496-7; discussion 497-8.

Saha, S., & Heber, S. (2006). In silico prediction of yeast deletion phenotypes. *Genet Mol Res*, *5*(1). 224-32.

Schmith, J., Lemke, N., Mombach, J. C. M., Benelli, P., Barcellos, C. K., & Bedin, G. B. (2005). Damage, connectivity and essentiality in protein-protein interaction networks. *Physica A: Statistical Mechanics and its Applications*, *349*(3-4), 675-684.

Seringhaus, M., Paccanaro, A., Borneman, A., Snyder M., & Gerstein, M. (2006). Predicting essential genes in fungal genomes. *Genome Res*, *16*(9), 1126-35.

Smalley, D. J., Whiteley, M., & Conway, T. (2003). In search of the minimal *Escherichia coli* genome. *Trends Microbiol*, *11*(1), 6-8.

Steinmetz, L. M., Scharfe, C., Deutschbauer, A. M., Mokranjac, D., Herman, Z. S., Jones, T., Chu, A. M., Giaever, G., Prokisch, H., Oefner, P. J., & Davis, R. W. (2002). Systematic screen for human disease genes in yeast. *Nat Genet*, *31*(4), 400-4.

Stephen, P. B. (2006). Identifying sets of key players in a social network. *Comput. Math. Organ. Theory*, *12*(1), 21-34.

Tew, K. L., Li, X. L., & Tan, S. H. (2007). Functional centrality: detecting lethality of proteins in protein interaction networks. *Genome Inform.*, *19*, 166-77.

The Gene Ontology Consortium (2001). Creating the gene ontology resource: design and implementation. *Genome Res*, *11*(8), 1425-33.

Ulitsky, I., & Shamir, R. (2007). Pathway redundancy and protein essentiality revealed in the *Saccharomyces cerevisiae* interaction networks. *Mol Syst Biol*, *3*, 104.

Vagner, V., Dervyn, E., & Ehrlich, S. D. (1998). A vector for systematic gene inactivation in *Bacillus subtilis*. *Microbiology*, *144* (Pt 11), 3097-104.

Vapnik, V. (1999). *The Nature of Statistical Learning Theory (Information Science and Statistics)*. Springer.

Wach, A., Brachat, A., Pohlmann R., & Philippsen, P. (1994). New heterologous modules for classical or PCR-based gene disruptions in *Saccharomyces cerevisiae*. *Yeast*, *10*(13), 1793-808.

Watts, D. J., & Strogatz, S. H. (1998). Collective dynamics of 'small-world' networks. *Nature*, *393*(6684), 440-2.

Wilson, A. C., Carlson, S. S., & White, T. J. (1977). Biochemical evolution. Annu Rev Biochem, *46*, 573-639.

Wu, X., Zhu, L., Guo, J., Zhang D. Y., & Lin, K. (2006). Prediction of yeast protein-protein interaction network: insights from the Gene Ontology and annotations. *Nucleic Acids Res*, *34*(7), 2137-50.

Yang, J., Gu, Z., & Li, W. H. (2003). Rate of protein evolution versus fitness effect of gene deletion. *Mol Biol Evol*, *20*(5), 772-4.

Yu, H., Greenbaum, D., Xin H., Lu, X. Z., & Gerstein. M. (2004). Genomic analysis of essentiality within protein networks. Trends Genet, *20*(6), 227-31.

Zhang, J., & He, X. (2005). Significant impact of protein dispensability on the instantaneous rate of protein evolution. Mol Biol Evol, *22*(4), 1147-55.

Chapter XII
Predicting Protein Functions from Protein Interaction Networks

Hon Nian Chua
Institute for Infocomm Research, Singapore

Limsoon Wong
National University of Singapore, Singapore

ABSTRACT

Functional characterization of genes and their protein products is essential to biological and clinical research. Yet, there is still no reliable way of assigning functional annotations to proteins in a high-throughput manner. In this chapter, the authors provide an introduction to the task of automated protein function prediction. They discuss about the motivation for automated protein function prediction, the challenges faced in this task, as well as some approaches that are currently available. In particular, they take a closer look at methods that use protein-protein interaction for protein function prediction, elaborating on their underlying techniques and assumptions, as well as their strengths and limitations.

INTRODUCTION

Since the completion of the Human Genome Project (HGP) in 2003, genomic and proteomic research has gained much momentum. Based on statistics from the Genome OnLine Database (GOLD) (Liolios et al. 2008), the number of genomes sequenced grew exponentially since 1995, with nearly 700 genomes completely sequenced by 2007 (See Figure 1). With the maturation of genomic data generation, the focus in biological research has shifted towards the understanding of the complex functional and interactive processes between proteins and multi-component molecular machines that contribute to the majority of operations in cells, as well as the transcriptional regulatory mechanisms and pathways that control these cellular processes (Frazier et al. 2003). There is also a pressing need for the functional characterization of genes in clinical research to better understand diseases (Hu et al. 2007).

In contrast to the unprecedented rate at which new genes are being discovered, the pace at which novel genes and their corresponding protein products are characterized pales in comparison. A recent survey on function

Figure 1. Number of completely sequenced genomes from year 2005 to 2007. Credit: Image adapted from http://www.genomesonline.org/gold_statistics.htm.

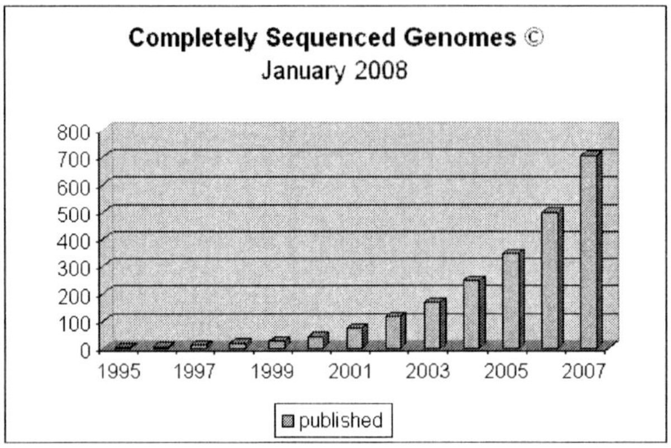

prediction techniques showed that out of 345 genomes listed in the KEGG Genome collection (Kanehisa et al. 2004), 222 have some ambiguous functional annotations assigned to half or more of its genes (putative, probable, and unknown) (Hawkins et al. 2007). This is may be attributed to the lack of reliable high-throughput method to identify the functional nature of proteins. Unlike genomic sequences, function is an abstract and complex notion, and can only be ascertained through the observation of multiple aspects of a protein, such as its sequence, structure, interaction behavior and changes in phenotype upon its mutation or removal.

Besides the influx of genomic sequence data, the maturation of high-throughput techniques for various other genomic analyses such as gene expression profiling (Eisen et al. 1998; Hughes et al. 2000), immuno-precipitation, genetic interactions, two-hybrid (Gietz et al. 1997), tandem-affinity purification, mass spectrometry, and more recently, flow cytometry and Protein-Fragment Complementation Assay (Tarassov et al. 2008), also makes available a wealth of other biological data. Advancements in computational techniques such as secondary and tertiary structure prediction also make it possible to generate computationally predicted data in large scale (Rost et al. 2003). This multitude of heterogeneous information presents to researchers a global perspective of the mechanisms behind genes and their protein products, and offers hope to elucidate the functions of proteins which cannot be easily characterized by sequence alone. However, this escalating rate of growth in biological data also makes manual annotation of protein function an increasingly daunting task. This paves the way to the emergence and popularization of automated function prediction. While it is unlikely that automated function prediction can produce authoritative annotations, it can provide systematic identification of potential novel annotations, which may be used to guide the prioritization of resource allocation for experimental verification. This can potentially improve the throughput of conventional functional characterization.

Objectives of the Chapter

In this chapter, we hope to provide a concise overview on the background, challenges and approaches taken in the task of automated protein function prediction, especially for methods that utilize protein-protein interactions. The chapter is organized as follows: First, we provide an overview of techniques used in automated protein function prediction. Second, we discuss the difficulties that are associated with this task. Finally, we describe in greater detail some of the techniques that use protein-protein interactions to predict protein function.

BACKGROUND

Protein Function Prediction using Sequence Information

Sequence homology is the classical methodology used to infer the function of a novel protein. A simple way to infer possible characteristics of a protein is to use an alignment software such as the Basic Local Alignment Search Tool (BLAST) (Altschul et al. 1990), PSI-BLAST (Altschul et al. 1997) and FASTA (Pearson et al. 1988) to find possible homologs in sequence databases such as TrEMBL (Boeckmann et al. 2003) and UniProt (Wu et al. 2006). Using standardized annotation schemes such as the Gene Ontology, some methods such as OntoBlast (Zehetner 2003), PFP (Hawkins et al. 2006), Gotcha (Martin et al. 2004), GOFigure (Khan et al. 2003) and GOPET (Vinayagam et al. 2006) further process such alignment results using statistical and machine learning techniques to predict functional annotations for the query protein. Other methods search for patterns in segments of amino acid sequences that are conserved in specific families of proteins. These conserved patterns are usually computed by first performing multiple sequence alignment (Thompson et al. 1994) on a group of similar protein sequences, and using the alignment results to build a representative model of these sequences. Some methods, such as eBLOCKs (Su et al. 2005) and PRINTS (Attwood et al. 2003), model these patterns as position-specific weight matrices. Others, such as PROSITE (Bairoch 1993) and EMOTIF (Huang et al. 2001), used regular expressions. PFAM (Finn et al. 2008) used a statistical model known as hidden Markov model to represent these conserved amino acid patterns. These models are commonly referred to as domains or motifs.

While sequence homology has proven to be effective and reliable for inferring protein function, its viability is limited to cases where substantial sequence similarity to annotated proteins can be found. Based on a study of over a million sequence alignments in (Rost 1999), it was shown that while 90% of alignments with at least 30% sequence identity correspond to alignments between homologous proteins, only 10% of alignments with 25% sequence identity or less do. Hence to maintain reasonably low false positives, the coverage of methods that utilize sequence alignments may be limited to some extent.

Protein Function Prediction using Non-Sequence Information

Sequence is only one aspect of a protein. Upon translation of a protein sequence, it may undergo post-translational modifications, and will form the tertiary structure by forming contacts between residues which are distant in the sequence. The final protein product will also interact with other proteins and form complexes. The functional behavior of a protein may hence be better understood when we also look at other aspects of its characteristics beyond sequence.

Some approaches analyze the secondary (Wang et al. 2005; Ferre et al. 2006) and tertiary structures (Pazos et al. 2004; Laskowski et al. 2005; Laskowski et al. 2005) of protein. Tertiary structures reflect the physical characters of translated proteins, and offer clues to the actual mechanism of protein function (Laskowski et al. 2005; Laskowski et al. 2005; Chiaraluce et al. 2007; Komander et al. 2008). However, tertiary structures are derived using relatively costly and time-consuming experimental techniques. The majority of structures are derived using X-ray crystallography (about 85.3%) and Protein nuclear magnetic resonance spectroscopy (NMR) (about 14.2%). A small fraction of protein structures (about 0.6%) are derived using less common techniques such as Electron Microscopy, Electron Crystallography, X-ray Fiber Diffraction and Fourier Transform Infrared Spectroscopy (statistics taken from http://www.pdb.org/pdb/statistics/holdings.do). The number of known tertiary structures is small relative to the number of protein sequences known. At the time of writing, there are 47,688 released structures in the Protein Data Bank (Berman et al. 2000), which is a far cry from the 390,696 and over 6 million proteins catalogued by Swiss-Prot and TREMBL respectively. Moreover, tertiary structures cannot be always reliably predicted from protein sequences, especially when appropriate template structures for homology modeling are not available. Secondary structures, on the other hand, can be effectively predicted from sequences (Rost et al. 2003) and used to complement sequence homology for function prediction (Wang et al. 2005; Ferre et al. 2006).

Other types of information have also been used for protein function prediction. These include experimentally derived data such as protein-protein interactions (Schwikowski et al. 2000; Brun et al. 2003; Deng et al. 2003; Letovsky et al. 2003; Samanta et al. 2003; Vazquez et al. 2003; Chen et al. 2004; Chua et al. 2006; Chua et al. 2007) and gene expression profile (Zhou et al. 2002), as well as computationally derived data such as phylogenetic profiles (Pellegrini et al. 1999; Wu et al. 2003), co-occurrence of proteins in operons or genome context (Dandekar et al. 1998; Overbeek et al. 1999; Salgado et al. 2000) and common domains in fusion proteins (Enright et al. 1999; Marcotte et al. 1999; Huynen et al. 2000). The availability of such diverse biological information also motivated approaches that combine multiple heterogeneous data sources to make better predictions (Troyanskaya et al. 2003; Chen et al. 2004; Deng et al. 2004; Karaoz et al. 2004; Lanckriet et al. 2004; Tsuda et al. 2005; Xiong et al. 2006; Chua et al. 2007). An excellent review on approaches in automated protein function prediction is provided in (Hawkins et al. 2007).

COMMON CHALLENGES IN PROTEIN FUNCTION PREDICTION

Regardless of the type of biological information used or the techniques involved, approaches to automated function prediction face several common challenges:

Incomplete Data

Due to the limitations in experimental techniques as well as resources, many biological data provide only partial information. Gene expression profiles from microarray experiments may not have constant time intervals. The conditions in which the profiles are extracted are also insufficient to distinguish between all genes. Consequently, expression profiles can be very similar for a large number of genes, such as household genes or cell cycle genes (Spellman et al. 1998). Protein-protein interaction information, as well as functional annotations are also incomplete, especially for the less well-studied genomes (Chua et al. 2007). As mentioned earlier, tertiary structure information is also far from being comprehensive. This poses two problems: First, using a single type of biological data for functional prediction will have limited coverage. Second, validation of prediction results can result in biased conclusions.

The first problem may be addressed by taking an integrative approach (Troyanskaya et al. 2003; Chen et al. 2004; Deng et al. 2004; Karaoz et al. 2004; Lanckriet et al. 2004; Tsuda et al. 2005; Xiong et al. 2006; Chua et al. 2007). It is shown in (Deng et al. 2004; Xiong et al. 2006; Chua et al. 2007) that predictions made using multiple sources of information are far more precise than that made from each individual source.

The second problem stems from the fact that function is an abstract concept which evolves as scientists learn more about proteins and their behavior. Most annotation schemes are hierarchical in nature, with annotation terms ranging from general (e.g. metabolic process) to specific (e.g. glycosinolate biosynthetic process). Well-studied functions tend to be specified in greater detail, while less well-studied ones tend to be more general. Terms associated with better understood functions are also more likely to be annotated more frequently. Using all annotation terms in the evaluation of a set of predictions may produce very biased conclusions. This is because a term has substantial overlap with its ancestor and descendant terms. Hence a method that predicts well for better understood functions with deeper hierarchies will tend to be over-estimated relative to one that predicts well for functions with a smaller hierarchies, since there will be substantially more terms associated with the former functions.

Several approaches have been taken to ensure that the evaluation of a predictions based on known annotations is meaningful. Deng et al. (2003), Lanckriet et al. (2004) and Tsuda et al. (2005) uses only the most general annotation terms from the Munich Information Center for Protein Sequences (MIPS) for validation, as shown in Table 1. A simple strategy to evaluate predictions using more specific terms is to simply use annotations from a fixed level in the hierarchy (Chua et al. 2007; Gabow et al. 2008) .

Another strategy is to use the concept of informative annotation terms (Zhou et al. 2002; Chua et al. 2006). A term is defined to be informative if there are at least 30 proteins annotated with it; and has no descendant terms annotated with at least 30 proteins. Using only informative terms to evaluate predictions allow more specific

Table 1. 13 functional classes from MIPS

	Category
1	Metabolism
2	Energy
3	Cell cycle & DNA processing
4	Transcription
5	Protein synthesis
6	Protein fate
7	Cellular transport & transport mechanism
8	Cell rescue, defense & virulence
9	Interaction with the cellular environment
10	Cell fate
11	Control of cellular organization
12	Transport facilitation
13	Others

terms to be used, whilst ensuring that terms used for validation: 1) appears in at least 30 instances and may be statistically evaluated; 2) do not overlap in description, since no descendent or ancestor of an informative term can be informative.

Lack of a Common Protein Naming Convention

Many biological databases contain overlapping or complementary information on the same proteins. In the task of function prediction, it is necessary to obtain data from multiple sources. For example, references to the same yeast protein may be found as a gene product in the Comprehensive Yeast Genome Database (CYGD) or the Saccharomyces Genome Database (SGD) (Cherry et al. 1998); as an interacting entity in the Biomolecular Interaction Network Database (BIND) (Bader et al. 2000) or the General Repository for Interaction Data (GRID) (Breitkreutz et al. 2002); as a sequence in the EMBL Nucleotide Sequence Database (EMBL-Bank) (Kulikova et al. 2007), GenBank (Benson et al. 2007), SwissProt or TREMBL (Boeckmann et al. 2003; Bairoch et al. 2005); or as an annotated protein in Gene Ontology (Ashburner et al. 2000). Each of these databases may refer to the same protein using different names.

The yeast gene product GIP4, for example, is identified by an EMBL accession number (U12980) in EMBL-Bank, a RefSeq accession number (NP_009371) in GenBank, an UniProt ID (P39732) in UniProt, a systematic name (YAL031C) in CYGD, and an SGD ID (S000000029) in SGD. Interaction databases may adopt some of these naming convention, e.g. GenBank accession numbers in BIND, and CYGD systematic name in GRID.

The adoption of different naming conventions may stem from various reasons, such as legacy, or the nature of the data being referenced (e.g. sequences vs. genes). Nonetheless, this poses some problems in protein function prediction when we need to combine data from different sources. Cross-referencing tables are sometimes provided in some of these databases, but these are often incomplete and not up-to-date. Without complete cross-referencing between different databases, automated function prediction using cross database information will face problems of redundancy and incomplete association between proteins.

Initiatives such as the International Protein Index (IPI) (Kersey et al. 2004) and the UniProt Universal Protein Resource (Bairoch et al. 2005) have been established to provide complete cross-referencing information as well as unique, non-redundant identifiers for distinct proteins. UniProt provides a unique identifier to every distinct protein sequence, while the IPI provides a unique identifier for every distinct annotated protein. These resources show great foresight, and are key to integrating all available biological databases into one coherent web of information that can work in synergy for many biological and bioinformatics problems including protein function prediction.

Noisy Data

Some biological data, such as high-throughput protein interaction assays and gene expression profiles, tend to be noisy (i.e. contain many false positives). Two-hybrid (Y2H) experiments, which are susceptible to sticky proteins that can activate the reporter genes of non-interacting proteins, are particularly notorious for noise. The level of false positives in yeast two-hybrid experiments has been estimated to be as high as 50% (Legrain et al. 2001; Deng et al. 2002; von Mering et al. 2002; Sprinzak et al. 2003). Affinity purification methods such as Tandem Affinity Purification (TAP) and Co-Immunoprecipitation detects a set of proteins (preys) that bind to a protein of interest (bait), instead of capturing pairwise interactions. Some of the prey proteins may not bind to the bait protein directly, but instead bind to other prey proteins that do so. However, these data are often treated as pairwise interactions, either based on a Spoke model (each prey protein interacts with bait protein) or a matrix model (each prey interacts with the bait and all other preys) (Bader et al. 2003). This also introduces false positive pairwise interactions that did not actually occur. Binary interactions detected by both Y2H and TAP are known to be noisy and incomplete (Shoemaker et al. 2007), and it is not clear which method is relatively more reliable.

Approaches that make use of such biological data will need to take noise into consideration to achieve consistent prediction performance. In the case of protein-protein interactions, several computational techniques have been proposed to reduce false positives in experimental datasets. A most intuitive and common way to assess the confidence of an interaction is to check if it is reproducible in multiple independent experiments (Nabieva et al. 2005; Chua et al. 2006). Deane et al. (2002) and Bader et al. (2004) proposed using auxiliary information such as the correlation in gene expression profiles of the participating proteins as a form of confidence indicator. Saito et al. (2003) and Chen et al. (2006) proposed using network topological measures to estimate the reliability of protein-protein interactions. A recent review on methods to assign reliability scores for protein-protein interaction data can be found in (Chua et al. 2008).

USING PROTEIN INTERACTIONS FOR FUNCTIONAL PREDICTION

Proteins do not work alone, but interacts with other biological entities such as DNA, RNA, as well as other proteins to perform their function. Hence the function of a protein may be inferred by looking at its interaction neighborhood. Chua et al. (2007) provided justification for the use of protein-protein interactions as a complementary approach to sequence homology by illustrating the maximum additional coverage which protein-protein interactions can potentially gain over using BLAST with an E-value threshold to infer protein functions. Figure 2, which is adapted from (Chua et al. 2007), shows that additional fraction of known functional annotations that can be discovered using protein-protein interaction (shown in light blue) but cannot be inferred by looking at homologs from BLAST search results using various log E-value thresholds. Further increase in coverage (shown in dark blue) can be obtained by also considering indirect interactions (relationships between proteins that do not interact by interacts with common proteins). The first column depicts results for *Molecular Function* terms from Gene Ontology, while the second and third columns depict results for the *Biological Process* and *Cellular Component* terms. The top row presents results for *Saccharomyces cerevisiae* (Bakers' Yeast) while the bottom row presents results for *Drosophila melanogaster* (Fruit Fly). Protein-protein interactions were obtained from the BioGRID database. The study provides support that protein-protein interactions can potentially infer a significant number functional annotations which would otherwise be missed by homology search alone.

Figure 2. Functional coverage of protein–protein interactions. The fraction of known functional annotations that can be suggested through BLAST homology search; and the additional annotations that can be suggested through: 1) direct protein interactions (PPI) and 2) indirect protein interactions. A range of BLAST E-value cutoffs between 1 to 1e-10 is used. BLAST is performed on sequences from the gene ontology database. Proteins with very close homologs (E-value ≤ 1e-25) are excluded from analysis. The top row shows the results from S. cerevisiae, and the bottom row shows the results from D. melanogaster. The three columns depict results on the biological process (left), molecular function (center) and cellular component (right) categories of the Gene Ontology. Credit: Figure adapted from (Chua et al. 2007)

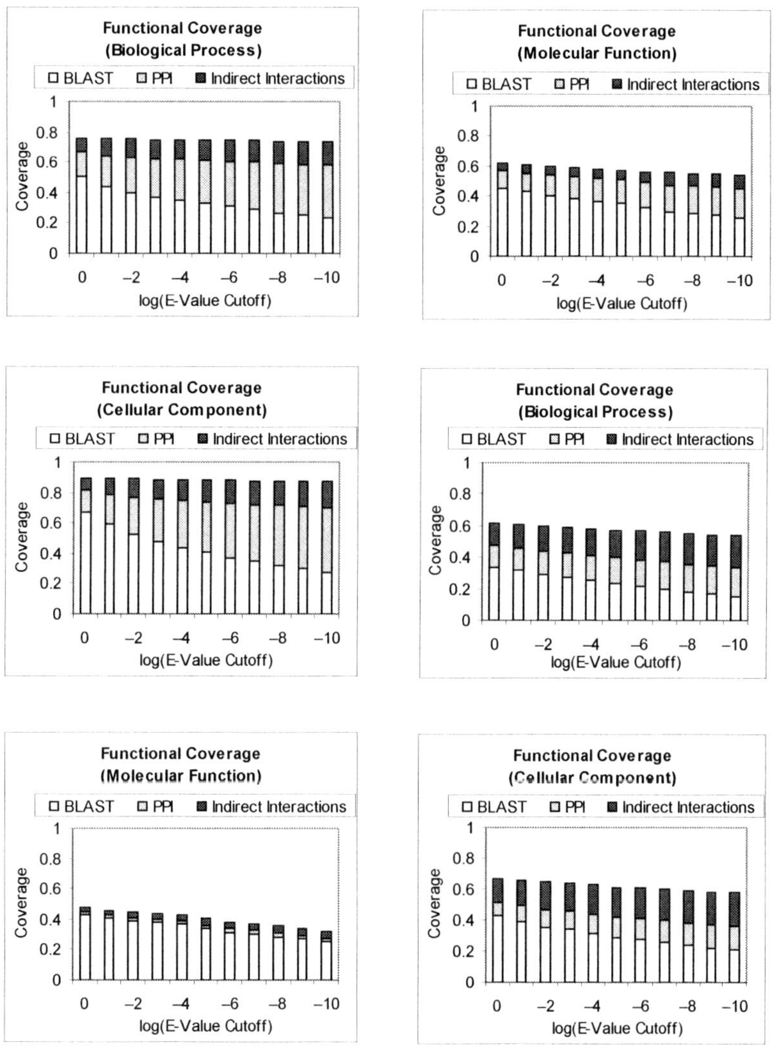

In this section, we will look in greater detail the techniques underlying several approaches that use protein-protein interaction for protein function prediction. To facilitate discussion on the various algorithms, we first introduce a popular graph-based representation for protein-protein interactions. A protein-protein interaction network can be represented as an undirected graph $G = (V, E)$ with a set of vertices V and a set of edges E. Each vertex $u \in V$ represents a unique protein, while each edge $(u, v) \in E$ represents an observed interaction between proteins u and v. Generally, methods that use protein-protein interactions to predict protein function can be categorized into two main groups. The first group, which we refer to as *local prediction* methods, infers the function of each protein based on the neighbors in its local interaction neighborhood. The second group, which we refer to as *global optimization* methods, derives functional annotations for each protein such that the final annotations for all proteins in the entire interaction network are optimal based on some measure.

Local Prediction Methods

Local prediction methods infer the functional annotations of a protein by looking at its interaction neighbors. The main advantage in such approaches lies in their simplicity, which makes such methods computationally more viable and hence scalable to large datasets. The straightforward nature of these methods also makes it easy to explain how a prediction is made, as well as list the neighbors that contributed to the prediction. This can help a biologist analyze whether a prediction is biologically plausible before using it further. Ironically, the main drawback of these approaches also lies in their simplicity. Since predictions are made based on the local interaction neighborhood, predictions made will be limited in both coverage and precision when the local interaction neighborhood is small, or when the proteins in the neighborhood are not well annotated.

Neighbor Counting

A simple yet effective local prediction approach assigns a protein with the function that occurs most frequently among its interaction partners (Schwikowski et al. 2000). The method is popularly referred to as Neighbor Counting. For each protein u, each function x is given a score based on the frequency of its occurrence in the neighbors of u.

Equation 1. Neighbor Counting scoring function

$$f_x(u) = \sum_{v \in N_u} \delta(v, x)$$

$\delta(v, x) = 1$ if v has function x, 0 otherwise;
N_x refers to the interaction neighbors of protein x.

The function k with the largest score $f_k(u)$ is predicted for protein u. To assign multiple functions to u, all functions that are associated with the neighbors of protein u can be sorted based on decreasing $f_x(u)$, and scored based on their rank (Deng et al. 2003).

Chi-Square

The Neighbor Counting approach simply considers the frequency which a function f is annotated to the neighbors of a protein, but do not consider the frequency with which f is annotated to all the proteins in the interaction network (i.e. its background frequency). If a function a is annotated to more proteins in the network relative to another function b, then a is relatively more likely than b to be found in the interaction neighborhood of a protein, and vice versa. Hence if both functions appear the same number of times in the interaction neighborhood of protein u, a has a lower likelihood than b to be a function of u. The Neighbor Counting method, however, would assign the same score to both functions. The Chi-Square method overcome this limitation by using the Chi-Square statistics instead of frequency as a scoring function (Hishigaki et al. 2001). The statistical measure computes the deviation of the observed occurrence of function x in the neighbors of u from its expected occurrence:

Equation 2. Chi-Square scoring function

$$S_x(u) = \frac{(f_x(u) - e_x(u))^2}{e_x(u)}$$

$e_x(u)$ is the expected number of proteins with function x among the interaction partners of u, and is computed by multiplying the number of annotated interaction partners of u with the fraction of all annotated proteins in the interaction map that are annotated with function x.

The function with the largest chi-square value is then predicted for protein u. To assign multiple functions to each protein, all functions that are associated with the neighbors of protein u can be sorted based on decreasing $S_x(u)$, and scored based on their rank (Deng et al. 2003).

PRODISTIN

Both the Neighbor Counting and the Chi-Square methods do not assign any weight to the edges of the interaction network. PRODISTIN (Brun et al. 2003) assigns a weight to each pair of proteins by using a graph theory measure known as the Czekanowski-Dice distance (CD-Distance). The CD-Distance between two proteins u and v is given by:

Equation 3. Czekanowski-Dice distance

$$D(u,v) = \frac{|N'_u \Delta N'_v|}{|N'_u \cup N'_v| + |N'_u \cap N'_v|}$$

N'_x refers to the set of proteins that contain x and its interaction neighbors
$X \Delta Y$ refers to the symmetric difference between two sets X and Y.
$D(u,v) < 1$ if u and v are neighbors. If $N_u = N_v$, $D(u,v)$ will be evaluated to 0. On the other extreme, if $N_u \cap N_v = \varnothing$, $D(u,v)$ will be evaluated to 1.

Each pair of proteins that share at least one interaction neighbor will have a distance less than 1. The distance will be large for protein pairs that share few neighbors, with the maximum distance being 1 when there are no common neighbors. The minimum distance between two proteins is 0 when both have identical set of neighbors.

Using the Czekanowski-Dice distance as the distance metric, proteins are clustered using the BIONJ algorithm (Gascuel 1997), which was designed to reconstruct phylogeny trees from sequences. Only the largest connected component in the protein interaction network is used. The BIONJ algorithm produces a hierarchical classification tree. PRODISTIN define a functional class for each function k as that largest possible subtree in the classification tree that: 1) contains at least three proteins annotated with k; and 2) has at least 50% of its annotated members annotated with k. Un-annotated proteins in the functional class are then predicted with the function.

We consider PRODISTIN to be a local prediction method since predictions are made based on a cluster within the reconstructed tree instead of using a global objective function to guide prediction. PRODISTIN has some advantages over Neighbor Counting and the Chi-Square methods. Firstly predictions for a protein are not limited to its immediate neighbors, since two proteins that share some neighbors can have a weight less than 1 even if they do not interact. Moreover, during the construction of the tree, proteins that do not share any neighbors may also be clustered together. Secondly, the algorithm assigns a weight to each protein pair using common interaction neighbors, while Neighbor Counting and Chi-Square simply consider protein pairs that interacts, and do not distinguish between them in any way. As discussed earlier in this chapter, protein-protein interaction data, especially those derived from high-throughput experimental assays tend to have a lot of false positives. Since the likelihood that two proteins are falsely detected as interacting is much higher that the likelihood that they are falsely sharing a number of interacting neighbors (especially if this is large), the assignment of the Czekanowski-Dice distance helps to reduce the effect of false positives on the prediction results.

Hypergeometric Distance

Like PRODISTIN, Samanta et al. (2003) also computed a distance for each pair of proteins, but using a different distance metric and clustering technique. The distance between a pair of proteins u and v is computed using the hypergeometric score based on each protein's interaction neighbors, which the authors refer to as the P-value:

Equation 4. Hypergeometric P-value

$$P(N,u,v,m)=\frac{\binom{N}{m}\binom{N-m}{n_1-m}\binom{N-n_1}{n_2-m}}{\binom{N}{n_1}\binom{N}{n_2}}$$

N refers to all proteins in the interaction network
$m = |N_u \cap N_v|$
$n_1 = |N_u|$
$n_2 = |N_v|$

The P-value reflects the likelihood that proteins u and v share m neighbors by chance in a network of N proteins, given that u has n_1 neighbors and v has n_2 neighbors. A smaller P-value indicates that the observation is less likely to occur by chance, and is more likely to be biologically significant. Using a P-value threshold of 10^{-8} for their dataset, the authors retain only protein pairs with P-value above or equal this threshold, and assign functions to un-annotated proteins based on majority vote similar to Neighbor Counting. Using the P-value as a distance metric, proteins are also clustered using a bottom-up hierarchical clustering approach to form functional modules. Beginning with each protein as a cluster, the two clusters with the smallest P-value are merged to form a cluster with two proteins. The P-value between two clusters is computed by the geometric mean of the P-value between each pair of their constituent proteins. This merging process is continued a certain P-value threshold is reached.

Functionalflow

Nabieva et al. (2005) proposed a network-based algorithm that simulates functional flow between proteins over d number of biological time steps. The amount of flow between each edge is derived based on the confidence of the experiments from which the interaction represented by the edge is observed. The confidence of proteins interactions from an experimental assay is estimated by the fraction of interacting protein pairs from the experiment that share at least one common function. A protein pair u and v that is observed in multiple experiments are weighted by combining the confidence of each experimental assay in a probabilistic manner:

Equation 5. Confidence computation for interacting protein pair u and v

$$r_{u,v} = 1 - \prod_{i \in E_{u,v}} (1 - r_i)$$

r_i is the confidence of experimental source i
$E_{u,v}$ is the set of experimental sources in which interaction between u and v is observed

Each protein is initially assigned to hold an infinite reservoir of function x if the protein is annotated with x, and hold an empty reservoir of x otherwise. At each time step, the reservoir of function x in each protein u is computed by considering the amount of flow of function x into, and out of the u:

Equation 6. Function reservoir of x in protein u at time step t

$$R_t^x(u) = R_{t-1}^x(u) + \sum_{(u,v)\in E} \left(g_t^x(v,u) - g_t^x(u,v) \right)$$

$R_t^x(u)$ refers to the reservoir of function x in protein u at time step t

$g_t^x(u,v)$ refers to the flow of function x from u to v at time step t

The flow of function at time t, $g_t^x(u,v)$, is 0 at time 0, and is defined as:

Equation 7. Functional Flow from u to v at time step t

$$g_t^x(u,v) = \begin{cases} 0, & R_{t-1}^x(u) < R_{t-1}^x(v) \\ \min\left(w_{u,v}, \dfrac{w_{u,v}}{\sum_{(u,y)\in E} w_{u,y}} \right) & \textit{otherwise} \end{cases}$$

After d time steps, the flow is terminated, and the score for each function x for protein u is computed by the amount of flow that has entered u:

Equation 8. FunctionalFlow scoring function of function x for protein u over d time steps

$$f_x(u) = \sum_{t=1}^{d} \sum_{(u,v)\in E} g_t^x(v,u)$$

One obvious advantage of FunctionalFlow over Neighbor Counting and Chi-Square is its ability to assign a protein with functions that are not annotated to proteins in its immediate neighborhood. Also, the estimated confidence of the experiments that contributed to an edge is used to decide the amount of functional flow, which helps to reduce the effect of noise in the interaction data. Nabieva et al. (2005) showed that FunctionalFlow can predict function with significantly better accuracy than Neighbor Counting and Chi-Square.

FS-Weighted Averaging

Chua et al. (2006) extend upon the ideas proposed by (Brun et al. 2003) and (Samanta et al. 2003) to include indirect neighbors (proteins that share common interaction partners but do not interact) when predicting the function of a protein, but explicitly distinguish these indirect interaction neighbors from direct ones (proteins that interacts). The authors proposed that function sharing between direct neighbors occurs in a different model from indirect neighbors, and coined the term *direct functional association* and *indirect functional association* to differentiate between the two. Proteins interact to form part of a functional pathway to perform a synergized biological function; hence direct interaction neighbors are likely to share common functions. Indirect interaction neighbors do not interact, but interact with some common proteins. These indirect neighbors may interact with the same protein due to some common physical or biochemical characteristics, especially if they share many common interaction neighbors.

The authors provided support for their hypothesis by gathering some statistics based on known interactions in *S.cerevisiae*. Among the 4,162 annotated yeast proteins in the dataset studied, only 48.0% share some function with its direct neighbors, while 22.7% share some function with its indirect neighbors but do not share any function with its direct neighbors. Less than 2% of the proteins share functions exclusively with direct neighbors.

The authors defined a new scoring function Functional Similarity Weight (FS-Weight), which is adapted from (Brun et al. 2003), to assign weights to protein pairs that interacts or share interaction neighbors. The main difference between FS-Weight and the CD-Distance is the inclusion of the confidence of protein interactions from different experimental assays similar to that derived in (Nabieva et al. 2005).

For each protein *u*, each function *x* is given a score using an averaging method based on the frequency of its occurrence in the direct and indirect neighbors of *u*, as well as the weight between each neighbor and *u*:

Equation 9. FS-Weighted Averaging function

$$f_x(u) = \frac{1}{Z}\left[\lambda r_{\text{int}}\pi_x + \sum_{v \in N_u}\left(S(u,v)\delta(v,x) + \sum_{w \in N_v} S(u,w)\delta(w,x)\right)\right]$$

$S(u,v)$ is the FS-Weight score for u and v

r_{int} is the fraction of all interaction pairs that share some function

$\delta(p, x) = 1$ if p has function x, 0 otherwise

π_x is the frequency of function x in annotated proteins

$0 \leq \lambda \leq 1$ is the weight representing the contribution of background frequency to the score

Z is the sum of all weights, given by:

$$Z = 1 + \sum_{v \in N_u}\left(S(u,v) + \sum_{w \in N_v} S(u,w)\right)$$

The equation confers greater weight on direct interactions, and also includes a component with the frequency of function *x* so that a prediction is made largely based on the background frequency of *x* when there are few or no annotated proteins in the interaction neighborhood of *u*.

The strength in FS-Weighted Averaging lies in the use of weighted edges and estimated confidence of experimental assays. This helps to reduce the effect of false positives in the dataset effectively. The inclusion of indirect interaction neighbors with a different weight from direct neighbors also seem more intuitive than considering them with similar weight. Experiments in (Chua et al. 2006) showed that the method achieved significant improvement over a number of existing methods including Neighbor Counting, Chi-Square and PRODISTIN.

Global Optimization Methods

Local prediction methods assign annotations to un-annotated proteins such that its annotations are the most consistent with its neighbors. By predicting an un-annotated protein with the most frequent annotations among its neighbors, Neighbor Counting ensures that the similarity between the predicted functions of the un-annotated protein and that of its neighbors are maximized. Chi-Square, FunctionalFlow and other local prediction methods employs similar principles, and differ mainly in the use of weighted edges, and varying neighborhood sizes. Global optimization methods, on the other hand, assign functional annotations such that global consistency in the final functional annotations is optimized. The differences between methods in this category lie mainly in the objective function used to define "consistency", and the technique used for optimization.

Simulated Annealing

Vazquez et al. (2003) defined an objective function based on the same principle in Neighbor Counting, that is, interacting proteins should share function, while non-interacting proteins should not. Each un-annotated protein u is assigned a function σ_u from all possible annotations such that the following objective function is minimized:

Equation 10. Objective Function for Simulated Annealing

$$E = -\sum_{u,v} J_{uv}\delta(\sigma_u, \sigma_v) - \sum_{u} h_u(\sigma_u)$$

J_{uv} is equal to 1 if protein u and v interact and are both un-annotated, 0 otherwise
$\delta(\sigma_u, \sigma_v) = 1$ if $\sigma_u = \sigma_v$, 0 otherwise
$h_u(\sigma_u)$ is the number of annotated neighbors of protein u with function σ_u.

The first summation reflects the consistency between the predicted functions for un-annotated proteins and un-annotated proteins in their interaction neighborhood. The second summation in the equation reflects the consistency between the predicted functions for un-annotated proteins and annotated proteins in their interaction neighborhood.

To find a set of functional assignments that can minimize the objective function, the authors used a popular technique known as Simulated Annealing (Kirkpatrick et al. 1983). The technique is inspired by the process in which solid materials are heated and allowed to cool gradually to alter its physical properties. Initially, each un-annotated protein is assigned a random annotation and the objective function E is computed. Temperature T is set to an initial high value. This is analogous to heating a material to a temperature such that atoms are freed from their initial configurations and wander around randomly. From this initial condition, multiple steps of "cooling" are simulated. At each step, a protein is randomly selected and its assigned annotation is replaced by another randomly picked annotation. The updated objective function E' is then re-computed. If the change in the objective function $\Delta E = E' - E$ is positive, the updated configuration is accepted with probability $r = exp(-\Delta E/T)$, and rejected otherwise. This step is repeated until ΔE becomes 0. Then T is decreased slightly, and the "cooling" step is repeated. The process is terminated when the protein annotations stabilize, and the final annotations are the predicted functions. The annealing process is repeated a number of time (the authors repeated for 100 times) using different initial functional assignments and the score with which each protein u is predicted with a function x is computed by the fraction of times in which x is annotated to u at the end of the simulated annealing process.

Markov Random Fields

Deng et al. (2003) proposed another global optimization method based on Random Markov Fields. For each function annotation f, $X = (X_1, X_2, \ldots, X_N)$ is defined as the annotations for all proteins $p_1, \ldots, p_N$, where X_i is a random variable such that $X_i = 1$ when protein p_i has function f, and 0 otherwise. Proteins $p_1, \ldots, p_n$ are un-annotated, while proteins $p_{n+1}, \ldots, p_{n+m}$ are annotated. $X_{n+1}, \ldots, X_{n+m}$ has corresponding values $\mu_1, \ldots, \mu_m$. The objective is to find the posterior distribution of the un-annotated proteins:

$$P(X_1, \ldots, X_n \mid X_{n+1} = \mu_1, \ldots, X_{n+m} = \mu_m)$$

To find the posterior distribution, a Gibbs Sampling algorithm is used. Gibbs Sampling is an algorithm for generating a sequence of samples from the joint distribution of multiple random variables when the distribution is unknown but the conditional distribution of each variable is known (Geman et al. 1990). In (Deng et al. 2003), the conditional distribution of each variable X_i is derived to be:

Equation 12. Conditional distribution each variable $\mathbf{X}_i$

$$P(X_i = 1 \mid X_{[-i]}, \theta) = \frac{e^{\alpha+(\beta-1)M_0^{(i)}+(\gamma-\beta)M_1^{(i)}}}{1+e^{\alpha+(\beta-1)M_0^{(i)}+(\gamma-\beta)M_1^{(i)}}}$$

$X_{[-i]} = (X_1, \dots, X_{i-1}, X_{i+1}, \dots, X_{n+m})$
$M_0^{(i)}$ is the number of neighbors of protein p_i that is not annotated with f
$M_1^{(i)}$ is the number of neighbors of protein p_i that is annotated with f
$\theta = (\alpha,\beta,\gamma)$ are parameters, where

$$\alpha = \log\left(\frac{\pi}{1-\pi}\right),$$

and the total probability of the functional labeling is proportional to $exp(-U(x))$, where
$U(x) = -\alpha N_1 - \beta N_{10} - \gamma N_{11} - N_{00}$, with
N_1 being the number of proteins annotated with function *f*,
N_{10} being the number of interactions in which only one protein is annotated with function *f*,
N_{11} being the number of interactions in which both proteins are annotated with function *f*,
N_{00} being the number of interactions in which both proteins are not annotated with function *f*.

The parameters $\theta = (\alpha,\beta,\gamma)$ are estimated using a quasi-likelihood estimation method described in (Li 1995). After the parameters are estimated, the probability that each un-annotated protein has function *f* is estimated by Gibbs Sampling. First, random variables $X_1, \dots, X_n$ representing the un-annotated proteins are randomly assigned with values 1 or 0 with probability π, where π is the probability that a protein has function *f*, computed by the fraction of all annotated protein that has function *f*. Next, $X_1, \dots, X_n$ are updated repeatedly using Equation 9 until all $P(X_i|X_{[-i]})$ stabilize. The authors discard results from the first 100 updates, which is referred to as the "burn-in-period". The results from the subsequent 10 updates, which is referred to as the "lag-period" are then averaged and used as the predicted probability that each un-annotated protein has functional annotation *f*. Deng et al. (2003) showed that this method significantly outperforms the Neighbor Counting and Chi-Square methods.

Support Vector Machines

Lanckriet et al. (2004) introduced an integrated Support Vector Machines classifier for function prediction, in which protein-protein interaction data was used to derive one of the kernels. Support vector machines (Vapnik 1998) are a set of popular classifiers that has been shown to perform well in many applications. Data points in a training set are first transformed by a kernel, and a multi-dimensional hyperplane that can achieve maximum separation between data points of different class is derived. This hyperplane can then be used to classify new data points. The kernel used by the authors is built based on similarities between the nodes in the interaction graph using a general method known as diffusion kernels proposed by (Risi Imre et al. 2002). This method efficiently computes similarity between all nodes in the graph based on random walks, and considers all possible paths connecting two nodes as well as their lengths. Two nodes that are connected by shorter paths or by multiple paths are more similar. For each functional annotation, a classifier is built based on the diffusion kernel using the annotated proteins, and used to predict whether each un-annotated protein has that function.

DISCUSSION AND FUTURE TRENDS

In this chapter, we have provided a background on the task of automated protein function prediction, and also covered in detail some popular methods of using protein-protein interactions to predict protein function. Local prediction methods are simpler and more efficient, while global optimization methods are able to predict protein functions that are more consistent in a global scale, taking into account both known and predicted annotations. Both approaches also have their drawbacks. Local prediction methods do not take into account the global optimality of the predicted functions and hence may not be able to infer the correct predictions when the local neighborhood is small or largely un-annotated. On the other hand, the fact that the interaction network for many genomes are largely incomplete, and the high level of noise in high-througput interaction data may mean that global methods, or a local method like FunctionalFlow may propagate erroneous predictions over multiple interactions. A study found that global optimization techniques may not yield significant improvement over local prediction methods (Murali et al. 2006). Some recent directions in function prediction using protein-protein interactions that are being actively pursued currently are not covered in this chapter. One approach is the identification of conserved network patterns or motifs (Chen et al. 2007; Kirac et al. 2008). Another interesting approach is the use of comparative analysis of protein-protein interactions from multiple species to gain further insight on protein function (Bandyopadhyay et al. 2006; Fionda et al. 2007; Singh et al. 2007). Unlike the methods described earlier, which utilize information from a single protein-protein interaction network, these approaches infer protein functions through the comparison of multiple protein-protein interaction networks from different species. Meanwhile, the influx of more biological data of different nature pushes the trend of automated protein function prediction towards integration methods (Troyanskaya et al. 2003; Chen et al. 2004; Deng et al. 2004; Karaoz et al. 2004; Lanckriet et al. 2004; Tsuda et al. 2005; Xiong et al. 2006; Chua et al. 2007).

REFERENCES

Altschul, S. F., Gish, W., et al. (1990). Basic local alignment search tool. *J Mol Biol,, 215*(3), 403-10.

Altschul, S. F., Madden, T. L., et al. (1997). Gapped BLAST and PSI-BLAST: A new generation of protein database search programs. *Nucleic Acids Res, 25*(17), 3389-402.

Ashburner, M., Ball, C. A., et al. (2000). Gene ontology: tool for the unification of biology. The Gene Ontology Consortium. *Nat Genet, 25*(1), 25-9.

Attwood, T. K., P. Bradley, et al. (2003). PRINTS and its automatic supplement, prePRINTS." Nucleic Acids Res 31(1): 400-2.

Bader, G. D., & Hogue, C. W. (2000). BIND--a data specification for storing and describing biomolecular interactions, molecular complexes and pathways. Bioinformatics, *16*(5), 465-77.

Bader, G. D., & Hogue, C. W. (2003). An automated method for finding molecular complexes in large protein interaction networks. *BMC Bioinformatics, 4*(2).

Bader, J. S., Chaudhuri, A. et al. (2004). Gaining confidence in high-throughput protein interaction networks. *Nat Biotechnol, 22*(1), 78-85.

Bairoch, A. (1993). The PROSITE dictionary of sites and patterns in proteins, its current status. *Nucleic Acids Res, 21*(13), 3097-103.

Bairoch, A., R. Apweiler, et al. (2005). The Universal Protein Resource (UniProt). *Nucleic Acids Res, 33*(Database issue), D154-9.

Bandyopadhyay, S., Sharan, R. et al. (2006). Systematic identification of functional orthologs based on protein network comparison. Genome Res, *16*(3), 428-35.

Benson, D. A., Karsch-Mizrachi, I. et al. (2007). GenBank. *Nucleic Acids Res, 35*(Database issue): D21-5.

Berman, H. M., J. Westbrook, et al. (2000). The Protein Data Bank. *Nucleic Acids Res, 28*(1), 235-42.

Boeckmann, B., Bairoch, A. et al. (2003). The SWISS-PROT protein knowledgebase and its supplement TrEMBL in 2003. *Nucleic Acids Res*, *31*(1), 365-70.

Breitkreutz, B. J., Stark, C. et al. (2003). The GRID: The General Repository for Interaction Datasets. *Genome Biol, 4*(3), R23.

Brun, C., Chevenet, F., et al. (2003). Functional classification of proteins for the prediction of cellular function from a protein-protein interaction network. *Genome Biol, 5*(1), R6.

Chen, J., Hsu, W., et al. (2007). Labeling network motifs in protein interactomes for protein function prediction. *Proceedings of the IEEE 23rd International Conference on Data Engineering* (pp. 546-555).

Chen, J., Hsu, W., et al. (2006). Increasing confidence of protein interactomes using network topological metrics. *Bioinformatics, 22*(16), 1998-2004.

Chen, Y., & Xu, D. (2004). Global protein function annotation through mining genome-scale data in yeast Saccharomyces cerevisiae. *Nucleic Acids Res*, *32*(21), 6414-24.

Cherry, J. M., Adler, C. et al. (1998). SGD: Saccharomyces Genome Database. *Nucleic Acids Res*, *26*(1), 73-9.

Chiaraluce, R., Florio, R. et al. (2007). Tertiary structure in 7.9 M guanidinium chloride--the role of Glu53 and Asp287 in Pyrococcus furiosus endo-beta-1,3-glucanase. *Febs J, 274*(23), 6167-79.

Chua, H. N., Sung, W.-K. et al. (2007). An efficient strategy for extensive integration of diverse biological data for protein function prediction. *Bioinformatics, 23*(24), 3364-3373.

Chua, H. N., Sung, W. K. et al. (2006). Exploiting indirect neighbours and topological weight to predict protein function from protein-protein interactions. *Bioinformatics*, *22*(13), 1623-30.

Chua, H. N., Sung, W. K. et al. (2007). Using indirect protein interactions for the prediction of Gene Ontology functions. *(BMC) Bioinformatics*, *8*(Suppl 4), S8.

Chua, H. N., & Wong, L. (2008). Increasing the reliability of protein interactomes. *Drug Discov Today, 13*(15/16), 652-658.

Dandekar, T., Snel, B. et al. (1998). Conservation of gene order: a fingerprint of proteins that physically interact. *Trends Biochem Sci, 23*(9), 324-8.

Deane, C. M., Salwinski, L. et al. (2002). Protein interactions: two methods for assessment of the reliability of high throughput observations. *Mol Cell Proteomics*, *1*(5), 349-56.

Deng, M., Chen, T. et al. (2004). An integrated probabilistic model for functional prediction of proteins. *J Comput Biol, 11*(2-3), 463-75.

Deng, M., Mehta, S., et al. (2002). Inferring domain-domain interactions from protein-protein interactions. *Genome, 12*(10), 1540-8.

Deng, M., Zhang, K., et al. (2003). Prediction of protein function using protein-protein interaction data. *J Comput Biol, 10*(6), 947-60.

Eisen, M. B., Spellman, P. T., et al. (1998). Cluster analysis and display of genome-wide expression patterns. *Proc Natl Acad Sci U S A, 95*(25), 14863-8.

Enright, A. J., Iliopoulos, I., et al. (1999). Protein interaction maps for complete genomes based on gene fusion events. *Nature, 402*(6757), 86-90.

Ferre, S., & King, R. D. (2006). Finding motifs in protein secondary structure for use in function prediction. *J Comput Biol, 13*(3), 719-31.

Finn, R. D., Tate, J., et al. (2008). The Pfam protein families database. *Nucleic Acids Res, 36*(Database issue), D281-8.

Fionda, V., Palopoli, L., et al. (2007). GRAPPIN: Bipartite GRAph Based Protein-Protein Interaction Network Similarity Search. *2007 IEEE International Conference on Bioinformatics and Biomedicine (BIBM 2007).*

Frazier, M. E., Johnson, G. M., et al. (2003). Realizing the potential of the genome revolution: the genomes to life program. *Science, 300*(5617), 290-3.

Gabow, A., Leach, S., et al. (2008). Improving protein function prediction methods with integrated literature data. *BMC Bioinformatics, 9*(1), 198.

Gascuel, O. (1997). BIONJ: an improved version of the NJ algorithm based on a simple model of sequence data. *Mol Biol Evol, 14*(7), 685-95.

Geman, S., & Geman, D. (1990). Stochastic relaxation, Gibbs distributions, and the Bayesian restoration of images. *Readings in uncertain reasoning* (pp. 452-472). Morgan Kaufmann Publishers Inc.

Gietz, R. D., Triggs-Raine, B., et al. (1997). Identification of proteins that interact with a protein of interest: Applications of the yeast two-hybrid system. *Molecular and Cellular Biochemistry, 172*(1), 67-79.

Hawkins, T., & Kihara, D. (2007). Function prediction of uncharacterized proteins. *J Bioinform Comput Biol, 5*(1), 1-30.

Hawkins, T., Luban, S., et al. (2006). Enhanced automated function prediction using distantly related sequences and contextual association by PFP. *Protein Sci, 15*(6), 1550-6.

Hishigaki, H., Nakai, K., et al. (2001). Assessment of prediction accuracy of protein function from protein--protein interaction data. *Yeast, 18*(6), 523-31.

Hu, P., Bader, G., et al. (2007). Computational prediction of cancer-gene function. *Nat Rev Cancer, 7*(1), 23-34.

Huang, J. Y., & Brutlag, D. L. (2001). The EMOTIF database. *Nucleic Acids Res, 29*(1), 202-4.

Hughes, T. R., Marton, M. J., et al. (2000). Functional discovery via a compendium of expression profiles. *Cell, 102*(1), 109-26.

Huynen, M., Snel, B., et al. (2000). Predicting protein function by genomic context: quantitative evaluation and qualitative inferences. *Genome Res, 10*(8), 1204-10.

Kanehisa, M., Goto, S., et al. (2004). The KEGG resource for deciphering the genome. *Nucleic Acids Res, 32*(Database issue), D277-80.

Karaoz, U., Murali, T. M., et al. (2004). Whole-genome annotation by using evidence integration in functional-linkage networks. *Proc Natl Acad Sci U S A, 101*(9), 2888-93.

Kersey, P. J., Duarte, J., et al. (2004). The International Protein Index: an integrated database for proteomics experiments. *Proteomics, 4*(7), 1985-8.

Khan, S., Situ, G., et al. (2003). GoFigure: automated Gene Ontology annotation. *Bioinformatics, 19*(18), 2484-5.

Kirac, M., & Ozsoyoglu, G. (2008). Protein Function Prediction Based on Patterns in Biological Networks. *Research in Computational Molecular Biology,* (pp. 197-213).

Kirkpatrick, S., Gelatt Jr., C. D., et al. (1983). Optimization by Simulated Annealing. *Science, 220*(4598), 671-680.

Komander, D., & Barford, D. (2008). Structure of the A20 OTU domain and mechanistic insights into deubiquitination. *Biochem J, 409*(1), 77-85.

Kulikova, T., Akhtar, R., et al. (2007). EMBL Nucleotide Sequence Database in 2006. *Nucleic Acids Res, 35*(Database issue), D16-20.

Lanckriet, G. R., Deng, M., et al. (2004). Kernel-based data fusion and its application to protein function prediction in yeast. *Pac Symp Biocomput*, (pp. 300-11).

Laskowski, R. A., Watson, J. D., et al. (2005). ProFunc: a server for predicting protein function from 3D structure. *Nucleic Acids Res, 33*(Web Server issue), W89-93.

Laskowski, R. A., Watson, J. D., et al. (2005). Protein function prediction using local 3D templates. *J Mol Biol, 351*(3), 614-26.

Legrain, P., Wojcik, J., et al. (2001). Protein--protein interaction maps: A lead towards cellular functions. *Trends Genet, 17*(6), 346-52.

Letovsky, S., & Kasif, S. (2003). Predicting protein function from protein/protein interaction data: a probabilistic approach. *Bioinformatics, 19*(Suppl 1)i197-204.

Li, S. Z. (1995). *Markov Random Field Modeling in Computer Vision*. Springer-Verlag New York, Inc.

Liolios, K., Mavromatis, K., et al. (2008). The Genomes On Line Database (GOLD) in 2007: status of genomic and metagenomic projects and their associated metadata. *Nucleic Acids Res, 36*(Database issue), D475-9.

Marcotte, E. M., Pellegrini, M., et al. (1999). Detecting protein function and protein-protein interactions from genome sequences. *Science, 285*(5428), 751-3.

Martin, D. M., Berriman, M., et al. (2004). GOtcha: a new method for prediction of protein function assessed by the annotation of seven genomes. *BMC Bioinformatics, 5*, 178.

Murali, T. M., Wu, C. J., et al. (2006). The art of gene function prediction. *Nat Biotechnol, 24(*12), 1474-5; author reply 1475-6.

Nabieva, E., Jim, K., et al. (2005). Whole-proteome prediction of protein function via graph-theoretic analysis of interaction maps. *Bioinformatics, 21(*Suppl 1), i302-10.

Overbeek, R., Fonstein, M., et al. (1999). The use of gene clusters to infer functional coupling. *Proc Natl Acad Sci U S A, 96*(6), 2896-901.

Pazos, F., & Sternberg, M. J. (2004). Automated prediction of protein function and detection of functional sites from structure. *Proc Natl Acad Sci U S A, 101*(41), 14754-9.

Pearson, W. R., & Lipman, D. J. (1988). Improved tools for biological sequence comparison. *Proc Natl Acad Sci U S A, 85*(8), 2444-8.

Pellegrini, M., Marcotte, E. M., et al. (1999). Assigning protein functions by comparative genome analysis: protein phylogenetic profiles. *Proc Natl Acad Sci U S A, 96*(8), 4285-8.

Risi Imre, K., & John, D. L. (2002). Diffusion Kernels on Graphs and Other Discrete Input Spaces. *Proceedings of the Nineteenth International Conference on Machine Learning.* Morgan Kaufmann Publishers Inc.

Rost, B. (1999). Twilight zone of protein sequence alignments. *Protein Eng, 12*(2), 85-94.

Rost, B., & Liu, J. (2003). The PredictProtein server. *Nucleic Acids Res, 31*(13), 3300-4.

Saito, R., Suzuki, H., et al. (2003). Construction of reliable protein-protein interaction networks with a new interaction generality measure. *Bioinformatics, 19*(6), 756-63.

Salgado, H., Moreno-Hagelsieb, G., et al. (2000). Operons in Escherichia coli: genomic analyses and predictions. *Proc Natl Acad Sci U S A, 97*(12), 6652-7.

Samanta, M. P., & Liang, S. (2003). Predicting protein functions from redundancies in large-scale protein interaction networks. *Proc Natl Acad Sci U S A, 100*(22), 12579-83.

Schwikowski, B., P. Uetz, et al. (2000). A network of protein-protein interactions in yeast. *Nat Biotechnol, 18*(12), 1257-61.

Shoemaker, B. A., & Panchenko, A. R. (2007). Deciphering protein-protein interactions. Part I. Experimental techniques and databases. *PLoS Comput Biol, 3*(3), e42.

Singh, R., Xu, J., et al. (2007). Pairwise Global Alignment of Protein Interaction Networks By Matching Neighborhood Topology. *Proceedings of the 11th Annual International Conference on Research in Computational Molecular Biology (2007): Lecture Notes in Computer Science, 4453*, 16-31.

Spellman, P. T., Sherlock, G., et al. (1998). Comprehensive identification of cell cycle-regulated genes of the yeast Saccharomyces cerevisiae by microarray hybridization. *Mol Biol Cell, 9*(12), 3273-97.

Sprinzak, E., Sattath, S., et al. (2003). How reliable are experimental protein-protein interaction data? *J Mol Biol, 327*(5), 919-23.

Su, Q. J., L. Lu, et al. (2005). eBLOCKs: enumerating conserved protein blocks to achieve maximal sensitivity and specificity. *Nucleic Acids Res, 33*(Database issue), D178-82.

Tarassov, K., Messier, V., et al. (2008). An in vivo map of the yeast protein interactome. *Science, 320*(5882), 1465-70.

Thompson, J. D., Higgins, D. G., et al. (1994). CLUSTAL W: improving the sensitivity of progressive multiple sequence alignment through sequence weighting, position-specific gap penalties and weight matrix choice. *Nucleic Acids Res, 22*(22), 4673-80.

Troyanskaya, O. G., Dolinski, K., et al. (2003). A Bayesian framework for combining heterogeneous data sources for gene function prediction (in Saccharomyces cerevisiae). *Proc Natl Acad Sci U S A, 100*(14), 8348-53.

Tsuda, K., Shin, H., et al. (2005). Fast protein classification with multiple networks. *Bioinformatics, 21(*Suppl 2), ii59-65.

Vapnik, V. (1998). *Statistical Learning Theory.* Wiley-Interscience.

Vazquez, A., Flammini, A., et al. (2003). Global protein function prediction from protein-protein interaction networks. *Nat Biotechnol, 21*(6), 697-700.

Vinayagam, A., del Val, C., et al. (2006). GOPET: a tool for automated predictions of Gene Ontology terms. *BMC Bioinformatics, 7*, 161.

von Mering, C., Krause, R., et al. (2002). Comparative assessment of large-scale data sets of protein-protein interactions. *Nature, 417*(6887), 399-403.

Wang, K., & Samudrala, R. (2005). FSSA: a novel method for identifying functional signatures from structural alignments. *Bioinformatics, 21*(13), 2969-77.

Wu, C. H., Apweiler, R., et al. (2006). The Universal Protein Resource (UniProt): an expanding universe of protein information. *Nucleic Acids Res, 34*(Database issue), D187-91.

Wu, J., Kasif, S., et al. (2003). Identification of functional links between genes using phylogenetic profiles. *Bioinformatics, 19*(12), 1524-30.

Xiong, J., Rayner, S., et al. (2006). Genome wide prediction of protein function via a generic knowledge discovery approach based on evidence integration. *BMC Bioinformatics, 7*, 268.

Zehetner, G. (2003). OntoBlast function: From sequence similarities directly to potential functional annotations by ontology terms. *Nucleic Acids Res, 31*(13), 3799-803.

Zhou, X., Kao, M. C., et al. (2002). Transitive functional annotation by shortest-path analysis of gene expression data. *Proc Natl Acad Sci U S A, 99*(20), 12783-8.

Chapter XIII
Protein Interactions for Functional Genomics

Pablo Minguez
Centro de Investigación Príncipe Felipe (CIPF), Spain

Joaquin Dopazo
Centro de Investigación Príncipe Felipe (CIPF), Spain

ABSTRACT

Here the authors review the state of the art in the use of protein-protein interactions (ppis) within the context of the interpretation of genomic experiments. They report the available resources and methodologies used to create a curated compilation of ppis introducing a novel approach to filter interactions. Special attention is paid in the complexity of the topology of the networks formed by proteins (nodes) and pairwise interactions (edges). These networks can be studied using graph theory and a brief introduction to the characterization of biological networks and definitions of the more used network parameters is also given. Also a report on the available resources to perform different modes of functional profiling using ppi data is provided along with a discussion on the approaches that have typically been applied into this context. They also introduce a novel methodology for the evaluation of networks and some examples of its application.

INTRODUCTION

The available data for protein-protein interactions (ppis) has increased enormously in the last few years with the emergence of high-throughput techniques that can report thousands of ppis in a short time span. The most used techniques in this field are: yeast two hybrid (y2h), tandem affinity purification (TAP) and high-throughput Mass Spectrometry techniques (MS). Reviews on these and related methodologies can be found in Drewes and Bouwmeester (2003), Cho *et al.* (2003), Falk *et al.* (2007) and Berggard *et al.* (2007).

The reliability of this data is not exempt of controversy. Studies comparing resulting data from several experiments demonstrate that the overlap between them is not as complete as desirable. This can be due to the fact that some methodologies do not reach the saturation point (Bader & Hogue, 2002) or because of the lack of accuracy and coverage on some of them (von Mering *et al.*, 2002). A conventional large-scale experiment can cover only 3-9% of the total interactome, so limited overlap should be expected (Han *et al.*, 2005). False positives are also a problem: in y2h these represent up to 50% of the total data (Ito *et al.*, 2001; Mrowka *et al.*, 2001). Moreover, there is a bias in the functional categories of the ppis each technique detects, e.g. y2h fails in detecting proteins involved in translation (von Mering *et al.*, 2002).

Beyond discussions about accuracy and coverage of this kind of experiments, the relevance of ppis in the cellular machinery has fostered an unprecedented interest in the exploration of the interactome of model organisms such as *Saccharomyces cerevisiae* (Uetz *et al.*, 2000; Ito *et al.*, 2001), *Drosophila melanogaster* (Gio et al., 2003; Formstecher *et al.*, 2005), *Caenorhabditis elegans* (Li *et al.*, 2004) or human (Stelzl *et al.*, 2005, Rual *et al.*, 2005), just to cite a few examples.

Actually, after years of intensive study, there is a high-quality, literature curated set of ppis free from false positives that probably represents the complete yeast interactome (Reguly *et al.*, 2006). In the case of human, the scenario is still far away from this degree of detail. The estimated size of the human interactome is of 650,000 ppis (Stumpf *et al.*, 2008). None of the public databases contain more than 10% of this number of ppis, and a compilation of all the known ppis would only cover about 10% of the interactions.

The interactome is an abstract scaffold that does not provide information about particular conditions, cell developmental stage or cell type in which a particular ppi occurs (if any). To infer a case-specific interactome it is necessary to integrate other types of data that provide information that allows inferring the active ppis at a particular condition. To achieve this, the transcriptome, defined as the set of transcripts that are expressed at a given moment in a particular cell type, can be used. An integrative study of the interactome filtered by the transcriptome will provide valuable information on the active ppis in a given cell state.

Actually, ppis play a central role at almost every level of cell activity: they are involved in the structure of organelles (structural proteins), transport machinery (nuclear pore importins), response to stimulus (signalling cascades), regulation of gene expression (transcription factors), protein modification (kinases) among many other processes. The proper use of this type of information is of crucial importance in order to understand cell behaviour.

However, the conventional methodologies used to understand the functional basis of the cell behaviour are almost restricted to functional profiling methods. Such methods exploit the differences observed in the comparison of transcriptomes among different experimental conditions to find over-representations of predefined functional modules of genes (see Dopazo 2006 for a recent review). Classically, standard annotations like Gene Ontology (GO) terms (Ashburner et al., 2000), KEGG pathways (Kanehisa et al., 2004) or Mesh terms, have been used to define such modules. Nevertheless, ppis have not extensively used for such puroses.

The combination of expression and interactions has been used to infer gene function (Ideker *et al.*, 2001), to extract signatures to predict disease phenotypes (Camargo & Azuaje, 2007; Lee *et al.*, 2007; Liu *et al.*, 2007; Chuang *et al.*, 2007) as well as to detect possible drug targets by inferring topological features of particular classes of genes (Wachi *et al.*, 2005; Johsson and Bates, 2006).

It is widely accepted that there are very few processes that can be explained by the action of a single protein. On the contrary, the units of activity involved in cellular processes seem to be modules composed by several interacting molecules (Hartwell *et al.*, 1999; Barabasi and Oltvai, 2004). Apart from classical definitions of these modules, such as proteins that share a GO term or proteins integrating the same biological pathway, ppi data is also used to define modules, as representative of units of action characterized by the interaction of their components.

INTERACTOME-RELATED DEFINITIONS

PPI Resources

In this new era of massive production of biological data, an important challenge is its storage in a standardised format with the proper annotation. This facilitate further queries, as simple as possible, to retrieve relevant information from the databases. Data from high-throughput technologies, such as DNA sequences or microarray experiments have developed structured formats to submit the data to the databases with annotations following an ontology-based vocabulary. Learning from those experiences, the Proteomic Standards Initiative (PSI) of the

Human Proteome Organization (HUPO) has established a Molecular Interaction (MI) group to develop a standard format to interchange information called PSI-MI (Hermjakob *et al.*, 2004).

At the time of writing this revision, there is not a common repository that stores all the ppis. Contrarily to other genomic data such as sequences, microarrays, protein structures, etc., ppi data are spread through several databases, among which a small overlapping exists. Moreover, there are strong differences in the type and depth of ppi annotations among the databases. The major repositories are the Human Protein Reference Database (HPRD, Peri *et al.*, 2003), IntAct (Kerrien *et al.*, 2006), the Bimolecular Interaction Network Database (BIND, Bader *et al.*, 2003), the Database of Interacting Proteins (DIP, Salwinski *et al.*, 2004), BioGRID (Breitkreutz *et al.*, 2008) and the Molecular INTeractions database (MINT, Chatr-aryamontri *et al.*, 2006); see Resources section for web addresses. Reviews on the resources dedicated to store and annotate ppis can be found in Xenarios and Eisenberg (2001) and Mathivanan *et al.* (2006).

Therefore, it is not a trivial task for the end user to obtain a reasonably complete and curated set of ppis to work with. Several methodologies have been proposed to solve this problem; see Methods for a small revision on them.

Recently, an ambitious initiative has been proposed by FEBS Letters journal (Ceol *et al.*, 2008), which consists on linking scientific manuscripts with protein interactions databases through a structured summary with controlled vocabulary that has to be filled by the authors. These kinds of approaches are going to be crucial in the quality and accessibility of biological data.

PPIs as Networks (The Interactome)

Whichever the set of ppis chosen, a subset of the interactome can be defined as a compilation of pairwise relationships that, taken all together, represent a network where the nodes are the proteins and the edges the interaction events. Apart from the elements of the network (nodes and edges), the topology of the networks is also of crucial importance when trying to understand their role in a cellular process (Yeger-Lotem *et al.*, 2004).

Graph theory has helped biology to study these networks and established the bases for their description. One of the first discoveries brought about by graph theory was that biological networks are scale-free networks (Barabasi & Albert, 1999; Barabasi & Bonabeu, 2003) instead of random networks. Scale-free networks are defined by a connections degree, number of connections of a node, and a distribution that approximates to a power law $P(k) = k^{\gamma}$, being $\gamma < 3$. This indicates that the network has a low number of highly connected nodes called hubs. In other words, there are a few proteins, the hubs, which connect at long distance much of the whole network. Indeed, identifying hubs is a hot topic in functional analysis (Batada *et al.*, 2006, He & Zhang, 2006, Sporns *et al.*, 2007).

Apart from degree, which identifies hubs, there are other network parameters that help to describe properties of these systems (Barabasi & Oltvai, 2004). The Betweenness Centrality of a node v (C_B (v)) is a parameter that accounts for the centrality of the node V within the graph. It is obtained from the expression:

$$C_B(v) = \sum_{s \neq v \neq t \in V} \frac{\sigma_{st}(v)}{\sigma_{st}}$$

being $\sigma_{st}(v)$, the number of shortest paths through a node and σ_{st}, the total number of shortest paths in the graph. Relative betweenness centrality (rC_B (v)) is calculated as:

$$rC_B(v) = \frac{2 * C_B(v)}{n^2 - (3n + 2)}$$

being n the total number of nodes in the graph.

A node with high betweenness centrality is a protein which has many shortest paths between any two other nodes passing through it. The action of removing that particular node from the network would cause a strong disconnection of the network. By this description, central nodes seem to be crucial in the global compactness of the network and in the definitions of boundaries of sub-networks seen as modules of action (Girvan and Newman, 2002; Wilkinson and Huberman, 2004; Joy *et al.*, 2005). Betweenness has also been shown to be bigger in modules formed by proteins associated with cancer (Hernandez *et al.*, 2007).

Clustering coefficient of a node v (C(v)) is a measure of connectivity that evaluates how connected is the node neighborhood. This parameter helps to distinguish among the highly connected proteins that form star-shaped sub-networks (classical hub configuration) and the proteins in a more connected area, e.g. complexes. The clustering coefficient is calculated as follows:

$$C(v) = \frac{2e_n}{n_v(n_v - 1)}$$

where e_n is the number of edges among the nodes connected to node v, and n_v is the number of neighbors of node v.

Other interesting features in the structure of a network are the concepts of components and bicomponents. A component is a group of nodes connected among them and a bicomponent is a group of nodes connected to another group of nodes by a single edge, which is called the articulation point. When analysing network parameters of a set of proteins related to some phenotype (such as gene/protein signatures of diseases or differentially expressed genes in a two conditions comparison in microarray experiments) components and bicomponents can be considered extreme examples of gene/protein modules if they can be defined as sub-networks with a higher internal connectivity than its connectivity to other modules.

Resources to Apply PPI Data to Functional Genomics

Historically, in the field of ppis, the majority of the resources available have been focused mainly on visualization aspects rather than in the proper analytical steps. Most of the databases have their own visualization tools that exceptionally provide some applications to carry out very simple analyses. DIP through its satellite project LiveDIP has a tool for finding the path between two proteins. This tool also performs a simple mapping of the proteins selected in microarray experiments onto the interactome. BIND can export directly to Cytoscape (Shannon et al., 2003), a popular visualization tool, and may perform some simple analysis of enrichment in GO clusters called OntoGlyphs. HPRD does not provide visualization although it does have transcriptome information. MINT estimates the Minimal Connected Network (MCN, see Methods for complete explanation) and has a java environment available for visualization purposes. The IntAct database includes an application called MINE that can calculate and represent the MCN. BioGRID does not provide any visualization yet although the application Osprey (Breitkreutz *et al.*, 2003) uses it as underlying support database and consequently can be used as an interface to BioGRID. The database STRING (von Mering, *et al.*, 2007) includes a visualization tool with a complete set of options including generating the MCN and displaying co-expression analysis of the set of proteins.

Besides the facilities provided by the databases there are programs, such as Osprey (Breitkreutz *et al.*, 2003), Cytoscape (Shannon *et al.*, 2003), VisANT (Hu *et al.*, 2007) and PATIKA (Dogrusoz *et al.*, 2006), that aim to provide a general framework for ppi data management. Cytoscape and VisANT allows the development of plug-ins that can be integrated into them to perform more specific tasks. Cytoscape is probably the most successful application in this field and it has an ample community of users and developers. At the time of writing this revision there were 48 plug-ins available. A good review about visualization and network management packages can be found in Suderman *et al.* (2007).

Other applications like the Agile Protein Interaction DataAnalyzer (APID) (Prieto *et al.*, 2006), Genes2Networks (Berger *et al.*, 2007) and PIANA (Aragues *et al.*, 2006) were developed with the aim of become a common repository for different ppi datasets. APID and Genes2Networks are web-based tools that make the datasets available. PIANA is more orientated to computer scientists as a working framework for ppi data management. It also can predict novel interactions and calculate some topological parameters.

In a more general functional profiling context, ppi data has quite recently been introduced into suites of programs like Babelomics (Al-Shahrour *et al.*, 2006, 2007, 2008) and DAVID (Dennis *et al.*, 2003) although with different grade of sophistication. DAVID simply reports the interactions associated to the genes of a list and performs a classical enrichment analysis for each of the interactions. Babelomics has included a new module called SNOW (Studying Networks in the Omic World) that calculates the MCN and evaluates the significance of its robustness as functional class comparing its topological parameters versus distributions of same sized lists of random genes or proteins (Al-Shahrour et al., 2008). It also evaluates the presence of hubs, central nodes and highly connected areas in the pre-selected genes or proteins versus a curated interactome. The methodologies that SNOW applies are explained in more detail in the Methods section.

METHODS

Interactome Generation Methodologies

As discussed in the previous section, obtaining a curated set of ppis as complete as possible to work with is not a trivial task. The databases' coverage, the depth and type of annotation and the lack of accuracy of some of the techniques constitute a limitation for this type of analysis. Nevertheless, Reguly *et al.* (2006) established a milestone in this field by generating what probably is the complete yeast interactome, free from false positives, via manual curation. The interactomes of the rest of species are far from this level of completion, basically due to their comparative bigger size. Even small differences in genome sizes can account for drastic increases in the number of ppis, suggesting that the final cause of organism's complexity must be a post-transcriptional event. For instance, human has 21,541 protein coding genes and 650,000 predicted interactions while *Caenorhabditis elegans* has 20,140 genes and three times less predicted interactions (gene counts taken from ensembl genome browser release 49 and interactome size predictions taken from Stumpf *et al.*, 2007).

There is a clear necessity of methodologies to filter ppis given that manual curation is not always a feasible task. Therefore, several approaches have been proposed (review on Badet *et al.*, 2004). We will point out some of them:

1. Promiscuity (Uetz *et al.*, 2000; Ito *et al.*, 2001; Gavin *et al.*, 2002; Ho *et al.*, 2002), that consists in removing proteins having many interaction partners (known as sticky proteins asd most of them presenting a binding without any biological meaning).
2. Topological criteria (Bader & Hogue, 2002), it is specific to Co-IP experiments, retaining the bait-hit (spoke) rather than the bait-hit and hit-hit (matrix) interactions.
3. Intersection of multiple high-throughput datasets (von Mering *et al.*, 2002; Deane *et al.*, 2002).
4. Selecting ppis detected with two different techniques (von Mering *et al.*, 2002), based on the idea that the combination of two methods will increase coverage and accuracy.
5. Intersection with other type of data, e.g. interacting proteins whose transcripts co-express are more likely to be real (Ge *et al.*, 2001; Deane *et al.*, 2002; Jansen *et al.*, 2002) or inferences of ppis across species due to protein homology (Deane *et al.*, 2002).
6. Logistic regression approach (Bader *et al.*, 2004), that uses statistical and topological descriptors to predict the biological relevance of protein-protein interactions obtained from high-throughput screens.

Our experience in compiling data to build an accurate set of human ppis showed us that the annotation in the different databases sometimes is not comparable. The approach proposed here is a modification of the point 4. To build a filtered interactome we have taken the six top categories of experimental methods described in the Molecular Interaction (MI) Ontology (Hermjakob *et al.*, 2004) plus the categories *in vivo* and *in vitro* from HPRD as reference. HPRD seems to be essential when approaching a human interactome (Mathivanan *et al.*, 2006). Every ppi in each of the datasets was annotated with these categories. Ppis verified by at least two of these methods were introduced in the filtered interactome. By using lower levels of depth in the ontology of "techniques annotation" we ensure that ppis extracted with experiments with similar basics (that may have same biases in the detection process) are not selected.

Network Features Evaluation

A classical experiment in Functional Genomics ends up with the assessment of a functional interpretation of the results of a genome-scale experiment such as a microarray analysis. Typically, these results are lists of genes or proteins potentially relevant to the case of study because they have a common behaviour in their expression (for instance, they are over or under expressed in a disease or they have a similar pattern of expression through time when a disease is treated with a drug, etc.).

Conventional methods for functional profiling use functional definitions provided by well known repositories (such as GO, KEGG pathways, etc.) and study the enrichment on these modules among the selected list of genes. Ppi data can also be used under a similar philosophy. An interesting analysis than can be conducted on a list of genes or proteins is to check whether it is enriched in any kind of nodes. That is, to check whether it has significantly more hubs, central proteins or proteins in a very connected area compared to the complete interactome. This can be done comparing the distribution of the connections degree, betweenness centrality and clustering coefficients, respectively, versus the distribution of these parameters in the set of ppis used as background. Quite useful information about the set of proteins can be obtained in this way. For example, it has recently been reported that cancer-related proteins exhibit a higher degree of connections and centrality than the nodes in complete interactome not associated to the disease (Johsson *et al.*, 2006; Hernandez *et al.*, 2007). Nevertheless, the final aim in this kind of experiments is to seek for modules of proteins with a cooperative activity. Therefore, a more holistic approach is needed.

Methodologies to Infer a Sub-Network

The introduction of ppi data into a functional profiling framework requires from different analytical methods due to the nature of this kind of annotation. The way in which ppis are defined do not conform discrete classes (as GOs, KEGGs, etc.) but they have an internal structure in the form of networks where proteins have different roles according to their position in the network and its global shape.

As previously mentioned in the introduction, a module in a network is a sub-network with an internal connectivity higher than its connectivity to other modules. Many attempts have been made to explore the interactome seeking for modules of action, most of them based on the application of clustering methods to weighted matrices. Pereira-Leal *et al.* (2004) proposed the number of experiments that support a given ppi as the index of the general ppi matrix. Rives and Galitski (2003) used the shortest paths among pairs of nodes to measure de relationship between nodes. There are also other approaches based for instance on topological features of the network such as the betweenness (Girvan and Newman, 2002; Wilkinson and Huberman, 2004). Central nodes, with a high betweenness, may define the boundaries of the sub-networks because this implies that many shortest paths pass through them and the action of removing them from the network would lead to the disconnection of some sub-networks.

Modules obtained by these methodologies may be enriched in proteins with related biological functionalities, shown by its significant enrichment in GO terms (Luo *et al.*, 2006) or by its co-occurrence within the literature (Wilkinson and Huberman, 2004). Indeed, it has also been shown that there are sub-networks associated to diseases (Badano and Katsanis, 2002; Brunner and van Driel, 2004; Gandhi *et al.*, 2006). Gandhi *et al.* (2006) found in the analysis of the human interactome that proteins encoded by genes mutated in inherited genetic disorders are likely to interact with proteins known to cause similar disorders.

When seeking for modules of action (sub-networks) within lists of genes or proteins, these have not to be considered as a mere collection of nodes but rather as a potential unit of functionality in cell activity, in a similar way than Gene Ontology terms or KEGG pathways are defined. It is not enough obtaining only the interactions associated to each of node and explore the function of the interacting proteins. Thereby, we need to test for specific sub-networks within the set of proteins analysed in order to assign a common putative functionality to the list. In other words, it is essential for the biological interpretation of a gene or protein list to seek for the sub-network that they might configure, or in other words, the module that has been activated.

A common approach to envisage this sub-network is to calculate the called Minimal Connected Network (MCN). The MCN is the minimal network that connects a set of nodes. It is generated by the calculation of the shortest path between any two proteins in the list. When generating the MCN for functional profiling of experiments, the resulting network should be representative of the list of proteins analysed. Therefore not all the paths should be integrated in the final graph but only the ones that connect directly any pair of proteins in the list. Such graphs can be completed using other proteins not contained in the list analysed that might have been missing (a typical problem in proteomics experiments) and could contribute with important connections to the resulting network. The number of proteins outside the list that connects pairs of proteins within the list should be small enough to keep equilibrium between the exploratory capabilities of this methodology and the maintenance of the accuracy of the assignment of a network to the list.

Indeed, proteins no pre-selected by expression profiling has been reported to be related to disease due to its inclusion into a network of ppis (Xu and Li, 2006; Liu et al., 2007; Chuang *et al.*, 2007). In microarray analysis it is a common practice to apply a threshold in the p-value (typically 0.05 considering multiple testing adjustment) to the selection of differentially expressed genes. This could be an additional problem in the selection of important genes because strong corrections are applied to the p-values, due to the multiple testing nature of the analysis, and some important genes might not be reported as differentially expressed when they actually are. Moreover, there are observations that point out that important proteins in the networks such as hubs and superhubs may not be differentially expressed (Camargo & Azuaje, 2007).

Methodologies to Evaluate a Sub-Network

When trying to assign a functional interpretation to a list of genes using classical annotations such as GO terms or KEGG pathways a simple inventory of the annotations found does not give significant information. The frequencies of the annotations found have to be compared to the background to test whether the annotations show a significant enrichment or not. There are several methodologies and resources available for functional profiling (review in Dopazo, 2006). The application of ppi data to this field is quite recent so there are not standard methodologies to be applied to the evaluation of the modules found yet.

A standard approach proposed has been to test if the proteins in a given network are enriched in any functional category (Wilkinson and Huberman, 2004; Luo *et al.*, 2006). There are specific tools for doing this task. BINGO (Maere *et al.*, 2005) is a java applet that can be integrated into Cytoscape visualization tool (Shannon *et al.*, 2003) that performs GO enrichment analysis to the nodes of a network. Although informative, this test does not guarantee in the case of negative results that the network is not a module of action yet unannotated or simply a module of proteins that do not share a functional category but they are indeed carrying out some cooperative function. In fact, functional analysis using ppis do not always overlap with functional label based analysis (Liu *et al.*, 2007).

Liu *et al.* (2007) proposed a systems biology oriented approach called Gene Network Enrichment Analysis (GNEA). The method evaluates the association of sub-networks to a determined disease. A limitation of this method is that it must start with a set of pre-defined gene signatures already associated to the disease, each of them with a particular annotation. The gene signatures are assembled and then, the relative expression in a microarray analysis, exploring the case of study, is mapped to a global network of ppis. From the interactomic and transcriptomic information a High Scoring Matrix (HSM) is extracted as a sub-network that is highly transcriptionally affected in the disease. Finally, they evaluate the hypothesis that a particular gene signature is enriched into the sub-network. Basically, ppis in this methodology substitutes the classical differentially expression analysis but it is not taking advantage of the structured data of the biological networks.

For the evaluation of a sub-network we propose to take into account the special topological features of the biological networks. Our hypothesis is that sub-networks associated to a specific cellular activity should have a compact topology. This type of topology can be characterised by having a higher distribution of connections degree and significantly less components than a sub-network integrated by random proteins that do not share any functionality. The distribution of two other parameters, clustering coefficient and betweenness centrality, that are more related to the special features of each functional type of active network, can also be evaluated. Thus, finding statistical significance in the different parameters points towards different possible topologies of the network and the combined study of some of them may reflect the actual shape of the network. In this way, obtaining a significantly higher connection degree but a non significant clustering coefficient would point to a star-shaped network. A significantly low number of components with a significantly high connections degree would reflect a compact network. Only a significant number of connections degrees might point to the presence of a group of small protein complexes. Finally, significance for betweenness centrality but not for connections degree could indicate the presence of a cascade signalling network.

This methodology uses as input a list of proteins selected from a genome scale experiment (e.g. proteins that co-express or that are differentially expressed under certain conditions). The aim is to find the active networks inside these lists and to evaluate if they are important in the cooperative behaviour of the list. In other words, the methodology proposed intends to highlight the sub-networks activated in response to new stimuli and to evaluate their importance within the entity selected as the unit of study, which is the list of proteins.

We first calculate the MCN of the pre-selected proteins or genes and then we test, using the Kolmogorov-Smirnov test, each of the distributions of the MCN node parameters (connections degree, betweenness centrality and clustering coefficient) against their corresponding reference distribution, which we take from generating MCNs of lists of the same size as the query list, populated with random proteins. The number of components of the MCN is compared to a 95% confidence interval generated from the random datasets.

This novel methodology is implemented and available in the SNOW module of Babelomics suite for functional profiling of genome-scale experiments.

CONNECTIVITY OF CONVENTIONAL FUNCTIONAL MODULES

We performed a massive analysis of lists of human genes and proteins taken from microarray experiments, co-expression modules in cancer, GO terms, KEGG pathways and Biocarta classes with the aim of studying how ppi networks are spread in different types of lists and in some classical sources of annotation normally used for functional profiling. Table 1 shows the percentages of the lists in each category that presented a positive result in one of the analysis performed. The procedure was to take every list and calculate its MCN allowing the inclusion of one non-listed node into the network. We used a curated human interactome as described in the Methods section. The distribution of the connections degree (degree), betweenness centrality (betweenness) and clustering coefficient of the nodes in each of the MCNs generated was compared using a Kolmogorov-Smirnov test versus

Table1. Network parameters evaluation in different types of sets of genes. Lists of genes taken from differential expression analysis of microarray experiments (cancer lists, non-cancer lists, up-regulated lists and down-regulated lists), modules of co-expression in cancer also taken from microarray experiments and genes belonging to the same annotation class (GO terms, KEGG pathways and Biocarta pathways). In rows we show the network topological parameters evaluated, bdcG (betweenness, degree and clustering coefficient have a significantly higher value than random sets), bG (betweenness significantly higher than random), dG (degree significantly higher than random), cG (clustering coefficient significantly higher than random), compL (number of components below the 95% confidence interval of random sets) and compL + dG (number of components below the 95% confidence interval of random sets plus degree significantly higher than random) . The value in the cells is the percentage of the set of lists with a p-value less than 0.05 compared to networks generated using same size random lists.

	Gene Ontology	Modules	Cancer Lists	Non-Cancer Lists	Up-regulated Lists	Down-regulated Lists	Biocarta	KEGGs
bdcG	15.11	10.5	6.67	1.9	1.91	5.6	9.27	26.9
bG	36.53	34.9	20.44	19.85	21.02	20	32.59	48.97
dG	71.52	52.1	29.33	31.23	30.57	30	55.91	59.31
cG	22.21	13.4	7.56	2.91	2.87	6.4	12.46	31.03
compL	51.92	38	18.22	18.4	15.61	19.2	33.87	52.41
compL + dG	47.72	34.9	14.67	14.29	13.38	14	32.91	48.28

the distribution of the same parameter in a set of 10,000 MCNs generated from a same size range set of lists populated with random proteins/genes.

This analysis reveals the global role of networks across biologically meaningful lists. The results show that generally the ppi networks seem to have a wide distribution in classes formed by functionally related proteins. We take a significant result in the compL + dG analysis as indicative of a compact network, where the nodes have more connections and fewer components than a network of random genes, hypothetically not functionally related. Betweenness and clustering coefficient seem to be parameters more related to the shape of the network that may point out its activity, e.g. a signalling cascade network is a network with a low clustering coefficient distribution because its nodes are not in very connected areas. Nevertheless, it should indeed have more connections and fewer components than a network coming from a random list.

Summarising, the results of the analysis indicate that networks are more widely spread in the annotation classes (GO, KEGG, BioCarta and modules of co-expression) than in the up- and down-regulated lists of genes (cancer and non-cancer lists are contained into this categories as well). GO terms and KEGG pathways appear to be the more connected entities. The classification of differentially expressed lists into 4 categories do not show differences among them but in the clustering coefficient comparison where cancer and down-regulated lists show a higher percentage indicating that their networks must have a higher interconnectivity.

CONCLUSIONS AND FUTURE TRENDS

In conclusion, we could say that ppi networks do have an important role in functionally related genes. Even in genes detected to be differentially expressed in an experiment that may not be involved in a single activity but in more than one, we could detect modules of action using ppi data. Therefore the applicability of interactomics as a source of annotation in functional profiling is more than justifiable.

Nevertheless, there are still three main challenges that have to be approached to be able to obtain whole capabilities from this kind of data:

1. High-throughput techniques produce a high proportion of false positives. Besides, there is still a low coverage of the interactome for the majority of the species. For filtering ppis according to their accuracy, literature curation does not seem to be a realistic approach, so new methodologies have to be proposed while the techniques do not overcome this limitation.
2. There is a clear necessity of applying the standard annotation developed by the HUPO for the ppi experiments. This should be enough to encourage the community to take a policy of sharing data to be able to have one or several repositories with all the available ppis.
3. The network nature of the functional classes that the ppis form requires more complex methodologies to study modules enrichment. The topology of the networks should be taken into account as an important parameter of the module. A protein in a network cannot be annotated just as part of the network but as a node with a special position that affects the rest of nodes. Moreover, the global shape of the network is characteristic of its functional activity.

The introduction of more structured data as the network concept in functional analysis is more and more required. Analysis of regulatory (Yeger-Lotem *et al.*, 2004), co-expression (Ghazalpou *et al.*, 2006) and genetic (Kelley & Ideker, 2005) networks are some examples of the applicability of graph theory to biological data.

CONCLUDING REMARKS

In this chapter we have highlighted the relevance of introducing interactomic data into Functional Genomics. Last discoveries show that cell complexity is more associated to post-transcriptional and post-translational events rather than to size of the genome. Hence, to understand cell behaviour is of crucial importance to introduce not only functional gene annotations but also information of post-transcriptional events such as protein-protein interactions.

We have given short but comprehensive review on the databases that store ppi data, methodologies to curate ppi data as well as to infer and evaluate networks within a functional genomics context. In addition, we have analysed the role of ppi networks in different gene modules build up using different functional criteria. Our observations suggests that conventional functional modules (GO, KEGG, Biocarta) seems to be fairly connected. This shows that the introduction of ppi data into Functional Genomics may provide both complementary and reinforcing results in the functional profiling of genome scale experiments.

RESOURCES

General Resources

A very complete revision on some available resources related to ppis:
http://www.imb-jena.de/jcb/ppi/jcb_ppi_databases.html

Protein Interaction Databases

Database of Interacting Proteins (DIP):
http://dip.doe-mbi.ucla.edu/

Biomolecular Interaction Network Database (BIND):
http://bond.unleashedinformatics.com/

Human Protein Reference Database (HPRD):
http://www.hprd.org/

Molecular INTeractions database (MINT):
http://mint.bio.uniroma2.it/mint/Welcome.do

Database of protein InterAction data (IntAct):
http://www.ebi.ac.uk/intact/

General Repository for Interaction Datasets (BioGRID):
http://www.thebiogrid.org/

Literature Curated (LC) Saccharomyces cerevisiae ppis database (Reguly *et al.*, 2006):
http://www.thebiogrid.org/ & http://yeastgenome.org

STRING
http://string.embl.de/

Visualization Tools

Agile Protein Interaction DataAnalyzer (APID):
http://bioinfow.dep.usal.es/apid/index.htm

APID2NET (APID applet for Cytoscape)
http://bioinfow.dep.usal.es/apid/apid2net.html

CYTOSCAPE
http://www.cytoscape.org

Integrative Visual Analysis Tool for Biological Networks and Pathways (VisAnt)
http://visant.bu.edu/

Osprey
http://biodata.mshri.on.ca/osprey/servlet/Index

STRING
http://string.embl.de/

Tools for Inferring (and Evaluate) Networks

MINE (IntAct interface)
http://www.ebi.ac.uk/intact/mine/do/welcome

MINT
http://mint.bio.uniroma2.it/mint/search/search.do

MEDUSA (interface to STRING database)
http://coot.embl.de/medusa/

Genes2Networks
http://actin.pharm.mssm.edu/genes2networks/

PIANA
http://sbi.imim.es/piana/

Gene Network Enrichment Analysis (GNEA), R package
http://genomics10.bu.edu/manwayl/

BINGO (Cytoscape plug-in)
http://www.psb.ugent.be/cbd/papers/BiNGO/

Babelomics (SNOW module)
http://www.babelomics.org

Packages for Functional Genomics Analysis

Babelomics
http://www.babelomics.org

DAVID
http://david.abcc.ncifcrf.gov/

OntoTools
http://www.bioconductor.org/packages/2.0/bioc/html/ontoTools.html

ACKNOWLEDGMENT

This work is supported by grants from projects BIO 2005-01078 from the Spanish Ministry of Education and Science and National Institute of Bioinformatics (www.inab.org), a platform of Genoma España. The CIBER de Enfermedades Raras (CIBERER) is an initiative of the ISCIII.

REFERENCES

Al-Shahrour, F., Díaz-Uriarte, R., & Dopazo, J. (2004). FatiGO: A web tool for finding significant associations of Gene Ontology terms with groups of genes. *Bioinformatics, 20*, 578-580

Al-Shahrour *et al.* (2006). BABELOMICS: a systems biology perspectve in the functional annotation of genome-scale experiments. *Nucleic Acids Research, 34*(Web Server Issue), W472-W476

Al-Shahrour, F., Minguez, P., Tárraga, J., Medina, I., Alloza, E., Montaner, D., & Dopazo, J. (2007). FatiGO+: a functional profiling tool for genomic data. Integration of functional annotation, regulatory motifs and interpretation data with microarray experiments. *Nucleic Acids Research, 35*(Web Server issue), W91-96

Al-Shahrour, F., Carbonell, J., Minguez, P., Goetz, S., Conesa, A., Tárraga, J., Medina, I., Alloza, E., Montaner, D., & Dopazo, J. (2008). Babelomics: Advanced functional profiling of transcriptomics, proteomics and genomics experiments. *Nucleic Acid Research, 36*, 341-346.

Aragues R., Jaeggi D., & Oliva, B. (2006). PIANA: Protein Interactions and Network Analysis. *Bioinformatics, 22*(8), 1015-7.

Ashburner, M., Ball, C. A., Blake, J. A., Botstein, D., Butler, H., Cherry, J. M., Davis, A. P., Dolinski, K., Dwight, S. S., Eppig, J. T. et al. (2000). Gene ontology: tool for the unification of biology. *The Gene Ontology Consortium. Nat Genet, 25*, 25-29.

Badano, J. L., & Katsanis, N. (2002). Beyond Mendel: an evolving view of human genetic disease transmission. *Nat. Rev. Genet., 3*, 779–789.

Bader, G. D. & Hogue C. W. V. (2002). Analyzing yeast protein–protein interaction data obtained from different sources. *Nature Biotechnology, 20*, 991–997.

Bader, G. D., Betel, D., & Hogue, C. W. V. (2003). BIND: The Biomolecular Interaction Network Database. *Nucleic Acids Research, 31*(1).

Bader, J. S. et al. (2004). Gaining confidence in high-throughput protein interaction networks. *Nature Biotechnology, 22*(1), 78-84.

Barabasi, A. L., & Albert, R. (1999). Emergence of scaling in random networks. *Science 286*(5439), 509-12.

Barabasi, A. L., & Bonabeu, E. (2003). Scale-Free Networks. *Scientific American 288*(5), 60-9.

Barabasi, A. L., & Oltvai, Z. N. (2004). Network Biology: Understanding the cell's functional organization. *Nature Reviews, 5*(2), 101-13.

Batada, N. N., Hurst, L. D., & Tyers, M. (2006). Evolutionary and Physiological Importance of Hub Proteins. *Plos Computational Biology, 2*(7), e88.

Berger, S. I., Posner, J. M., & Ma'ayan, A. (2007). Genes2Networks: Connecting lists of gene symbols using mammalian protein interactions databases. *BMC Bioinformatics, 8*, 372.

Berggard, T., Linse, S., & James, P. (2007). Methods for the detection and analyisis of protein-protein interactions. *Proteomics, 7*, 2833-2842.

Breitkreutz, B., Stark, C., & Tyers, M. (2003). Osprey: A network visualization system. *Genome Biology, 4*, R22.

Breitkreutz, B., et al. (2008). The BioGRID Interaction Database: 2008 update. *Nucleic Acids Research, 36*(Database issue)D637-D640.

Brunner, H. G., & van Driel, M. A. (2004). From syndrome families to functional genomics. *Nat. Rev. Genet., 5*, 545–551.

Camargo A., & Azuaje, F. (2007). Linking Gene Expression and Functional Network Data in Human Heart Failure. *PloS One, 12*, e1347.

Ceol, A., Chatr-Aryamontri A., Liacata L., & Cesareni G. (2008). Linking entries in protein interaction database to structured text: The FEBS Letters experiment. *FEBS Lettwers, 585*, 1171-1177.

Chatr-aryamontri, A., Ceol, A., Palazzi L., M., Nardelli, G., Schneider, M. V., Castagnoli L., & Cesareni, G. (2006). *MINT: the Molecular INTeraction database Nucleic Acids Research.*

Cho, S. et al. (2003). Protein-protein Interaction Networks: from Interactions to Networks. *Journal of Biochemistry and Molecular Biology, 37*, 1, 45-52.

Chuang, H., Lee, E., Liu, Y., Lee, D., & Ideker, T. (2007). Network-based classification of breast cancer metastasis. *Molecular Systems Biology, 3*, 140

Deane, G. D. *et al.* (2002). Protein Interactions: two methods for assessment of the realiability of high throughput observations. *Mol. Cel. Proteomics, 1*, 349-356.

Dennis Jr., G., Sherman, B. T., Hosack, D. A., Yang, J., Gao, W., Lane, H. C., &Lempicki, R. A. (n/d). DAVID: Database for Annotation, Visualization, and Integrated Discovery. *Genome Biology, 2003, 4*(5):P3.

Dopazo, J. (2006). Functional interpretation of microarray experiments. *Omics, 10*, 398-410.

Drewes, G., & Bouwmeester, T. (2003). Global approaches to protein-protein interactions. *Current Opinion in Cell Biology, 15*, 199-205.

Falk, R., *et al.* (2007). Approaches for systematic proteome exploration. *Biomolecular Engineering, 24*, 155-168.

Drogrusoz, U., *et al.* (2006). PATIKAweb: A Web interface for analyzing biological pathways through advanced querying and visualization. *Bioinformatics, 22*(3), 374-375.

Formstecher, E., *et al.* (2005). Protein interaction mapping: A Drosophila case study. Genome Research, *15*, 376-384.

Gandhi, T. K., *et al.* (2006). Analysis of the human protein interactome and comparison with yeast, worm and fly interaction datasets. *Nat. Genet., 38*, 285–293.

Gavin, A. C. *et al.* (2002). Functional organization of the yeast proteome by systematic analysis of protein complexes. *Nature, 415*, 141-147

Ge, H., *et al.* (2001). Correlation between transcriptome and interactome mapping data from Saccharomyces cerevisiae. *Nature Genetics, 29*, 482-486

Ghazalpour, A., et al. (2006). Integrating Genetic and Network Analysis to Characterize Genes Related to Mouse Weight. *Plos Genetics, 2*(8), e130.

Giot, L. *et al.* (2003). A protein interaction map of Drosophila melagaster. *Science, 302*, 1727-1736.

Girvan, M., & Newman, M. E. J. (2002). Community structure in socila and biological networks. *Proc. Natl. Acad. Sci., 99*(12), 7821-7826.

Han, J-DJ., *et al.* (2005). ffect of sampling on topology predictions of protein-protein interaction networks. *Nature Biotechnology, 23*, 839-844.

Hartwell, H., Hopfield, J., Leibler, S., & Murray, A. W. (1999). From molecular to modular biology. *Nature, 402*, C47-C52.

He, X., & Zhang, J. (2006). Why do hubs tend to be essential in protein networks? *Plos Genetics*, *2*(6), e88.

Hermjakob, H., et al. (2004). The HUPO PSI's Molecular Interaction format-a community standard for the representation of protein interaction data. *Nature Biotechnology, 22*(2), 177-183.

Hernandez, P., et al. (2007). Evidence for system-level molecular mechanisms of tumorgenesis. *BMC Genomics, 8*, 186

Hernandez-Toro, J., Prieto C., & De Las Rivas, J. (2007). APID2NET: unified interactome graphic analyzer. *Bioinformatics, 23*(18), 2495-2497

Ho, Y. et al. (2002). Systematic identifiaction of protein complexes in Saccharomyces cerevisiae by mass spectrometry. *Nature, 415*, 180-183

Hu, Z. et al. (2007). VisANT 3.0: new modules for pathway visualization, editing, prediction and construction. *Nucleic Acids Research*, (pp. 1-8).

Ideker, T. et al. (2001). Integrated Genomic and Proteomic Analyses of a Systematically Perturbed Metabolic Network. *Science, 292*, 929-933

Ito, T. et al. (2001). A comprehensive two-hybrid analysis to explore the yeast protein interactome. *Proc. Natl. Acad. Sci. 98*, 4569-4574.

Jansen, R. et al. (2002). Relating whole-genome expression data with protein-protein inetractions. *Genome Research, 12*, 37-46

Johsson, P. F., & Bates, P. A. (2006). Global topological features of cancer proteins in the human interactome. *Bioinformatics, 22*(18), 2291-2297.

Joy, M. P., Brock, A., Ingber, D. E., & Huang, S. (2005). High-Betweenness Porteins in the Yeast Protein Interaction Network. *Journal of Biomedicine and Biotechnology, 2*, 96-103

Kanehisa, M., Goto, S., Kawashima, S., Okuno, Y., & Hattori, M. (2004). The KEGG resource for deciphering the genome. *Nucleic Acids Res, 32*, D277-280.

Kerrien, S. et al. (2006). IntAct – Open Source Resource for Molecular Interaction Data. *Nucleic Acids Research, 2006.*

Kelley, R., & Ideker, T. (2005). Systematic interpretation of genetic interactions using protein networks. *Nature Biotechnology, 23*, 5, 561-566

Lee et al. (2007). Asingle gene network accurately predicts phenotypic effects of gene perturbation in Caenorhabditis elegans. *Nature Genetics, 40*(2), 181-188

Li, S. (2004). A map of the interactome network of the metazoan C. elegans. *Science, 303,* 540-543

Liu, M., Liberzon, A., Kong, S. W., Lai, W. R., Park, P. J., Kohane, I. S., & Kasif, S. (2007). Network-Based Analysis of Affected Biological Proceses in Type 2 Diabetes Models. *Plos Genetics*, *3*(6), e96.

Luo, F. et al. (2006). Modular organization of protein interaction networks. *Bioinformatics, 23*(2), 207-214 .

Maere, S., Heymans, K., & Kuiper, M. (2005). BiNGO: a Cytoscape plugin to assess overrepresentation of gene ontology categories in biological networks. *Bioinformatics, 21*(16), 3448-9.

Mathivanan, S. et al. (2006). An evaluation of human protein-protein inetraction data in the public domain. *BMC Bioinformatics, 7*(5), S19.

Mrowka, R., Patzak, A., & Herzel, H. (2001). Is there a bias in proteome research? *Genome Research, 11*, 1971-1973

Pereira-Leal, J. B. et al. (2004). Detection of functional modules from protein-protein interaction networks. *Proteins: Struct. Func. Bioinformatics, 54*, 49-57.

Peri, S. et al. (2003). Development of human protein reference database as an initial platform for approaching systems biology in humans. *Genome Research, 13*, 2363-2371.

Prieto, C., & De Las Rivas, J. (2006). APID: Agile Protein Interaction DataAnalyzer. *Nucl. Acids Res., 34*, W298-W302.

Reguly T. et al. (2006). Comprehensive curation and analysis of global interaction networks in Saccharomyces cerevisiae. *J. Biol, 5*(11).

Rives, A. W., & Galitski, T. (2003). Modular organization of cellular networks. *Proc. Natl. Acad. Sci., 100*(3), 1128-1133.

Rual, J. F., Venkatesan, K., Hao, T., Hirozane-Kishikawa, T., Dricot, A., Li, N., Berriz, G. F., Gibbons, F. D., Dreze, M., Ayivi-Guedehoussou, N., Klitgord, N., Simon, C., Boxem, M., Milstein, S., Rosenberg, J., Goldberg, D. S., Zhang, L. V., Wong, S. L., Franklin, G., Li, S., Albala, J. S., Lim, J., Fraughton, C., Llamosas, E., Cevik, S., Bex, C., Lamesch, P., Sikorski, R. S., Vandenhaute, J., Zoghbi, H. Y., Smolyar, A., Bosak, S., Sequerra, R., Doucette-Stamm, L., Cusick, M. E., Hill, D. E., Roth, F. P., & Vidal, M. (2005). Towards a proteome-scale map of the human protein-protein interaction network. Nature, 20, 437(7062), 1173-8.

Salwinski, L., Miller, C. S., Smith, A. J., Pettit, F. K., Bowie, J. U., & Eisenberg, D. (2004). The Database of Interacting Proteins: 2004 update. *NAR, 32*, Database issue D449-51

Shannon, P. et al. (2003). Cytoscape: A software environment for integrated models of biomolecular interaction networks. *Genome Research, 13*(11), 2498-504.

Sporn, O., Honey, C. J., & Kötter, R. (2007). Identification and classification of hubs in brain networks. *Plos One, 10*, e1049.

Stelzl, U., Worm, U., Lalowski, M., Haenig, C., Brembeck, F. H., Goehler, H., Stroedicke, M., Zenkner, M., Schoenherr, A., Koeppen, S., Timm, J., Mintzlaff, S., Abraham, C., Bock, N., Kietzmann, S., Goedde, A., Toksöz, E., Droege, A., Krobitsch, S., Korn, B., Birchmeier, W., Lehrach, H., & Wanker, E. E. (2005). A human protein-protein interaction network: A resource for annotating the proteome. *Cell* 2005, *122*(6), 957-68

Stumpf, M. P. H. et al. (2008). Estimating the size of the human interactome. *Proc. Natl. Acad. Sci., 105*(19), 6959-6964

Suderman, M., & Hallet, M. (2007). Tools for Visually Exploring Biological Networks. *Bioinformatics, 23*(20), 2651-9.

Uetz, P. et al. (2005). A comprehensive analysis of protein-protein interactions in Saccharomyces cerevisiae. *Nature, 403*, 603-627.

von Mering, C. et al. (2002). Comparative assessment of large-scale data sets of protein-protein interactions. *Nature, 417*, 399-403.

von Mering, C. et al. (2007). STRING 7 – recent developments in the integration and prediction of protein interactions. *Nucleic Acids Research, 35*(Database issue), D358-D362.

Wachi, S., Yoneda, K., & Wu, R. (2005). Interactome-trancriptome analysis reveals the high centrality of genes differentially expressed in lung cancer tissues. *Bioinformatics, 21*(23), 4205-4208.

Wilkinson, D. M., & Huberman, B. A. (2004). A method for finding communities of related genes. *Proc. Natl. Acad. Sci., 101*(1), 5241-5248.

Xenarios, I., & Eisenberg, D. (2001). Protein interaction databases. *Current Opinion in Biotechnology, 12*, 334-339.

Xu, J., & Li, Y. (2006). Discovering disease-genes by topological features in human protein-protein interaction network. *Bioinformatics, 22*, 2800-2805.

Yeger-Loten, E., Sattah, S., Kashtan, N., Itzkovitz, S., Milo, R., Pinter, R., Alon, U., & Margalit, H. (2004). Networks motifs in integrated cellular networks of transcription-regulation and protein-protein interactions. *PNAS, 101*(16), 5934-5939.

Zeeberg, B. R., Feng, W., Wang, G., Wang, M. D., Fojo, A. T., Sunshine, M., Narasimhan, S., Kane, D. W., Reinhold, W.C., Lababidi, S. et al. (2003). GoMiner: A resource for biological interpretation of genomic and proteomic data. *Genome Biol, 4*, R28.

Chapter XIV
Prioritizing Disease Genes and Understanding Disease Pathways

Xiaoyue Zhao
Bionovo Inc., USA

Lilia M. Iakoucheva
Rockefeller University, USA

Michael Q. Zhang
Cold Spring Harbor Laboratory, USA

ABSTRACT

Genetic factors play a major role in the etiology of many human diseases. Genome-wide experimental methods produce an increasing number of genes associated with such diseases. This chapter introduces data sources, bioinformatics tools, and computational methods for prioritizing disease candidate genes and identifying disease pathways. The main strategy is to examine the similarity among the candidate genes and known disease genes at the functional level. The authors review different similarity measures and prevailing methods for integrating results from different functional aspects. They hope this chapter will help advocate many useful resources that the researchers can use to investigate diseases of their interest.

INTRODUCTION

Genetic factors play a major role in the etiology of many diseases, including cancers and neurological disorders. Identifying genes that confer increased risk to the disease, and elucidating cellular and molecular processes in which these genes participate are very important problems in biomedical research. Genome-wide experimental methods, such as linkage, association (Botstein & Risch, 2003) and recently copy number variation (CNV) studies (McCarroll & Altshuler, 2007; Sebat, 2007), are all aimed at narrowing down genomic regions containing candidate disease genes. However, due to the linkage disequilibrium and the limited resolution of genome-wide technologies, the disease-associated regions could contain hundreds of candidate genes. The list of genes produced from such studies is constantly growing. The traditional one-gene-at-a-time approach is a time-consuming and expensive step to validate the disease-causing genes using experimental methods. Therefore it is of great importance and also a challenging task to use computational methods to prioritize disease gene candidates. Computational methods could greatly speed up the efforts directed towards elucidating disease mechanisms and ultimately translating genetic findings into effective prevention, diagnosis and treatment.

The recent availability of a large variety of genomic data and modern high-throughput technologies provide unique opportunities and complementary powerful resources for this purpose. Although disease-gene relationships are not simple (such as different diseases may be caused by mutations in the same gene, and the same disease may be caused by mutations in different genes), disease genes usually share at least some common characteristics including sequence features, expression patterns, involvement in the same protein-protein interaction sub-network, common gene ontology annotations, shared pathways and others (Goh et al., 2007; Oti & Brunner, 2007). For example, it was shown that genes involved in the same disease share up to 80% of their annotations in the GO and InterPro databases (Mulder et al., 2007). The similarity among disease genes is not restricted to the sequences and annotations; the similarity in their functions could also be noted. This leads to the main strategy in prioritizing disease genes, that is, to examine the similarity among candidate genes and known disease genes at the functional level (Han, 2008).

This functional approach is starting to yield new insights into the underlying biology of more and more diseases. For example, Mootha et al. integrated gene co-expression data and tandem mass spectra proteomics data to pinpoint a single candidate gene in a physical map of candidate disease loci for a familial cytochrome c oxidase deficiency LSFC (Mootha et al., 2003). In another study, the PPI network for Huntington's disease (HD) was generated to identify many new interactions and to pinpoint several novel proteins which may be involved in HD (Goehler et al., 2004). Further insights into neuronal toxicity in HD were revealed by another recently developed huntingtin-centered network (Kaltenbach et al., 2007). In the latter study, the authors identified a large number of new proteins that bind to normal and mutant forms of the huntingtin protein using two complementary approaches, high-throughput yeast two-hybrid screening and affinity pull down followed by mass spectrometry. A recent study by Lim et al. found a high degree of connectivity between different ataxia-causing proteins by constructing an ataxia-centered protein interaction network (Lim et al., 2006). Genes associated with breast cancer and a potential link between breast cancer susceptibility and centrosome dysfunction were identified via a network modeling strategy employed by Pujana et al. (Pujana et al., 2007). In the above study, known breast cancer susceptibility genes BRCA1 and BRCA2 were used as baits to generate hundreds of potential functional associations by combining gene expression with functional genomic and proteomic data. In our recent study of autism spectrum disorders (ASD), we tested whether newly identified ASD candidate loci are enriched in genes that are functionally related in transcriptional networks, protein-protein interaction networks, pathways and biological processes (manuscript in preparation) . We found evidence for increased connectivity between autism copy number variant genes and a set of known ASD genes, allowing us to prioritize ASD candidate genes, biological processes and cellular pathways for further studies. Taken together, all these examples suggest that the combination of experimental and computational approaches at a systems biology level could provide important insights into mechanisms of various diseases.

In this chapter, we describe a variety of useful data sources and bioinformatics tools and methods that can help prioritize disease genes and identify disease pathways. A selection of tools and resources is given in Table 1. We present the principles underlying different methods to help understand their advantages and disadvantages, and emphasize the importance of integrating results from different functional aspects. We use a few detailed studies to illustrate the power of this approach.

RESOURCES AND METHODS

When provided with a diverse list of genes associated with a particular disease, there are two frequently asked questions: what genes in this list are functionally related to each other? And what genes are functionally related to the reference genes that have already been implicated in the same disease or diseases with similar phenotypes?

The genes that are similar to other candidate genes or reference genes are more likely to be involved in the same disease pathway.

There are a variety of ways to define the similarity between genes at the functional level. Direct interaction in the PPI network, co-expression in the transcriptional network, involvement in the same metabolic or signaling pathways, shared GO annotations, and many others, each may imply functional relatedness between genes. Below, we are going to first describe the resources available for detecting the functional involvement of genes and appropriate resource-specific gene similarity measures. We will also introduce tools and methods for using these resources collectively to prioritize candidate genes for better understanding disease-relevant pathways.

RESOURCES AND SIMILARITY MEASURE

Phenotypic Similarity

Description from the literature can be useful for measuring the similarity between two genes. If the literature descriptions for two genes have many terms in common, it is likely that these genes are functionally related. The computational method to capture this similarity would be to build a vocabulary vector for each gene, where each entry in the vector is the frequency of the corresponding term in the relevant literature, and then use cosine function between these two vectors to measure the similarity. The application of this approach was demonstrated by Aerts et al (Aerts et al., 2006) who performed text mining of MEDLINE abstracts and built literature profiles for each gene based on the GO vocabulary. A similarity score of a candidate gene to a set of known disease genes was calculated by the cosine between the literature profile of the candidate gene and the average literature profile of the known disease genes.

Phenotypically similar diseases tend to be caused by similar functionally annotated genes (Oti, Huynen, & Brunner, 2008). Genes causing a similar disease can be used to help prioritize disease genes. Feature vectors were used to describe diseases and utilized to measure the phenotypic similarity among them (Lage et al., 2007; van Driel et al., 2006; Wu et al., 2008). Phenotype vectors could be constructed based on OMIM (Hamosh et al., 2005) using the vocabularies from Medical Subject Headings (MeSH) terms (Lowe & Barnett, 1994) or the Unified Medical Language System (UMLS) (Bodenreider, 2004). Phenotype similarity (overlap) scores can then be calculated between phenotype vectors.

PPI Network

The increasing knowledge about protein networks is being utilized to identify new genes and genetic mechanisms behind diseases. Certain properties of protein-protein interaction data can be used to prioritize disease genes. In the binary PPI network, the direct connection between two proteins being examined suggests their functional relevance. Even when two proteins are not directly connected, but share a large number of first-order neighbors, this may also suggest their functional similarity. Furthermore, network proteins with shorter mean shortest path length could also be considered more similar to each other.

Although experimentally derived PPI networks for many organisms including human have now become available (Giot et al., 2003; Ito et al., 2001; Li et al., 2004; Rain et al., 2001; Rual et al., 2005; Stelzl et al., 2005; Uetz et al., 2000), they still suffer from limited coverage, and high error rates, mainly due to their high-throughput nature. Furthermore, the majority of them still correspond to only a small fraction of the entire organism-specific interactomes. Computational approaches could be helpful to address at least some of these shortcomings.

To reduce the effect of false positive interactions resulting from high-throughput methods, confidence scores could be assigned for each edge in weighted PPI network. Different types of weights have been introduced. One score is based on the topology of the local network around the interaction (de Lichtenberg et al., 2005). It was found that the reliability of an interaction was correlated with the number of non-shared interaction partners for the yeast two-hybrid experiments. The raw quality score was then defined as

$S = -\log((N_1 + 1)(N_2 + 1))$,

where N_1 and N_2 are the number of non-shared interaction. The closer the raw score is to zero, the fewer the number of non-shared interaction partners is. If an interaction is reported in a greater number of independent experiments and the experiments are at a smaller scale, the interaction is considered as more reliable. Therefore, the raw quality score could be further scaled based on the number of publications reporting a particular interaction and on the number of reported interactions in each publication (Lage et al., 2007).

Another type of score borrows information from other functional relationship data to further expand the existing PPI networks and to extract more accurate information from the existing maps. Xia et al. built a Bayesian classifier to predict human PPI networks based on 27 heterogeneous genomic, proteomic and functional annotation datasets (Xia, Dong, & Han, 2006). The constructed database, IntNetDB, includes more than 180,000 predicted protein-protein interactions among more than 9,900 human proteins. Since the Bayesian classifier approach is similar to the one used for integrating different evidence for prioritizing disease genes in the section below, we will introduce the detailed method there.

The most likely candidates would be the ones that interact with known disease-causing genes. There is often limited knowledge of the known genes for a given disease, but if the gene in question interacts with a gene involved in phenotypically similar disorder, this may be suggestive of its relevance to the disease. For instance, one is interested in finding proteins involved in disorder type 1, and protein D is known to be involved in phenotypically similar disorder type 2. Suppose a list of four candidate proteins, C1, C2, C3 and C4, is given. If protein C1 interacts with protein D, then C1 is also likely to be involved in the disorder type 1. Based on this idea, one study constructed a disease gene predictor that utilizes protein phenotype information in the protein interaction network. For each protein in the neighborhood, a phenotype vector showing weighted frequencies of a list of selected words in OMIM reports was constructed and then phenotype similarity (overlap) score was calculated via the cosine distance between the vector of the protein and the vector of the disease under study (Lage et al., 2007). Another study went beyond the first order neighbor and established a mapping between phenotype network and PPI network (Wu et al., 2008). It used the correlation between gene–gene relatedness and phenotype–phenotype similarity to rank the candidate genes. Please refer to the section of integration of different evidence in this chapter for more detailed description of the method.

Leading databases containing human protein-protein interactions are listed in Table 1 and include Biomolecular Interaction Network Database (BIND), IntAct, Human Protein Reference Database (HPRD), BioGrid, Database of Interacting Proteins (DIP), Molecular Interaction Database (MINT) and STRING. They contain interaction data from high-throughput experiments (for example, yeast two-hybrid assays) and from the literature-based curation of physical interactions. Cytoscape is an easy-to-use tool for visualizing networks and producing network plots (Shannon et al., 2003). Other free and useful network tools include Osprey, VisANT, Graphviz and Bioconductor packages, some of which are reviewed in (Suderman & Hallett, 2007).

Transcriptional Network

Besides PPI networks, gene expression (or transcriptional) networks that carry different types of information (e.g. mRNA expression patterns) could also be built and used for disease gene prioritization. Genes of unknown function that are co-expressed with the genes already implicated in some disorder may be functionally relevant to the disease.

Pearson correlation coefficient (PCC) and, more recently, topological overlap (TO) (Oldham, Horvath, & Geschwind, 2006; Zhang & Horvath, 2005) are used to search for clusters of co-expressed genes. Genes having high PCC or high TO with each other are more likely to be co-regulated. TO measures the gene-gene relatedness via the similarity of their patterns of connection strength to other genes in the expression network and thus can exclude spurious or isolated connections during network construction. A recent study has demonstrated its power in identifying several modules of co-expressed genes that are conserved between Alzheimer disease (AD) and normal aging (Miller, Oldham, & Geschwind, 2008). Furthermore, the authors identified several novel genes as hubs of the transcriptional network that were not previously implicated in AD.

To compute TO, correlation of genes is first calculated and then converted into measures of connection strength a_{ij} by raising either the absolute value of correlation or the positive correlation to a power β. The topological overlap matrix $\Omega = [\omega_{ij}]$ is calculated as follows:

$$\omega_{ij} = \frac{l_{ij} + a_{ij}}{\min\{k_i, k_j\} + 1 - a_{ij}},$$

where

$$l_{ij} = \sum_u a_{iu} a_{uj},$$

and

$$k_i = \sum_u a_{iu}$$

is the node connectivity.

In our autism study, we computed TO between all pairs of genes using mRNA expression data across a variety of tissues from healthy individuals. Using TO, we constructed an autism-centered transcriptional network based on co-expression of autism CNV genes and known autism reference genes, and discovered that autism gene candidates form three distinct functional modules. Significantly increased connectivity was observed between autism CNV genes and autism reference genes in the brain module (manuscript in preparation).

Before calculating correlation or TO, it is important to remove genes whose expression levels do not change across all the experiments and also genes which have consistently low expression values. For the latter case, the high variation of the expression values of a gene could be simply due to the noise. Coefficient of variation (CV) is often used for this purpose. CV is defined as the ratio of the standard deviation to the mean. Typically threshold 0.1 is applied, and it can also be selected from random control data. Without this filtering step, the gene expression network will be overwhelmed by the noisy uninformative genes.

The Gene Expression Omnibus (GEO) at NCBI currently represents the largest single resource for public gene expression data, deposited by more than 2000 laboratories from around the world (Barrett et al., 2007). Since there is a big variety of data in GEO, researchers need to choose datasets appropriate for their purpose. The datasets used for building autism gene expression network in our ASD study was extracted from SymAtlas (Su et al., 2004), also deposited in GEO. SymAtlas contains the gene expression data for 33,698 human transcripts across 79 human tissues, 27 of which were derived from samples of postmortem brain.

Cluster and TreeView software (Eisen et al., 1998) are two most popular tools for analyzing expression data. Cluster performs the hierarchical clustering based on Pearson correlation coefficients (PCC) to find clusters of genes with similar expression patterns and TreeView visualizes the outputted clusters. The tools introduced for visualizing PPI network can also be used for the gene expression network.

GO Annotation and Pathway

Functional similarity of genes can also be measured by looking whether they share functional annotation or whether they are involved in a common biochemical or signaling pathway.

Gene Ontology Analysis

The Gene Ontology (GO) (Ashburner et al., 2000) is a database of vocabularies for the functional annotation of genes in three domains, molecular function, biological process and cellular component. Gene ontology database provides a useful resource to analyze the functions of a large number of genes. GO terms that are highly enriched in a list of candidate genes (or reference genes) may tell us something about the biology of the studied disease.

Candidate genes that are annotated by shared and enriched GO terms have higher probability of being functionally related and involved in the same disease.

The most commonly used statistical test that indicates the enrichment of a particular property/term involves the hypergeometric distribution. It describes the number of successes of N draws from a finite population without replacement. For example, one has an urn with M_1 black balls and M_2 white balls. Define drawing a black ball as a success and drawing a white ball as a failure. If N balls are drawn from the urn without replacement, the hypergeometric distribution calculates the probability of having n black balls drawn as following:

$$\frac{\binom{M_1}{n}\binom{M_2}{N-n}}{\binom{M_1+M_2}{N}}$$

In the context of GO analysis, the event of success is having a gene associated with the GO term under examination. Suppose a list of candidate genes contains N genes. The probability of having n number of genes in a certain GO term containing M_1 genes can be calculated using the above formula. For each term, the number of overlapping genes from the input is calculated and then compared with the one expected. Fisher's Exact Test using the hypergeometric distribution is performed to judge whether the observed difference is significant or not. For each GO category, this will result in a p-value, a probability that the observed counts or more extreme counts could have been obtained just due to chance. Since a large number of GO terms are examined, the calculated p-values need to be corrected for multiple testing. A number of tools have been developed based on the above procedure, most of which are listed at the GO website (http://www.geneontology.org/), including GOstat (Beissbarth & Speed, 2004), GOTM (Zhang et al., 2004) and DAVID (Dennis et al., 2003), three web-based user-friendly tools.

The above approach gives a straightforward and simple measure for the overrepresentation of an individual GO term. The GO terms are structured as a directed acyclic graph (DAG), where each term is represented as a node connected to more specific nodes by directed edges. Therefore, the GO terms are not independent due to the overlapping GO terms and overlapping annotations. That is, each GO term shares the annotations of its descendants and one gene could have multiple different GO annotations. Better tools that take into account these parent-child relationships have been developed such as Ontologizer (Grossmann et al., 2007). Tools like AmiGO can be used for visualizing the GO hierarchical structures. Detailed information about these tools is listed in Table 1.

Pathway Analysis

Human body is essentially a network of molecular interactions. In this network, molecules are synthesized, transported to other locations, interact with other molecules, undergo different types of modifications, and participate in other activities. Pathway information describes how different components work together to maintain some function of an entity. Pathways significantly over-represented in a list of candidate genes and known disease genes may reflect the underlying pathology of the disease.

Similarly to GO enrichment analysis, Fisher exact test that is based on hyper-geometric distribution is commonly used to calculate the probability of observing a certain number of overlapping genes for each pathway as compared to a list of random genes. Multiple testing correction is usually needed.

Only limited knowledge about metabolic and signaling pathways currently exists. The scientific community has made enormous effort to curate and assemble the pathways described in the literature and to organize them into databases. The leading freely accessible pathway databases include KEGG, Panther, HumCyc and Reactome. Two other useful commercially available pathway databases include Ingenuity Pathway Analysis (IPA) and BioCarta. DAVID is also a useful tool for detecting relationships among genes utilizing pathway databases. More detailed information about these pathway databases and tools are listed in Table 1.

Table 1. Resources for prioritizing disease genes and identifying disease pathways

Resource Name	URL	References
Protein-protein interaction & protein domain databases		
HPRD	http://www.hprd.org	(Peri et al., 2003)
BIND	http://www.bind.ca	(Bader, Betel, & Hogue, 2003)
BioGrid	http://www.thebiogrid.org	(Stark et al., 2006)
DIP	http://dip.doe-mbi.ucla.edu	(Xenarios et al., 2000)
IntAct	http://www.ebi.ac.uk/intact	(Hermjakob et al., 2004)
MINT	http://mint.bio.uniroma2.it/mint	(Zanzoni et al., 2002)
IntNetDB	http://hanlab.genetics.ac.cn/sys/intnetdb	(Xia et al., 2006)
STRING	http://string.embl.de	(von Mering et al., 2007)
InterPro	http://www.ebi.ac.uk/interpro	(Mulder et al., 2007)
Pathway databases		
KEGG	http://www.genome.jp/kegg	(Kanehisa & Goto, 2000)
Panther	http://www.pantherdb.org	(Thomas et al., 2003)
HumCyc	http://HumanCyc.org	(Romero et al., 2005)
Reactome	http://www.reactome.org	(Vastrik et al., 2007)
IPA	http://www.ingenuity.com	Commercial
BioCarta	http://www.biocarta.com	Commercial
Gene expressions databases		
GEO	http://www.ncbi.nlm.nih.gov/geo	(Barrett et al., 2007)
SymAtlas	http://symatlas.gnf.org/SymAtlas	(Su et al., 2004)
Tools for gene expression analysis		
Cluster, TreeView	http://rana.lbl.gov/EisenSoftware.htm	(Eisen et al., 1998)
Tools for GO analysis		
AmiGO	http://amigo.geneontology.org/cgi-bin/amigo/go.cgi	
GOstat	http://gostat.wehi.edu.au/cgi-bin/goStat.pl	(Beissbarth & Speed, 2004)
GOTM	http://bioinfo.vanderbilt.edu/gotm	(B. Zhang et al., 2004)
Ontologizer	http://www.charite.de/ch/medgen/ontologizer	(Grossmann et al., 2007)
DAVID	http://david.abcc.ncifcrf.gov/home.jsp	(Dennis et al., 2003)
Genome browsers and gene structure analysis		
UCSC genome browser	http://genome.ucsc.edu	(Hinrichs et al., 2006)
Ensembl	http://www.ensembl.org	(Birney et al., 2006)
Tools for producing and visualizing networks		
Cytoscape	http://www.cytoscape.org	(Shannon et al., 2003)
Osprey	http://biodata.mshri.on.ca/osprey/servlet/Index	(Breitkreutz, Stark, & Tyers, 2002)
VisANT	http://visant.bu.edu	(Hu et al., 2005)
Graphviz	http://www.graphviz.org	

continued on following page

Table 1. continued

Resource Name	URL	References
Bioconductor packages	http://bioconductor.org/packages/release/GraphsAndNetworks.html	
Tools for prioritizing disease genes		
ENDEAVOR	http://homes.esat.kuleuven.be/~bioiuser/endeavour/endeavour.php	(Aerts et al., 2006)
SUSPECT	http://www.genetics.med.ed.ac.uk/suspects	(Adie et al., 2006)
Prioritizer	http://humgen.med.uu.nl/~lude/prioritizer	(Franke et al., 2006)
G2D	http://www.ogic.ca/projects/g2d_2	(Perez-Iratxeta et al., 2005)

Sequence Similarity

Disease and pathway genes often share a high degree of sequence similarity with other members of the same disease or pathway. Sequence similarity could be defined at several levels. The most obvious is the coding sequence similarity: if two genes have parts of coding sequences in common, they may share important functional domains and thus manifest functional similarity. For example, Aerts et al used the blastp algorithm to score each candidate gene against each known disease gene, by using the longest coding sequence available in the Ensembl database (Aerts et al., 2006). They then ranked each candidate gene based on the lowest e-value when compared to each known gene.

Another level of similarity is domain similarity which might be also indicative of functional similarity between proteins. If a candidate gene carries a protein domain that is over-represented in the known disease genes, this gene is likely to be involved in the same disease. InterPro database (Mulder et al., 2007) could be used to assess the similarity between proteins at the level of domains or functional sites.

Yet another level of similarity is the similarity at the level of cis-regulatory elements in the promoters of two genes. This type of similarity may indicate that two genes are co-regulated by the same set of transcription factors or other regulatory molecules. Promoter is commonly referred to as the DNA region that is required for regulating the transcription initiation of the gene immediately downstream from it. Binding of the transcription factors (TFs) to specific sites in the promoter sequences is one of the first steps determining whether a gene is under the control of such TFs. Binding sites of a transcription factor are often represented as position weight matrix (PWM) where weights are proportional to the frequency (or log frequency) of each base at each position among the known binding sites. Other more complex models have been developed to improve the prediction accuracy, such as high-order Markov models (Roulet et al., 2002; Zhang & Marr, 1993), the maximum entropy model (Yeo & Burge, 2004), Bayesian networks (Barash, et al., 2003), and generalizations of Bayesian networks (Ben-Gal et al., 2005; Zhao, Huang, & Speed, 2005), when more binding sites are available. TF motifs determined experimentally have been collected into databases such as TRANSFAC (Matys et al., 2003) and JASPAR (Sandelin et al., 2004) . De-novo motif finding methods have been developed to discover novel enriched motifs in a set of sequences. We refer to (Zhang, Zhao, & Zhang, 2007) for details on the topic of motif finding. A recent study used TF binding sites similarity to prioritize disease genes (Aerts et al., 2006). The authors organized the predicted sites of all transcription factors from TRNASFAC into a vector, whose entry represents the best score of each PWM in the sequence examined. Then, the vector of each candidate gene was compared with all the vectors of known reference genes using Pearson correlation coefficient. Aerts et al. also searched for the best combination of five transcription factors upstream of known disease genes, and assigned each candidate gene a module score of these five PWMs as the summations of the best score of each matrix. In summary, these examples show successful application of the similarity measure in sequence motifs towards disease genes prioritization.

INTEGRATION OF DIFFERENT EVIDENCE (DISEASE GENE PREDICTOR)

Large-scale high-throughput data are inevitably noisy, and the functional data are incompletely annotated and biased towards certain biological functions. For example, protein-protein interaction data covers only a small fraction of the full network and contains a high proportion of false positives. GO and pathway databases are not completely annotated and usually biased towards better-studied genes. Therefore, data integration is needed to provide a more accurate and informative selection of candidate genes, and to obtain a more comprehensive and less biased view of network via combination of the strengths of different sources. There are several methods to combine the data. We will demonstrate the integrative approach based on two different scenarios.

In one scenario, if some number of the high-confident disease-causing genes is available, they can be used as baits to pull out additional candidate genes from a larger list of potential targets. If a candidate gene shares a greater number of functional annotations with the set of known disease genes, it is more likely to be involved in the pathogenesis of the same disease. This "guilt-by-association" logic has enabled many researchers to build disease gene predictors, which take a list of candidate genes and the information of the disease (such as known disease genes and disease phenotype) as an input, and then produce a ranked list of candidate genes as well as significantly over-represented functional annotations (GO, pathways, networks, etc) related to the disease. Several methods have been developed in this direction to build a bioinformatics framework capable of comparing and integrating different gene features to rank test genes according to their similarity with known training genes (Adie et al., 2006; Aerts et al., 2006; Dalkilic et al., 2008; De Bie et al., 2007; Franke et al., 2006; Kann, 2007; Lage et al., 2007; Wu et al., 2008).

It is often the case that one does not have enough known high-confident disease genes available. In this scenario, simple intersection or presence/absence analysis can be applied. For each type of source, score 1 is assigned to a gene if it carries certain property similar to known disease genes. Then the total score from different sources is used for ranking genes. This scoring system is simple and easy-to-use. Although the scores are not weighted for different genes or different types of evidence, it has been successfully and efficiently applied to identify a novel breast cancer gene, HMMR (Pujana et al., 2007). We also used this ranking procedure in the autism network study (manuscript in preparation). Firstly, we assigned score of 1 to a gene if (1) it has high topological overlap with a known autism gene in human gene expression network, or (2) it has high topological overlap with a known autism gene in mouse gene expression network, or (3) it is directly connected to a known autism gene or an autism CNV gene in the PPI network, or (4) it is involved in a significantly enriched GO term, or (5) it participates in a significantly enriched Ingenuity pathway. Secondly, the autism CNV genes were sorted based on the total score of the above five lines of evidence, and then resorted based on the number of direct connections with autism reference genes in the human expression network. This procedure allowed for identifying a list of highly ranked autism CNV genes with compelling evidence for autism development.

Below, several representative methods are introduced for integrating different data sources to prioritize disease genes.

Combining P-Values and Rank Ratio Scores

The basic idea of this method is to obtain a score for each data source and then integratively combine these scores. Since different data sources are heterogeneous, it is important to derive a comparable score across these sources. If p-values from each evidence are available, one way is to combine the p-values using

$$-2\sum_{i=1}^{n}\log(P_i),$$

where n is the number of data sources used, and P_i, $i=1,\dots,n$ are the corresponding p-values. It is not difficult to show that

$$-2\sum_{i=1}^{n}\log(P_i)$$

follows a chi-square distribution with degree of freedom equal to $2n$ under the assumption that P_i, i = 1,…,n are independently and identically distributed uniform random variables on the unit interval [0,1]. P-value thus can be calculated for the combined p-value score using the chi-square distribution.

In cases when it is not straightforward to derive the p-values, order statistic may be helpful for the integration. For each type of evidence, candidate genes are first ranked based on the data source-specific statistic, and are then assigned a score of rank ratio by dividing each rank by the total number of ranked genes. The probability of getting the observed rank ratio by chance can be calculated using the following formula (Stuart et al., 2003):

$$Q(r_1, r_2, ..., r_N) = N! \int_0^{r_1} \int_{s_1}^{r_2} \cdots \int_{s_{N-1}}^{r_N} dS_N dS_{N-1} \cdots dS_1$$

where r_i is the rank ratio for data source i, N is the number of data sources used, and $r_0 = 0$. In this way, one study integrated information from different heterogeneous sources including PPI network, GO, KEGG, expression, InterPro, sequence, regulatory motifs and literature description, and a software called ENDEAVOR was developed as a result of this integration (Aerts et al., 2006). The authors were able to identify a novel gene involved in craniofacial development from a 2-Mb chromosomal region, deleted in some patients with DiGeorge-like birth defects. The order-statistic method becomes computationally very expensive when the number of integrated data types increases. A new more efficient approach developed by the same group applied the kernel method to fuse data for hunting disease genes (De Bie et al., 2007). The kernel method finds a hyperplane to separate the disease genes from the origin with the largest margin (distance). A gene is more likely to be a disease gene if it lies farther in the direction of this hyperplane on the same side as disease genes. Each data source was represented as a kernel matrix. Kernel matrices based on different views of genes are then combined into a single kernel matrix in the algorithm.

Bayesian Classifier

Another way of data integration is to use Bayesian classifiers to integrate all individual prioritizations. Using Bayesian theorem, one can compute the probability that a gene is disease-related given the existing evidence for that gene. High probabilities should be assigned to candidate genes that are more similar to the set of known disease genes. Suppose we have n genes in an interval associated with some disease. If each of n genes can be disease genes, the odds of candidate gene i being the disease gene is computed as follows:

$$\frac{P(X_i = 1 \mid AllData)}{P(X_i = 0 \mid AllData)} = \frac{\prod_{J=1}^{N} P(dataJ \mid X_i = 1)P(X_i = 1)}{\prod_{J=1}^{N} P(dataJ \mid X_i = 0)P(X_i = 0)} \quad (1)$$

where X_i takes values 1 or 0, with 1 for protein i being a disease protein and 0 otherwise, and N is the total number of evidence used. $P(X_i = 1)$ is the prior probability and $P(X_i = 1 \mid AllData)$ is the posterior probability. These two probabilities tell us how much we believe candidate gene i being a disease gene before and after seeing all the data, respectively.

Under the assumption that one and only one gene in the disease loci is the disease gene, the probability that the disease gene is candidate gene i is computed as follows:

$$P(D = i \mid AllData) = \frac{P(AllData \mid D = i)P(D = i)}{\sum_{j=1}^{n} P(AllData \mid D = j)P(D = j)} \quad (2)$$

where

$$P(AllData \mid D = i) = \prod_{k=1}^{i-1} P(dataJ \mid X_k = 0)P(dataJ \mid X_i = 1)\prod_{k=i+1}^{N} P(dataJ \mid X_k = 0).$$

Here $P(D = i)$ is the prior probability and $P(D = i \mid AllData)$ is the posterior probability. These two probabilities tell us how much we believe the disease gene is candidate gene i before and after seeing all the data, respectively.

The difference between model (1) and model (2) is that the prior probability is specified differently. In the first model, any gene could be disease gene and $P(X_i = 1) + P(X_i = 0) = 1$, while in the second model, one and only one gene from the disease loci is the disease gene and $P(D = 1) + P(D = 2) + ... + P(D = n) = 1$. In order to calculate the posterior probability, we need to estimate the parameters in the model. For prior probability, we can use the uninformative (uniform) prior, that is, to assign $P(X_i = 1) = P(X_i = 0) = 1/2$ and $P(D = i) = 1/N$ for each candidate i. $P(AllData \mid X_i = 1)$ and $P(AllData \mid D = i)$ is called likelihood, which is the probability of obtaining the observed evidence if gene i is the disease gene. In order to improve the coverage and accuracy of the PPI network, Bayesian classifier was also applied in several studies to predict protein-protein interactions by combining a variety of biological data (Jansen et al., 2003; Rhodes et al., 2005; Xia et al., 2006).

To discriminate the disease protein from the non-disease protein, two sets of training data are needed. One is the positive set, consisting of the proteins known to be involved in the investigated disease. The other one is the negative set, consisting of the proteins not involved in the disease. An appropriate model is the one that captures the main characteristics of positive and negative data. The parameters of the models are different for the positive and negative sets, which allows for distinguishing between them. The parameters for calculating the likelihood can be estimated for both disease-associated and non-disease-associated genes from the interaction data, transcription network data, and other data described above. It is important to keep in mind that the goal for mathematical modeling is not to construct a model that fully describes the biological reality, but to construct a model that is able to explain the most important aspects of the system and make good predictions for unseen conditions. There is usually insufficient data available to reliably estimate a large number of parameters and therefore a complex model will usually have poor performance in predictions.

Combining PPI Network and Disease Network

Based on the assumption that phenotypically similar diseases are caused by functionally related genes, human protein–protein interactions, disease phenotype similarities, and known gene–phenotype associations could be integrated to capture the complex relationships between phenotypes and genotypes (Lage et al., 2007; Wu et al., 2008). The study by Lage et al. assigned high probabilities to candidates that interact with one or more proteins involved in disorders that are phenotypically similar to the one being investigated (Lage et al., 2007). Two of their main parameters include (i) the probability of a protein having any known disease (or similar disease) genes as interaction partners and (ii) the probability that the number of interaction partners that are involved in similar disorders falls in a certain range. Combining different probabilities via Bayesian classifier introduced above, the authors were able to provide confident predictions of novel candidate disease genes for intervals involved in disorders such as retinitis pigmentosa, epithelial ovarian cancer, inflammatory bowel disease, etc.

Based on the global concordance between the human protein network and the phenotype network, CIPHER program has been developed and shown to reliably predict disease genes (Wu et al., 2008). Figure 1 shows the scoring scheme of CIPHER. Suppose we have n phenotypes $p_1, p_2, ..., p_n$ in the disease network. For a disease p of interest, the concordance score CS_{pg} between disease p and a candidate gene g was defined as the Pearson correlation coefficient between phenotype similarity vector $S_p = (S_{pp_1}, S_{pp_2}, ..., S_{pp_n})$ and gene closeness vector $\Phi_g = (\Phi_{gp_1}, \Phi_{gp_2}, ..., \Phi_{gp_n})$:

$$CS_{pg} = \frac{\operatorname{cov}(S_p, \Phi_g)}{\sigma(S_p)\sigma(\Phi_g)},$$

where S_{ppi} is the similarity between phenotype p and p_i, and Φ_{gpi} is the similarity between gene g and phenotype p_i. The closeness Φ_{gpi} of gene g to phenotype p_i was defined as gene-gene closeness from gene g to all disease genes $G(p_i)$ of phenotype p_i:

Figure 1. Scoring scheme of CIPHER for a particular phenotype-gene pair (p, g) (from Figure 1 Wu et al. 2008). (1) The human phenotype network, protein network, and gene–phenotype network are assembled into an integrated network. (2) The phenotype similarity vector for p and the gene closeness vector for g are constructed from the integrated network. (3) The linear correlation of the two profiles is calculated and assigned as the concordance score between the phenotype p and the gene g.

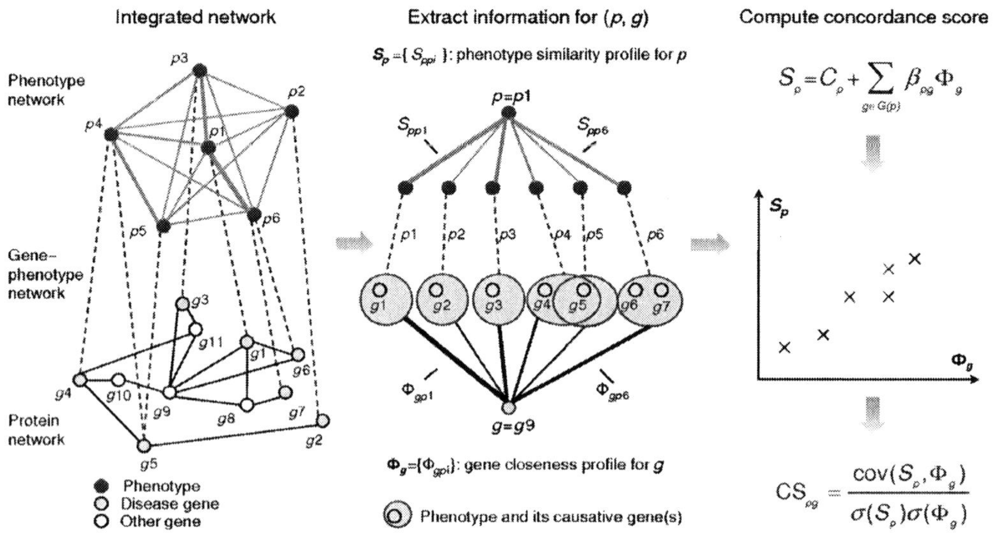

$$\Phi_{gpi} = \sum_{g' \in G(pi)} e^{-L^2(g,g')},$$

where $L(g, g')$ is the topological distance in the PPI network. The shortest path length (SP) or directed neighbor ($L(g, g') = 1$ for directed neighbors, $= \infty$ otherwise) was used to define $L(g, g')$. If the disease of interest is similar to some diseases in the phenotype network and a candidate gene is close to the genes causing those diseases in the PPI network, this gene is likely to cause the disease under the study. The concordance measure CS_{pg} between gene g and disease p makes use of the information implicated in the entire protein network and disease network, that is, the relationship to all the other phenotypes or other disease genes causing those phenotypes. This global nature makes the concordance score more robust to the individual imprecision and false information in both networks. CIPHER was shown to reproduce many of the known gene–disease associations from genetic databases. In a case study of breast cancer, 10 out of 16 well-established breast-cancer genes were given high association scores. The program was also used to assign high rankings to 15 genes that have more recently been proposed as linked to breast cancer. This highlights the effectiveness of CIPHER as a tool for facilitating the discovery of disease genes. CIPHER has been shown to outperform ENDEAVOR (Aerts et al. 2006) and the method introduced in Lage et al. (Lage et al., 2007) which prioritizes candidate genes using phenotypes associated with the first order neighbor in the PPI network.

In this section, we have introduced several representative methods for combining different sources. The rank statistics used in ENDEAVOR and the kernel method applied by De Bie et al. (De Bie et al., 2007) have been successfully used for prioritizing disease genes. Compared to ENDEAVOR, kernel method is more efficient in computational time, has the flexibility in weighting different sources and outperforms ENDEAVOR. No comparison has been carried out between the kernel method and the methods taking into account the phenotypic information of diseases such as the Bayesian predictor developed by Lage et al and CIPHER. CIPHER has shown to perform better than the Bayesian predictor. One of the reasons for explaining its better performance is that the Lage et al. used local PPI network (up to first order neighbor), while CIPHER used the global concordance of the whole PPI and disease network. This global property makes sufficient use of information from the entire networks and enables CIPHER to be more robust to annotation errors in the networks. Both methods only used PPI data for calibrating gene-gene interactions. Extracting information from other sources to increase the coverage of gene network might help to further improve the performance. For the accessibility, ENDEAVOR is available from its website and very flexible to use, while there is no released software for the kernel method, the Bayesian predictor and CIPHER. However, the genome-wide prioritization of candidate genes by CIHPER for over 5000 human phenotypes is publicly released at its website.

DISCUSSIONS AND CONCLUSION

In this chapter, we describe the data sources and methods for prioritizing disease candidate genes and exploring disease pathways. In the methods section we discuss disease genes prioritization approaches in the context of the biological networks. As shown above, using computational methods for prioritization as well as various data integration approaches could facilitate the understanding of the disease mechanisms and the diagnosis and treatment of human disease. This represents a step forward towards the emerging field of network medicine (Barabasi, 2007; Loscalzo, Kohane, & Barabasi, 2007).

Due to the noisy nature of large scale high-throughput data and the limited coverage of each data source, methods capable of comparing and integrating available gene characteristics could be a powerful tool to rank unknown candidate genes according to their similarity with known "reference" genes. The purpose of data integration is to obtain a more comprehensive, less technically biased and more accurate view of the underlying true network. It is not necessarily true that the more data one integrate, the better precision one will get. It also depends on what kinds of data and what methods to use. Only those data related to the characteristics of disease genes can help. It will be also useful if weights could be assigned to different data so that irrelevant or very noisy data sources have lower weights.

Studies have shown that diseases sharing similar symptoms (phenotypes) are often caused by abnormalities in functionally related genes (genotypes) (Oti et al., 2008). The idea of integrating similarity-based disease network and functional-relatedness-based genetic network has shown great promise in this emerging field (Lage et al., 2007; Wu et al., 2008). However, most studies are limited to the available PPI data. Expanding the current PPI data may help to improve the power and accuracy in detecting disease genes. The ongoing large-scale mapping of human interaction networks (Aloy, 2007) and systematic collection of human phenotypic data (Freimer & Sabatti, 2003) represent the efforts for increasing coverage and quality of human interaction network, as well as for expanding more standardized and objective phenotype descriptions.

To identify what functional characteristics are significantly over-represented in a list of genes, the list of the control genes is often needed. Due to the nature of the data, most lists of candidate genes are derived from a set of regions detected by linkage or association studies, or by copy number variation analysis. It was observed that genes in close genomic neighborhood may have similar expressions and functional annotations (Su et al. 2002). Thus appropriate control datasets need to be chosen to correct for such bias. In our study of identifying pathways and networks using copy number variable regions in autism patients, two types of control sets were used for an unbiased comparison: (1) randomly permuted CNV regions across human genome; (2) naturally occurring human CNVs in normal individuals extracted from Database of Genomic Variants (DGV, http://projects.tcag.ca/variation/). Using these two types of control sets, an increased connectivity among brain-expressed genes from autism CNVs but not from "normal" CNVs or random genomic regions was observed (manuscript in preparation).

Traditional linkage or association studies have limited power in detecting disease genes for genetically heterogeneous diseases due to the need of unrealistically large sample sizes. However, they still could be directly used for disease gene prioritization if appropriate methods are applied. A recent study combined standard genetic linkage analysis with molecular interaction data to predict networks or pathways that contribute to the common diseases (Iossifov et al. 2008). Instead of looking at individual intervals, the authors tested gene sets that encode interacting molecules and assigned gene-specific significance score. Although the developed model is limited by its simplified assumptions, this new thinking can help to further explore genetic studies for identifying disease genes and pathways.

The most likely new disease candidates are those that are similar to the known disease causative genes. The similarity is examined not just at the sequence level, but at a more broad functional level. Despite the marked cytogenetic and phenotypic heterogeneity of the patients in some diseases, the same specific biological processes and molecular functions might be compromised in many of them. At the level of biological processes and pathways, the pathology of many diseases is less divergent than its etiologies appear to be. The newly emerging functional approaches have already provided insight into the underlying biology of many diseases. They will continue to provide useful information about the molecular bases of diseases as technologies are improving and ever more functional data are generated in the post-genomic era.

REFERENCES

Adie, E. A., Adams, R. R., Evans, K. L., Porteous, D. J., & Pickard, B. S. (2006). SUSPECTS: Enabling fast and effective prioritization of positional candidates. *Bioinformatics, 22*(6), 773-774.

Aerts, S., Lambrechts, D., Maity, S., Van Loo, P., Coessens, B., De Smet, F., et al. (2006). Gene prioritization through genomic data fusion. *Nat Biotechnol, 24*(5), 537-544.

Aloy, P. (2007). Shaping the future of interactome networks. *Genome Biol, 8*(10), 316.

Ashburner, M., Ball, C. A., Blake, J. A., Botstein, D., Butler, H., Cherry, J. M., et al. (2000). Gene ontology: Tool for the unification of biology. The Gene Ontology Consortium. *Nat Genet, 25*(1), 25-29.

Bader, G. D., Betel, D., & Hogue, C. W. (2003). BIND: The Biomolecular Interaction Network Database. *Nucleic Acids Res, 31*(1), 248-250.

Barabasi, A. L. (2007). Network medicine--from obesity to the "diseasome". *N Engl J Med, 357*(4), 404-407.

Barash, Y., Kaplan, T., Friedman, N., & Elidan, G. (2003). *Modeling dependencies in protein-DNA binding sites.* Paper presented at the the 7th International Conference on Research in Computational Molecular Biology (RECOMB), San Diego.

Barrett, T., Troup, D. B., Wilhite, S. E., Ledoux, P., Rudnev, D., Evangelista, C., et al. (2007). NCBI GEO: Mining tens of millions of expression profiles--database and tools update. *Nucleic Acids Res, 35*(Database issue), D760-765.

Beissbarth, T., & Speed, T. P. (2004). GOstat: find statistically overrepresented Gene Ontologies within a group of genes. *Bioinformatics, 20*(9), 1464-1465.

Ben-Gal, I., Shani, A., Gohr, A., Grau, J., Arviv, S., Shmilovici, A., et al. (2005). Identification of transcription factor binding sites with variable-order Bayesian networks. *Bioinformatics, 21*(11), 2657-2666.

Birney, E., Andrews, D., Caccamo, M., Chen, Y., Clarke, L., Coates, G., et al. (2006). Ensembl 2006. *Nucl. Acids Res., 34*(suppl_1), D556-561.

Bodenreider, O. (2004). The Unified Medical Language System (UMLS): Integrating biomedical terminology. *Nucleic Acids Res, 32*(Database issue), D267-270.

Botstein, D., & Risch, N. (2003). Discovering genotypes underlying human phenotypes: past successes for mendelian disease, future approaches for complex disease. *Nat Genet, 33 Suppl*, 228-237.

Breitkreutz, B. J., Stark, C., & Tyers, M. (2002). Osprey: A network visualization system. *Genome Biol, 3*(12), PREPRINT0012.

Dalkilic, M. M., Costello, J. C., Clark, W. T., & Radivojac, P. (2008). From protein-disease associations to disease informatics. *Front Biosci, 13*, 3391-3407.

De Bie, T., Tranchevent, L. C., van Oeffelen, L. M., & Moreau, Y. (2007). Kernel-based data fusion for gene prioritization. *Bioinformatics, 23*(13), i125-132.

de Lichtenberg, U., Jensen, L. J., Brunak, S., & Bork, P. (2005). Dynamic complex formation during the yeast cell cycle. *Science, 307*(5710), 724-727.

Dennis, G., Jr., Sherman, B. T., Hosack, D. A., Yang, J., Gao, W., Lane, H. C., et al. (2003). DAVID: Database for Annotation, Visualization, and Integrated Discovery. *Genome Biol, 4*(5), P3.

Eisen, M. B., Spellman, P. T., Brown, P. O., & Botstein, D. (1998). Cluster analysis and display of genome-wide expression patterns. *Proc Natl Acad Sci U S A, 95*(25), 14863-14868.

Franke, L., van Bakel, H., Fokkens, L., de Jong, E. D., Egmont-Petersen, M., & Wijmenga, C. (2006). Reconstruction of a functional human gene network, with an application for prioritizing positional candidate genes. *Am J Hum Genet, 78*(6), 1011-1025.

Freimer, N., & Sabatti, C. (2003). The human phenome project. *Nat Genet, 34*(1), 15-21. Giot, L., Bader, J. S., Brouwer, C., Chaudhuri, A., Kuang, B., Li, Y., et al. (2003). A protein interaction map of Drosophila melanogaster. *Science, 302*(5651), 1727-1736.

Goehler, H., Lalowski, M., Stelzl, U., Waelter, S., Stroedicke, M., Worm, U., et al. (2004). A protein interaction network links GIT1, an enhancer of huntingtin aggregation, to Huntington's disease. *Mol Cell, 15*(6), 853-865.

Goh, K. I., Cusick, M. E., Valle, D., Childs, B., Vidal, M., & Barabasi, A. L. (2007). The human disease network. *Proc Natl Acad Sci U S A, 104*(21), 8685-8690.

Grossmann, S., Bauer, S., Robinson, P. N., & Vingron, M. (2007). Improved detection of overrepresentation of Gene-Ontology annotations with parent child analysis. *Bioinformatics, 23*(22), 3024-3031.

Hamosh, A., Scott, A. F., Amberger, J. S., Bocchini, C. A., & McKusick, V. A. (2005). Online Mendelian Inheritance in Man (OMIM), a knowledgebase of human genes and genetic disorders. *Nucleic Acids Res, 33*(Database issue), D514-517.

Han, J. D. (2008). Understanding biological functions through molecular networks. *Cell Res, 18*(2), 224-237.

Hermjakob, H., Montecchi-Palazzi, L., Lewington, C., Mudali, S., Kerrien, S., Orchard, S., et al. (2004). IntAct: an open source molecular interaction database. *Nucleic Acids Res, 32*(Database issue), D452-455.

Hinrichs, A. S., Karolchik, D., Baertsch, R., Barber, G. P., Bejerano, G., Clawson, H., et al. (2006). The UCSC Genome Browser Database: update 2006. *Nucl. Acids Res., 34*(suppl_1), D590-598.

Hu, Z., Mellor, J., Wu, J., Yamada, T., Holloway, D., & Delisi, C. (2005). VisANT: data-integrating visual framework for biological networks and modules. *Nucleic Acids Res, 33*(Web Server issue), W352-357.

Ito, T., Chiba, T., Ozawa, R., Yoshida, M., Hattori, M., & Sakaki, Y. (2001). A comprehensive two-hybrid analysis to explore the yeast protein interactome. *Proc Natl Acad Sci U S A, 98*(8), 4569-4574.

Jansen, R., Yu, H., Greenbaum, D., Kluger, Y., Krogan, N. J., Chung, S., et al. (2003). A Bayesian networks approach for predicting protein-protein interactions from genomic data. *Science, 302*(5644), 449-453.

Kaltenbach, L. S., Romero, E., Becklin, R. R., Chettier, R., Bell, R., Phansalkar, A., et al. (2007). Huntingtin interacting proteins are genetic modifiers of neurodegeneration. *PLoS Genet, 3*(5), e82.

Kanehisa, M., & Goto, S. (2000). KEGG: kyoto encyclopedia of genes and genomes. *Nucleic Acids Res, 28*(1), 27-30.

Kann, M. G. (2007). Protein interactions and disease: computational approaches to uncover the etiology of diseases. *Brief Bioinform, 8*(5), 333-346.

Lage, K., Karlberg, E. O., Storling, Z. M., Olason, P. I., Pedersen, A. G., Rigina, O., et al. (2007). A human phenome-interactome network of protein complexes implicated in genetic disorders. *Nat Biotechnol, 25*(3), 309-316.

Li, S., Armstrong, C. M., Bertin, N., Ge, H., Milstein, S., Boxem, M., et al. (2004). A map of the interactome network of the metazoan C. elegans. *Science, 303*(5657), 540-543.

Lim, J., Hao, T., Shaw, C., Patel, A. J., Szabo, G., Rual, J. F., et al. (2006). A protein-protein interaction network for human inherited ataxias and disorders of Purkinje cell degeneration. *Cell, 125*(4), 801-814.

Loscalzo, J., Kohane, I., & Barabasi, A. L. (2007). Human disease classification in the postgenomic era: a complex systems approach to human pathobiology. *Mol Syst Biol, 3*, 124.

Lowe, H. J., & Barnett, G. O. (1994). Understanding and using the medical subject headings (MeSH) vocabulary to perform literature searches. *Jama, 271*(14), 1103-1108.

Matys, V., Fricke, E., Geffers, R., Gossling, E., Haubrock, M., Hehl, R., et al. (2003). TRANSFAC: transcriptional regulation, from patterns to profiles. *Nucleic Acids Res, 31*(1), 374-378.

McCarroll, S. A., & Altshuler, D. M. (2007). Copy-number variation and association studies of human disease. *Nat Genet, 39*(7 Suppl), S37-42.

Miller, J. A., Oldham, M. C., & Geschwind, D. H. (2008). A systems level analysis of transcriptional changes in Alzheimer's disease and normal aging. *J Neurosci, 28*(6), 1410-1420.

Mootha, V. K., Lepage, P., Miller, K., Bunkenborg, J., Reich, M., Hjerrild, M., et al. (2003). Identification of a gene causing human cytochrome c oxidase deficiency by integrative genomics. *Proc Natl Acad Sci U S A, 100*(2), 605-610.

Mulder, N. J., Apweiler, R., Attwood, T. K., Bairoch, A., Bateman, A., Binns, D., et al. (2007). New developments in the InterPro database. *Nucleic Acids Res, 35*(Database issue), D224-228.

Oldham, M. C., Horvath, S., & Geschwind, D. H. (2006). Conservation and evolution of gene coexpression networks in human and chimpanzee brains. *Proc Natl Acad Sci U S A, 103*(47), 17973-17978.

Oti, M., & Brunner, H. G. (2007). The modular nature of genetic diseases. *Clin Genet, 71*(1), 1-11.

Oti, M., Huynen, M. A., & Brunner, H. G. (2008). Phenome connections. *Trends Genet, 24*(3), 103-106.

Perez-Iratxeta, C., Wjst, M., Bork, P., & Andrade, M. A. (2005). G2D: a tool for mining genes associated with disease. *BMC Genet, 6*, 45.

Peri, S., Navarro, J. D., Amanchy, R., Kristiansen, T. Z., Jonnalagadda, C. K., Surendranath, V., et al. (2003). Development of human protein reference database as an initial platform for approaching systems biology in humans. *Genome Res, 13*(10), 2363-2371.

Pujana, M. A., Han, J. D., Starita, L. M., Stevens, K. N., Tewari, M., Ahn, J. S., et al. (2007). Network modeling links breast cancer susceptibility and centrosome dysfunction. *Nat Genet, 39*(11), 1338-1349.

Rain, J. C., Selig, L., De Reuse, H., Battaglia, V., Reverdy, C., Simon, S., et al. (2001). The protein-protein interaction map of Helicobacter pylori. *Nature, 409*(6817), 211-215.

Rhodes, D. R., Tomlins, S. A., Varambally, S., Mahavisno, V., Barrette, T., Kalyana-Sundaram, S., et al. (2005). Probabilistic model of the human protein-protein interaction network. *Nat Biotechnol, 23*(8), 951-959.

Romero, P., Wagg, J., Green, M. L., Kaiser, D., Krummenacker, M., & Karp, P. D. (2005). Computational prediction of human metabolic pathways from the complete human genome. *Genome Biol, 6*(1), R2.

Roulet, E., Busso, S., Camargo, A. A., Simpson, A. J., Mermod, N., & Bucher, P. (2002). High-throughput SELEX SAGE method for quantitative modeling of transcription-factor binding sites. *Nat Biotechnol, 20*(8), 831-835.

Rual, J. F., Venkatesan, K., Hao, T., Hirozane-Kishikawa, T., Dricot, A., Li, N., et al. (2005). Towards a proteome-scale map of the human protein-protein interaction network. *Nature, 437*(7062), 1173-1178.

Sandelin, A., Alkema, W., Engstrom, P., Wasserman, W. W., & Lenhard, B. (2004). JASPAR: an open-access database for eukaryotic transcription factor binding profiles. *Nucleic Acids Res, 32*(Database issue), D91-94.

Sebat, J. (2007). Major changes in our DNA lead to major changes in our thinking. *Nat Genet, 39*(7 Suppl), S3-5.

Shannon, P., Markiel, A., Ozier, O., Baliga, N. S., Wang, J. T., Ramage, D., et al. (2003). Cytoscape: a software environment for integrated models of biomolecular interaction networks. *Genome Res, 13*(11), 2498-2504.

Stark, C., Breitkreutz, B. J., Reguly, T., Boucher, L., Breitkreutz, A., & Tyers, M. (2006). BioGRID: a general repository for interaction datasets. *Nucleic Acids Res, 34*(Database issue), D535-539.

Stelzl, U., Worm, U., Lalowski, M., Haenig, C., Brembeck, F. H., Goehler, H., et al. (2005). A human protein-protein interaction network: a resource for annotating the proteome. *Cell, 122*(6), 957-968.

Stuart, J. M., Segal, E., Koller, D., & Kim, S. K. (2003). A gene-coexpression network for global discovery of conserved genetic modules. *Science, 302*(5643), 249-255.

Su, A. I., Wiltshire, T., Batalov, S., Lapp, H., Ching, K. A., Block, D., et al. (2004). A gene atlas of the mouse and human protein-encoding transcriptomes. *Proc Natl Acad Sci U S A, 101*(16), 6062-6067.

Suderman, M., & Hallett, M. (2007). Tools for visually exploring biological networks. *Bioinformatics, 23*(20), 2651-2659.

Thomas, P. D., Campbell, M. J., Kejariwal, A., Mi, H., Karlak, B., Daverman, R., et al. (2003). PANTHER: a library of protein families and subfamilies indexed by function. *Genome Res, 13*(9), 2129-2141.

Uetz, P., Giot, L., Cagney, G., Mansfield, T. A., Judson, R. S., Knight, J. R., et al. (2000). A comprehensive analysis of protein-protein interactions in Saccharomyces cerevisiae. *Nature, 403*(6770), 623-627.

van Driel, M. A., Bruggeman, J., Vriend, G., Brunner, H. G., & Leunissen, J. A. (2006). A text-mining analysis of the human phenome. *Eur J Hum Genet, 14*(5), 535-542.

Vastrik, I., D'Eustachio, P., Schmidt, E., Joshi-Tope, G., Gopinath, G., Croft, D., et al. (2007). Reactome: a knowledge base of biologic pathways and processes. *Genome Biol, 8*(3), R39.

von Mering, C., Jensen, L. J., Kuhn, M., Chaffron, S., Doerks, T., Kruger, B., et al. (2007). STRING 7--recent developments in the integration and prediction of protein interactions. *Nucleic Acids Res, 35*(Database issue), D358-362.

Wu, X., Jiang, R., Zhang, M. Q., & Li, S. (2008). Network-based global inference of human disease genes. *Mol Syst Biol, 4*, 189.

Xenarios, I., Rice, D. W., Salwinski, L., Baron, M. K., Marcotte, E. M., & Eisenberg, D. (2000). DIP: the database of interacting proteins. *Nucleic Acids Res, 28*(1), 289-291.

Xia, K., Dong, D., & Han, J. D. (2006). IntNetDB v1.0: an integrated protein-protein interaction network database generated by a probabilistic model. *BMC Bioinformatics, 7*, 508.

Yeo, G., & Burge, C. B. (2004). Maximum entropy modeling of short sequence motifs with applications to RNA splicing signals. *J Comput Biol, 11*(2-3), 377-394.

Zanzoni, A., Montecchi-Palazzi, L., Quondam, M., Ausiello, G., Helmer-Citterich, M., & Cesareni, G. (2002). MINT: a Molecular INTeraction database. *FEBS Lett, 513*(1), 135-140.

Zhang, B., & Horvath, S. (2005). A general framework for weighted gene co-expression network analysis. *Stat Appl Genet Mol Biol, 4*(17).

Zhang, B., Schmoyer, D., Kirov, S., & Snoddy, J. (2004). GOTree Machine (GOTM): a web-based platform for interpreting sets of interesting genes using Gene Ontology hierarchies. *BMC Bioinformatics, 5*(16).

Zhang, C., Zhao, X. & Zhang, M.Q. (2007). Functional in silico analysis of gene regulatory polymorphism. In M. R. Barnes (Ed.), *Bioinformatics for Geneticists: A Bioinformatics Primer for the Analysis of Genetic Data* (2nd ed., pp. 281-310). England: Wiley.

Zhang, M. Q., & Marr, T. G. (1993). A weight array method for splicing signal analysis. *Comput Appl Biosci, 9*(5), 499-509.

Zhao, X., Huang, H., & Speed, T. P. (2005). Finding short DNA motifs using permuted Markov models. *J Comput Biol, 12*(6), 894-906.

Chapter XV
Dynamics of Protein–Protein Interaction Network in Plasmodium Falciparum

Smita Mohanty
Indian Institute of Science, India

Shashi Bhushan Pandit
Georgia Institute of Technology, USA

Narayanaswamy Srinivasan
Indian Institute of Science, India

ABSTRACT

Integration of organism-wide protein interactome data with information on expression of genes, cellular localization of proteins and their functions has proved extremely useful in developing biologically intuitive interaction networks. This chapter highlights the dynamics in protein interaction network across different stages in the lifecycle of Plasmodium falciparum, a malarial parasite, and the implication of the network dynamics in different physiological processes. The main focus of the chapter is the integration of information on experimentally derived interactions of P.falciparum proteins with expression data and analysis of the implications of interactions in different cellular processes. Extensive analysis has been made to quantify the interaction dynamics across various stages, as well as correlating it with the dynamics of the cellular pathways involving the interacting proteins. The authors' analysis demonstrates the power of strategic integration of genome-wide datasets in extracting information on dynamics of biological pathways and processes.

INTRODUCTION

A living cell comprises of several protein assemblies such as DNA polymerases, spliceosomes and scaffolding protein complexes involved in signaling pathways. These machineries can be viewed as multi-protein complexes (Spirin & Mirny, 2003) comprising of interacting proteins as it is illustrated in the example of RNA polymerase which is responsible for mRNA synthesis. The core complex of RNA polymerase is made up of five subunits namely, two α, β, β' and ω subunits with σ subunit included in one of the functional forms. Together all these subunits interact to give rise to an active RNA polymerase. Presence of each of these subunits allows the polymerase to recognize a specific promoter, initiate transcription, polymerize RNA on a DNA template, continue polymerization till the end of a gene and finally terminate polymerization. Thus evolution of large multi-protein complexes enables a system to carry out multiple processes with greater fidelity in a highly regulated manner. Apart from the existence of such multi-protein assemblies transient protein-protein interactions are also very commonly seen during biological processes. Thus, a cellular event is most often an outcome of several protein molecules interacting in an ordered manner (Alberts, 1998). Given the significance and widespread occurrence of protein-protein interactions in cellular processes it is natural to expect a large proportion of proteome of an organism is involved in mediating protein-protein interactions and carry out different biological processes in an efficient and regulated manner (Legrain, Wojcik & Gauthier, 2001). Apart from using protein-protein interactions to study the complexity of an organism it can also be immensely helpful in identifying suitable drug targets in pathogens based on the importance of the interactions for the survival of the pathogen. Abrogation of critical interactions could lead to successful elimination of the pathogen. In addition to deciphering interactions existing within an organism, protein-protein interactions are also now being identified between pathogens and their hosts (Davis et al., 2007; Krishnadev & Srinivasan, 2008) since the process of invasion of a host cell by pathogen depends on series of protein-protein interactions between the host and the pathogen.

In the recent past several experimental strategies have been developed to study protein-protein interactions (Shoemaker & Panchenko, 2007a). Identification of protein interactions can be performed either at a small scale leading to identification of a few protein-protein complexes or at a large scale identifying interactions at genomic scale. Techniques such as yeast-two-hybrid (Y2H) (Fields & Song, 1989; Golemis, Serebriiskii & Law, 1999) aim towards building the whole organism protein-protein interaction map, synonymously referred as "interactome" (Schwikowski, Uetz, & Field, 2000). Such interactomes have been constructed for many model organisms such as *C.elegans* (Li et al., 2004) and *D.melanogaster* (Giot et al., 2003). Apart from Y2H there are other experimental approaches such as affinity purification and protein chips (Zhu et al., 2001) to obtain high throughput protein interaction data. Although each of these methods has many advantages and shortcomings, Y2H is one of the popular means for obtaining interaction data at genomic level. The Y2H approach is simple whereby ability for two proteins to interact is checked *in vivo* in the yeast system by coupling protein interaction with the expression of a reporter gene. Thus the approach allows for identifying interactions in a manner they would occur in a cell. However Y2H method itself has several limitations. For example, proteins initiating transcription by themselves cannot be targeted in Y2H experiments; and the use of fused sequence products can impose difficulties as fusion can change the structure of a target protein (Shoemaker & Panchenko, 2007a). In addition, protein folding and posttranslational modifications can differ between yeast and other organisms. This makes it difficult to screen proteins from mammalian and prokaryotic cells using Y2H and it is difficult to employ Y2H for cytoplasmic and membrane proteins. Apart from inherent inability to detect all the interactions, not all the interactions reported by Y2H may occur *in vivo*. Such false positive interactions may exist due to indirect interactions between proteins which are spatially proximal but physically not interacting. Thus extensive efforts have been made to reduce false positives and increase the coverage of the interactions detected by Y2H (Bader et al., 2004; Deane et al., 2002; Wojcik, Boneca & Legrain, 2002).

Experimental approaches to identify interactions are difficult and time consuming and hence have not been successfully adopted in obtaining interaction maps for many organisms of interest. Till date no interactome map exists for a large majority of organisms including important pathogens like *M.tuberculosis*. Thus there is a need for computational exploration of protein – protein interactions which in many cases are of reasonable accuracy (Schachter, 2002; Gomez & Rzhetsky, 2002). Interaction maps have been derived computationally for various organisms based on the homology of a pair of proteins in the target organism with an interacting pair of proteins in a template organism (Interologue mapping) (Itzhaki et al., 2006). The approach can work well in the case of two closely related organisms since the physiologies of such organisms are similar and hence the protein-protein interaction patterns would be similar (Mika & Rost, 2006). In the case of distantly related organisms, additional filtering is necessary to assess the biological feasibility of proposed interactions. Many groups have successfully made use of integrated databases for computational detection of protein-protein interactions (Lee et al., 2006;

Rhodes et al., 2005). The criteria used are usually correlated expression, similar function and co-localization. Use of different statistical tools to combine results from various databases leads to robust prediction of protein-protein interactions in an organism.

With the increase in information on interactions from both experimental studies and computational predictions it is eminent to device tools in-order to study these interactions to reveal their biological relevance. Visualization of biological process as networks or graphs is one such method commonly employed in different systems like genome scale protein-protein interaction analysis (La Count et al, 2005; Schwikowski et al., 2000), gene transcription regulation etc (Luscombe et al, 2004; Bansal et al., 2007). Graph theory provides a platform for methodical analyses of such networks (Albert & Baraba'si 2002). Here the interacting proteins or genes are viewed as nodes and the interaction between them as edges. A protein-protein interaction network is typically considered as an undirected graph with weighted edges. The edges can be weighted using appropriate confidence measures such as conservation of an interacting pair across other genomes, reproducibility of the interaction from other experimental approaches, support from other non-interaction data such as co-expression, co-localization and functional similarity. Subsequent to the generation of a graph there are several features of a graph that can be estimated. Most of these features are dependent on the density of interactions/connections as well as their distribution in the graph. Table 1 enlists some of these features and their implication in protein-protein interaction network analysis.

In this chapter, we have analyzed the protein–protein interactions occurring in an important human pathogen, *Plasmodium falciparum* which is the causative agent of malaria. The aim of the chapter is to integrate data sets from different biological sources along with protein-protein interactions to derive meaningful and biologically feasible hypothesis regarding the physiology of this organism.

Malaria is one of the most serious infectious diseases, cause by *Plasmodium sp.*, responsible for millions of death annually. *Plasmodium falciparum* is responsible for the most lethal form of the malaria and contributes maximally to the mortality rate (Breman, Egan & Keusch, 2001). The development of resistance to antimalarials and inherent complexity of parasite biology necessitates an urgent need to understand better the biological roles of *P. falciparum* genes in order to develop new chemotherapeutic and vaccine strategies (Miller et al., 2002). The life cycle of *P. falciparum* involves a complex process involving two hosts, a vertebrate host (the human) and invertebrate host (the female *Anopheles* mosquito). The parasite spends most of its asexual phase (ring, trophozoites, schizont, and merozoite stages) and a part of sexual life cycle (gametocytes stage) in the human host and it completes rest of the sexual life cycle in the mosquito.

Y2H based interactome information is available for *Plasmodium falciparum* (LaCount et al., 2005) and it gives us an overview of the complexity existing at the molecular level of the parasite. However protein interaction data alone is not enough and it is essential to abstract information about the different cellular process in which these interactions occur, in order to be able to understand the effect an interaction has on the organism. Apart from integration of the interaction data with appropriate cellular process, added level of complexity arises as a result of the multiple morphologically distinct stages in the life cycle of Plasmodium. As the parasite progresses through different stages in its life cycle there is concomitant change in expression of genes corresponding to various proteins. Such variations in expression of genes will also alter the interaction pattern obtained for the gene products as the parasite progresses through different stages. Thus a dynamic interactome built from integration of expression information and interaction data along with the information about the cellular processes in which these interactions takes place can provide a powerful tool to detect interacting modules active in different stages and their implication on the survival of the parasite.

In the following section we discuss in details the approach used to obtain a dynamic interaction map for *Plasmodium falciparum* and use this interactome further to explore various properties exhibited by the network at the global and modular level.

METHODOLOGY AND DATA INTEGRATION

In order to visualize the dynamics of the interaction network across the various stages of the 48 hour life cycle of the parasite within the human host, the interaction dataset has been integrated with microarray information from four independent sources (Ben Mamoun et al., 2001; DeRisi, Iyer & Brown, 1997; Rhian et al., 2000; Roch et al., 2004) along with mass spectrometry information from two different sources (Florens et al., 2002; Lasonder et al., 2002). All of these together reflect the level of expression of various genes in the seven different stages of *Plasmodium falciparum* namely: Gaemetocyte, Gamete, Sporozoite, Merozoite, Ring, Trophozoite and Schizoint stage. Each of these stages exemplifies different events that take place in the invasion and pathogenesis process of the parasite. Thus each of the stage requires a completely different inventory of cellular process, which in turn leads to differential expression of genes depending on their requirement in a particular stage. Use of all the relevant information allows one to view pathways as snapshots of a dynamical process where the links between proteins change depending on the regulation of the gene.

The processes occurring at every stage in the life cycle of the parasite are mapped on the interaction map. The pathway information has been obtained from different sources such as KEGG (Kanchisa et al., 2004) and Plasmocyc (Caspi et al., 2008). Out of a total of 576 proteins known to participate in 85 physiological processes 138 are involved in mediating protein interactions. These 138 proteins encompass 48 different physiological processes.

Table 2 enlists pathways where more than 40% of proteins are involved in mediating protein interactions and the numbers of interactions made across different stages in the life cycle of the parasite. Apart from altering expression of genes within a pathway, altered expression of the interacting partners may also aid in the regulation process in a stage specific manner depending on the interacting partners.

In-order to traverse from one node to other through the interaction network the BFS algorithm (Breadth First Search) available in MATLAB was used. This enabled us to determine the shortest paths between any two nodes as well as disjoint sub-graphs (clusters) within the interaction network. Different features of the graph were calculated using stage specific interaction networks to study their variations in a stage specific manner.

To estimate the prevalence of a pathway in a stage, feasibility index was calculated. This index depends on the number of genes actively expressed in a pathway. Also the index incorporates weight to individual genes such that a pathway is more feasible if it progresses through majority of the steps in it. If in a pathway a particular gene is not expressed then the weight is readjusted to 1 at that step, so that the pathway has low overall feasibility. Also the weighting scheme is sensitive to the position of the steps in a pathway, such that higher weights are given to steps that occur later in the pathway.

$$F.I = \sum_{i=1}^{n} 2^i C_i / \sum_{i=1}^{n} 2^i \qquad (1)$$

$$F.I = \left[\sum_{i=1}^{j-1} 2^i * C_i + 2^0 + \sum_{j+1=1}^{n} 2^{(j+1)} * C_i \right] / \sum_{i=1}^{n} 2^i \qquad (2)$$

F.I= Feasibility index, n= no of steps in a pathway, i=ith step of the pathway, C_i= 0 for gene expressed and 1 for not expressed, j steps of pathway not identified

Equations 1 and 2 summarize the feasibility index calculation with equation 2 incorporating the reset of the value of C_i at steps (jth) where the protein involved is found not expressed. The feasibility index values range from 0 to 1, with 1 characterizing a fully functional pathway. This can be further exemplified by considering the example of Phosphatidylinositol (PI) signaling pathway. This pathway comprises of 5 steps mediated by 7 proteins with step 2 progressing in 3 different directions resulting in the involvement of 7 proteins. All the proteins acting in the considered pathway are annotated in *Plasmodium falciparum*. However for one of the proteins involved in the second step of the pathway, MAL1P2.32, there is no evidence for expression. Hence in the calculation of feasibility indices for 3 directions of this pathway the value of i at step 2 is reset to zero. Therefore the value of the denominator and 2^i (indicated within brackets) calculated at each step are as follows: - 2(1), 1(0), 2(1), 2(1), 4(2), 8(3) and 16(4). Similarly in the Merozoite stage PF10_0132 (step 2) shows no expression [2(1), 1(0), 1(0), 2(1), 4(2), 8(3), 16(4)]. Hence there is a decrease in the value of feasibility index. If any of the proteins involved in the latter steps of the pathway are absent then the reduction in the value of the feasibility index would be greater. For example if step 4 showed lack of expression in any of the stages then the feasibility index value would have been 0.28 [2(1), 4(2), 4(2), 4(2), 8(3), 1(0), 2(1)], assuming in all other steps all the proteins involved are expressed.

ANALYZING GLOBAL PROPERTIES EMANATING FROM INTERACTION MAP

In-Depth Analysis of Interaction Dynamics during the Course of Parasite's Lifecycle

In order to study the variations in the interaction network, as it progress through the different stages of the lifecycle of the parasite certain gross level parameters describing the interaction network were estimated (Table 1). Most of these parameters highlight the density and distribution of connectivity within the network. Thus estimation of these parameters across the 7 stages aids in capturing the dynamic behavior of the interaction graph. The parameters estimated are as follows:

1. Shortest path between any two nodes
2. Clustering coefficient
3. Cluster size and function
4. Distribution of Degree of connectivity of nodes
5. Hub identification and their dynamics

The parameters *viz,* shortest path and clustering coefficient aid in understanding the extent of connectivity of networks as well as its modularity. Table 3 enlists the values for average shortest path for the 7 stages as well as the average shortest path obtained for the co-expression dataset and the full core dataset. Co-expression dataset comprises of all the interactions co-expressed in at least one stage and the full core dataset comprises of all the interactions. As we progress through the various stages we observe difference in the average shortest path value depending on the variation in the underlying connectivity. Figure 1 depicts the frequency distribution of path length for different stages/network. This is further corroborated by the degree distribution of nodes obtained across all the seven stages along with the co-expressed dataset (Figure 2). Degree distribution also varies across stages; Gamete stage comprises of proteins with low connectivity (maximum value=6), while the Schizoint stage comprises of nodes forming hubs with connectivity as high as 29. Other stages also comprise of nodes with high connectivity like in the Ring stage 10% of the nodes have connectivity varying from 4-16. In the Gametocyte stage 11% of nodes show connectivity between 4 and 16. Majority of connection for any of the stage lies between 1 and 2 as is evident from Figure 2. Many proteins with high connectivity act as key regulators and thus mediated direct or indirect interaction amongst different process (A detailed discussion to follow in the later section). Large differences in the frequency of nodes with high connectivity across stages indicate active regulation of these proteins as well as the interactions mediated by them leading to fluctuation in the number of interactions (connection) made by them across different stages.

Upon comparing the clustering coefficient values (Table 3) it is seen that the Schizoint stage shows relatively higher clustering than the Gametocyte stage followed by the other stages which show even lower clustering coefficients. This is also supported by the presence of nodes with higher degree of connection in the Schizoint and Gametocyte stages. Gamete stage shows a very poor coverage of the interactions as is depicted by an average clustering coefficient of almost zero.

Table 3 lists the number of disjoint clusters obtained for each of the stages and their average sizes (BFS algorithm). Both the number of clusters and cluster size varies across stages depending on the presence and absence of the interacting members (Figure 3) for e.g. largest cluster obtained from the co-expressed dataset could be broken into 4 non-disjoint clusters in the Tropohozoite stage (Figure 3e). Such comparisons aid us in understanding how the interaction graph is evolving as we progress from one stage to the other. Despite the overall variation in sizes of the clusters across stages, most of the clusters have identical functional composition across stages. The largest disjoint cluster of the full dataset obtained (Figure 3) encompass 33 different processes as described in

Table 1. Different features of graph analyzed during network analysis

Feature	Definition	Description	Implications for protein interaction network
Degree	k_i - Number of nodes connected to node i	Degree varies from node to node for a non-random graph	Number of interactions mediated by proteins occur in a non-random manner and thus vary from one protein to the other.
Degree distribution	Frequency distribution of the degrees obtained for the nodes	Typically this is non-random and follows a power law distribution	Few proteins show above average number of interactions (Hubs). Form connectors between different cellular modules
Clustering coefficient	$C_i = 2 \times (n_i)/(k_i - 1)k_i$ C_i - clustering coefficient of node i. n_i – interaction observed between the nodes connected to i k_i - Number of nodes connected to i	Depicts the modularity of a graph. Value ranges from 0 to 1 corresponding to completely isolated and fully connected cliques. Value obtained is higher for a non-random graph.	Highly connected modules are observed amongst functionally related or functionally similar proteins.
Shortest path	Shortest path between any two nodes	For a non-random graph the value obtained is lower than a random graph with same number of connections. For nodes with no connecting path a value of infinity is obtained.	Depends on the density of connectivity in a network. Interaction networks with higher connectivity show relatively short path. Plausible modes of regulation between proteins through the shortest path as well as a protein-protein interaction mediated mode of relating different cellular processes.
Number of clusters	Number of independent sub-graphs (disjoint clusters) obtained from a large network.	Sparse networks containing nodes with infinite path show higher probability of collapsing into sub-graphs.	Presence of such interaction islands depict closely knit process which may have direct or indirect relations amongst them mediated by protein interactions.
Compactness of clusters	Extent of connectivity within a cluster	A given cluster may have extensive interactions within the cluster or the interactions could be sparse.	Highly compact clusters are functionally homogenous.

Table 2. List of pathways with at least 40% of the constituent proteins participating in mediating protein interactions. Column 2-4 contain the total number of proteins present in a pathway, number of proteins involved in mediating interactions and total number of interactions made respectively. Column 5 onwards lists the number of interactions in different stages where both interacting partners are expressed. Stages where no co-expression amongst interacting proteins is found are left blank. $^+$ Ga- Gametocyte, G-Gamete, Sp-Sporozoite, M-Merozoite, R-Ring, T-Trophozoite, S-Schizont

Pathway	Total number of proteins	Number of proteins involved in interactions	Total number of interactions	Ga$^+$	G$^+$	Sp$^+$	M$^+$	R$^+$	T$^+$	S$^+$
Purine metabolism	25	17	63	45	1	23	33	33	39	42
Pentose phosphate pathway	12	5	9	6	1	3	5	6	5	6
Peptidoglycan biosynthesis	1	1	1	1	1	1	1	1	1	1
Starch and sucrose metabolism	5	5	8	6		3	6	6	6	7
Galactose metabolism	5	2	2	1	1					
Phosphatidylinositol signaling system	9	5	31	12	3	8	8	10	12	13
Selenoamino acid metabolism	5	2	14	14	5	11	13	13	14	14
Ribosome	71	34	104	74	22	54	83	76	89	84
Oxidative phosphorylation	12	5	5	4		3	3	2	5	5
Inositol phosphate metabolism	6	4	19	3		1	1	2	2	2
RNA polymerase	1	1	3	3		1	2	2	2	2
Fructose and mannose metabolism	10	4	7	6	1	4	4	5	5	5
Carbon fixation	11	5	16	8		5	6	7	7	7
Glycolysis / Gluconeogenesis	18	9	38	24	9	18	21	21	24	26
SNARE interactions in vesicular transport	10	4	10	7		2				3
Fatty acid metabolism	3	3	5	3		3	4	2	3	4
Folate biosynthesis	7	4	7	7		4	6	6	7	7
Ubiquitin mediated proteolysis	18	8	60	46	4	40	42	40	47	
Glutamate metabolism	15	7	33	22	2	18	25	20	24	48
Phenylpropanoid biosynthesis	2	1	1	1		1	1	1	1	24

Table 3. Various parameters estimated for the interaction network and their variation across the different stages. Average path length is the average distance obtained between any two nodes in a network.

Parameter	STAGES							Co-exp+ dataset	Core dataset*
	Gamete	Gaemetocyte	Sporozoite	Merozoite	Trophozoite	Ring	Schizoint		
Average path length	0.2757	3.101	1.587	3.49	3	2.74	3.22	3.35	3.48
Average clustering coefficient	0	0.00415	0.0039	0.00243	0.004	0.0026	0.00438	0.0035	0.021
No. of clusters	9	11	12	9	10	14	13	8	2
Average cluster size	6.7	27.7	16.1	31.11	31.1	19.14	26.58	44.87	635
Average compactness of cluster	0.445	0.384	0.438	0.379	0.3	0.451	0.439	0.323	0.335

+ *Dataset of interactions where both proteins are co-expressed in at least one stage*

* *Dataset of all known interactions*

Figure 1. Frequency distribution of shortest path length between any two nodes and their variations across the seven stages. Columns represent the interactions from different stages (1: Gamete, 2: Gametocyte, 3 Sporozoite, 4: Merozoite, 5: Trophozoite, 6: Ring, 7: Schizoint, 8: Coexpressed datset, 9: Full interactions), rows indicate the path length and the color gradation is based on the frequency occurrence. Large portions of the matrix appear blue due to very low frequencies observed, especially in case of the Gamete stage very few interactions are observed. Most of the nodes have path lengths between 2-8.

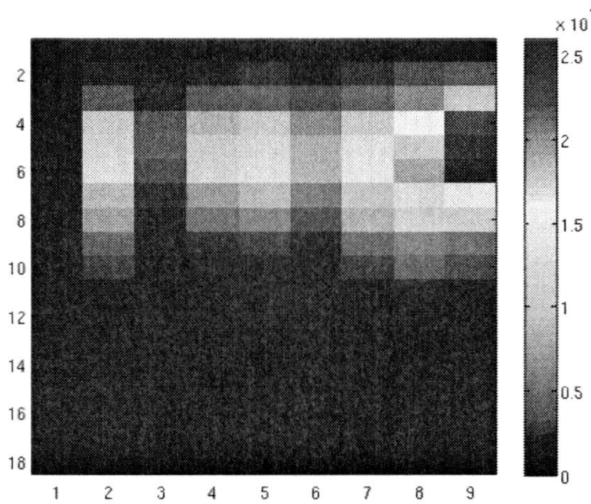

Figure 2. Degree distribution matrix depicting the frequency of distribution of degree of a node i.e the number of interactions made by a node with respect to the various stages (columns 1: Gamete, 2: Gametocyte, 3 Sporozoite, 4: Merozoite, 5: Trophozoite, 6: Ring, 7: Schizoint,). Rows indicate the different degrees, most nodes show very few interactions there are few proteins interacting with large number of other proteins.

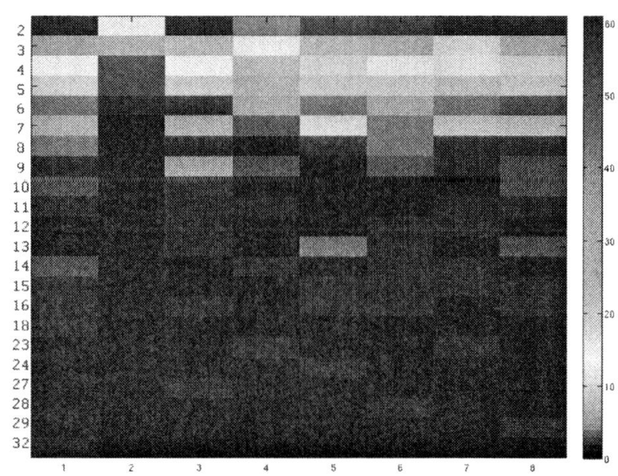

the KEGG database, with Glycolysis (KEGG code 00010), Purine metabolism (00230), Ribosomal (03010) and Proteosomal (03050) assembly containing a large number of proteins mediating many interactions. However, few interactions could be observed in the Gamete stage. Both Sporozoite and Gamete stage show relatively fewer interactions (indicated by the dark blue regions in Figure 4).

Many of the proteins in the interaction network show high connectivity also known as hubs as enlisted in Table 4. Several of these proteins perform regulatory roles and are connected to different processes undergoing

Figure 3. The largest disjoint cluster obtained from interaction network comprising of all those interactions which are co-expressed in at least one stage. The light grey circles indicate genes not expressed in a particular stage and dark highlighted circles are genes being expressed in a stage. As we progress from one stage to other one observes variations in the expression of genes, leading to formation of different disjoint clusters(encircled) and thus affecting the overall properties of the network.

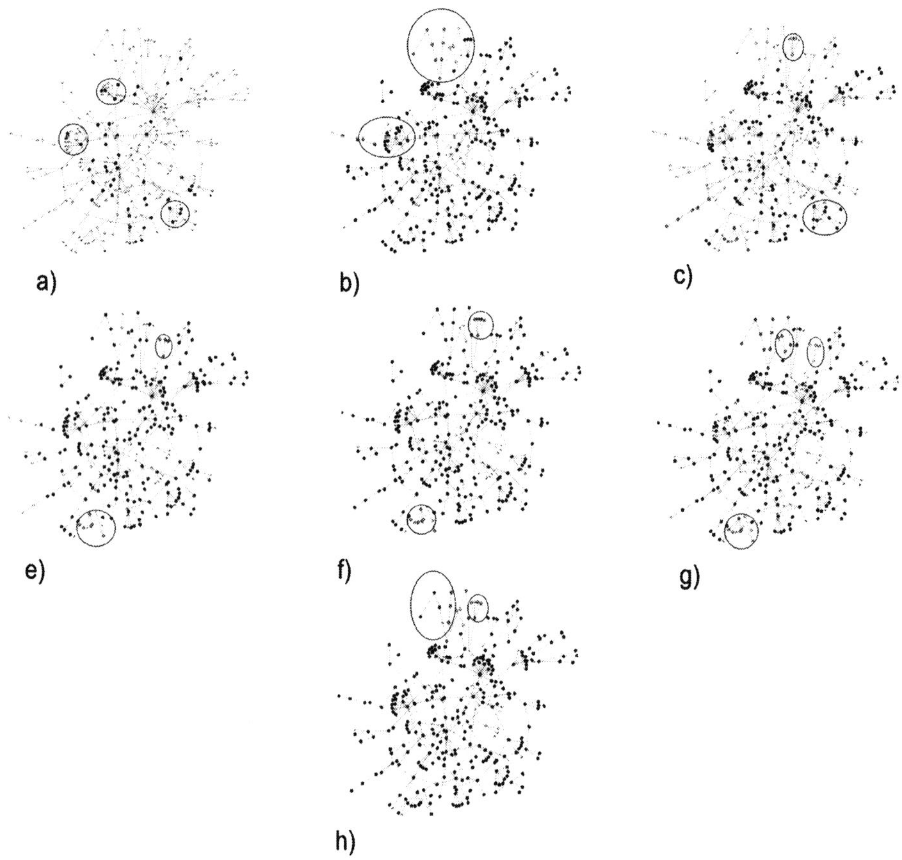

Figure 4. Matrix showing the distribution of different functional classes (33) observed in the largest independent cluster obtained. There are few pathways which contain significant number of interacting proteins as highlighted.

regulation, For example PF14_0323 is a Calmodulin which is a Ca^{2+} sensor (Bahl et al., 2003) that regulates different pathways depending on Ca^{2+} concentration. The parasite protein PF11_0053 (PfSNF2L) (Bahl et al., 2003) contains SNF2 domain involved in a variety of transcription regulatory processes (e.g., SNF2, STH1, brahma, MOT1), DNA repair (e.g., ERCC6, RAD16, RAD5), DNA recombination among other functions. Apart from proteins involved in generic cellular functions involving interactions with several proteins such as regulation, transcription and proteosomal degradation there are also examples of enzymes catalyzing specific reactions, which also tend to show a large connectivity. For example, GAPDH, a glycolytic enzyme has been shown to be involved in many different cellular processes aside from glycolysis and thus mediates interaction with many different proteins (Sirover, 1999; Kim & Dang, 2005).

Understanding and Classifying Interactions Underlying the Integrated Interaction Network

Interaction network when coupled with biological processes can be grouped into three major categories:

Table 4. P. falciparum proteins showing large number of interactions in Y2H interaction map subsequent to integration with pathway information such that only those interactions were considered where one of the interacting proteins is involved in a pathway.

S.No	Protein	Number of interacting proteins	Function	Number of interactions at various stages						
				Ga^+	G^+	Sp^+	M^+	T^+	R^+	S^+
1	PF14_0323	10	Calmodulin	3	9	7	7	8	6	9
2	PF14_0598	11	glyceraldehyde-3-phosphate dehydrogenase	5	9	7	7	9	7	10
3	PF10_0111	15	20S proteasome beta subunit putative	4	11	9	13	13	9	13
4	PF10_0155	10	Enolase	6	3	6	7	8	7	8
5	PFD0665c	16	26s proteasome aaa-ATPase subunit Rpt putative	5	16	14	15	15	14	15
7	PF13_0228	14	40S ribosomal subunit protein S6 putative	1	10	9	13	14	12	13
8	PF11_0053	10	PfSNF2L	9	0	4	7	8	7	9
9	PFE1345c	14	Minichromosome maintenance protein putative	2	12	6	13	12	11	11
10	MAL8P1.23	32	ubiquitin-protein ligase putative	0	27	23	24	28	23	29
11	PF08_0132	18	Glutamate dehydrogenase putative	0	13	11	16	16	15	16

1. Interactions amongst proteins acting in the same pathway.
2. Interactions amongst proteins acting in different pathways.
3. Interactions between proteins involved in different metabolic processes with apparently extraneous proteins not directly involved in the metabolic process.

In the cell these interactions have different implications. Interactions mediated amongst proteins within a pathway are usually correlated to co-expression and co-localization of proteins which is seen to increase the efficiency of the pathway. Such complexes are analogous to multifunctional enzymes formed by gene fusion. The second kind of interactions, occurring amongst proteins across different pathways promotes crosstalk between different pathways essential for cellular maintenance. The last mode of interaction involving proteins typically not participating in the metabolic process largely point toward series of regulatory proteins controlling the flux of pathways. These regulatory proteins most often interact with more than one process thus bringing relevant pathways under the control of the same physiological cue. Final category of interactions also involves structural proteins like actin acting as scaffolds for different pathways thus mimicking large multimeric protein complexes (Harris & Winzor, 1990). Table 5 enlists the distribution of interactions for *Plasmodium falciparum*.

Total number of interactions encompassed in different processes (as described in KEGG database) is 469, from a total of 2848 interactions reported in Y2H. These 469 interactions involve 449 proteins. Out of the 311 proteins are not involved in any of the above considered biological processes and 167 are hypothetical proteins. Rest of the 144 proteins are involved in many other functions such as membrane traversing, antigens, kinases, DNA binding and transcription activators. Several of these proteins may have implications in regulating the underlying biological process. Also, few proteins may have been annotated recently and thus missed in KEGG. One has to consider all these possibilities during the analysis and study them in great detail to understand the biology of the organism.

In *P. falciparum* we note that the number of interactions mediated amongst proteins within a pathway is very few (as mentioned in the previous section). Most of the interactions are mediated with proteins not involved in metabolic pathways. This sparseness of interactions within pathways could be due to various reasons. It is possible that the inherent tendency to form interactions between proteins occurring within a pathway is rare. To test this, we documented the three types of interactions (described above) in other genomes having Y2H data like Yeast (Ito et al., 2001; Schwikowski, Uetz & Fields, 2000) and *C.elegans* (Li at al., 2004), and compared it with *P. falciparum*. It is apparent from Table 6 that interactions within a pathway are few as compared to the other types of interactions for any organism. However, the number of interactions occurring between proteins within a pathway is rare in *P. falciparum* as compared to Yeast and *C.elegans*. Thus, our assumption that the inherent tendency to form interaction within a pathway is rare is not applicable for all the organisms studied. The reason for such variability could be due to one or more of the following reasons:

1. Limitation of Y2H in detecting all the interactions and thus several interactions remain undetected.
2. *P. falciparum* genome is highly diverged and hence orthologues of many of its proteins in other organisms is undetected (Aravind et al., 2003; Suthram, Sittler & Ideker, 2005). Hence no function annotation and information on functional link is available.
3. *P. falciparum* is an obligate intracellular parasite and it does not require many pathways. So, many interactions are abrogated during the course of evolution.
4. Also, some of the interactions reported in other organisms can be false positives and thus does not take place *in vivo*.

Upon comparing the number of proteins in some of the common pathways between *P. falciparum* and the other organisms (Table 7) we note examples of pathways which show massive deterioration in *P. falciparum* such as TCA cycle and oxidative phosphorylation. Since the parasite largely depends on rapid anaerobic conversion of glucose to lactic acid for energy (Roth, 1990), these pathways are not essential and hence some of the proteins might have been lost from the genome. There are also examples of pathways which are quite important for *P. falciparum* and hence, by and large, the proteins concerned are conserved and yet we note large differences in the number of interactions mediated in the organisms (for example, pyrimidine biosynthesis pathway). The life

Table 5. Distribution of Y2H data for P. falciparum. Row 1 contains all interactions as reported in Y2H while row 2 includes only those interactions where the interacting partners are co-expressed in at least one stage. Table enlists the total number of protein interactions occurring in different pathways, the number of proteins mediating them and their distribution, in-terms of the number of proteins involved in pathways directly, those involved indirectly and hypothetical proteins.

Y2H interaction data	Total number of interactions involved in different biological processes	Total number of proteins	Proteins involved in pathways	Hypothetical proteins	Other functions	Number of Pathways involved
Core data (2848)	469	449	138	167	144	48
Interactions involving co-expressed proteins (2358)	410	402	127	151	124	44

Table 6. A Comparison between interactions occurring in different pathways in Yeast and C.elegans with those found in P. falciparum. Table lists the number of interactions participating in different pathways and their sub-classification into different subtypes namely-: Intra pathway: between protein in the same pathway; Inter-pathway: between proteins from different pathways and finally interactions mediated by proteins involved in pathways with non-metabolic proteins.

Organism	Total number of proteins	Total number of interactions	Total number of Pathways	Intra pathway	Inter pathway	Interaction with non metabolic proteins
Yeast	311	422	108	552	412	3632
C.elegans	332	1014	120	446	32	536
P.falciparum	449	469	85	8	15	446

cycle and environment of an organism plays an important role in remodeling the protein interaction network of the organism. Finally, there are certain pathways (e.g., folate biosynthesis) which show nearly the same amount of interactions with few interactions missing.

In *P.falciparum*, protein interactions within a pathway giving rise to multimeric complexes show rare occurrences. Most of the interactions are important in regulating pathways. There are also interactions linking different pathways thus facilitating cross-talk between pathways. Many of the other interactions involve cytosolic or membrane proteins acting as scaffolds for pathways for stabilization and enhanced activity. These observations are consistent with the power law distribution which is observed for protein interaction network (Wuchty, 2007).

Table 7. Examples of pathways occurring in Yeast, C.elegans and P. falciparum along with the number of proteins known to be involved in a pathway for an organism. The numbers of intra-pathway interactions reported are mentioned within brackets ().

Pathway	**Yeast (interactions)**	***C.elegans* (interactions)**	***P.falciparum* (interactions)**
Citrate Cycle (TCA)	30 (2)	29 (6)	15 (0)
Oxidative phosphorylation	73(10)	103(33)	33
Pyrimidine biosynthesis	70(40)	60(8)	58(1)
Folate biosynthesis	31 (1)	16(1)	11(1)
Pyruvate metabolism	33(2)	22(5)	22(0)

The power law distribution implicates the presence of fewer highly connected hub proteins which tend to connect many different functional modules thus giving rise to more complex and tightly regulated cellular physiology.

Correlating Functioning of Pathways with the Interactions Mediated by Them

Apart from studying the nature and dynamics of protein interaction across various stages and their effect on the global properties of the network, we also studied the effect of gene expression on the overall functioning of a pathway in a stage in order to understand the underlying biological relevance of the protein interactions. Feasibility index for a few pathways (11) and the number of interactions made across the 7 stages for the same pathways are shown in Figure 5. The glycolytic pathway is an example of a pathway where significant number of interactions can be seen and this is constant in Trophozoite, gametocyte and schizoint stage but declines in gamete and sporozoite stages despite high feasibility index values in all stages. Most pathways contain few interacting proteins and all of them show few/no interactions in the gamete stage. On comparing the feasibility of pathways across stages with the number of interactions made by the pathway, no definite correlation is obtained thus most of the interactions are not necessary for the progress of the pathway.

Abstracting information from protein interaction network in such a manner has been useful in understanding some of the general principles of protein interaction network especially in the context of metabolic pathways. Apart from understanding the global properties of protein interaction network it is also relevant to highlight some of the biological pathways where protein interactions play a critical role.

ANALYSIS OF LOCAL PROPERTIES OF INTERACTION NETWORK

In the following sections we explore some of the pathways in *Plasmodium falciparum* in detail in order to investigate the effects of dynamic protein interactions on these processes and their implications on the survival and propagation of the parasite.

Glycolysis / Gluconeogenesis

Glycolysis is one of the primary modes of energy production releasing two moles of ATP per glucose molecule. It is typified by ten steps where it converts one molecule of glucose to two molecules of pyruvate and in the process releases two ATP molecules. The same pathway in reverse order with few modifications is also involved in glucose synthesis (Gluconeogenesis). In *P. falciparum*, glycolysis is a critical source for energy production. Parasites increase glucose consumption 50-100 fold as compared to uninfected red cells. Most of the glucose is metabolized to lactic acid (Roth, 1990). Energy production is ubiquitous in the parasite's lifecycle. Despite being the primary source of energy production glycolytic pathway is interesting because of the presence of several enzymes which

Figure 5. Feasibility index of pathways across different stages correlated with the interactions made by proteins acting in a pathway. X axis enlists the pathways (11), stages of expression is shown along Y-axis and the number of interactions shown along Z-axis. The color coding is based on the feasibility index values obtained for each pathway across various stages. Pathways show increase or decrease in the feasibility index values across stages depending on the expression of genes as well show variation in the number of interactions made, shown by the height. However both features are not essentially correlated for many pathways one observes poor interaction with high feasibility and vice versa.

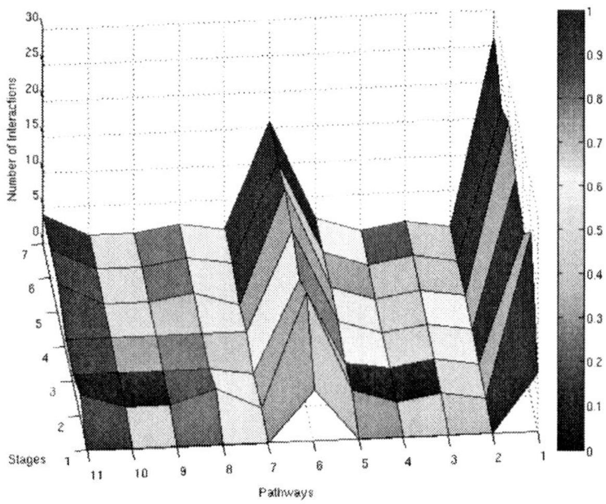

exhibit many non-glycolytic functions (Kim & Dang, 2005). Such proteins therefore form interactions with many other proteins involved in different processes (Details explained in subsequent sections).

All of the Glycolytic enzymes are present in *P.falciparum*. All except one of the proteins are involved in mediating protein interactions. Table 8 shows the integration of interaction data with microarray and mass spectrometry data and the distribution of the interactions into different functional classes.

Figure 6 represents all the glycolytic enzymes and the interactions mediated by them. Light grey circle indicate glycolytic enzymes (e.g. PFI1105w) and white indicates hypothetical proteins. This network enlists all the 9 enzymes and their neighboring nodes. All the protein interactions made have been catalogued in Table 8 depending on the functions of the proteins interacting with the various glycolytic enzymes. Using this information one can glean into the entire network and its physiological implication. Upon a closer look over the glycolysis sub-network four main features emerge

1. Inter-pathway interactions are made between enzymes within the pathway.
2. Interactions are also mediated with membrane proteins such as MSP-9 and RESA like protein PF70.
3. Interactions are mediated between pathway proteins and regulatory proteins.
4. Interactions are mediated with proteins in other pathways.

Interaction between Enolase and GAPDH

Y2H data suggests interaction between Enolase (PF10_0155) and GAPDH (PF14_0598) the 9th and 5th steps respectively of the glycolytic pathway. Both these enzymes are constitutively expressed. There is only one report till date of the two enzymes associating together on the cell wall of Lactobacillus (Antikainen et al., 2007). Both the proteins are shown to interact with human plasminogen. Enolase and GAPDH are multifunctional proteins involved in bacterium-host interactions and the relevance to potential ability of these proteins in Plasmodium's invasion of the host cell is an open question.

Figure 6. Glycolytic enzymes are shown along with the immediate interacting partners. Circles in light grey indicate glycolytic enzymes. Circles in white indicate hypothetical proteins, black circles indicate other proteins involved in various processes. This figure and all the other figures depicting protein interaction networks have been generated using Cytoscape (Shannon et al., 2003).

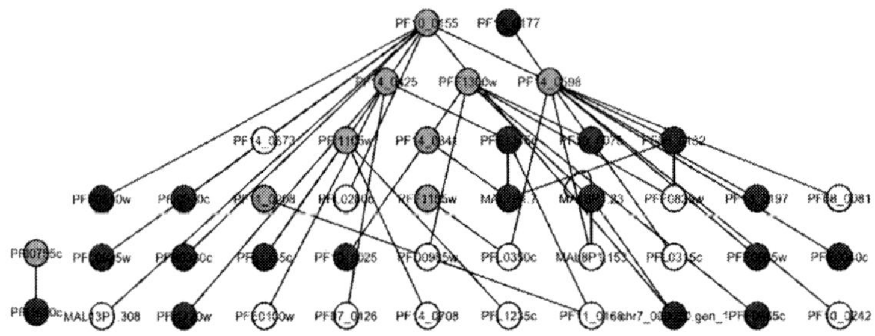

Interaction with Membrane Proteins

At least five of the glycolytic enzymes show interactions with membrane proteins such as MSP-9, PF70 and RESA-like protein in *P.falciparum* (Figure 7). However proteins such as MSP-9 and MSP-7 are surface localized and they do not have any detectable transmembrane region. Thus it is difficult to envisage such interactions between cytosolic proteins and surface proteins unless the former also localize to the membrane. Membrane localization of glycolytic enzymes has been reported in several prokaryotic and eukaryotic organisms (Barbosa et al., 2006; Gil-Navarro et al., 1997). Especially in case of GAPDH in yeast (Delgado at al., 2001) it has been shown that it is a native member of the membrane though its role is not clear. Similar reports also exist for FBP aldolase, Phosphoglycerate kinase and enolase in other organisms (Delgado et al., 2001). However none of these proteins which are secreted out have a typical signal peptide and they probably employ a novel mode of translocation not yet characterized (Delgado et al., 2001). In a few organisms, these surface proteins aid in recognition of the host and also in the adhesion process of the parasites (Barbosa et al., 2006).

Figure 8 (a-g) shows the stage specific expression of all the glycolytic enzymes and their interaction with different membrane proteins. Genes shown in dark grey are expressed. Out of the 8 glycolytic enzymes for which expression information is available 5 are ubiquitously expressed in the 7 stages. PF14_0425 is not expressed in the gametocyte stage (Figure 8a) and PF14_0341 is not expressed in the sporozoite stage (Figure 8c). Figure 8a-c depict Gametocyte, Gamete and Sporozoite stages where one notices inconsistent expression of the other interacting proteins. It is only from the early ring stage, most of the interacting partners begin to be expressed (Figure 8d-g). Thus most of the interactions between the glycolytic enzymes and the membrane protein occur in the intra-erythrocytic stages. The implication of such interactions on the pathogenecity of the organism needs further exploration.

Regulation of Pathway Proteins

6-Phospho fructose kinase (PFK), a bifunctional enzyme, is one of the checkpoints regulating the glycolytic pathway (Kurland & Pilkis, 1995). The alteration of the stability of an enzyme is one of the ways in which an enzyme can be regulated as in PFK(PFI0755c). *P. falciparum* PFK shows interaction with PFI1610c which is annotated as putative calcylin binding domain (Bahl et al., 2003). The latter contains a SGS domain which is responsible for binding to S100A6, a Ca^{2+} binding protein. Besides binding to S100A6, the SGS domain is also responsible for binding to Hsp90/80 chaperone (Zabka et al., 2008). However, both the interactions are complementary such that increased Ca^{2+} decreases the binding of the chaperone (Zabka et al., 2008). It has been seen that PFK mediates interaction with cytosolic proteins like Hsp60 for stability (Schwock et al., 2004). The parasite PFK binds to SGS domain containing protein (PFI1610c) which binds to Hsp86 belonging to Hsp90 family. Thus PFI1610c recruits in the chaperone which aids in the stability of PFK1. However yet another direction of regulation exists

Table 8. Proteins of glycolysis pathway involved in mediating protein interactions. Table enlists the stages in which the individual proteins are expressed, the number of interactions made by them and their further classification depending on the interacting partner whether it is hypothetical, membrane localized etc. Categories where no interactions were found have been left blank. + Ga- Gametocyte, G-Gamete, Sp-Sporozoite, M-Merozoite, R-Ring, T-Trophozoite, S-Schizont

S.no	Name and Function (EC number)	Expression+ Ga,G,Sp,M,R,T,S	Total number of interactions	Inter pathway interaction	Extra pathway interaction			
					Intra-pathway	Regulatory	hypothetical	Membrane Or Cytoskeletal
1.	PFF1155w (2.7.1.1)	No expression	1	0			1	
2.	PF14_0341 (5.3.1.9)	Ga, G, M, R, T, S	1	0				1
3.	PFI0755c (2.7.1.11)	Ga, G, Sp, M, R, T, S	1	0		1		
4.	PF14_0425 (4.1.2.13)	Ga, Sp, M, R, T, S	4	0	1		3	
5.	PF14_0598 (1.2.1.12)	Ga, G, Sp, M, R, T, S	11	1	3		5	2
6.	PFI1105w (2.7.2.3)	Ga, G, Sp, M, R, T, S	3	0			2	1
7.	PF11_0208(5.4.2.1)	Ga, G, Sp, M, R, T, S	1	0			1	
8.	PF10_0155 (4.2.1.11)	Ga, G, Sp, M, R, T, S	9	1	2	1	3	2
9.	PFF1300w (2.7.1.40)	Not expressed	6			1	2	3

Figure 7. Interactions between glycolytic proteins and membrane localized proteins. Black circles highlights proteins involved in glycolysis; Grey circles indicate hypothetical proteins bridging interaction between glycolytic enzymes. White circles indicate different membrane proteins.

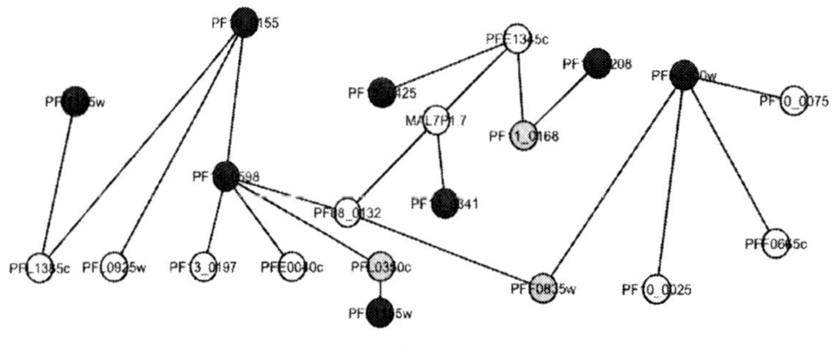

Figure 8. Stage specific expression of the glycolytic enzymes and the interacting proteins. Dark circles indicate proteins expressed in a stage. Stages a) Gametocyte, b) Gamete, c) Sporozoite, d) Merozoite, e) Ring, f) Trophozoite, g) Schizoint.

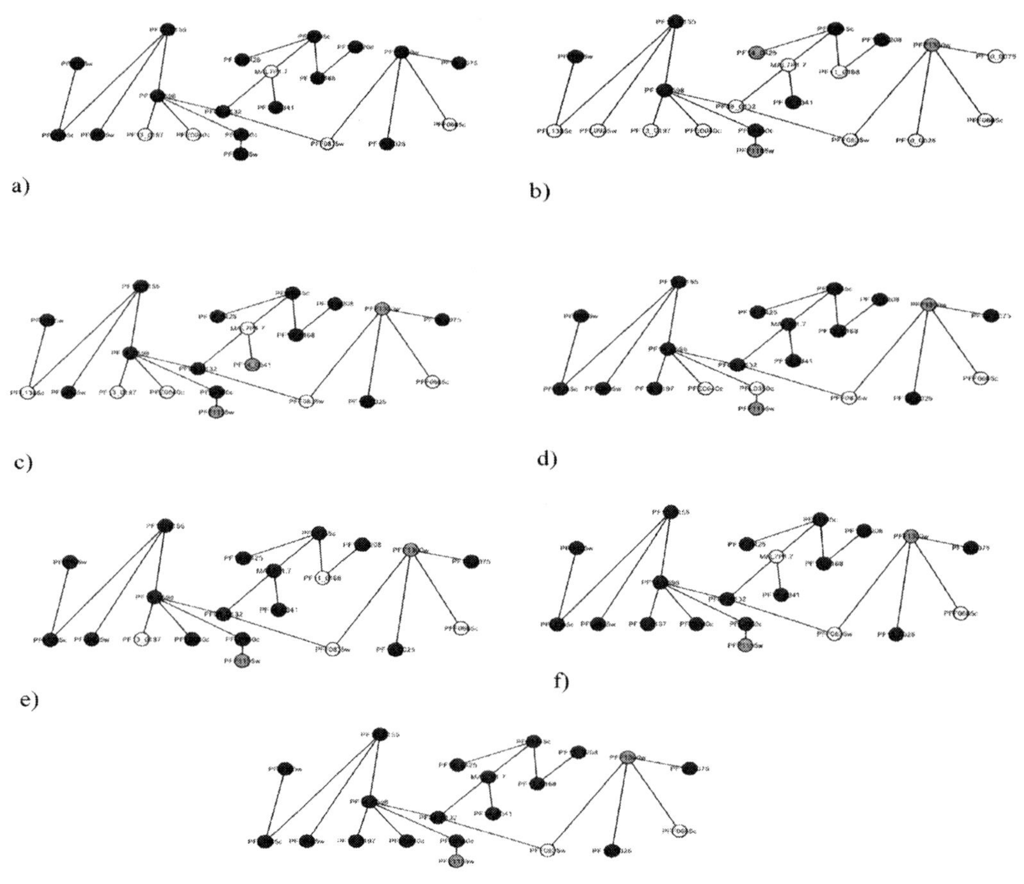

that is imparted by the ability of the SGS domain to bind to S100A6 in the presence of Ca^{2+} which in turn prevents binding of the chaperone and therefore affect the stability of PFK-1. The latter mode of regulation is similar to Ca^{2+} calmodulin dependent destabilization of PFK-1 (Lan & Steiner, 1991). In *P.falciparum* PFK and PFI1610c are reported to interact. However the latter is expressed only during gametocytogenesis and during this phase requirement of glycolysis is least. Thus PFI1610c destabilizes PFK by preventing the binding of chaperone. However in all the other stages there is no expression for PFI1610c thus there is no apparent regulation of PFK by PFI1610c.

Intra-Pathway Interactions

Glyceraldehyde -3-posphate dehydrogenase mediates interactions between glycolysis and other cellular processes. Figure 9a shows all the interactions mediated by the protein. Out of the 11 proteins with which it makes interactions, five are hypothetical proteins with no function annotation available.

This protein mediates interaction with membrane localized proteins such as MSP-7 and surface antigen PFE0040c, Both the proteins are expressed in the trophozoite and schizoint stages. MSP-7 is also expressed in the merozoite stage while PFE0040c is exclusively expressed in the ring stage (Figure 9f). The surface localization and interaction with membrane proteins might aid in the pathogenecity of the parasite.

Glycolyitc enzymes are popularly known to moonlight in many other functional contexts and one such protein is GAPDH (Kim & Dang, 2005; Sirover, 1999). Glyceraldehyde -3-phosphate dehydrogenase interacts with MCM2-DNA replication licensing factor thus linking the glycolytic process with DNA replication. By establishing such a linkage some aspects of DNA synthesis can be modulated in response to energy provided by the environment (Jannière et al., 2007; Figure 9)

Glyceraldehyde -3-phosphate dehydrogenase is a multifaceted enzyme; apart from its glycolytic function it also carries out other non-glycolytic functions as is quite clearly shown by its interaction with protein from other process like interaction with PFE0835w, a ubiquitin carboxy terminal hydrolase,

Finally, we discuss the implications of interactions mediated between GAPDH and Glutamate dehydrogenase. The latter is an enzyme involved in Glutamate metabolism and it catalyzes the conversion of α-ketoglutate (derived from TCA cycle) and ammonia to Glutamate along with release of $NAD/NADP^+$. Glutamate provides the central link between carbon and nitrogen metabolisms, between catabolism and anabolism. It has been shown during the recent years that several regulatory mechanisms in bacteria have evolved that ensure a balanced expression of genes at the interface of carbon and nitrogen metabolism (Commichau, Forchhammer, & Stulke, 2006). As such there is no literature evidence to support a direct physical interaction between these two proteins. However glutamate dehydrogenase is a reversible enzyme and when catalyzing the synthesis of glutamate it also releases NAD^+. This can later be used by GAPDH which binds with NAD^+ to catalyze the conversion of Glyceraldehyde-3 phosphate to 1,3 bisphospho glycerate. Thus the α-ketoglutate derived from TCA cycle can be utilized by glutamate dehydrogenase to release NAD^+. However this link needs to be further explored to understand its functional significance. Both proteins are co-expressed in all the 6 stages except for the gamete stage where glutamate dehydogenase is not expressed (Figure 9).

Phosphatidylinositol Signaling

Figure 10 shows the phosphatidylinositol signaling pathway. It is involved in producing poly phosphorylated inositols some of which are involved in regulating cellular processes such as actin remodeling, release of Ca^{2+} and other cellular processes. Five proteins of the pathway are involved in mediating protein interactions and details about their expression and interactions are given in Table 9.

Interactions Mediated by Calmodulin

Calmodulin is a calcium binding protein and it is responsible for calcium mediated activation of several proteins. Calmodulin activates PfPKB in the erythrocyte stages of the parasite and this is required for the erythrocyte

Figure 9. Stage specific expression of the GAPDH (PF14_0598) and its interacting partners. a) Light grey circles indicate glycolytic enzymes and black circles indicate all the hypothetical proteins. Section b-h depict stage specific expression of interacting proteins here dark circles indicate being proteins expressed in a stage Stages b) Gametocyte, c) Gamete, d) Sporozoite, e) Merozoite, f) Ring, g) Trophozoite, h) Schizoint.

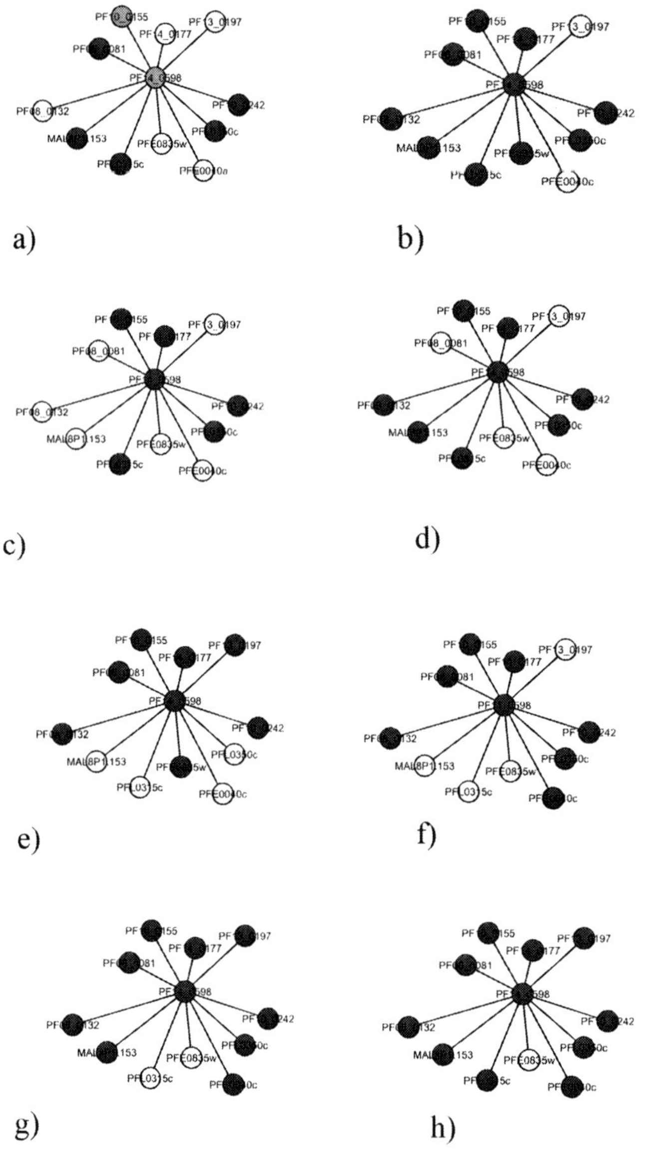

invasion process (Vaid & Sharma, 2006). Expression of Calmodulin is constitutive. Figure 11 shows the various interacting partners (hypothetical proteins are highlighted in grey). Large numbers of proteins mediating interactions are hypothetical proteins of yet unknown function. Calmodulin also mediates interaction with surface proteins like MSP-1.

Calmodulin interacts with PF08_0129, a protein phosphatase known as calceneurine. Calceneurine has an inactivation domain and a Calmodulin binding domain, binding of Calmodulin activates calceneurine, active

Table 9. Proteins involved in phophatidylinositol signaling pathway. All the proteins listed are involved in mediating protein interactions. Table enlists the total number of interactions made by the proteins and their distribution into different types depending on the interacting partner. Categories where no interactions were found have been left blank.

S.no	Name and Function(EC number)	Expression⁺ Ga,G,Sp,M,R,T,S	Total number of interactions	Extra pathway interaction			
				Intra-pathway	Regulatory	hypothetical	Membrane Or Cytoskeletal
1.	PFE0485w (2.7.1.67)	Ga, Sp, M, R, T, S	1	1			
2.	PF10_0132 (3.1.4.11)	Ga, Sp, R, T, S	1		1		
3.	PFF1370w (2.7.1.-)	Not expressed	15	4	1	8	2
4.	PFI1485c (2.7.1.107)	Ga, Sp, M, R, T, S	2				2
5.	PF14_0323	Ga, G, Sp, M, R, T, S	11	1	2	6	2

calceneurine causes dephosphorylation of target proteins such as transcription factors and thus their subsequent activation. Role of calceneurine has been established previously in *P. falciparum* (Kumar, Musiyenko & Barik, 2005; Dobson et al., 1999). Calmodulin also mediates interactions with cysteine protease PFB0340c belonging to papain family of protease. It has been documented previously that calceneurine is a substrate for cyteine proteases like calpain (Wu et al., 2004), a Ca^{2+} dependent protease. Further processing relieves calceneurine from inhibition and it is constitutively active in a Ca^{2+} independent manner. Similar observations are also made in *P. falciparum* where it has been shown that Calmodulin along with an atypical calpain are involved in activation of calceneurine (Kumar et al., 2004). The interaction data supports the binding of papain instead of calpain to calmodulin and thus one can speculate the replacement of calpain by papain in *Plasmodium falciparum* for complete activation of calceneurine. Calcenurine is constitutively expressed; however papain is activated predominantly in the intra-erythrocytic stages. Thus it is possible that prior to expression of papain, activation of calceneurine was mediated by calmodulin alone. However in the erythrocyte it is coupled with papain causing proteolytic degradation of the inhibitory regions of calceneurine.

Other proteins involved in the phosphatidylinositol signaling pathway also mediate interactions with many other proteins (Figure 11) Several of these interactions are mediated with hypothetical proteins with yet unknown function. Involvement of these hypothetical proteins in interactions with phosphatidylinositol signaling pathways provide clues about the biological context in which these hypothetical proteins work. Interestingly, according to the interaction dataset, several of these interactions are found to be unique in *Plasmodium falciparum*.

In the above sections two examples of pathways were discussed where individual members showed extensive protein interactions. There are several other examples of entire pathways e.g. Folate biosynthesis where many members within the pathway mediate interactions. It is important to explore such pathways to study the implication of protein-protein interaction in their function and regulation. It should be noted however that the interaction network comprises of several hypothetical proteins for which no function can still be ascribed and thus interaction mediated by them still need to be understood better. Also there are several interactions which are uniquely found in *P. falciparum* . Such interactions are organism specific and they should be validated by further experiments.

Figure 10. Schematic representation of the phophatidylinositol signaling pathway. PLC (phospholipase C) leads to Ca^{2+} release as well as PKC (protein kinase C) activation which in turn activate several other cellular processes. There is also production of poly phosphorylated Inositols (IP4,IP5 etc.) which in turn are thought to participate in the regulatory processes.

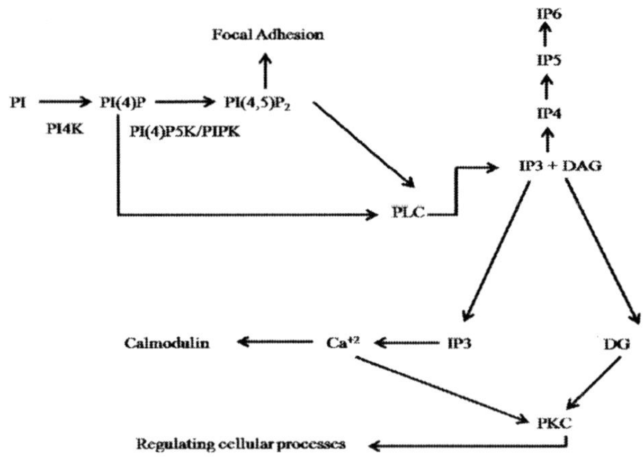

Figure 11. Interaction network of proteins involved in Phophatydilinositol signaling. Nodes in dark grey are proteins involved in signaling pathway, nodes in light grey are hypothetical proteins, while the rest are involved in other processes.

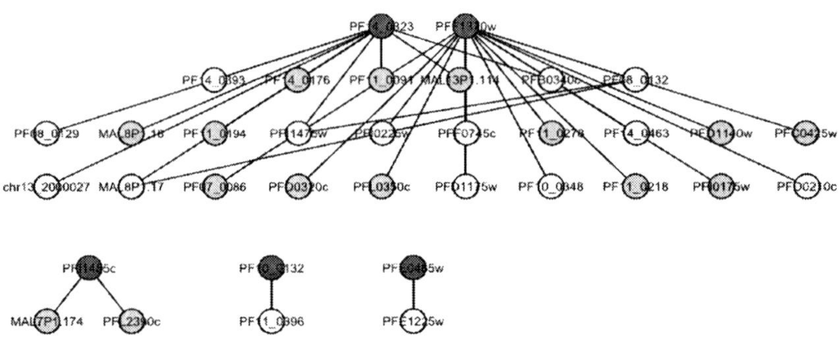

CONCLUSION

Analyzing interaction network in the light of different cellular process enables us in learning many new facets of different metabolic pathways. It allows us in learning about the different regulatory modes and how it differs in the different stages of the lifecycle of the parasite and about interactions mediated between pathways and their effect on each other. All this information pinpoints towards a closely knit meshwork of various metabolic pathways involving various regulatory and structural proteins which are critical for parasites survival and maintenance. Use of expression information available from genome scale microarray and mass spectrometry has added yet another dimension to these static networks. One learns about the changes brought about at the molecular level which may have profound implications at the cellular level. However use of such an integrated approach in understanding the cellular physiology has some limitations:

1. Regarding quality of the Y2H data, there is some concern on the exhaustiveness and accuracy of the Y2H data (Sprinzak, Sattah & Margalit, 2003). The experimental procedures have their limitations and thus many interactions are not reported and some of the reported interactions could be false positives (Shoemaker & Panchenko, 2007a). Various steps have been taken to further refine the quality of the data such as repeating experiments to check the extent of reproducibility of the interaction and confirming interactions via other *in vivo* and *in silico* approaches. *In silico* measures have been taken to validate interaction such as analysis of the extent of conservation of interactions in other organisms, level of functional similarity between interacting partners by taking aid from predicted functions for the proteins from Gene Ontology consortia, co-expression of the proteins (Ge et al., 2001; Leach et al., 2007). Use of 3-D structures to validate interactions is also an effective step. All of these measures with a series of other statistical tools weighting interactions based on different features such as co-expression and average connectivity have been used in identifying reliable interactions (Deane et al., 2002).
2. The exhaustiveness of Y2H data is questionable as explained previously. In the recent past large scale experimental data for protein-protein interaction has been obtained for many organisms such as Yeast, *C.elegans, D.melanogaster, H.pylori* and *P.falciparum*. Upon comparing the interaction networks across organisms however one notes very little overlap (von Mering et al., 2002). Such absence of interaction could mean that the organisms are quite variable and thus their cellular processes and underlying interactions also change. However it could also mean that several interactions have not yet been reported by such large scale experiments. The latter argument is further strengthened by the fact that many interactions reported within an organism through different approaches do not overlap (von Mering et al., 2002) and thus the interaction data contains many false positives and many real interactions are not recognized (false negatives). *In silico* methods have been employed to predict interactions by combining it with the experimental procedure with an ambition of generating an exhaustive inventory of protein-protein interactions (Shoemaker & Panchenko, 2007b). Several algorithms have been developed using different aspects of protein-protein interactions like interaction conservation, structural compatibility, co-evolution, phylogenetic profiling, functional similarity, genomic mapping etc. Many more sophisticated algorithms like SVM, naïve-Bayesian classifiers etc have employed which do not just rely on any one aspect of protein-protein interaction but use multiple signatures in predicting highly accurate interactions.
3. Apart from quality of interaction the other concern lies in the quality of functional annotation available for the genes within an organism. Apart from tedious and slow biochemical assays used for characterization of proteins several high-throughput approaches such as microarrays and structural genomics initiative are now being employed in order to ascertain function of proteins directly or indirectly. Computational approaches have been very useful in successful functional annotations of proteins using sensitive approaches such as PSI-BLAST, RPS-BLAST and HMMER. All these approaches establish relation between sequences with known function or protein families. Despite all these efforts there still remain a large proportion of genome for which no functional annotation is available (e.g. in *P. falciparum* approximately 50% of the genome has no functional assignment). This lack of information greatly deters network analysis where no functional link can be established between interacting proteins. However network analysis has also proven to be helpful in some cases to predict interaction for unannotated proteins based on the interacting partners and their function.
4. Need for more biological data sources, to be integrated with interaction network for a more comprehensive and detailed analysis to reveal different aspects of cellular physiology and effect of interaction on them. There is a need for biological information in terms of cellular localization, expression, 3-D structure etc which will all aid in further probing the interaction networks

Despite these short comings, use of high throughput interaction information coupled with other biological data is a very useful and is a necessary tool for bridging the relation between the molecular interactions and their dynamics to cellular process and phenotype observed at the organism level. An exhaustive network of protein–protein interaction coupled with information from other biological resources such as gene and protein expression, sub-cellular localization and function within a cell together map a working cell. Use of such tools has found immense use in different aspects of biology such as rational drug design, by allowing us in choosing appropriate targets and also aid in rational design of experiments to study different aspects of regulation.

ACKNOWLEDGMENT

This research is supported by Council of Scientific and Industrial Research, Government of India.

REFERENCES

Alberts, B. (1998). The cell as a collection of protein machines: preparing the next generation of molecular biologists. *Cell, 92*(3), 291-294.

Albert, R., & Baraba'si, A-L. (2002). Statistical mechanics of complex networks. *Reviews of Mordern Physics, 74 (1),* 47-96.

Antikainen, J., Kupannen, V., Lähteenmäk, K., & Korhonen, T. K. (2007). pH-Dependent Association of Enolase and Glyceraldehyde-3-Phosphate Dehydrogenase of Lactobacillus crispatus with the Cell Wall and Lipoteichoic Acids. *J Bacteriol., 189*(12), 4539-4543.

Aravind, L., Iyer, L. M., Wellems, T. E., & Miller, L. H. (2003). Plasmodium Biology: Genomic Gleanings. *Cell, 115*, 771-785.

Bader, J. S., Chaudhuri, A., Rothberg, J. M., & Chant, J. (2004). Gaining confidence in high-throughput protein interaction networks. *Nat. Biotechnol., 22*, 78–85.

Bahl, A., Brunk, B., Crabtree, J., Fraunholz, M., Gajria, B., Grant, G., et al. (2003). PlasmoDB: the Plasmodium genome resource. A database integrating experimental and computational data. *Nucleic Acids Res, 31*, 212-215.

Bansal, M., Belcastro, V., et al. (2007). How to infer gene networks from expression profiles. *Molecular Sytems Biology, 3***,** 1-10.

Barbosa, M. S., Báo, S. N., Andreotti, P. F., Faria, F. P. d., Felipe, M. S. S., Feitosa, L. d. S., et al. (2006). Glyceraldehyde-3-Phosphate Dehydrogenase of Paracoccidioides brasiliensis Is a Cell Surface Protein Involved in Fungal Adhesion to Extracellular Matrix Proteins and Interaction with Cells. *Infection and Immunity, 74*(1), 382-389.

Ben Mamoun, C., Gluzman, I. Y., Hott, C., MacMillan, S. K., Amarakone, A. S., Anderson, D. L., et al. (2001). Co-ordinated programme of gene expression during asexual intraerythrocytic development of the human malaria parasite Plasmodium falciparum revealed by microarray analysis. *Mol Microbiol., 39*, 26-36.

Breman, J., Egan, A., & Keusch, G. (2001). The intolerable burden of malaria: a new look at the numbers. *Am J Trop Med Hyg., 64*, 4-7.

Caspi, R., Foerster, H., Fulcher, C., Kaipa, P., Krummenacker, M., Latendresse, M., et al. (2008). The MetaCyc Database of metabolic pathways and enzymes and the BioCyc collection of Pathway/Genome Databases. *Nucleic Acids Res., 36*, D623-631.

Commichau, F. M., Forchhammer, K., & Stulke, J. (2006). Regulatory links between carbon and nitrogen metabolism. *Curr Opin Microbiol, 9*(2), 167-172.

Davis, F. P., Barkan, D. T., Eswar, N., McKerrow, J. H., & Sali, A. (2007). Host pathogen protein interactions predicted by comparative modeling. *Protein Sci, 16*(12), 2585-2596.

Deane, M., Salwinski, L., Xenarios, I., & Eisenbergh, D. (2002). Protein Interactions two methods for assessment of the reliability of high throughput observations. *Mol. Cellular Proteomics, 1*, 349-356.

Delgado, M. L., O'Connor, J. E., Azorín, I., Renau-Piqueras, J., Gil, M. L., & Gozalbo, D. (2001). The glyceraldehyde-3-phosphate dehydrogenase polypeptides encoded by the Saccharomyces cerevisiae TDH1, TDH2 and TDH3 genes are also cell wall proteins. *Microbiology, 147*, 411-417.

DeRisi, J. L., Iyer, V. R., & Brown, P. O. (1997). Exploring the metabolic and genetic control of gene expression on a genomic scale. *Science, 278*, 680-686.

Dobson, S., May, T., Berriman, M., Del, V. C., Fairlamb, A., Chakrabarti, D., et al. (1999). Characterization of protein Ser/Thr phosphatases of the malaria parasite, Plasmodium falciparum: inhibition of the parasitic calcineurin by cyclophilin-cyclosporin complex. *Mol Biochem Parasitol., 99*(2), 167-181.

Fields, S., & Song, O. (1989). A novel genetic system to detect protein-protein interactions. *Nature, 340*(6230), 245-246.

Florens, L., Washburn, M., Raine, J., Anthony, R., Grainger, M., Haynes, J., et al. (2002). A proteomic view of the Plasmodium falciparum life cycle. *Nature, 419*(6906), 520-526.

Ge, H., Liu, Z., Church, G. M., & Vidal, M. (2001). Correlation between transcriptome and interactome mapping data from Saccharomyces cerevisiae. *Nat. Genet., 29*, 482–486.

Gil-Navarro, I., Gil, M., Casanova, M., O'Connor, J., Martinez, J., & Gozalbo, D. (1997). The glycolytic enzyme glyceraldehyde-3-phosphate dehydrogenase of Candida albicans is a surface antigen. *J Bacteriol., 179*(16), 4992-4999.

Giot, L., Bader, J. S., Brouwer, C., Chaudhuri, A., Kuang, B., & Li, Y. (2003). A protein interaction map of Drosophila melanogaster. *Science, 302*, 1727-1736.

Golemis, E. A., Serebriiskii, I., & Law, S. (1999). The yeast two-hybrid system: criteria for detecting physiologically significant protein-protein interactions. *Curr Issues Mol Biol., 1*(1-2), 31-45.

Gomez, S. M., & Rzhetsky, A. (2002). Towards the prediction of complete protein--protein interaction networks. *Pac Symp Biocomput*, 413-424.

Harris, S. J., & Winzor, D. J. (1990). Interactions of glycolyltic enzymes with erythrocyte membrane. *Biochemica et Biophysica Acta, 1038*, 306-314.

Ito, T., Chiba, T., Ozawa, R., Yoshida, M., Hattori, M., & Sakaki, Y. (2001). A comprehensive two-hybrid analysis to explore the yeast protein interactome. *Proc. Natl Acad. Sci. USA., 98*, 4569–4574.

Itzhaki, Z., Akiva, E., Altuvia, Y., & Margalit, H. (2006). Evolutionary conservation of domain-domain interactions. *Genome Biologyx, 7*, R125-137.

Jannière, L., Canceill, D., Suski, C., Kanga, S., Dalmais, B., & Lestini, R. (2007). Genetic evidence for a link between glycolysis and DNA replication. *PLoS ONE 2*, e447.

Kanehisa, M., Goto, S., Kawashima, S., Okuno, Y., & Hattori, M. (2004). The KEGG resource for deciphering the genome. *Nucleic Acids Res., 32*, D277-D280.

Kim, J., & Dang, C. V. (2005). Multifaceted roles of glycolytic enzymes. *Trends Biochem Sci, 30*, 142–150.

Krishnadev, O., & Srinivasan, N. (2008). A data integration approach to predict host-pathogen protein-protein interactions: application to recognize protein interactions between human and a malarial parasite. *In Silico Biology, 8*, 0020.

Kumar, R., Musiyenko, A., & Barik, S. (2005). Plasmodium falciparum calcineurin and its association with heat shock protein 90: mechanisms for the antimalarial activity of cyclosporin A and synergism with geldanamycin. *Mol Biochem Parasitol.*, *141*(1).

Kumar, R., Musiyenko, A., Oldenburg, A., Adams, B., & BMC, B. S. (2004). Post-translational generation of constitutively active cores from larger phosphatases in the malaria parasite, Plasmodium falciparum: implications for proteomics. *Mol Biol.*, *5*(6).

Kurland, I. J., & Pilkis, S. J. (1995). Covalent control of 6-phosphofructo-2-kinase/bifunctional enzyme fructose-2,6-bisphosphatase: Insights into autoregulation of a bifunctional enzyme. *Prorein Sci.*, *4*, 1023-1027.

LaCount, D., Vignali, M., Chettier, R., Phansalkar, A., Bell, R., Hesselberth, J., et al. (2005). A protein interaction network of the malaria parasite Plasmodium falciparum. *Nature*, *438*(7064), 103-107.

Lan, J., & Steiner, R. F. (1991). The interaction of troponin C with phosphofructokinase comparision with calmodulin. *Biochem. J.*, *274*, 445-451.

Lasonder, E., Ishihama, Y., Andersen, J., Vermunt, A., Pain, A., Sauerwein, R., et al. (2002). Analysis of the Plasmodium falciparum proteome by high-accuracy mass spectrometry. *Nature*, *419*(6906), 537-542.

Leach, S., Gabow, A., Hunter, L., & Goldberg, D. S. (2007). Assessing and combining reliability of protein interaction sources. *Pac Symp Biocomput*, (pp. 433-444).

Lee, H., Deng, M., Sun, F., & Chen, T. (2006). An integrated approach to the prediction of domain-domain interactions. *BMC Bioinformatics*, *7*, 269-284.

Legrain, P., Wojcik, J., & Gauthier, J. M. (2001). Protein--protein interaction maps: a lead towards cellular functions. *Trends Genet*, *17*(6), 346-352.

Li, S., Armstrong, C. M., Bertin, N., Ge, H., Milstein, S., Boxem, M., & Vidalain, P. O. (2004). A map of the interactome network of the metazoan C.elegans. *Science*, *303*, 540-543.

Luscombe, N.M., Babu, M.M. et al. (2004). Genomic analysis of regulatory network dynamics reveal large topological changes. *Nature*, *431*, 308-312.

Mika, S., & Rost, B. (2006). Protein-protein interactions more conserved within species than across species. *PLoS Comput Biol.*, *7*, e79-91.

Miller, L. H., Baruch, D. I., Marsh, K., & Doumbo, O. K. (2002). The pathogenic basis of malaria. *Nature*, *415*, 673-679.

Pancholi, V., & Fischetti, V. (1992). A major surface protein on group A streptococci is a glyceraldehyde-3- phosphate-dehydrogenase with multiple binding activity. *Journal of Experimental Medicine*, *176*, 415-426.

Rhian, E. H., Joseph, L. D., Suad, A., David, C. K., Patrick, O., & Pradipsinh, K. R. (2000). Shotgun DNA microarrays and stage-specific gene expression in Plasmodium falciparum malaria *Molecular Microbiology*, *35*(1), 6-14.

Rhodes, D., Tomlins, S., Varambally, S., Mahavisno, V., Barrette, T., Kalyana-Sundaram, S., et al. (2005). Probabilistic model of the human protein-protein interaction network. *Nat. Biotechnol.*, *23*(8), 951-959.

Roch, K. G. L., Zhou, Y., Blair, P. L., Grainger, M., Moch, J. K., Haynes, J. D., et al. (2004). Discovery of Gene Function by Expression Profiling of the Malaria Parasite Life Cycle. *Science*, *301*(5639), 1503-1508.

Roth , E. J. (1990). Plasmodium falciparum carbohydrate metabolism: a connection between host cell and parasite. *Blood Cells*, *16*(2-3), 453-460.

Schachter, V. (2002). Construction and prediction of protein-protein interaction maps. *Ernst Schering Res Found Workshop,* (38), 191-220.

Schwikowski, B., Uetz, P., & Fields, S. (2000). A network of protein–protein interactions in yeast. *Nat. Biotechnol., 18*, 1257-1261.

Schwock, J. o., Kirchberger, J. u., Edelmann, A., Kriegel, T. M., & Kopperschl¨ager, G. (2004). Interaction of 6-phosphofructokinase with cytosolic proteins of Saccharomyces cerevisiae. *Yeast, 21*, 483-494.

Shannon, P., Markiel, A., Ozier, O., Baliga, N. S., Wang, J. T., Ramage, D., et al. (2003). Cytoscape: a software environment for integrated models of biomolecular interaction networks. *Genome Res, 13*(11), 2498-2504.

Shoemaker, B., & Panchenko, A. (2007a). Deciphering protein-protein interactions. Part I. Experimental techniques and databases. *PLoS Comput Biol., 3*(3), e42.

Shoemaker, B. A., & Panchenko, A. R. (2007b). Deciphering Protein–Protein Interactions. Part II. Computational Methods to Predict Protein and Domain Interaction Partners. *PLoS Comput Biol., 3*(3), e43.

Sirover, M. A. (1999). New insights into an old protein: the functional diversity of mammalian glyceraldehyde-3-phosphate dehydrogenase. *Biochemica et Biophysica Acta, 1432*, 159-184.

Spirin, V., & Mirny, L. A. (2003). Protein complexes and functional modules in molecular networks. *Proc Natl Acad Sci U S A, 100*(21), 12123-12128.

Sprinzak, E., Sattath, S., & Margalit, H. (2003). How reliable are experimental protein-protein interaction data? *J Mol Biol., 327*(5), 919-923.

Suthram, S., Sittler, T., & Ideker, T. (2005). The Plasmodium protein network diverges from those of other eukaryotes. *Nature, 438*(7064), 108-112.

Schwikowski, B., Uetz, P., Fields, S. (2000). A network of protein–protein interactions in yeast. *Nature Biotechnology, 18*, 1257-1261.

Uetz, P., Giot, L., Cagney, G., Mansfield, T. A., Judson, R. S., & Knight, J. R. (2000). Comprehensive analysis of protein-protein interactions in Saccharomyces cerevisiae. *Nature, 403* 623-627.

Uetz, P., & Goll, J. (2006). The elusive yeast interactome. *Genome Biology, 7*, 223-229.

Vaid, A., & Sharma, P. (2006). PfPKB, a protein kinase B-like enzyme from Plasmodium falciparum: II. Identification of calcium/calmodulin as its upstream activator and dissection of a novel signaling pathway. *J Biol Chem., 281*(37), 27126-27133.

von Mering, C., Krause, R., Snel, B., Cornell, M., Oliver, S. G., Fields, S., et al. (2002). Comparative assessment of large-scale data sets of protein–protein interactions. *Nature, 417*, 399-403.

Wojcik, J., Boneca, I. G., & Legrain, P. (2002). Prediction, assessment and validation of protein interaction maps in bacteria. *J Mol Biol, 323*(4), 763-770.

Wu, H.-Y., Tomizawa, K., Oda, Y., Wei, F.-Y., Lu, Y.-F., Matsushita, M., et al. (2004). Critical Role of Calpain-mediated Cleavage of Calcineurin in Excitotoxic Neurodegeneration. *J. Biol. Chem., 279*(6), 4929-4940.

Wuchty, S. (2007). Rich-club phenomenon in the interactome of P. falciparum--artifact or signature of a parasitic life style? *PLoS ONE, 2*(3), e335.

Zabka, M., Les'niak, W., Prus, W., nicki, J. K., & Filipek, A. (2008). Sgtl has co-chaperone properties and is up-regulated by heat shock. *Biochemical and Biophysical Research Communications, 370*, 179-183.

Zhu, H., Bilgin, M., Bangham, R., Hall, D., Casamayor, A., Bertone, P., et al. (2001). Global analysis of protein activities using proteome chips. *Science, 293*(5537), 2101-2105.

Section V
Tools for Analysis of PPI Networks

We will conclude the book by three chapters that discuss state-of-the-art software tools that provide visualization and analysis for protein interaction networks.

Chapter XVI
Graphical Analysis and Visualization Tools for Protein Interaction Networks

Sirisha Gollapudi*
MyCIB, University of Nottingham, UK

Alex Marshall*
MyCIB, University of Nottingham, UK

Daniel Zadik*
MyCIB, University of Nottingham, UK

Charlie Hodgman
MyCIB, University of Nottingham, UK

* *Joint first authors*

ABSTRACT

Software for the visualization and analysis of protein-protein interaction (PPI) networks can enable general exploration, as well as providing graph-theoretic algorithms for specific tasks. Analyses can include reduction of complexity or the scope of the network in order to make it more manageable, or increase in complexity by integration with other datasets, to represent biology more accurately. Two software approaches are outlined in this chapter: desktop applications and web services. Desktop applications have attractive user interfaces with a wide range of analysis tools, and often capabilities for integration of other bio-molecular data. Web services provide a newer approach to network analysis. They have the advantages of a broader range of potential functionalities and a more extensible framework than standalone desktop tools. However, their relative infancy means that they are not as well developed. This chapter provides an evaluation of some common desktop applications, compared to and contrasted with several examples of web services.

INTRODUCTION

In recent years, a predominant theme in biological research has been a shift from the reductionist approach to an integrative systems approach (Kitano *et al.*, 2002; Oltvai & Barabási, 2002; Han, 2008), which views the components of a cell as acting together in a network of reactions and interactions. This change has been aspired to for many years, as it is widely understood that observed cell behaviour is rarely attributed to one component acting alone. The continued growth in the processing power of computers and the increase in availability of biological data from high throughput technologies have allowed such an approach to become realistic.

This chapter describes software that uses algorithms taken from the field of graph theory to draw biological conclusions about protein-protein interaction (PPI) networks. It contains background on the subject, a method for the representation of networks, including bio-molecular networks, as a set of nodes connected by edges. In PPI networks, the nodes represent proteins, while the edges represent the interactions between them. Very often, PPI networks aim to represent the whole proteome of a species. As an example, Figure 1 shows a holistic PPI network for the budding yeast, *Saccharomyces cerevisiae* (Schwikowski, Uetz & Fields, 2000). Holistic PPI networks are usually constructed from two-hybrid screening (Fields & Song, 1989) and other experimental data, much of which is available in public databases (see Background).

Two-hybrid technology is particularly powerful as it identifies large numbers of putative interactions. To test if two proteins interact, the DNA-binding domain from a transcription factor is spliced onto the end of the gene for one, and the activating domain from the transcription factor is spliced onto the other. When the genes are transformed into a cell (usually in *S. cerevisiae* or *E. coli*) and expressed, two hybrid proteins are produced. The protein with the DNA-binding domain binds to an upstream activating sequence (UAS) of a reporter gene and the protein with the activating domain binds the remaining transcriptional machinery. If the two proteins bind each other then the transcriptional machinery will be brought into close proximity with the UAS, and the reporter gene will be expressed. Whole libraries of genes can be transformed into a cell suspension, to create a population with a very diverse set of combinations of hybrid genes. If a reporter gene is used that is essential for survival under particular conditions, then those conditions can be used to select only those cells that contain

Figure 1. An S. cerevisiae holistic PPI network, containing 1548 proteins and 2358 interactions, from a study carried out by Schwikowski, Uetz and Fields, (2000). (Permission to reproduce figure kindly provided by Peter Uetz, Institute of Genetics, University of Karlsruhe.)

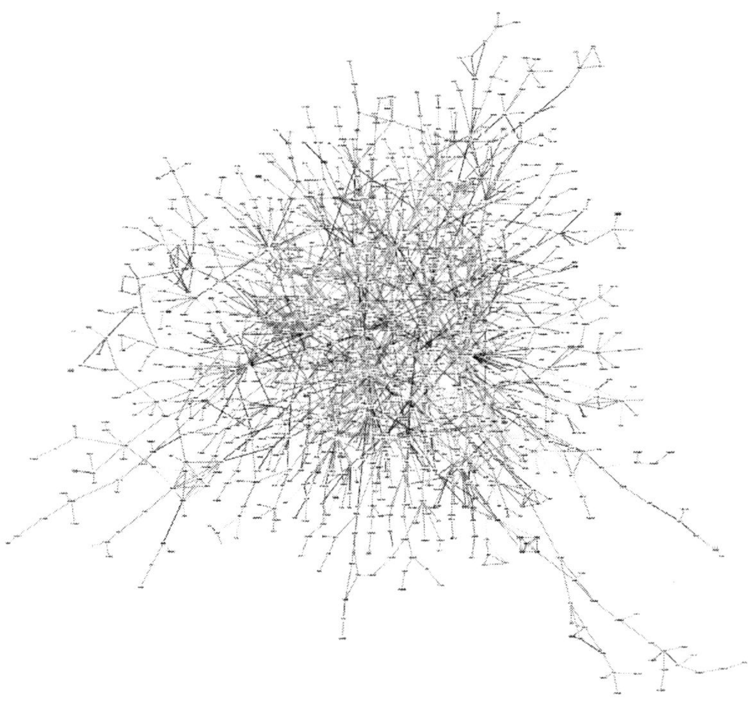

a pair of interacting hybrid proteins, and the genes from these can then be sequenced to identify them. The artificial context in which the interactions take place can result in both false positive and false negative results. For example, animal proteins may fold incorrectly in a yeast cell, or proteins that can interact but would never come into contact *in vivo* (because they are found in different cellular compartments or at different times) are reported as interacting.

It is often desirable to visualise PPI networks in order to see their structure and therefore provide further insight into the system. Visualizations of large networks are often complicated and dense, making it impossible to discern structure by eye. However, the use of analysis techniques can help elucidate structure and therefore function. The results of such analyses may be interesting in their own right, or may lead to further visualization steps on selected subsets or simplifications of the network. For these reasons, the software for visualising networks usually contains various network analysis functions.

Network properties can tell us about the biology of the system. They can help to elucidate the mechanisms of signal transduction, complex formation, protein localisation and protein modification, which in turn can reveal the control mechanisms of many molecular and cellular processes. The network topology around a protein of interest can be used to estimate its essentiality to the functioning of the system. Knocking out proteins that have many interacting partners, or act as bridging nodes which connect different parts of the network, is much more likely to be fatal to the organism. Such measures can be used to explain the extreme variation of the severity of phenotypes of gene knockouts, as well as to identify potential drug targets.

Some of the more important visualization and analysis software are presented below. The major advantages and disadvantages of each tool are given, along with the analyses offered, and the capability for integration of other types of data (such as gene expression). Much of the existing software for the construction, visualization and analysis of protein interaction networks has been developed as standalone applications, which are installed and run on a local computer. Details of several of these tools are given in this chapter. They tend to be well-developed applications with intuitive user interfaces, capable of a wide variety of visualization and analysis tasks. There are several such tools in the public domain and many of these are open-source. Some can be used to study networks in any subject area, while others have been specifically developed to work with bio-molecular networks. These two approaches have a slightly different focus, and therefore different advantages.

Generic network software tends to have a powerful set of graph-theoretic algorithms, but is less specialised for biological tasks such as incorporating experimental data. Two examples of such software are detailed in this chapter, Network Workbench (http://nwb.slis.indiana.edu) and yEd (http://www.yworks.com/products/yed). Software designed for the analysis of bio-molecular networks has several advantages conferred by its specificity. Three examples of such software are Cytoscape (Shannon *et al.,* 2003; Cline *et al.,* 2007), BiologicalNetworks (Baitaluk, Sedova, Ray & Gupta, 2006) and VisAnt (Hu *et al.,* 2005). It tends to support a variety of types of bio-molecular network: PPI as well as metabolic, signal transduction and gene regulatory. It is often possible to integrate these different network types, along with biological information from databases or the user's experiments. Most notably, the overlaying of gene expression data onto networks can help to elucidate the control mechanisms of the systems.

More recently, a type of software called web services has been increasingly employed by the bioinformatics community. Web services are a distributed technology, involving remote tools that can be called from a local computer. Web services have several advantages over standalone applications: they bring together a range of geographically diverse resources for the user, and they harness remote computing power, so they are not limited by the power of the local computer. They are theoretically independent from platform and programming language constraints, and therefore a wide variety of client applications can access them. These include command-line scripts, websites and specific software tools. Most powerful, however, are workflow enactment engines, such as Taverna (Oinn *et al.,* 2004), which provide a graphical interface for the construction, storage, sharing and execution of workflows.

A fully automated workflow can take input from a user and/or databases, run one or several chains of web services (with each web service using the input data or the results of a previous web service) and return to the user the required result or set of results. Workflows amount to reproducible processes, which can be saved, reused and shared, with minimal potential for human error, and recording of experimental provenance. Provenance, in bioinformatics, is the recording of experimental steps carried out on data. This could be the source database of a piece of data, or the version of a tool used to carry out some analysis on the data. It can be thought of as how and why a piece of data was produced. In this chapter, three example workflows for PPI network analysis are used to demonstrate a variety of network analysis tasks, and the usefulness of the results is discussed. "Network Properties" runs many graph-theoretic web services in parallel, to give an array of statistics that help to characterise the topology of a PPI network. "Node Locality" extracts the local sub-network around a protein of interest from within a larger network in which the amount and density of information can become a barrier to understanding. This can suggest which proteins may have related functions, and can help to elucidate the biological function of unknown genes. "Annotating Metabolic Pathways with PPIs" adds to a metabolic pathway of interest, any PPIs from the IntAct database (Kerrien *et al.*, 2007) that involve the enzymes of that pathway, and can suggest control mechanisms.

This chapter aims to give background information on the types of visualization and analysis tools useful in the study of PPI networks, inform potential users of a number of useful tools and demonstrate the functions, advantages and disadvantages of each.

BACKGROUND

The field of systems biology has shown much promise as a means to analyse and better understand the interactions within the complex cell machinery. Bio-molecular networks can be constructed using many different types of experimental or predicted data. These include descriptions of metabolic reactions, gene regulation, signal transduction and the interactions between proteins. This book focuses on the interactions between one protein and another, but proteins also interact with many other molecules, including nucleic acids and metabolites.

Protein Interaction Data

PPIs are involved in many different biological functions: one protein can bind another and modify it; proteins can form complexes with different functionalities from their constituent parts; carrier proteins may bind other proteins to ensure their correct localisation, and pathways of protein interactions can transduce signals within the cell.

Since the development of high throughput technologies such as two-hybrid screening, large PPI datasets have been generated. These can be visualised as networks and analysed with a variety of tools to provide an understanding of the flow of information and mechanisms of control within the system. Where a network aims to represent every protein expressed in a particular species, it can be described as **holistic**. Conclusions can be drawn about such networks at several levels: global conclusions can be drawn from statistical analysis of the whole network, interesting local structures can be identified within the holistic network and local areas of previous specific interest can be examined for novel structure.

The first large-scale identification of protein interactions was carried out by Uetz *et al.* (2000), for *S. cerevisiae*, which identified 957 putative interactions between 1004 proteins. Ito *et al.* (2001) expanded on this knowledge by identifying a further 4549 interactions between 3728 proteins, again via the yeast two-hybrid method. Despite being prone to a high rate of false-positives, this high-throughput technique is still the source of most PPI data.

A number of protein interaction databases are publicly available (Bader, Cary & Sander, 2006). These are populated largely from two-hybrid data, as well as other experimental data, from the literature and predicted using a variety of methods for computational inferences. For example, Figure 2a shows the proportions of data generated by different techniques in IntAct. Some examples of prominent repositories are given in Table 1, together with information on export formats and database statistics including the Database of Interacting Proteins (DIP, Salwinski *et al.*, 2004), the Bio-molecular Object Network Database (BOND, Alfarano *et al.*, 2005), IntAct (Kerrien *et al.*, 2007), the Molecular INTeractions database (MINT, Chatr-Aryamontri *et al.*, 2007), the Biological General Repository for Interaction Datasets (BioGRID, Breitkreutz *et al.*, 2008), the Human Protein Reference Database (HPRD, Mishra *et al.*, 2006), and the *Arabidopsis thaliana* Protein Interaction Database (AtPID, Cui *et al.*, 2007). Extensive study of model organisms over recent years has led to an increase in available data. Figure 2b shows the rise in PPI data in the IntAct database, showing that the amount of PPI data generated is continuing to rise at a steady rate.

Figure 2. IntAct database statistics (http://www.ebi.ac.uk/intact/statisticView/do/statistics). (a) The proportions of PPIs generated by different identification methods. (b) The rise in the number of binary PPIs in the database. (Permission to reproduce figures kindly provided by IntAct.)

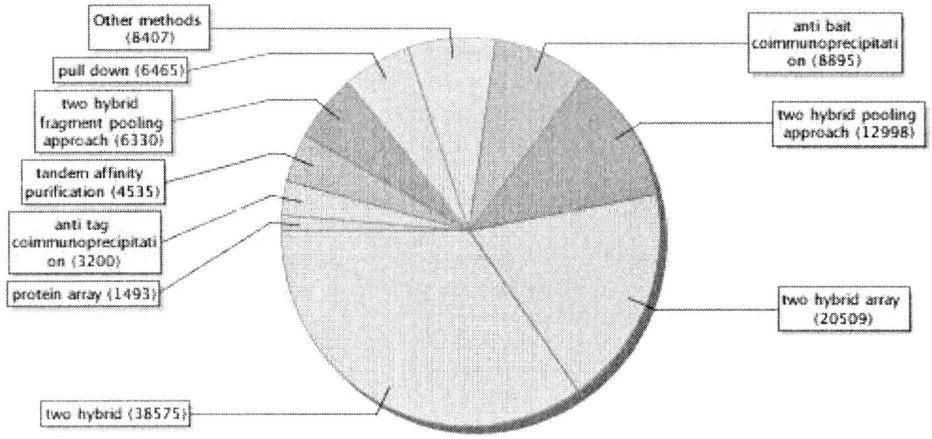

(a)

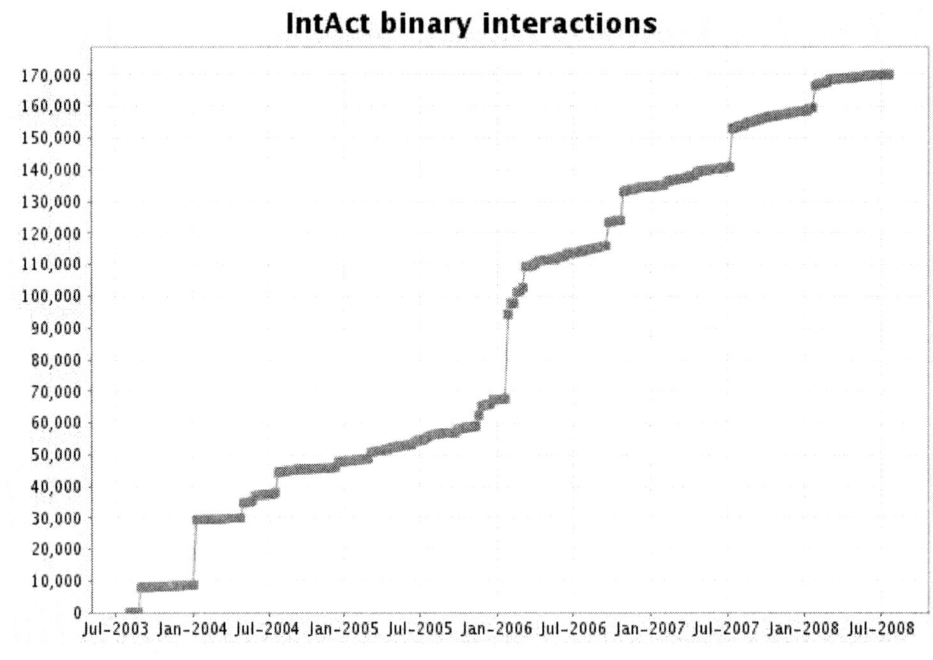

(b)

Table 1. A selection of widely used PPI databases (accessed as of May 2008). The number of interactions given is for non-redundant (i.e. the same interaction suggested by two techniques counts only as one interaction), binary (i.e. the interactions are represented as pairs of proteins, and of not of any larger groups) interactions only. "Curation from literature" implies the collation of experimental results from published sources. This can be at any scale, from the verification of individual interactions to the suggestion of thousands of interactions by two-hybrid techniques. Many more PPI databases, for a variety of species are available. Many of these are listed at http://www.pathguide.org (Bader, Cary & Sander, 2006).

Database	**Organism**	**Number of proteins**	**Number of interactions**	**Data Sources**	**Download formats**
DIP (Database of Interacting Proteins)	Various	19,765	56,493	Manual curation from literature (with computational aids for selecting papers for curation).	MIF, MI-TAB
BOND (Bio-molecular Object Network Database. Frozen 2005. Proprietary version, BONDplus, also available)	Various	82,218	188,517	Manual curation from literature.	MIF, SIF, many others
IntAct	Various	63,121	170,063	Manual curation from literature. Direct user submission.	MIF, MI-TAB
MINT (Molecular INTeractions database)	Various	28,723	105,537	Manual curation from literature. Direct user submission.	MIF, MI-TAB
BioGRID (Biological General Repository for Interaction Datasets)	Various	529,018	136,376	Manual curation from literature (with computational aids for selecting papers for curation)	MIF, tab-delimited
HPRD (Human Protein Reference Database)	*H. sapiens*	25,661	38,167	Manual curation from literature, Data deposited in Human Proteinpedia.	XML, MIF, tab-delimited
AtPID (*Arabidopsis thaliana* Protein Interactome Database)	*A. thaliana*	14,791	32,728	Manual curation from literature. IntAct, BIND, TAIR and KEGG databases. Computational prediction using a variety of methods.	Tab-delimited

Graph-Theoretical Concepts

Although usually stored as a list of interactions, such data can be more usefully thought of as a network. To aid the computational analysis of cellular networks, it is necessary to decide on appropriate graph frameworks to represent entities and relationships within the network. A network can be described formally as a graph that consists of a set of nodes and a set of edges connecting the nodes. The nature of the nodes and edges is different depending on the network type. Edges may be directed or undirected to determine the direction, if any, of information flow. Both nodes and edges have related metadata to describe their characteristics further. In most PPI networks, nodes represent proteins, and undirected edges represent interactions.

The network approach to biological systems borrows heavily from the wider fields of graph theory and complex network theory. This allows for algorithms originally designed to obtain information from, for example, social networks or food webs, to be applied to bio-molecular networks with the aim of increasing understanding of cellular behaviour at the systems level. It remains an important task to ensure that any such analyses are biologically meaningful and that their results can be identified as genuine biological phenomena.

Within a network, a variety of topological structures can be detected. Areas where nodes are more highly interconnected than elsewhere are of interest. These groupings are often called **clusters**, although this is a term used inconsistently within the field. Such areas where all nodes are connected to all other nodes are called **cliques**. In PPI networks, such structures represent related functions of the proteins involved and strongly suggest that they could form a complex. The extent to which the proteins adjacent to a protein of interest are also adjacent to each other, is called the **clustering coefficient**, and can suggest that the protein forms part of a complex.

Specific arrangements of nodes and edges are called **motifs**. These are better conserved than the network in general, suggesting that they represent important functions. The simplest of these are **self-loops**, where an edge joins a node to itself. In PPI networks, these suggest the formation of a homodimer or larger complexes of the same protein. An illustration of motifs and cliques can be found in Figure 3.

Figure 3. Example to demonstrate cliques, clustering coefficients and motifs. (a) In this network, nodes A, B and C form a clique (i.e. all nodes are connected to all others). The extent to which proteins adjacent to a protein of interest are also adjacent to each other is called the clustering coefficient. Each of the neighbors of node A is connected to each of the other neighbors of node A. Therefore, node A has a clustering coefficient of 1 (the maximum value). The same is true of node C. The clustering coefficient of node B is 1/3, because, of three possible edges required to connect all of its neighbours, only one exists. (b) Specific arrangements of nodes and edges are called motifs. In this network, the three nodes are connected to each other via directed edges. This is an example of an important motif known as a feed-forward loop, where node E influences node G indirectly via node F, but can also influence node G directly.

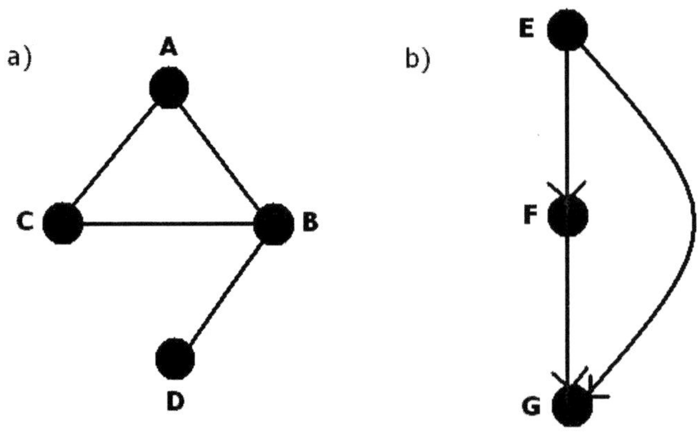

PPI networks often contain disconnected sub-networks called **components**. These are sets of interlinking proteins that do not connect to other parts of the network. These can either represent real biologically unconnected interaction networks or can indicate missing data. The least number of consecutive edges that connect two nodes of interest is called the **shortest path**. The **diameter** is the longest of all the shortest paths in a network or component, and is an estimate of how related the functions of the all proteins are.

Jeong, Mason, Barabási and Oltvai (2001) carried out the first large-scale graph-theoretic analysis of an *S. cerevisiae* network built from the interactions found by Uetz *et al.*, and additional interactions determined as part of a study carried out by Eisenberg, Marcotte, Xenarios and Yeates (2000). This led to the construction of an interaction network consisting of 1870 proteins and 2240 direct physical interactions. The **degree** of a node is the number of connections it has to other nodes, and the **degree distribution** of a network is the proportion of the nodes that have each degree. Analysis of the topology of the network for the detection of significant features revealed that the degree distribution followed a power law. This indicates a **scale-free** network (Barabási & Albert, 1999), *i.e.* a few proteins are highly connected, while most proteins have only a few links to others in the network.

The scale-free nature of this network led to the hypothesis that the removal of the proteins with the highest degree, also known as **hubs**, would disrupt the network the most, while removal of proteins with lower degree would be less likely to have a significant effect. This is known as the 'fragile yet robust' property of PPI and other biological networks (Barabási & Oltvai, 2004), which causes them to be susceptible to targeted attacks yet resilient in the face of random deletions. The position of a potential target in a network can be used to predict the outcome of deactivating it, and therefore methods to predict protein essentiality have several potentially significant applications, such as drug target identification. Jeong *et al.* (2001) conclude that highly connected proteins play a central role in the network's architecture and are three times more likely to be essential than proteins with a small degree.

Because of their generic function, hubs can sometimes reduce visibility for the more specific interactions. It is sometimes desirable to remove them from visualizations to draw more subtle conclusions. Having established the importance of hubs in PPI networks, a subsequent study investigated the importance of temporal and spatial factors (Han *et al.*, 2004). This study established two categories of protein hubs in the *S. cerevisiae* PPI network: 'party hubs', which interact with most of their partners simultaneously, and 'date hubs', which bind their partners at different times or locations.

A topological metric that has come into prominence recently is **betweenness centrality**. For a given node, this denotes its importance in the network based on the proportion of shortest paths between other nodes that go through this node. The higher the betweenness centrality, the more likely the node is to play a role in connecting or 'bridging' parts of the network. This has particular relevance when ranking nodes based on their degree (*i.e.* the number of connections).

A study carried out by Bader and Madduri (2007) analysed the PPI network of human proteins as obtained from various public interaction databases, comprising almost 18000 proteins and 44000 interactions. Having calculated degrees and betweenness centralities for all the proteins, it was discovered that the high degree proteins also tended to have high betweenness, an intuitive result as there are many proteins connected to hubs, resulting in them appearing on many shortest paths between pairs of proteins. However lower degree nodes displayed a greater amount of variation in their betweenness scores, indicating that some of these may also be globally important.

As betweenness and degree are correlated, it is unclear whether the essentiality of a protein is due to its high betweenness, high degree or a combination of the two. A study by Yu *et al.* (2007) attempted to address this issue by dividing proteins in the *S. cerevisiae* PPI network into four categories: those that were neither hubs nor bottlenecks (where a bottleneck is a protein with high betweenness), those that were hubs but not bottlenecks, those that were bottlenecks but not hubs, and those that were both. Analysis of the network established that both hub and non-hub bottlenecks tend to be essential. Interestingly, only hubs that were also bottlenecks were essential, leading to the conclusion that betweenness is a better predictor of essentiality than degree.

Another centrality measure to rank nodes is **closeness centrality**, which measures how fast information flows from a node of interest to all the reachable nodes. This measure is of particular relevance as PPI networks can include signal transduction pathways. Related to closeness centrality is the concept of the **network centre**. This is the node or set of nodes with the highest closeness centrality, and therefore the shortest average path length to all other nodes.

Complexity of data is a major issue with PPI networks. At different times, both reducing complexity and increasing it (to include as much information as possible) can be desirable. Reducing the complexity can improve visibility to the point where relevant structure becomes clear. Increasing the complexity by integrating PPI networks with other network types, databases, experimental data and manual annotation can provide insights on a far more profound level than the PPI network alone.

One way to decrease complexity is to concentrate on a subset of the network. The sub-network can then be analysed or visualised, and from this, its role and the functions of its constituent proteins and interactions can be suggested. Sub-networks for isolated study can be selected in several different ways, including isolation of the cliques, clusters, components and motifs described above, or the local network for a given distance around a protein of interest. The biological functions of uncharacterised proteins are very likely to be related to those of their better-annotated neighbours (Deng, Zhang, Mehta, Chen & Sun, 2003). As well as experimentally determining interactions, the study carried out by Ito *et al.* (2001) cites examples of biologically significant sub-networks extracted from the overall set of interactions. Another way to decrease complexity is to condense some or all of the proteins in a network into a smaller set of **metanodes**, based on commonality of any one of several different factors. This can be biochemical function, localisation, process or any other factor for which annotation is available. The resulting network is called a **metagraph**, and can elucidate how the more general categories of cellular machinery interact.

Data Integration

To add deeper meaning to PPI networks, they can be integrated with a variety of other data sets. These can include other bio-molecular networks, such as metabolic and gene regulatory networks, which provide greater understanding of the functions of the proteins and the relationships between those functions. In return, PPI networks suggest possible control mechanisms. Attributes, annotation and database links can also be assigned to proteins or interactions, either manually, or by using the existing name or annotation. This process provides deeper biological context for analyses carried out.

Amongst the most common network enrichment tasks carried out on PPI networks is the overlaying of data from bio-molecular profiling, such as metabolomic, polymorphism or most commonly gene expression data. This is often used to colour nodes according to levels of expression or fold change. Where two proteins are expressed together, it suggests shared control mechanisms and therefore related cellular function. Another common way to enrich PPI networks is by assigning weightings to the edges. These often denote either the reliability of the evidence or the strength of the interaction, which are usually determined from experimental data.

Software Considerations

Many network visualization and analysis tools are available. Visualization and analysis are separate computational tasks, but are often implemented together. The number of available tools has grown considerably in the last few years: Saraiya, North and Duca identified 16 software tools in 2005, and by 2007, Suderman and Hallett identified over 35. Both reviews emphasised visualization over analysis, though the majority of the tools surveyed contained basic analysis functions as well.

Several factors are important when developing analysis software. A well-designed visualization component is important, as bioinformatics software should aim to be intuitive to non-expert computer users. The ease of utilising such software is increased by visually guiding users through network analyses. The acceptance of standard data formats gives interoperability between a wide variety of software. Large networks should be supported, although this may be limited by local processing power. Some programming languages now have graph-theoretic libraries such as NetworkX for Python, the Graph module for Perl and the Boost Graph Library for C++, and these will assist greatly in the future development of network tools.

DESKTOP APPLICATIONS

There are many desktop applications for graph visualization and analysis available in the public domain. They tend to combine a wide range of network visualization and analysis functions, and use graph-theoretic techniques to extract interesting sub-networks. Many have the advantage that they are open-source and freely available, and are generally cross-platform. Where an Application Programming Interface (API) is available, extensions can be developed by fellow scientists to improve further the analytical techniques. Software packages for the analysis of holistic PPI networks tend to be standalone desktop applications. Some common applications and their properties are described in Tables 2 and 3.

Cytoscape

Cytoscape (Shannon *et al.,* 2003) is a freely available, Java-based application developed by a team funded by the National Institutes of Health (NIH), the National Science Foundation (NSF) and Unilever. It was originally written to visualise and analyse bio-molecular networks imported by the user. More recently, it has been expanded to support networks in any subject area or to access publicly available interaction data via a web service interface. The core software provides basic network visualization and integration functions, and there are also plugins available that extend the functionality and allow for the addition of more complex network analyses. Cytoscape is open-source, and benefits from an active development community who maintain bug-tracking facilities. New versions are often released. Cytoscape offers a wide range of analyses including sub-network selection, finding modules or clusters, network filtering and searching target nodes and edges.

Networks in Cytoscape can be integrated with many other data types. PPI networks can be integrated with other bio-molecular networks. Each node contains links to several external databases including IntAct, BioMart (http://www.biomart.org), Pathways Commons (http://www.pathwaycommons.org), PICR (http://www.ebi.ac.uk/Tools/picr), and Entrez (http://www.ncbi.nlm.nih.gov/) via a built-in web service plugin. Additionally, bio-

Table 2. The topological metrics supported by a selection of widely used network visualization and analysis tools (assessed as of May 2008).

Application	Simple topological metrics	Complex topological metrics
Cytoscape	Number of nodes and edges.	(calculated using various plug-ins such as NetworkAnalyzer) Node degree distributions, shortest paths, shared neighbours, topological coefficients, betweenness centrality and closeness centrality.
Network Workbench	Number of nodes and edges, isolated nodes, self-loops, weakly connected components, strongly connected components, number of nodes in the largest connected component, graph density.	Node degree distribution, k-nearest neighbour, clustering coefficients, diameter, average shortest path, shortest path distribution and betweenness centrality, multi-state discrete model dynamics analysis, extract and annotate k-cores, page-rank. GUESS visualization allows for colorization based on any numeric attribute.
yEd	Number of nodes and edges, number of self-loops, components; directed cycles check, planarity check,	Colouring graphs based on centrality, number of connected edges, graph centrality, node betweenness centrality and closeness centrality, Clustering graphs based on connectivity, bi-connectivity, edge betweenness centrality
VisANT	Number of nodes and edges.	Node degree distribution, clustering coefficients, average clustering coefficient and shortest paths, cycles, feed-forward loops and self-loops.
Biological Networks	Number of nodes and edges, average vertex degree, average distance, network diameter and network complexity.	Cycles and network blast, shortest paths, page rank, betweenness centrality, clustering coefficient, strong connected component.

Table 3. The data capabilities of a selection of widely used network visualization and analysis tools (assessed as of May 2008). The "Data Integration" column specifies whether tabular data can be mapped onto the nodes and edges of the network. This can be bio-molecular profile data such as gene expression, or annotations such as Gene Ontology.

Application	Network file import formats	Network file export formats	Integral datasets	Data Integration
Cytoscape	SIF, GML, XGMML, BioPax, PSI-MI, tab delimited and Excel files and OBO	Textual: SIF, GML, XGMML,PSI-MI Graphical: PDF, SVG, EPS, JPEG, PNG, BMP	Connections via web services to: IntAct, Pathway Commons, NCBI Entrez Gene and BioMart synonyms	Yes
Network Workbench	GraphML, XGMML, Pajek, NWB, TreeML, edge lists, CSV	Textual: Pajek, XGMML, GraphML, NWB, edge list, CSV Graphical: JPG, PNG,PPM,RAW,CGM, EMF, PDF, EPS, PS, SVG, SWF, GIF	Fly, Worm, and Yeast from MINT, TF_DNA_ regulonDB, and WI5, Boolean representation of a colorectal cancer model and drosophila embryo model.	No
yEd	GML, TGF, XGML, YGF, XML	Textual: GML, XGML, YGF, GraphML Graphical: PNG, GIF, JPG, PDF, EPS, EMF, SWF, SVG.	No	No
VisANT	VISML, BioPAX, edge list	Textual: CSV, VISML, BioPAX, tab-delimited file Graphical: JPG, SVG	405,000 functional/physical/ genetic associations, from 108 species including yeast, fly, human, from database such as Entrez Gene, MINT, BioGrid etc.	Yes
Biological Networks	BNM, SIF, BioPax,	Textual: BNM	14 curated data sources from model organisms	Yes

molecular profile data (such as gene expression or metabolomic) can be overlaid onto a network. This approach has been used with great success in breast cancer studies by Chuang *et al.* (2007) and Wang *et al.* (2005) (See Figure 4). The sub-networks shown are from these studies, which predicted patient survival based on expression profiles (van de Vijver *et al.,* 2002).

Cytoscape provides the ability to reduce or increase network complexity, depending on user requirements. To reduce complexity, nodes can be collapsed into metanodes, thus reducing the amount of data to render pictorially while keeping the same quantity of information. The user can increase the complexity by adding node and edge attributes, edge weightings and links to many databases, including Entrez (http://www.ncbi.nlm.nih.gov/sites/gquery), UniProt (UniProt Consortium, 2008) and Reactome (Joshi-Tope *et al.,* 2005). Cytoscape offers many different visual styles, colour schemes and layout algorithms to aid understanding of the network. Different layout algorithms can be used, including spring embedded, organic, hierarchical, random and edge weighted. There is a zoom function allowing both a bird's eye view of the whole network, and closer examination of areas of interest.

One of Cytoscape's major strengths is the capacity for development of external plugins. There are 45 plugins currently available, with widely differing functions for network analysis or enrichment (http://chianti.ucsd.edu/cyto_web/plugins/index.php). 13 plugins are for network analysis, 13 for network attributes, 4 for network inference, 3 for functional enrichment, 5 for communication and scripting and 10 for other functions. A set of the most universally useful plugins are supplied with the Cytoscape download and further plugins can be added using the plugin manager.

Networks can be analysed in light of functional information from the Gene Ontology (Ashburner *et al.,* 2000), using a plugin called Biological Network Gene Ontology (BiNGO, Maere *et al.,* 2005). The Gene Ontology is a controlled, hierarchical vocabulary to describe a gene or its product. To demonstrate the power of Cytoscape plugins, BiNGO and Molecular COmplex DEtection (MCODE, Bader & Hogue. 2003) analyses were carried out on an *S. cerevisiae* PPI network from the Biological General Repository for Interaction Datasets (BioGRID, Breitkreutz *et al.,* 2008 – see Figure 5a). First, the network was analysed using the BiNGO plugin, which groups proteins together into nodes that represent categories from the Gene Ontology (see Figure 5b). The more proteins from the original network there are in a particular category, the larger the node. Categories that are over-represented in this network, when compared to the genome-wide proportion, are likely to be of particular relevance.

Figure 4. Grey-scaling of nodes with expression data. In subgraphs A-E, the nodes are grey-scaled using the expression data from van de Vijver et al. (2002). Gene expression signatures were analysed as a predictor of the survival of breast cancer patients. Subgraphs F-K are grey-scaled using the expression data from Wang et al. (2005). Highly up-regulated nodes are black, highly down-regulated nodes are white, and the expression levels between are various shades of grey. Diamond nodes indicate that the gene is significantly up- or down-regulated (P < 0.05), and circular nodes indicate that it is not. (Permission to reproduce this figure from Chuang et al. (2007) has been kindly provided by Trey Ideker, UCSD. The colour scheme has been altered to allow for monochrome printing.)

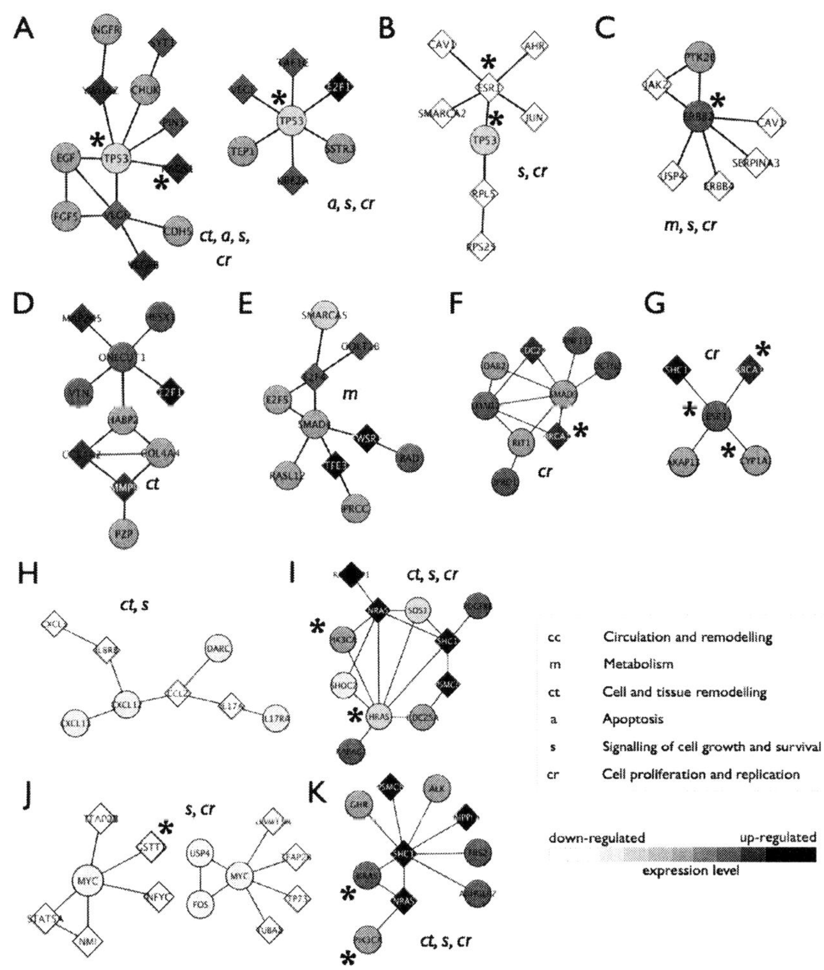

MCODE returns the most interconnected clusters within a network, which can represent complexes. When used on the same network, MCODE found 159 clusters. The most highly interconnected cluster contained just 39 proteins connected by 1058 interactions (see Figure 5c). Subsequent BiNGO analysis of the Gene Ontology terms of this cluster showed it largely consisted of proteins involved in "responses to DNA damage" (see Figure 5b). Figure 5d shows those genes involved in transcription.

Figure 5. Examples of analyses using Cytoscape. (a) An S. cerevisiae holistic PPI network, generated from the BioGRID database. This has 5388 nodes and 75749 edges. (b) BiNGO analysis of the network shown in Figure 4a, showing the representation of different Gene Ontologies. (c) MCODE cluster analysis of the network shown in Figure 4a, showing the most highly interconnected cluster, which has 59 nodes and 1499 edges. (d) The subgraph of Figure 4b, generated only from those proteins with Gene Ontology annotation suggesting that they are involved in transcription. This has 543 nodes and 578 edges. In Cytoscape, the zoom feature allows the user to read node labels easily.

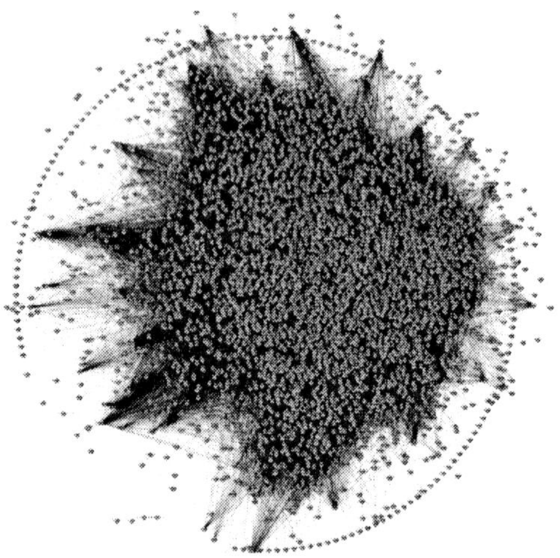

(a)

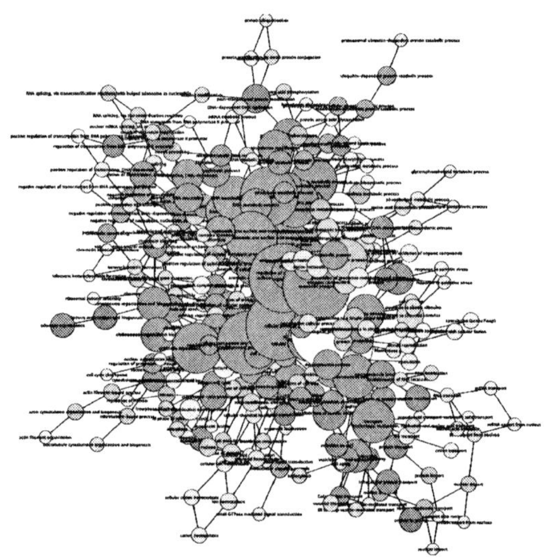

(b)

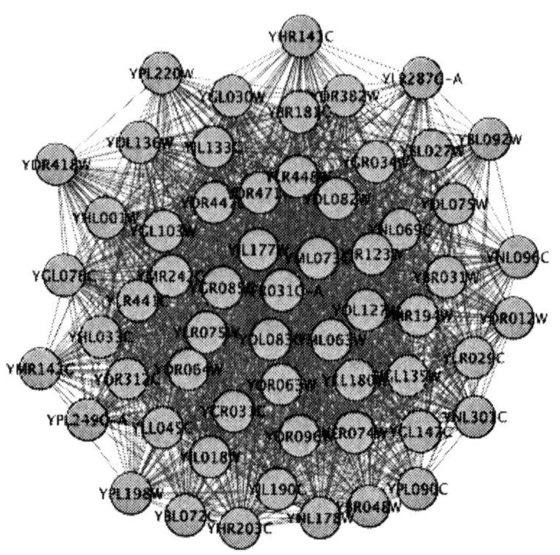

(c)

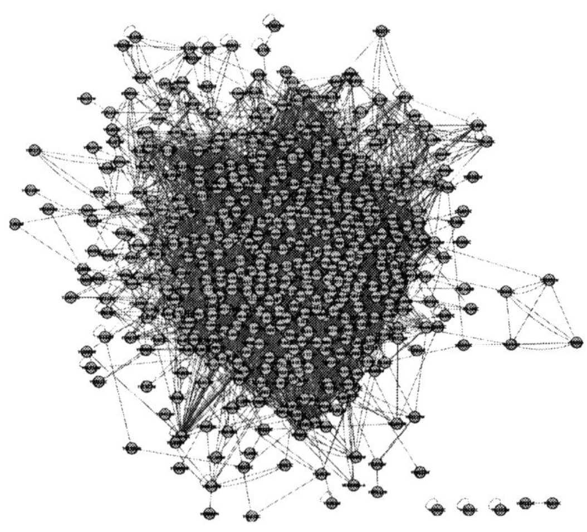

(d)

NetworkAnalyzer (Assenov *et al.*, 2008) is a plug-in that measures network properties including the clustering coefficient, network diameter, centralisation, shortest paths, average number of neighbours, network density and self-loops. Another plugin, called JActiveModules (Ideker *et al.*, 2002), extracts the "active" sub-networks, which are connected regions whose gene expression is significantly different from that of the whole network. These sub-networks are statistically scored and those that score highly are returned.

Network Workbench

Network Workbench (NWB) (http://nwb.slis.indiana.edu) is written mainly in Java and FORTRAN by a team at Indiana University collaborating with many other institutions, and is freely available and open-source. It was developed for visualization and analysis of networks imported by the user, from various subject areas. However, the NWB group do support network biology and provide access to major PPI datasets.

NWB can carry out a wide range of graph-theoretic analyses including node degree distribution, shortest path distribution, node betweenness centrality, connected components and clustering coefficients. NWB's strength lies in its array of complex graph-theory algorithms and its flexibility of subject area. It provides a range of PPI datasets from the MINT including fruit fly (*Drosophila melanogaster* - see Figure 6), the nematode (*Caenorhabditis elegans*) and *S. cerevisiae*. There are several layout algorithms including circular, spring embedded and tree map.

To demonstrate the power of NWB, a *D. melanogaster* network was visualised using a Kamada-Kawai layout (Kamada & Kawai, 1989) (see Figure 6a), and analysed to determine network properties including average clustering coefficient, network diameter and clustering, as well as degree distribution and site betweenness (see Figure 6b-e).

yEd

yEd (http://www.yworks.com) is another tool for the visualization and analysis of user-imported networks in any subject area. It is a Java-based application, produced by a company called yWorks. yEd is a front end to the yFiles software library.

yEd offers a range of network analysis features, including graph centrality, node betweenness centrality and closeness centrality. Nodes with a high degree of centrality are shown larger than those with low degree. yEd also supports a wide range of import and export formats including Graph Markup Language (GraphML), Y Graph Framework (YGF) and Graph Modelling Language (GML).

yEd can reduce network complexity by using nested graph hierarchies, which can be condensed into individual nodes. The software has a variety of layout algorithms including hierarchical, organic, orthogonal, tree and circular.

VisANT

VisANT (Hu *et al.*, 2007a,b) is a Java-based tool developed at Boston University, in collaboration with the KEGG group (Kyoto Encyclopaedia of Genes and Genomes, Kanehisa *et al.*, 2008), the Center of Cancer Systems Biology at Harvard University and University of California, San Diego. It is a freely available application for the visualization and analysis of bio-molecular networks. Its strength lies in its integration of PPI and pathway data. VisANT provides access to the KEGG metabolic dataset and Predictome PPI dataset (Mellor, Yanai, Clodfelter, Mintseris & DeLisi, 2002) or can analyse data imported by the user.

VisANT offers a range of statistical topological network analyses, including clustering coefficients, shortest paths, motif and self-loop detection. Proteins are annotated with web links to the relevant KEGG or Predictome web page. Functional associations are provided for 103 species and more detail is given for model species such as *D. melanogaster*, *S. cerevisiae* and human (*Homo sapiens*). Bio-molecular profile data (such as gene expression) can be integrated into networks and used to fill in the gaps in pathways. The software also provides links to the Stanford Microarray Database and Gene Expression Omnibus (GEO) database.

Figure 6. A D. melanogaster network from MINT, analysed in Network Workbench. (a) A network visualization, using the Kamada-Kawai layout. (b) The network properties provided by Network Workbench for this network. The low average clustering coefficient means the proteins in this network do not tend to cluster together. (c) The node degree list for 20 proteins of interest, showing the number of incident edges. Protein CG10032 has a higher degree than any other protein in this list and is therefore the most highly connected. (d) The degree distribution probability gives the proportion of nodes in the network with the given degree. (e) The node betweenness centrality for each protein is a way of ranking the importance of proteins based on the proportion of shortest paths they appear on. A relatively high betweenness value denotes a "bridging" node that connects parts of the network together. (f) The "Guess Visualization" graphical user interface, here visualising the network using a circular layout algorithm.

(a) *Drosophila* network

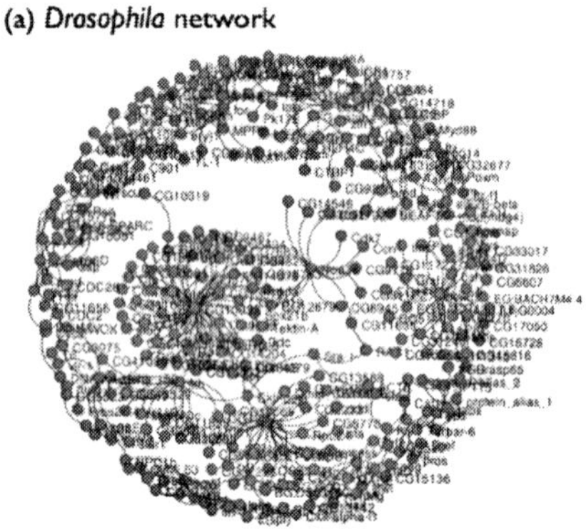

(b) Node degree

	Node	Number of Edges
1	"ACT8"	1
2	"AKAP1"	1
3	"ATPsyn-beta"	1
4	"Act57B"	9
5	"BEAF-32"	1
6	"BEST:CK01296"	1
7	"BG4"	1
8	"BG:DS00180.8"	1
9	"BM-40-SPARC"	1
10	"C901"	1
11	"CDC2"	1
12	"CDC28"	3
13	"CDK3"	3
14	"CDK4"	1
15	"CG10014"	3
16	"CG10017"	7
17	"CG10019"	1
18	"CG10032"	54
19	"CG10039"	1
20	"CG10051"	2

(c) network properties

330 nodes

295 edges

The average clustering coefficient is
0.07384615384615383

The diameter of the network is 9

There are 48 connected components!

Weak Component Clustering was selected.
48 clusters found, generating graphs for the top 48 clusters

(d) Degree distribution

Degree	Probability
1	0.803030E+00
2	0.787879E-01
3	0.606061E-01
4	0.151515E-01
5	0.121212E-01
6	0.303030E-02
7	0.303030E-02
8	0.303030E-02
9	0.606061E-02
10	0.000000E+00
11	0.303030E-02
12	0.000000E+00
13	0.000000E+00
14	0.000000E+00
15	0.303030E-02
16	0.000000E+00
17	0.000000E+00
18	0.000000E+00
19	0.000000E+00
20	0.000000E+00

(e) Node betweenness centrality

Node	Site betweenness
1	0.328000E+03
2	0.135000E+03
3	0.329000E+03
4	0.530400E+04
5	0.218000E+03
6	0.328000E+03
7	0.183000E+03
8	0.321000E+03
9	0.321000E+03
10	0.275000E+03
11	0.319000E+03
12	0.550000E+03
13	0.715000E+03
14	0.199000E+03
15	0.152400E+04
16	0.270700E+04
17	0.329000E+03
18	0.336150E+05
19	0.287000E+03
20	0.512000E+03

(f) Guess GUI network visualisation

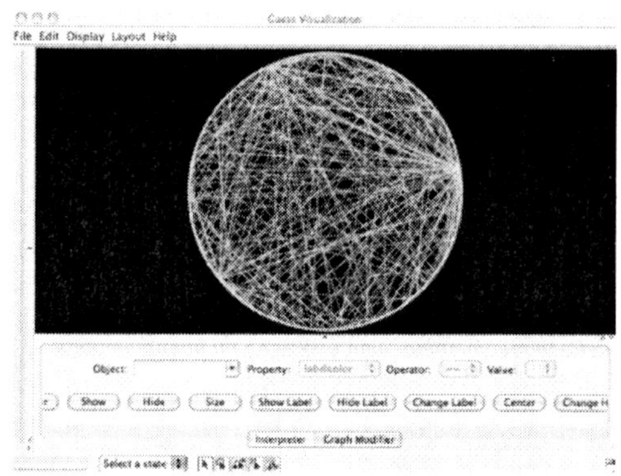

To reduce complexity, nodes can be grouped into metanodes. Available layout algorithms are circle, spoke, random, relaxing and elegant relaxing. Additionally, node properties and colours can be edited. The network can be exported in VisML format, which can be published online as a visualization containing clickable links to KEGG and Predictome.

BiologicalNetworks

BiologicalNetworks (Baitaluk *et al.*, 2006a) is a tool for visualising and analysing bio-molecular networks and integrating them with other resources, such as the Gene Ontology and taxonomies. The software is Java-based, freely available for academic use and can be executed locally or run as a web application. It is developed at the San Diego Supercomputer Centre at UCSD.

BiologicalNetworks can carry out shortest path analysis and calculate a variety of network statistics including number of nodes and edges, average node degree, network diameter and cycle identification. The BiologicalNetworks server provides a system where the user can query the PathSys (Baitaluk *et al.,* 2006b) integration server, comprising 14 curated genomic databases covering 8 species, including *S. cerevisiae, D. melanogaster, H. sapiens,* thale cress (*Arabidopsis thaliana*) and mouse (*Mus musculus*), in addition to the retrieval of multiple datasets. PPI networks can be integrated with other bio-molecular networks, as well as gene expression data, and the nomenclature is well integrated. BiologicalNetworks can then add further information, such as cellular compartment (important in filtering out false-positives from PPI networks), correlation networks and Gene Ontology data. Pathways can be identified based on such integrated networks.

BiologicalNetworks supports a variety of layout algorithms, including grid, circle, spring-embedded and hierarchical. Node colour can be changed by the user to highlight proteins of interest. BiologicalNetworks also has facilities to carry out dynamical modelling and simulations.

WEB SERVICES AND WORKFLOWS

A web service is a type of software application, defined formally by the W3C standards body as 'a software system designed to support interoperable machine-to-machine interaction over a network. It has an interface described in a machine-processable format' (http://www.w3.org/TR/ws-arch/). Web services are a distributed technology, originally developed by the e-business community as a solution to the interoperability issues of data exchange. Crucially, they were designed to be platform- and language-independent.

Web services use standard Internet protocols, and offer the necessary architecture for flexible and expandable integration of diverse scientific tools. More simply they can be thought of as callable routines made available over a network. This is the most common style of web service use, particularly in the bioinformatics domain. Web service protocols stress interoperability between services and imply a certain level of location transparency. This means that resources that are located in geographically distinct locations can be accessed by a user from a single location.

Web services are made up of a set of standards, or protocols, which are used to implement, define and locate them. These protocols each have a specification defined by the W3C. The core protocols are Service Oriented Architecture Protocol (SOAP), Web Services Description Language (WSDL) and Universal Description, Discovery and Integration (UDDI). Extensible Markup Language (XML) forms the basis of all three protocols, and is a platform-independent, general-purpose markup language with user-defined tags, designed to facilitate the sharing of data between heterogeneous systems. The web services protocol stack in Table 4 illustrates the relationship between the different protocols.

There are a number of ways to deploy web services, and almost every major programming language has now incorporated a web service toolkit. There are toolkits available for Java (Apache Axis), Perl (the SOAP::Lite module), Python (Zolera Soap Infrastructure) and C++ (gSOAP) amongst others. Support for automatic generation of WSDL, client-side stubs and server-side skeletons vary across the different toolkits. The language-independent nature of web services allows a developer to write a service in their favourite programming language, and, theoretically, a client program written in any other language can invoke it.

In order to make use of a web service, a client application must be designed to be able to send and receive data to and from the service. A client to a web service can be anything from a simple command-line script that takes information such as service location and parameters to be passed as arguments, to a more sophisticated graphical workbench that allows for the discovery of many different services from many locations. The latter is also known as a workflow enactment engine. A workflow is a chain of web services, where the output of one service is the input to the next. One such workbench is Taverna (Oinn *et al.,* 2004). Taverna was developed as part of the myGrid consortium, an initiative that seeks to develop a loosely coupled, service-based suite of open-source middleware components for bioinformaticians. This facilitates *in silico* research by pulling together resources from various providers, releasing the user from the need to install a variety of applications on their own computer, and instead harnesses the power of remote computers for the running of analyses.

Table 4. The web services protocol stack

Layer Name	Description	Protocols used
Service Transport	Responsible for transporting messages between network applications	HTTP, SMTP, FTP
Messaging	Responsible for encoding messages in XML format	SOAP, XML-RPC
Service Description	Used to describe a public interface to a specific web service	WSDL
Service Discovery	Centralises services into a common registry such that network web services can publish their locations and descriptions. Facilitates the discovery of services	UDDI

Figure 7 is a screenshot of Taverna, and comprises the Available Services window, which lists web services that can be included in a workflow, the Advanced Model Explorer, which enables workflow creation, and the workflow diagram pane, which depicts the workflow. By selecting services, or processors, from the list of those available, workflows can be built up by connecting inputs and outputs from one processor to the next. Processors are independent of each other, which means that different processors can be integrated in a workflow despite heterogeneous implementation techniques. The workflow enactment engine then handles the actual running of the workflow, and both the workflow itself and any results generated from running it may be saved.

To demonstrate the power of this approach, web services for performing various network analysis tasks have been deployed using the Soaplab tool (Senger, Rice & Oinn, 2003), which wraps command-line applications as web services and allows them to be discovered and invoked using the Taverna workbench. The services are constructed using implementations of graph-theoretic algorithms from the NetworkX package (http://networkx.lanl.gov) and the Perl Graph library (http://search.cpan.org/~jhi/Graph-0.84). The following examples show how these services, together with those produced by other providers, can be used to analyse PPI networks for significant network features.

Network Properties

Various graph-theoretic measures can be applied to PPI networks, which help characterise network topology and elucidate basic organisation. A workflow was developed to take a PPI network as input and calculate several topological parameters (see Figure 8).

The parameters calculated are the total number of nodes, which is the total number of proteins in the network, and the total number of edges or interactions linking them. The total number of disconnected sub-networks is calculated, and the largest sub-network is identified and returned as a network object to enable the calculation of the diameter and radius of the network (as these cannot be calculated for a whole, unconnected network). The proteins in the network are ranked by their degree, to identify hubs, and by their betweenness centrality, to determine which proteins are important in terms of bridging parts of the network. They are also ranked by their clustering coefficients, which measure the extent to which proteins that are adjacent to any one protein are also adjacent to each other. The workflow also contains a nested workflow that compares the list of proteins showing high betweenness centrality and high degree (*i.e.* the hubs) and pulls out the proteins that show either one or the other property, or both.

Node Locality

As holistic PPI networks are not easy to interpret by eye, it can be useful to study the local network around a protein of interest. To do this, the protein is selected, along with all the other proteins up to a given number of

Figure 7. The Taverna Workbench graphical user interface showing an example workflow, built up within the Advanced Model Explorer pane and displayed pictorially in the right hand pane.

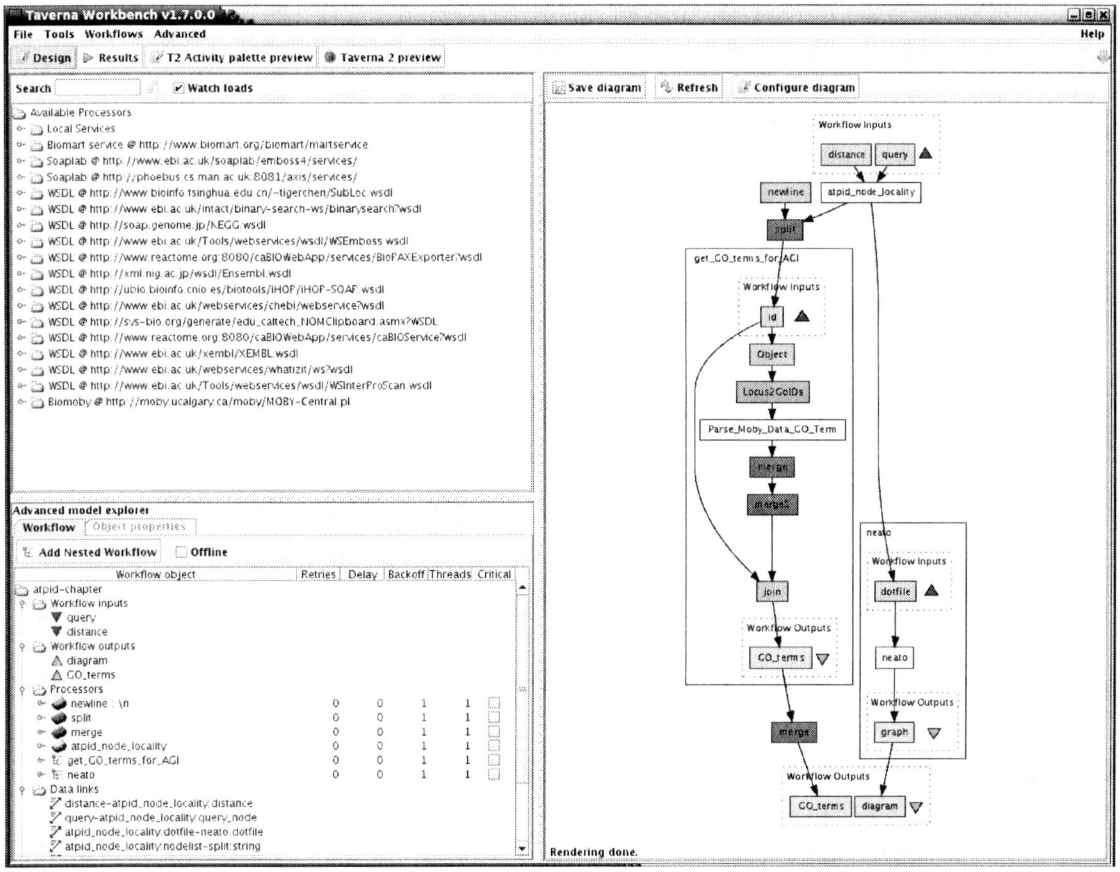

edges (or interactions) away, and any interconnections between these neighbouring proteins are identified. Due to the scale-free property of PPI networks the number of edges rarely exceeds three, as any larger would return most of the network. The resulting local network may be biologically more relevant than studying or visualising the entire network at once.

The workflow shown in Figure 9 has been constructed to view the local networks of selected proteins from AtPID. This database contains inferred interactions that have been computationally determined using a number of methods. The workflow takes as input the *Arabidopsis* Genome Initiative (AGI) accession number of the gene of interest, and an integer that determines how far away from the gene of interest the local network should extend.

This workflow is useful for characterising unknown genes and finding neighbours that are likely to have related cellular functions. This approach can be used to suggest the biological functions of uncharacterised proteins. In the example shown in Figure 9, the Gene Ontology terms associated with the query protein AT5G24150 show a common theme, which may have helped to characterise it, if its role had been unknown.

Annotating Metabolic Pathways with PPIs

The motivation for developing this workflow was to establish if, for a given metabolic pathway, any enzymes are mediated by PPIs, as this could generate hypotheses about the regulation of that pathway. The workflow accepts

Figure 8. (a) The network properties workflow and (b) the results obtained from running this workflow on the S. cerevisiae PPI dataset obtained from DIP.

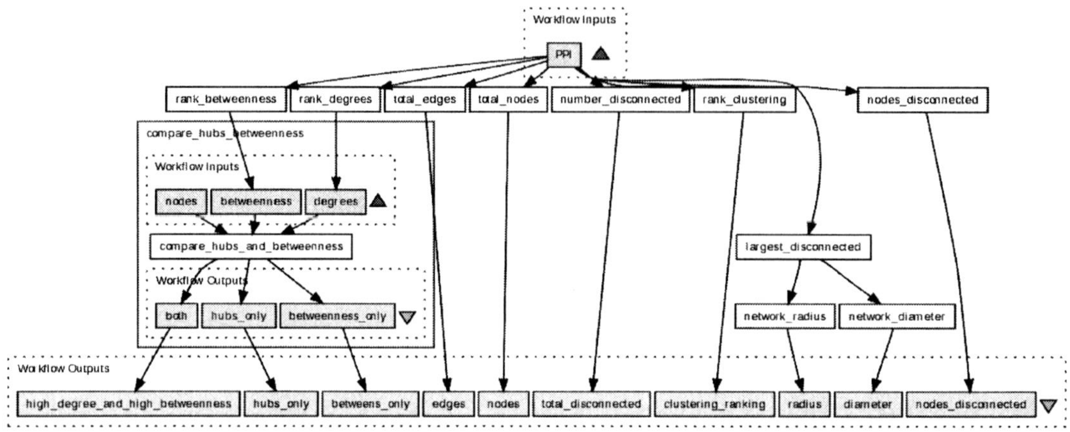

(a)

(b)

The basic topological metrics calculated:	
Total edges	17285
Total nodes	4966
Total disconnected subgraphs	32
Network diameter	11
Network radius	6

as input a KEGG pathway identifier, and then retrieves the corresponding pathway and its constituent enzyme identifiers. These are then queried against the IntAct PPI database, to discover if any proteins interact with more than one of these pathway enzymes.

The workflow was executed using the *S. cerevisiae* glycolytic pathway (see Figure 10a). The top 20 results in figure 10b show which proteins, identified by UniprotKB/Swiss-prot codes, interact with the most enzymes along the pathway ordered from the greatest number of interactions to least. The first identifier on each line is the interacting protein, and the following proteins are enzymes that this protein interacts with. The results revealed that one protein (with the UniprotKB/Swiss-prot identifier P53917) interacts with six of the nine enzymes listed in KEGG as involved with glycolysis (phosphoglycerate mutase, pyruvate decarboxylase, phosphopyruvate hydratase, triose-phosphate isomerase, pyruvate kinase and fructose-bisphosphate aldolase) and this protein was selected to be investigated further.

The literature identifies this protein as being Far11, a membrane-bound protein involved in pheromone-induced G1 cell cycle arrest (Kemp & Sprague, 2003). To answer the question of how exactly Far11 mediates cell cycle arrest, from the results of the workflow we may hypothesise that Far11 binds the six glycolytic enzymes to slow down glycolysis by removing them from the cytosol, where glycolysis would otherwise take place. This is depicted in Figure 10c. Glycolysis provides energy for many cellular functions, including cell division. If glycolysis were slowed down then it may be that sufficient energy would not be available for this to occur.

FUTURE DIRECTION

Currently even "holistic" networks represent the workings of a single, generic cell type for a species. An anticipated major movement within systems biology is the move from generic cell types, to specific cell types, to several interacting cells and eventually to whole organs and organisms.

Integration of different classes of bio-molecular network will more accurately represent the real biology of the cell. The combination of PPI, metabolic and gene regulatory networks will require new types of analysis,

Figure 9. (a) The node locality workflow. (b) The gene list of the local network of the A. thaliana gene AT5G24150, which encodes Squalene monooxygenase 1. (c) A visualization of this local network.

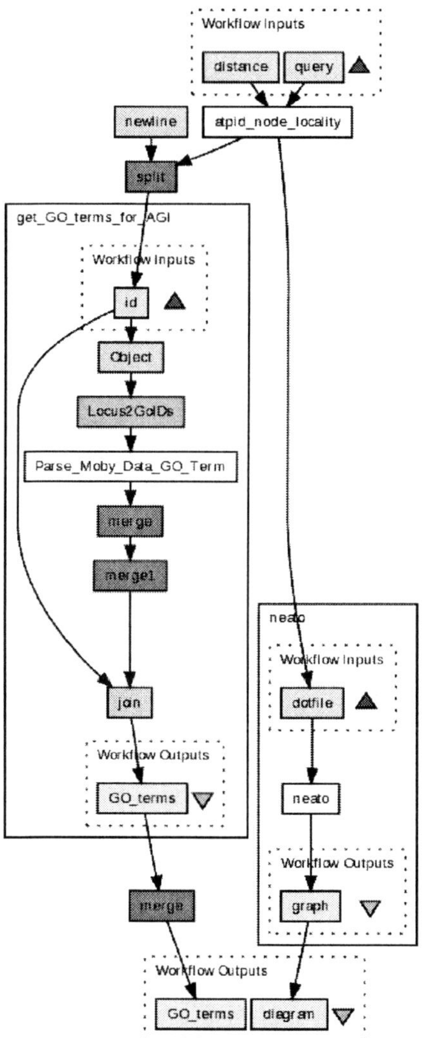

(a)

AT1G58440
sterol biosynthetic process
endomembrane system
oxidoreductase activity

AT4G37760
sterol biosynthetic process
response to jasmonic acid stimulus
response to wounding
endomembrane system
oxidoreductase activity

AT2G22830
sterol biosynthetic process
oxidoreductase activity

AT5G24140
sterol biosynthetic process
endomembrane system
oxidoreductase activity

AT5G24160
sterol biosynthetic process
endomembrane system
oxidoreductase activity

AT5G24150
squalene monooxygenase activity
sterol biosynthetic process
endomembrane system

(b)

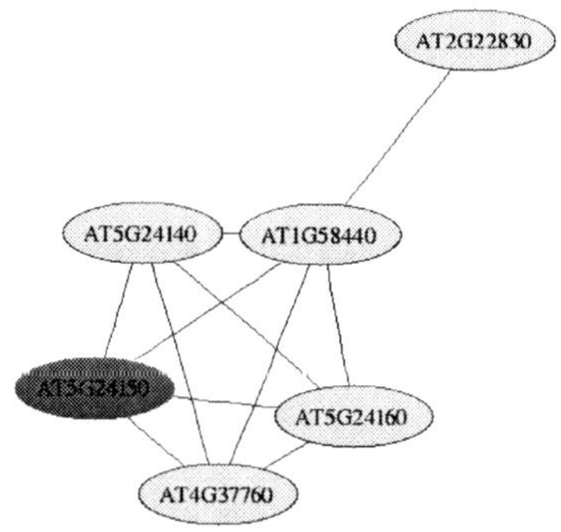

(c)

and will increase the size of the networks. The closer the networks are to reality, the more powerful the results of analyses will become, as actual biological phenomena can be more easily identified. The attempts to resolve nomenclature issues are likely to result in better coverage of synonyms, and therefore allow integration of datasets with greater ease.

In time, PPI datasets should become better annotated. Currently, there is rarely enough detailed information regarding temporal, compartmental or even tissue specific expression to replicate accurately the biological system when analysing PPI networks generated from two-hybrid data. It is very important for such factors to be taken into account so improved annotation will be increasingly used to filter two-hybrid datasets, and more thorough PPI detection techniques will add weight to the credibility of interactions that are reported as taking place.

A major future development in PPI network analysis will be the increase in use of high-performance computing. The algorithmic complexity of some graph-theoretic operations means that for very large PPI networks the time taken to find a solution becomes prohibitively large. In recent years, distributed, high-performance computing (HPC) has emerged as a solution in bioinformatics for dealing with the analysis of such large datasets. Graph-theoretic operations can be parallelised in order to take advantage of multi-core machines to speed up analysis, or software can be deployed across a 'grid' or computing cluster such as Condor (Litzkow, Livny & Mutka, 1988). The types of tasks suitable for such a cluster are those that can be partitioned into independent parts, also known as "embarrassingly parallel" jobs. These are loosely coupled segments of the job, which can be run separately, with the results combined at the end of the computation. Web services are a useful extension of this, as distributed workflows can be built where each web service or workflow component can be executed on a different machine, thereby reducing the local load and increasing efficiency and overall computation time of the workflow.

Network analysis functions are frequently duplicated over many available tools. As new functions are added to an application, by its own developers and the community, the disadvantages that it currently has (compared to other tools) may disappear. Most importantly, it is likely that the powerful graph-theoretic analyses currently implemented in generic network software will be integrated into tools specifically designed for bio-molecular network analysis. Over time, a few tools may dominate, and development effort may be focused on these.

CONCLUSION

The properties of bio-molecular networks can tell us much about the biology of a system. PPI networks represent protein modification, complex formation, protein localisation and signal transduction. Analysis of PPI networks can help to mine these large datasets for control mechanisms of molecular and cellular processes, and much of this can be achieved using graph-theoretic algorithms.

Several standalone desktop tools for the visualization and analysis of PPI networks are available. Each has its own combination of visualization and analysis functions, and some extend into dynamic modelling and simulation. Many are in the public domain and are open-source, allowing the community to extend their functionality. Different types of tool are most appropriate for different tasks. Standalone tools tend to be well-developed applications for doing a wide variety of visualization and analysis tasks, so are well suited to general exploration of a PPI network. Some have optimised their graph-theoretical analysis algorithms to be able to analyse very large networks on standard desktop machines. They tend to have attractive user interfaces, enabling ease of use for non-expert computer users, and require little or no programming knowledge. The more advanced tools are usually supported by a website where the software is available for download, together with comprehensive documentation and support.

Some standalone software is designed specifically for use on bio-molecular networks, and tends to be particularly good at specialised biological analyses such as the integration of biological datasets. Software for use on networks in any field often has a more advanced set of graph-theoretical operations. The very nature of this type of software means that a particular network analysis task cannot easily be integrated with external operations. However, the functionality of Cytoscape and VisANT can be extended by using plug-ins.

Remote invocation of a web service is beneficial, as a user can harness remote computing power. However, this does depend on the presence of an active network connection, and any failure to maintain the connection would result in the web service being unusable, a problem that does not affect independently running standalone

Figure 10. (a) A workflow for annotating metabolic pathways with PPIs. (b) Some of the top results when running the workflow on the S. cerevisiae glycolytic pathway. The first column is the interacting protein, while the remaining proteins on each line are the enzymes along the pathway that each protein interacts with. The protein investigated further, P53917, and the enzymes it interacts with are highlighted. Other proteins that interact with more enzymes along the pathway, such as P10592 and Q03330 are not investigated further as they interact with many other proteins (967 and 439 respectively). (c) A schematic of a cell showing a hypothetical role for Far11 in mediating pheromone-induced cell-cycle arrest. (i) Glycolysis continuing as normal in the cell: the black circles are enzymes involved with glycolysis and the arrows between them denote the metabolites produced in one reaction and used in the next. (ii) In the presence of the pheromone, the membrane protein Far11 may bind the enzymes, removing them from the cytosol and slowing down glycolysis.

(a)

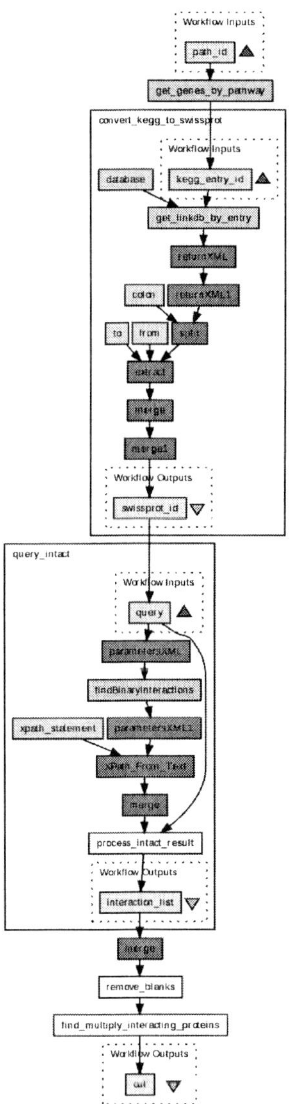

(b)

P10592	P32473 P00560 P40047 P09624 P00924 P00360 P00358 P33401 P52910 P07246 P47771 P12695
Q03330	P00560 P00359 P16861 P00925 P00360 P00358 P14540 P52910 P07246 P16862 P00331 P00330
P41940	P00549 Q07471 P40047 P16387 P09624 P33401 P52910 P07246 P12695 P52489 P46367
P02994	P32473 Q07471 P40047 P09624 P00358 P52910 P07246 P16862 P12695 P00330 P52489
P07259	P32473 Q07471 P40047 P16387 P33401 P52910 P07246 P12695 P52489 P46367
P40359	P12709 P00942 P26263 P16861 P00924 P14540 P00950 P00330 P52489
P38720	P00549 P00560 P00924 P00925 P00360 P14540 P06169 P16467 P00330
P02557	Q07471 P40047 P52910 P07246 P47771 P12695 P52489 P46367
P43555	P00560 P00359 P00925 P14540 P00950 P06169 P00330
P38962	P00560 P00359 P00925 P14540 P00950 P06169 P00330
P53917	P00549 P00942 P00925 P14540 P00950 P06169 P00330
P07866	P17709 P00560 P00942 P09624 P00924 P00925 P06169
Q02821	Q01574 P17709 P00359 P00360 P00358 P09201 P00330

(c)

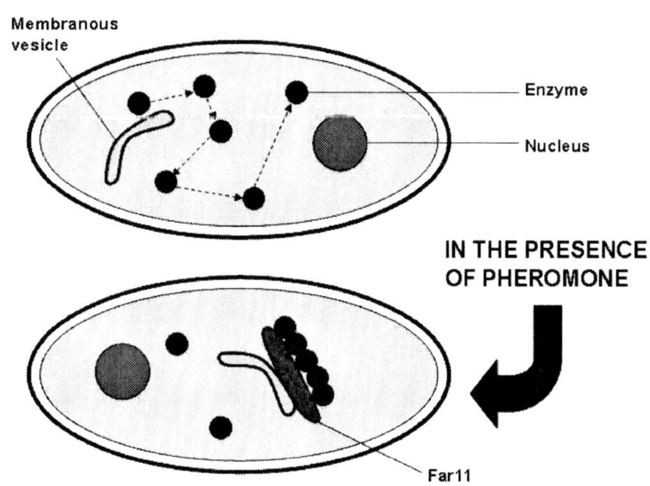

tools. A further problem with web services is that, as the number of simultaneous users connected to a service increases, accessing and returning data will be slower.

Web services are maintained on the server side by the service provider, who is responsible for patching and maintaining software, and therefore need not be updated by the user. However, this has the disadvantage that if the web service changes behaviour, the user has no choice but to update. Due to their relative novelty, web services tend to suffer from poor documentation. If a service provider chooses to change the location of their web services, then the user may be responsible for finding the new location, even if this has been poorly publicised.

The functionality of web services may be built upon by creating workflows, which comprise a set of web services, chained together to automate analyses. Workflows in bioinformatics are therefore transformed into a fully automated sequence of steps, starting with inputs provided by the user, and transformed by a sequence of web services into an end result. While each workflow carries out a very specific operation, as an approach they are very flexible, because their constituent web services can be combined in a number of ways. They also offer the advantage of explicitly capturing experimental provenance and increasing reproducibility. As long as different researchers have different tasks to complete, none of the existing single applications will be the best for everyone.

Very often, the PPI networks that are constructed are holistic, and therefore extremely large and complex. In order to concentrate on drawing a particular conclusion, simplifications have to be made. A smaller simplified network can then be analysed or visualised. Networks can be simplified either by isolating local areas of interest, or by reducing the level of detail. There are several different ways of doing each of these. The sub-network around a protein of interest can be isolated. This can be used to suggest biological function for uncharacterised proteins, which is likely to be related to that of their better-annotated neighbours. In addition, sub-networks with particular properties can be isolated. These can be clusters of highly interconnected nodes, suggesting a related function. To reduce complexity, groups of nodes can be condensed into metanodes, simplifying the graph for visualization purposes while preserving the complexity of the data for further analysis. This can be done by grouping on any one of several properties, depending on what is relevant for the current analysis.

Even though the analysis of PPI networks alone can be quite powerful, results that are far more meaningful can be gained when these networks are integrated with other data types. These can either be other network types (*e.g.* metabolic, gene regulatory or signal transduction *etc.*), links to databases (*e.g.* Gene Ontology) or experimental data (*e.g.* transcriptomic, metabolomic or Single Nucleotide Polymorphism (SNP) data). However, the inconsistency of nomenclature between data sets can often limit integration, but there are many ongoing attempts to remedy this. Amongst the more common integration tasks is the colouring of networks with gene expression data. This can suggest shared gene regulation mechanisms and functions between proteins.

Holistic PPI networks are constructed from two-hybrid data, as well as other experimental data. Two-hybrid screens are valuable for their volume of data, but are notorious for both false-positive and false-negative results, due to the artificial context in which each interaction does or does not take place. As with any computation, the quality of the input data absolutely limits the quality of any results. Integration of the network with spatial and temporal data can allow the filtering out of interactions that would never have the chance to occur. Spatial data is found in a number of databases, whereas temporal data is much less common.

This chapter has outlined some issues surrounding PPI network analysis. It has demonstrated two different types of software used to analyse a network, desktop applications and web services, and described their advantages and disadvantages.

REFERENCES

Alfarano, C., Andrade, C. E., Anthony, K., Bahroos, N., Bajec, M., Bantoft, K., et al. (2005). The Bio-molecular Interaction Network Database and related tools 2005 update. *Nucleic Acids Research. 33*(Database Issue), D418-24.

Ashburner, M., Ball, C. A., Blake, J. A., Botstein, D., Butler, H., Cherry, J. M., Davis, A. P., Dolinski, K., Dwight, S. S., Eppig, J. T., Harris, M. A., Hill, D. P., Issel-Tarver, L., Kasarskis, A., Lewis, S., Matese, J.C., Richardson,

J. E., Ringwald, M., Rubin, G. M., & Sherlock, G. (2000). Gene ontology: tool for the unification of biology. *Nature Genetics, 25*, 25-9.

Assenov, Y., Ramírez, F., Schelhorn, S. E., Lengauer, T., & Albrecht, M. (2008). Computing topological parameters of biological networks. *Bioinformatics, 24*(2), 282-284.

Bader, G. D., & Hogue, C. W. (2003). An automated method for finding molecular complexes in large protein interaction networks. *BMC Bioinformatics, 4*(2), doi:10.1186/1471-2105-4-2.

Bader, G. D., Cary, M. P., & Sander, C. (2006). Pathguide: A pathway resource list. *Nucleic Acids Research, 34*(Database issue), D504-6.

Bader, D. A., & Madduri, K. (2007). A Graph-Theoretic Analysis of the Human Protein-Interaction Network Using Multi-core Parallel Algorithms. *In Proceedings Parallel and Distributed Processing Symposium* (pp. 1-8).

Baitaluk, M., Sedova, M., Ray, A., & Gupta, A. (2006). BiologicalNetworks: visualization and analysis tool for systems biology. *Nucleic Acids Research, 34*(Webserver Issue), W466-71.

Baitaluk, M., Qian, X., Godbole, S., Raval, A., Ray, A., & Gupta, A. (2006). PathSys: integrating molecular interaction graphs for systems biology. *BMC Bioinformatics, 7*, 55.

Barabási, A. L., & Albert, R. (1999). Emergence of scaling in random networks *Science, 286*(5439), 509-12.

Barabási, A. L., & Oltvai, Z. N. (2004). Network biology: understanding the cell's functional organization. *Nature Reviews Genetics, 5*, 101-113.

Breitkreutz, B. J., Stark, C., Reguly, T., Boucher, L., Breitkreutz, A., Livstone, M., Oughtred, R., Lackner, D. H., Bähler, J., Wood, V., Dolinski, K., & Tyers, M. (2008). The BioGRID Interaction Database: 2008 update. *Nucleic Acids Research, 36*(Database Issue), D637-40.

Chatr-Aryamontri, A., Ceol, A., Palazzi, L. M., Nardelli, G., Schneider, M. V., Castagnoli, L., & Cesareni, G. (2007). MINT: the Molecular INTeraction database. *Nucleic Acids Research, 35*(Database Issue), D572-4.

Chuang, H-Y., Lee, E., Liu, Y-T., Lee, D., & Ideker, T. (2007) Network-based classification of breast cancer metastasis. *Molecular Systems Biology, 3*, 140.

Cline, M. S., Smoot, M., Cerami, E., Kuchinsky, A., Landys, N., Workman, C., et al. (2007). Integration of biological networks and gene expression data using Cytoscape. *Nature Protocols, 2*(10), 2366 - 2382.

Cui, J., Li, P., Li, G., Xu, F., Zhao, C., Li, Y., Yang, Z., Wang, G., Yu, Q., Li, Y., & Shi, T. (2008) AtPID: *Arabidopsis thaliana* protein interactome database--an integrative platform for plant systems biology. *Nucleic Acids Research, 36*(Database issue), D999-1008.

Deng, M., Zhang, K., Mehta, S., Chen, T., & Sun, F. (2003) Prediction of protein function using protein-protein interaction data. *Journal of Computational Biology, 10*(6), 947-960.

Eisenberg, D., Marcotte, E. M., Xenarios, I., & Yeates, T. O. (2000). Protein function in the post-genomic era. *Nature, 405*(6788), 823-6.

Fields, S., & Song, O. (1989). A novel genetic system to detect protein-protein interactions. *Nature, 340*(6230), 245-246.

Han, J-D.J., Bertin, H., Hao, T., Goldberg, D. S., Berriz, G. F., Zhang, L. V. Dupuy, D., Walhout, A. J. M., Cusick, M. E., Roth, F. P. & Vidal, M. (2004). Evidence for dynamically organized modularity in the yeast protein–protein interaction network. *Nature, 430*, 88-93.

Han, J-D.J. (2008). Understanding biological functions through molecular networks. *Cell Research, 18*, 224–237.

Hu, Z., Ng, D. M., Yamada, T., Chen, C., Kawashima, S., Mellor, J., Linghu, B., Kanehisa, M., Stuart, J. M., & DeLisi, C. (2007). VisANT 3.0: New modules for pathway visualization, prediction and construction. *Nucleic Acids Research, 35* (Webserver issue), W625-W632.

Hu, Z., Mellor, J., Wu, J., Kanehisa, M., Stuart, J., & DeLisi, C. (2007). Towards zoomable multidimensional maps of the cell. *Nature Biotechnology, 25*, 547-554.

Ideker, T., Ozier, O., Schwikowski, B., & Siegel, A. F. (2002). Discovering regulatory and signalling circuits in molecular interaction networks. *Bioinformatics, 18*(Supplement 1), S233–S240.

Ito, T., Chiba, T., Ozawa, R., Yoshida, M., Hattori, M., & Sakaki, Y. (2001) A comprehensive two-hybrid analysis to explore the yeast protein interactome. *Proceedings of the National Academy of Sciences, 98*(8), 4277-8.

Jeong, H., Mason, S., Barabási, A. L., & Oltvai, Z. N. (2001). Lethality and centrality in protein networks. *Nature, 411*, 41-42.

Joshi-Tope, G., Gillespie, M., Vastrik, I., D'Eustachio, P., Schmidt, E., de Bono, B., Jassal, B., Gopinath, G. R., Wu, G. R., Matthews, L., Lewis, S., Birney, E., Stein L. (2005). Reactome: a knowledgebase of biological pathways. *Nucleic Acids Research, 33*(Database issue), D428-32.

Kamada, T., & Kawai, S. (1989). An algorithm for drawing general undirected graphs. *Information Processing Letters, 31*(1), 7-15.

Kanehisa, M., Araki, M., Goto, S., Hattori, M., Hirakawa, M., Itoh, M., Katayama, T., Kawashima, S., Okuda, S., Tokimatsu, T., & Yamanishi, Y. (2008). KEGG for linking genomes to life and the environment. *Nucleic Acids Research, 36*(Database issue), D480-4.

Kemp, H. A., & Sprague, G. F. (2003). Far3 and five interacting proteins prevent premature recovery from pheromone arrest in the budding yeast *Saccharomyces cerevisiae. Molecular and Cellular Biology, 5*(23), 1750-63.

Kerrien, S., Alam-Faruque, Y., Aranda, B., Bancarz, I., Bridge, A., Derow, C., et al. (2007). IntAct - open-source resource for molecular interaction data. *Nucleic Acids Research, 35*(Database Issue). D561-5.

Kitano, H. (2002). Computational systems biology. *Nature, 420*(6912), 206-10.

Litzkow, M., Livny, M., & Mutka, M. W. (1988). Condor - A Hunter of Idle Workstations. *Proceedings of the 8th IEEE Distributed Computing Conference*, (San Jose, CA) (pp. 104-111).

Maere, S., Heymans, K., & Kuiper, M. (2005). BiNGO: a Cytoscape plugin to assess overrepresentation of Gene Ontology categories in biological networks. *Bioinformatics. 21*. 3448-3449.

Mellor, J. C., Yanai, I., Clodfelter, K. H., Mintseris, J., & DeLisi, C. (2002). Predictome: A database of putative functional links between proteins. *Nucleic Acids Research, 30*(1), 306-9.

Mishra, G. R., Suresh, M., Kumaran, K., Kannabiran, N., Suresh, S., Bala, P., et al. (2006). Human Protein Reference Database—2006 update. *Nucleic Acids Research, 34*(Database issue), D411-4.

Oinn, T., Addis, M., Ferris, J., Marvin, D., Senger, M., Greenwood, M., Carver, T., Glover, K., Pocock, M. R., Wipat, A., & Li, P. (2004) Taverna: A tool for the composition and enactment of bioinformatics workflows. *Bioinformatics, 20*(17), 3045-54.

Oltvai, Z. N., & Barabási, A. L. (2002). Life's complexity pyramid. *Science, 298*, 763-764.

Salwinski, L., Miller, C. S., Smith, A. J., Pettit, F. K., Bowie, J. U., & Eisenberg, D. (2004). The Database of Interacting Proteins: 2004 update. *Nucleic Acids Research. 32*(Database issue), D449-51.

Saraiya, P., North C., & Duca, K. (2005). Visualizing biological pathways: requirements analysis, systems evaluation and research agenda. *Information Visualization* (pp. 1-15).

Schwikowski, B., Uetz, P., & Fields, S. (2000). A network of interacting proteins in yeast. *Nature Biotechnology, 18*(12), 1257-1261.

Shannon, P., Markiel, A., Ozier, O., Baliga, N. S., Wang, J. T., Ramage, D., Amin, N., Schwikowski, B., & Ideker, T. (2003). Cytoscape: a software environment for integrated models of bio-molecular interaction networks. *Genome Research. 13*(11), 2498-504.

Suderman, M., & Hallett, M. (2007). Tools for Visually Exploring Biological Networks. *Bioinformatics, 23*(20), 2651-9.

Uetz, P., Giot, L., Cagney, G., Mansfield, T. A., Judson, R. S., Knight, J. R., et al. (2000). A comprehensive analysis of protein-protein interactions in *Saccharomyces cerevisiae. Nature*, *403*(6770), 601-3.

van de Vijver, M. J., He, Y. D., van't Veer, L. J., Dai, H., Hart, A. A., Voskuil, D. W., et al. (2002). A gene-expression signature as a predictor of survival in breast cancer. *The New England Journal of Medicine*, *347*(25), 1999-2009.

Wang, Y., Klijn, J.G., Zhang, Y., Sieuwerts, A. M., Look, M. P., Yang, F., et al. (2005). Gene-expression profiles to predict distant metastasis of lymph-node-negative primary breast cancer. *Lancet*, *365*, 671–679.

Yu, H., Kim, P. M., Sprecher, E., Trifonov, V., & Gerstein, M. (2007). The Importance of Bottlenecks in Protein Networks: Correlation with Gene Essentiality and Expression Dynamics. *PLoS Computational Biology, 3*(4), e59.

Chapter XVII
Network Querying Techniques for PPI Network Comparison

Valeria Fionda
Università della Calabria, Italy

Luigi Palopoli
Università della Calabria, Italy

ABSTRACT

The aim of this chapter is that of analyzing and comparing network querying techniques as applied to protein interaction networks. In the last few years, several automatic tools supporting knowledge discovery from available biological interaction data have been developed. In particular, network querying tools search a whole biological network to identify conserved occurrences of a query network module. The goal of such techniques is that of transferring biological knowledge. Indeed, the query subnetwork generally encodes a well-characterized functional module, and its occurrences in the queried network probably denote that this function is featured by the associated organism. The proposed analysis is intended to be useful to understand problems and research issues, state of the art and opportunities for researchers working in this research area.

INTRODUCTION

Biological data about molecular interactions are growing quickly. This fast growth has rendered the design and the development of automatic tools necessary for retrieving interesting information and discovering new knowledge. Interaction data are usually represented through the *biological networks* model. The comparison of such biological networks across species or different conditions is useful to understand the mechanisms underlying life processes (Zhang, 2008). Generally, biological networks are represented as graphs that can be fed as input to techniques suitable for topological and functional comparison. Such techniques analyze those networks by exploiting specialized algorithms and methodologies in order to infer new information about cellular activity and evolutive processes of the species.

A graph is a set of objects called *nodes* or *vertices* and connected by links called *edges*. More formally, a graph is an ordered pair $G = (V, E)$, in which V is the set of nodes and E is the set of edges, so that the elements

Figure 1. (a) An example of undirected graph and (b) an example of directed graph

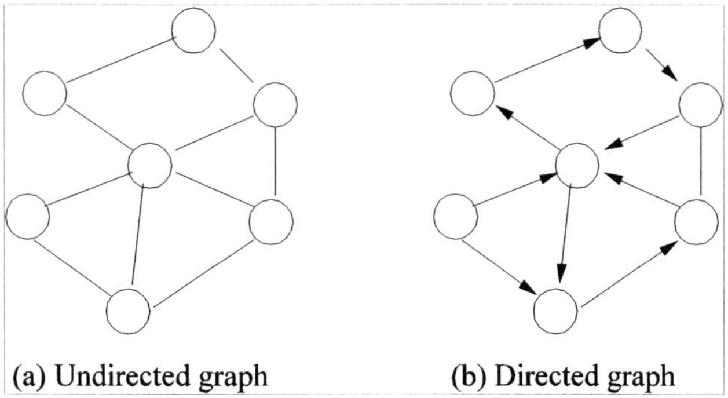

from *E* are pairs of elements from *V*. In an *undirected* graph, an edge linking nodes *A* and *B* can be traversed in both directions. In a *directed* graph, each edge is intended to be traversable in just one direction.

Different types of graphs are used to represent different types of biological networks. In fact, several kinds of biological networks have been defined, each of which stores interaction information related to specific entities or molecules. Main types of networks thereof are: transcriptional regulatory networks, signal transduction networks, metabolic networks and protein interaction network (or PIN). In the *transcriptional regulatory networks* the nodes of the graph represent genes and the edges are directed. An edge connects a source gene to a target gene if the source gene produces an RNA or protein molecule that functions as a transcriptional activator or inhibitor of the target gene. If the gene is an activator, then it is the source of a positive regulatory connection; if it is an inhibitor, then it is the source of a negative regulatory connection. In the *signal transduction networks*, the graph vertices represent proteins and the edges are directed. This type of network stores information about the processes by which a cell converts one kind of signal or stimulus into another. In particular, the signal transduction corresponds to the relaying of molecular signals or physical ones (for example, sensory stimuli) from a cell's exterior to its intracellular response mechanisms. In the *Metabolic networks*, the nodes represent metabolites and the edges are directed. These networks store the set of metabolic and physical processes of the cell and comprise the chemical reactions underlying the metabolism as well as the regulatory interactions that guide these reactions. In the *Protein Interaction Networks*, instead, the nodes represent proteins and the edges are undirected. They store information about the set of interactions between pairs of proteins in a proteome.

In the last few years, due to the large amount of experimentally discovered interaction data, many databases have been devised and made free accessible online. These databases allow storing and retrieving molecular interaction information (Kanehisa, 2000; Salwinski, 2004; Chatr-aryamontri, 2006). Many of them allow the user to search both online for interaction data and to download database files containing all stored interaction information.

Moreover, several automatic tools supporting knowledge discovery from available interaction data have been developed. Among them, the most related to the topic of this chapter are those tools designed to compare biological networks (see also, in this volume, the chapter *Discovering Interaction Motifs from Protein Interaction Networks*).

To illustrate, there are several ways to compare biological networks, but three of them, *network alignment*, *network integration* and *network querying*, can be regarded as the most significant ones (Sharan, 2006). Network alignment is the process of globally comparing two or more networks of the same type in different species, identifying similarity and dissimilarity regions. Network alignment is commonly applied to detect, across species, conserved sub-networks which are likely to represent common functional modules (Singh, 2007). Network integration is the process of combining several networks of the same species and different kinds, to study their interrelations. Network integration can be used to predict protein interactions and identify protein modules supported by interactions of different types (Kelley, 2005; Zhang, 2005). Finally, network querying techniques, which are the main topic of this chapter, search a whole biological network to identify conserved occurrences of a query network module, the goal thereof being the biological knowledge is transferred from one species to another (or, possibly, the same). Indeed, the query sub-network generally encodes a well-characterized functional module, and its occurrences in the queried network probably denote that this function is featured by the associated organism. For instance, using network querying techniques, a module involving cell-cycle transitions in the human network might be identified by exploiting a corresponding functional module of the yeast as the query network (Figure 15).

Notably, in solving network querying problems, two important issues must be taken into account. The first one is that sub-graph isomorphism checking is a sub-problem of network querying, and it is a well-known NP-complete problem (Garey, 1979), which limits the applicability of exact techniques in this context. The second issue is that any effective approach should look for approximated, rather than exact, occurrences of the query sub-network, since this takes into account the role of evolution in having possibly introduced biological variations (Berg, 2004) (see, in this volume, the chapter *Evolutionary Analyses of Protein Interaction Networks*).

In the last few years, the problem of querying biological networks has been studied by several researchers and some tools have been developed (Dost, 2007; Durand, 2006; Ferro, 2007; Fionda, 2008; Kelley, 2003; Pinter, 2005; Shlomi, 2006; Tian, 2007; Wernicke, 2007; Yang, 2007). However, available methods for network querying do not yet fully satisfy the requirements of system biology (Zhang, 2008), thus, this research area is still open and worth investigating. In this context, the goal of this chapter is to analyze network querying systems as applied to protein interaction networks.

To illustrate more formally, a protein interaction network N is commonly represented by an undirected graph $G^N = \langle P^N, I^N \rangle$ (Figure 2), in which the set of nodes P^N denotes a set of proteins and the edge set I^N encodes the interactions between pairs of proteins. In the most general definition, each edge $e^N \in I^N$ takes the form of a triplet

$$e^N = \langle p_i^N, p_j^N, w_{i,j}^N \rangle$$

where

$$p_i^N, p_j^N \in P^N$$

are the interacting proteins and

$$w_{i,j}^N \in [0,1]$$

Figure 2. An example of protein interaction network graph G^N

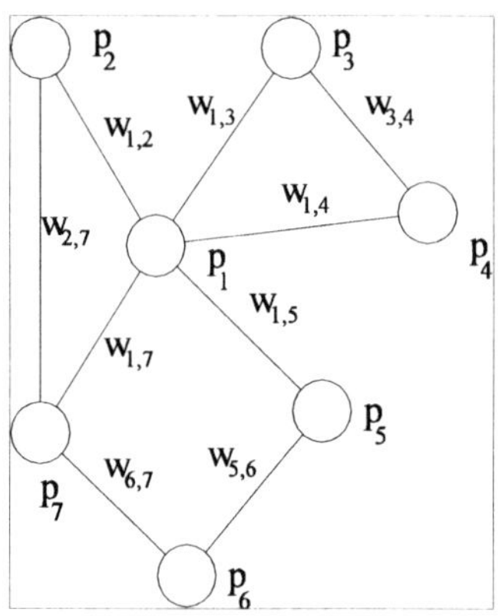

Figure 3. An example of alignment using BLAST

```
  Score =  413 bits (1061),  Expect = 6e-113
  Identities = 263/827 (31%), Positives = 429/827 (51%), Gaps = 89/827 (10%)

Query  11   LEATWNFIEPGINQILGNEKNQASTSKRVYKILSPTMYMEVYTAIYNYCV-----NKSRS  65
            L+  W+ +  GI Q+   +             ++ + YME+YT +YNYC      N++R
Sbjct  17   LDQIWDDLRAGIQQVYTRQS------------MAKSRYMELYTHVYNYCTSVHQSNQARG  64

Query  66   SGHFSTDSRTGQS---TILVGSEIYEKLKNYLKNYILNFKQSNS----ETFLQFYVKRWK  118
            +G   + S+ GQ+      VG E+Y++LK +LKNY+ N  +       E+ L+FY ++W+
Sbjct  65   AGVPPSKSKKGQTPGGAQFVGLELYKRLKEFLKNYLTNLLKDGEDLMDESVLKFYTQQWE  124
```

denotes the reliability for that interaction to be actually held under the particular experimentally measured condition. This formalism has been adopted in all of the presented approaches, but only in some of them the edge labels were actually used to incorporate reliability information (Dost, 2007; Fionda, 2008; Shlomi, 2006). Techniques that do not take into account reliability information can be assumed to manage graphs with all the edge weights set to 1. The interaction reliability information can be downloaded from some databases such as the MINT (Molecular INTeraction) database (Salwinski, 2004).

Generally, the similarity between pairs of proteins is computed by exploiting the score obtained by aligning their amino acid sequences. The alignment of two protein amino acid sequences is a way of arranging the two sequences to identify similarity regions (deriving from functional, structural, or evolutionary relationships between the sequences). One of the most widely used programs to align protein sequences is BLAST (Basic Local Alignment Search Tool) (Tatusova, 1999). The output of a BLAST alignment is an expectation value (the so called E-value) (see Figure 3). The lower the E-value is, the more significant the alignment is.

The network querying problem is defined on a pair of protein interaction networks, hereafter denoted by $G^Q = \langle P^Q, I^Q \rangle$ (Figure 4(a)) and $G^T = \langle P^T, I^T \rangle$ (Figure 4(b)), corresponding to the query sub-network and the queried target network, respectively. A potential solution of the query problem could be a sub-graph of G^T (Figure 4(c)), hereafter denoted by σ, which represents a possibly approximated occurrence of G^Q in G^T. In particular, approximation handling in the result sub-network is needed to deal with the possible occurrence through the evolution of *events* modifying the network structure. This also allows suitably to take into account the significant number of both false negative and false positive interactions found in databases for several organisms. Those events are *gene duplication*, that causes the addition of new nodes and *link dynamics*, corresponding to gain or loss of interactions through mutations in proteins (Berg, 2004). In its turn, gene duplication may be associated to both *node insertions*, corresponding to additions in the target network, and *node deletions*, corresponding to additions in the query network (Figure 4(c)). Moreover, node mismatches should be taken into account: this situation occurs when two proteins, characterized by low similarity, perform the same biological function. As a result, some of the nodes p^Q of G^Q may not correspond to any node of G^T, some of the nodes of σ may not correspond to any node of G^Q, and some corresponding pairs of nodes may have low similarity , but still the retrieved (approximated) occurrence σ of G^Q within G^T is biologically meaningful.

Of course, approximations should penalize the ranking of the specific potential solution in the overall set of solutions. However, not all the approaches developed for network querying take into account those approximations. Rather, some of them (e.g., Durand, 2006; Ferro, 2007) search for sub-graphs that satisfy all the structural constraint imposed by the query.

Given a query sub-network G^Q and a target network G^T, to solve the query problem an algorithm computes a score associated with solutions that also embodies node similarities and a penalty score for node insertions, deletions and mismatches. The best solution is the one with the maximum score.

The aim of this chapter is to analyze network querying techniques *as applied to protein interaction networks*. In particular, the comparison shall be illustrated along the following directions: (a) query structure restrictions,

Figure 4. (a) the query sub-network; (b) the target network; (c) a solution of the query problem. In the target network the black nodes represent the matching nodes while the white nodes represent the unmatched ones.

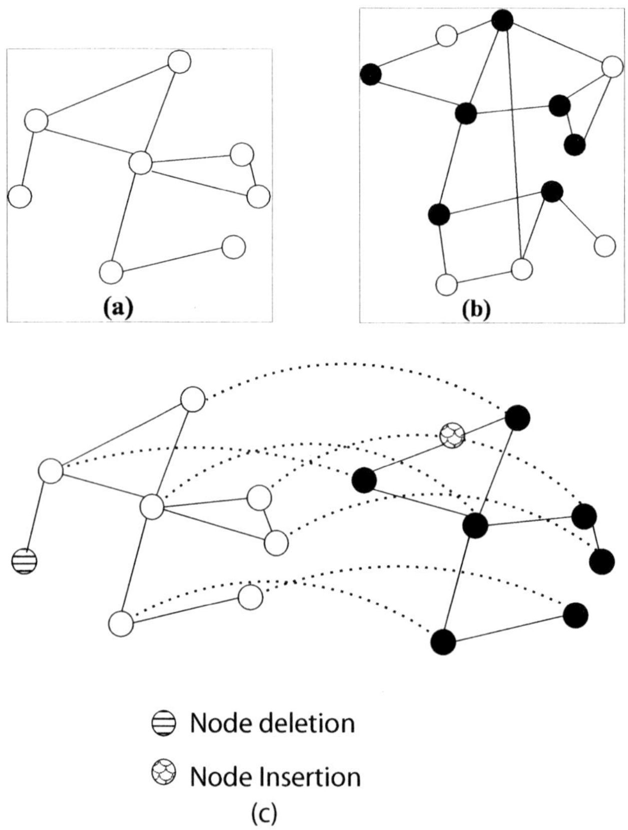

(b) general versus specific type of network analysis, (c) exploitation of additional information (e.g. sequence similarity, interaction reliabilities, etc.), (d) exact versus approximate occurrences, (e) types of approximation, (f) exact versus heuristic algorithms, (g) complexity, (h) delivered results (that is, the results as obtained by the authors of the described systems in the evaluation tests they illustrate).

The proposed analysis is intended to provide a comparative overview on the network querying techniques developed in the last few years, useful to understand problems and research issues, state of the art and opportunities for researchers working in this research area.

METHODS

In the last few years, the problem of querying biological networks has been studied by several researchers and some tools have been developed (Dost, 2007; Durand, 2006; Ferro, 2007; Fionda, 2008; Kelley, 2003; Pinter, 2005; Shlomi, 2006; Tian, 2007; Wernicke, 2007; Yang, 2007). Some of these tools were developed with particular focus on protein interaction networks and also tested on such networks (Dost, 2007; Fionda, 2008; Kelley, 2003; Shlomi, 2006; Yang, 2007), while others were intended to be applied to biological networks, different from the protein interaction ones, but maybe easily extended to work on them (Pinter, 2005; Tian, 2007; Wernicke, 2007) and further, other ones were designed to be *generally* applicable as they are able to query any type of biological graph (Durand, 2006; Ferro, 2007). We'll next survey all of them.

Methods Developed to Work on Protein Interaction Networks

In this section, a survey on the systems developed for querying PINs is depicted. Moreover, for the freely available tools, some operational details on the use of such softwares will be also provided.

PATHBLAST

PATHBLAST (Kelley, 2003) is a method for comparing and aligning two protein interaction networks by considering both sequence similarity and topology information to identify conserved pathways across the two species, and subsequently extended to identify protein interaction complexes and pathways by aligning more than two networks (Kelley, 2005; Kalaev, 2008). Originally, PATHBLAST was designed to align two entire networks. However, it can also be exploited to query a whole network against a specific pathway, merely using that pathway as one of the two input networks. The procedure takes into account evolutionary variations allowing for gaps (occurring when a protein interaction in one path skips over a protein in the other) and mismatches (occurring when aligned protein do not share a significant sequence similarity) to appear in the resulting alignment. The method builds a global alignment graph in which each vertex represents a pair of proteins, one from each of the input networks, that have a sufficiently large sequence similarity (a BLAST E-value $\leq 10^{-2}$), and each edge represents a conserved interaction, a gap or a mismatch (see Figure 5). Each pathway in such a graph corresponds to a sequence of conserved interactions across the two species.

The discovery of the best pathway solutions is based on the computation of a log probability score $S(P)$, computed over the vertices v and the edges e of the pathway G^Q:

$$S(G^Q) = \sum_{v \in P} \log_{10} \frac{p(v)}{p_{random}} + \sum_{e \in P} \log_{10} \frac{q(e)}{q_{random}}$$

where $p(v)$ is the probability of a true homology to hold for the protein pair encoded as v, $q(e)$ is the probability that the protein interactions denoted by e are real, p_{random} and q_{random} are the expected values of $p(v)$ and $q(e)$, resp., over all the vertices and the edges in the alignment graph.

Since the problem of finding the highest scoring paths of length q in acyclic graphs can be solved in linear time in the number of edges of the graph, and since the global alignment graph may contain some cycles, the method generates $5q!$ random acyclic sub-graph, where q is the length of the query pathway, and performs the search of conserved pathways on such graphs. Note that in this approach, the same protein pair cannot occur more than once in a pathway and neither gaps nor mismatches can occur consecutively.

The PATHBLAST system is accessible on-line at http://www.pathblast.org. To run the system the user should insert the proteins in the pathway query, identifying them by their amino acid sequences or protein Id, to a maximum of five proteins, which is the maximum query length. Then, the user must select the target organism and the search starts by clicking on the BLAST! button. The user is then directed to a page in which the Id of the submitted query and an estimate of the requested execution time are shown. The results, if any, will be displayed by clicking on the Result button in this web page.

QPath and QNet

QPath (Shlomi, 2006) is a tool proposed to query gene and protein interaction networks based on the colour coding technique. The query structure is limited to paths and the potential solutions are ranked according to the sequence similarity of matching nodes, number of node insertions, deletions and reliability of the involved edges.

Figure 5. Example pathway alignment with PATHBLAST. p'_i represents a generic protein of the query path. p''_j represents a generic protein of the path extracted from the target network.

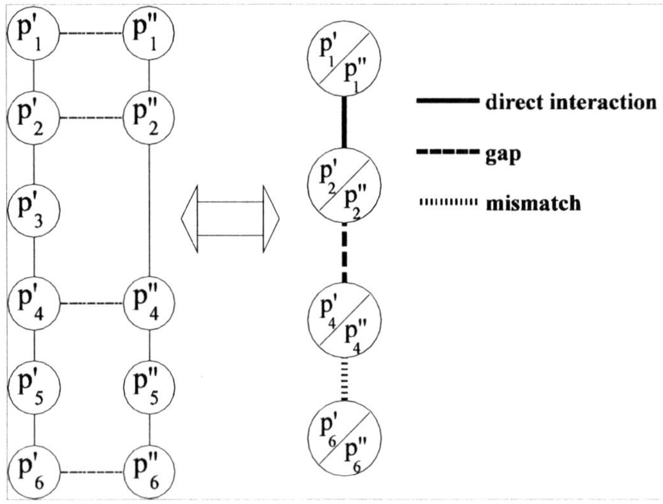

This technique guarantees that each resulting path consists of distinct proteins. The similarities between pairs of proteins, the first from the query pathway and the second from the target network, are computed by exploiting the BLAST E-Value obtained by aligning the associated protein sequences. The number of insertions and deletions in the potential solutions is bounded by two threshold values N_{ins} and N_{del}, respectively. The pathway search module is based on the colour coding technique, commonly used to find simple paths of length k in a graph. According to such technique, a random chosen colour from $\{1,...,k\}$ is assigned to each node and then the problem to find a path of length k that spans distinct colour is solved. In the matching approach, $k + N_{ins}$ distinct colours from $\{1,...,k + N_{ins}\}$ are used to take into account the N_{ins} allowed node insertions. The algorithm starts by assigning a colour to each node, and for a given colouring, a dynamic programming technique is exploited to find an optimal matching pathway involving a maximum number of N_{del} node deletions. Actually, several random colouring trials of the graph are to be executed since any particular path may be assigned non-distinct colours and, hence, may fail to be discovered. The algorithm outputs the set of high scoring matches for each exploited combination of the allowed number of insertions and the allowed number of deletions. The software QPath is available from the authors of the system upon request.

Notably, QPath was extended by QNet (Dost, 2007), this latter system being able to deal with queries shaped as trees or graphs having bounded treewidth. In this case, the basic employed technique is also colour coding, and the number of node insertions and deletions is also bounded. Moreover, the score of a potential solution is computed by applying insertion and deletion penalties and using matching node similarities.

PathMatch and GraphMatch

PathMatch and GraphMatch (Yang, 2007) are two algorithms developed for querying biological networks and, in particular, protein interaction networks and metabolic pathways to search for paths and graphs, respectively. The peculiarity of these approaches is that each node of the target network may correspond to more that one node of the query sub-network.

PathMatch takes advantage of the linearity of the query pattern and reduces the problem to the one of finding the longest weighted path in a directed acyclic graph. In particular, for each node v in the query path G^Q, it groups together all the nodes of the target graph related to v and chooses at most one node from each group to form the resulting path. In more detail, starting from the query path G^Q and the target network G^T, PathMatch builds, for each node $p_i \in G^Q$, a set of correspondences $V_i = \{v_{i1},\ldots,v_{i_{l_i}}\}$ with $v_{i1},\ldots,v_{i_{l_i}} \in G^T$. Exploiting these sets, it then builds a directed graph $G^{T}{}'$ in which each level of the graph, that is the set of nodes having the same distance from the graph roots, is related to a protein $p_i \in G^Q$ and contains the nodes in V_i. An edge between proteins $v_{ij}, v_k \in G^{T}{}'$ is added to $G^{T}{}'$ if the number of nodes in the path connecting v_{ij}, v_k in G^T is lower than the imposed constraints on the maximum number of node gaps allowed between two nodes (see, Figure 6).

The weight of each edge is less than 0 and proportional to the number of mismatches and gaps in the path represented by such an edge. The above construction thus reduces the path matching problem to the one of finding a path $G^{Q}{}'$ in $G^{T}{}'$ with the maximum sum of vertex and edge weights.

PathMatch and its user manual are downloadable from http://faculty.cs.tamu.edu/shsze/Pathmatch/.

The GraphMatch algorithm, instead, is an exact algorithm which takes in input a query graph G^Q and a target graph G^T. As PathMatch does, it starts by building, for each node $v_i \in G^T$, the set of correspondences V_i. GraphMatch fixes the maximum number of node insertions for each direct edge in G^Q, and it associates to each potential solution a score computed by exploiting the similarity values of the matching nodes and the number of gaps. The system enumerates all the potential solutions, so this search is effective if the correspondence lists are not particularly large.

GraphMatch and its user manual are downloadable from http://faculty.cs.tamu.edu/shsze/ graphmatch/.

Fionda et al.

The algorithm by Fionda et al. (Fionda, 2008), hereafter denoted by Fionda, is an heuristic approach to search for approximated occurrences of a query network into a target one. The system exploits node similarity information, sub-graph structural similarity and edge reliabilities to find a solution that satisfies user requests. The user can specify a threshold value on the score values associated to the potential solutions examined during the search, and the algorithm stops either when a solution with a score lower than the threshold is found, or when a maximum number of execution attempts has been reached. In the latter case, no solution is returned. At each iteration, a potential solution is computed by solving a minimum bipartite weighted matching involving the query sub-network nodes and the target network nodes, weighted using node distances, which are obtained by exploiting the BLAST E-value. The method computes the score associated to a solution on the basis of matching node distances and number of node insertions and deletions. If the computed score is lower than the threshold value set out by the user, the method returns that solution, otherwise it refines the distances of the involved nodes by exploiting a measure of satisfaction of structural constraints and starts a new iteration. As already mentioned, if a maximum number of iterations is reached, and no acceptable solution has been found, the method stops and no solution is returned.

The software is available from the authors of the system upon request.

Figure 6. Example of the $G^{T}{}'$ building. The same node letter in the figure is used to represent similar protein in the query path and in the target network.

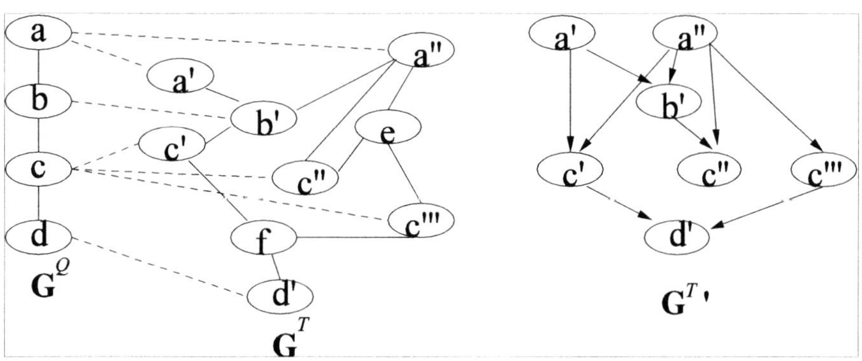

Methods Developed for Other Types of Biological Networks

In this section, a survey on the techniques developed for querying biological graphs, and tested on biological networks different from PINs is provided.

MetaPathwayHunter

MetaPathwayHunter (Pinter, 2005) is a tool developed to query protein interaction or metabolic networks, but actually optimised for these latter ones. The resulting alignments are computed both according to similarity between enzymes and to topological similarity between structures. Within this system, metabolic networks are modelled as graphs, with a structure that actually closely resembles the one of the protein interaction networks. Here, nodes represent enzymes and directed edges connect pairs of enzymes for which the product of one enzyme serves as a substrate to the other. The system takes advantage of the particular topology of most of the metabolic pathways, that can usually be represented as multi-source trees. A multi-source tree is a directed acyclic graph whose underlying undirected graph is a tree. Indeed, allowed queries are ones shaped as multi-source trees and the search is performed in a forest of such trees. MetaPathwayHunter does not permit deletions from the pattern, but deletions may be done from the target trees. The similarity between pairs of enzymes is measured according to the EC (Enzyme Commission) classification. EC numbers give a functional classification that does not necessarily reflect sequence similarity. For each potential solution, a score is computed to take into account the similarity between pairs of matching enzymes and node deletions (gap penalty).

SAGA

SAGA (Substructure Index-based Approximate Graph Alignment) (Tian, 2007) is a general graph matching system for searching for approximated occurrences of a query sub-graph in a graph database. The approximations taken into account are node gaps, node mismatches and graph structural differences. In particular, node gaps may denote both node insertions and deletions, while graph structural differences allow differences in node connectivity relationships. The search is based on the construction of an index on graph substructures extracted from graphs in the database. The query sub-graph is divided into fragments of size *k* (where *k* is the number of nodes of the fragments), these fragments are then used to query the index, and the matching fragments are then combined into larger matches. A label is associated with each node in the query and in the graphs in the database in order to identify node mismatches. Indeed, if two matching nodes have different labels, they correspond to a mismatch. SAGA was evaluated using metabolic networks and exploiting the EC classification to assign labels to nodes. The presented framework is general and applicable to all types of biological networks. In particular, to apply this system to PINs, the SAGA's authors suggest to take node labels from databases like COG (Clusters of Orthologous Groups) or KEGG (Kyoto Encyclopedia of Genes and Genomes). These databases organize proteins into orthologous groups, so that the same label can be assigned to similar proteins (and to graph nodes) belonging to the same group.

MetaPAT

MetaPAT (Wernicke, 2007) is a graph matching technique that searches for approximated occurrences of query sub-graphs in metabolic networks. The approach examines all the sub-graphs of the homeomorphic target network to the query sub-graph. Two graphs are homeomorphic if their edges can be subdivided (i.e. edges can be replaced by paths of arbitrary length in the same direction) in such a way that the resulting graphs are isomorphic. MetaPAT computes a score for each such potential solution, based on the similarity of the enzymes involved in the matching and the number of gaps. The algorithm starts out by aligning a vertex of the pattern with a vertex of the target graph, and then it uses a recursive sub-procedure to consider all possible extensions of the currently built partial solution.

Other Tools

This section presents a survey on techniques developed for querying general graphs, but also tested on some types of biological networks.

GenoLink

GenoLink (Durand, 2006) is a software platform developed for the exploration and querying of graphs, also applicable to biological networks. A query consists of a graph pattern in which vertices and edges are constrained, and the result is the set of all sub-graphs, from the target graph, which are isomorphic to the pattern and satisfy all the constraints.

In building the result set, the algorithm performs a depth-first search, which guarantees to find all the matching sub-graphs. Firstly, GenoLink uses the constraints on vertices and edges of the pattern to select a seed for the depth-first search; in particular, it chooses the node or edge with the smallest number of correspondences. Then, the system uses the constraints on nodes and edges during the search, progressively adding vertices and edges to the potential solution in such a way to satisfy the constraints.

NetMatch

NetMatch (Ferro, 2007) is a Cytoscape plug-in to search for patterns in biological networks. Queries may be approximated by leaving some parts of them unspecified. Thus, some elements of the query sub-graph are marked as constant, while others are not specified. In particular, a node or an edge labelled with a wildcard symbol '?' may correspond to any single value, while an approximated path (identified by a dashed edge in the pattern) may correspond to a path of length $\leq n$ or $\geq n$, where n is a positive integer parameter (Figure 7). The resulting sub-graphs are connected according to the same structure as the query graph.

COMPARING TOOLS

In this paragraph, a thorough comparison of the reviewed approaches for network querying is carried out. These systems, as applied to protein interaction networks, are analyzed along the following directions: (a) query structure restrictions, (b) general versus specific type of network analysis, (c) exploitation of additional information (e.g. sequence similarity, interaction reliabilities), (d) exact versus approximate occurrences, (e) types of handled approximation, (f) exact versus heuristic algorithms, (g) complexity, and (h) a comparative analysis of the experimental results obtained on protein interaction networks reported by the authors of those systems for the validation purposes (Dost, 2007; Fionda, 2008; Kelley, 2003; Shlomi, 2006; Yang, 2007). Some relevant data pertaining the comparisons are listed in Table 1 (for points (a) – (g)) and Table 2 (for point (h)).

Figure 7. Examples of NetMatch query sub-graphs

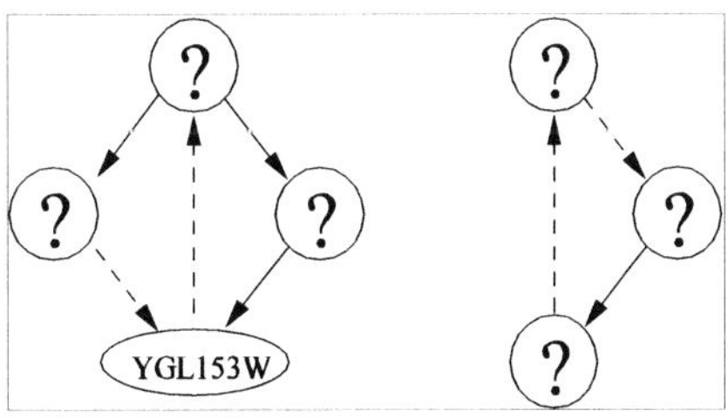

Query Structure Restrictions (a)

The techniques developed for querying PINs can be classified with reference to the structural constraints imposed on the query networks. Some of the proposed approaches, in particular PATHBLAST, QPath and PathMatch are only able to handle queries shaped as paths. Other approaches, such as MetaPathwayHunter and QNet, were developed to manage tree queries, also if QNet, in addition, is able to search for sub-graphs shaped as graphs with a bounded treewidth. The most general techniques, such as GenoLink, NetMatch, SAGA, GraphMatch, MetaPAT and Fionda, are able to handle queries shaped as general graphs.

The constraints imposed by some algorithms allow heuristic techniques to be used to perform the search efficiently, such as colour coding (Alon, 1995). In some cases, e.g., for MetaPathwayHunter, the restrictions imposed to the query structure are dictated by the particular topology of the most interesting biological substructures in the biological networks of interest (e.g., metabolic networks).

General versus Specific Type of Network Analysis (b)

The second direction, along which the querying techniques can be compared, is the contrast between general and specific network analysis. Some of the proposed systems, namely GenoLink and NetMatch, were proposed to query general graphs, and therefore also applicable to PINs , but they do not exploit specific information about proteins and their interactions because of their generality. Other proposed systems, instead, were developed to be applicable to particular type of biological networks. Indeed, QNet, PATHBLAST, QPath and Fionda were designed to be applied to PINs, and thus, they exploit specific information useful to characterize protein complexes (see next section). Other approaches, namely MetaPathwayHunter, SAGA and MetaPAT, were intended for the application to metabolic networks. However, these approaches can be used to query protein interaction networks just by changing the method used to compute the similarity between nodes and edges. Two of the reviewed approaches were oriented on the multi-type biological networks, namely PathMatch and GraphMatch, that were tested both on protein interaction networks and metabolic networks, highlighting their rather general applicability.

Exploitation of Additional Information (c)

Some of the reviewed systems exploit several types of information useful to characterize the particular type of network they analyze and which specific information used depends on the type of network these approaches were tested on. In this respect, some of the techniques we are reviewing limit themselves to exploit only structural knowledge, while others take advantage of additional information as well. In particular, two of the examined approaches, namely GenoLink and NetMatch, do not exploit additional information. These are methods designed to handle generic graphs, and they mostly focus on the structural, rather than the domain oriented, similarity issues. Those techniques tested on metabolic networks (MetaPathwayHunter, SAGA, MetaPAT, PathMatch and GraphMatch) exploit the functional classification of enzymes to define a similarity score between pairs of aligned enzymes. All the approaches tested on PINs (QPath, QNet, PATHBLAST, PathMatch, GraphMatch and Fionda), instead, compute the similarity between pair of proteins by exploiting the BLAST E-value obtained by aligning their amino acid sequences. Some of the approaches developed for querying PINs (QPath, QNet and Fionda) use weighted edges to represent protein interactions. This takes into account the reliability information associated with such interactions.

Exact versus Approximate Occurrences and the Types of Approximation Handled (d, e)

The identification of exact versus approximated occurrences, and the types of approximation handled in the result construction are worth analyzing for the purpose of our comparison. The two generic systems, GenoLink and NetMatch, require an exact correspondence of the query graph into the target one. However, in NetMatch (Ferro, 2007) the user is allowed to require a loose form of matching through specifying wildcards in the query graph which can be thought of as a "weak" form of approximation handling. All the approaches solely focused on querying particular types of biological networks, rather than generic graphs, namely PATHBLAST, MetaPathwayHunter, QPath, QNet, SAGA, PathMatch, GraphMatch, MetaPAT and Fionda, search for approximated occurrences. However, these techniques differ from one another with respect to the types of approximations they are able to handle. In particular, MetaPathwayHunter is able to handle only node insertions in the result sub-graphs, instead of QPath, QNet, MetaPAT, GraphMatch and Fionda which are able to handle both node insertions and deletions in the identified occurrences of the query graph within the target one, while PATHBLAST, SAGA, PathMatch manage node insertions, deletions and mismatches in the search of the best query occurrences.

Exact versus Heuristic Algorithms (f)

A further direction to compare the biological network querying systems is the exploitation of exact rather than heuristic algorithms. Only four of the 11 approaches, namely GenoLink, NetMatch, MetaPathwayHunter and GraphMatch, exploit exact algorithms and only one of them, GraphMatch (Yang, 2007), has been proposed to query protein interaction networks. Due to the complexity of the sub-graph matching problem (sub-graph isomorphism is NP-complete (Garey, 1979)), exact algorithms can be applied only to small problem instances. The other methods (QPath, QNet, PATHBLAST, SAGA, MetaPAT, PathMatch and Fionda), instead, exploit heuristic algorithms and therefore they do not guarantee to return the optimal solution.

Complexity (g)

The last direction along which the developed techniques were analyzed is time complexity. Let n and m be the numbers of nodes and edges in the target network, respectively; let q be the number of nodes of the query sub-network; let N_{ins} and N_{del} be the number of the maximum allowed node insertions and deletions, respectively, in those approaches in which they are used. For three techniques, namely GenoLink, SAGA and MetaPAT, the study of the time complexity was not carried out.

Polynomial Time Techniques

Three of the approaches (MetaPathwayHunter, PathMatch and Fionda) run in polynomial time. In particular, MetaPathwayHunter's time complexity (Pinter, 2005) is $O(q^2 n/\log q + qn + \log n)$, thus the running time is polynomial both in n and q. Fionda's running time is $O(MAXITERATION \cdot n^3)$, where $MAXITERATION$ is the allowed number of iterations the algorithm may perform, and the factor n^3 is implied by the computation of the minimum bipartite weighted matching problem. Finally, the PathMatch's time complexity (Yang, 2007) is $O(m + n + k)$, where k is the number of best scoring hits returned by the algorithm, so it is linear both in the number of nodes and edges of the target network.

Exponential Time Approaches

Five of the discussed systems run, instead, in exponential time in the number of nodes of the query sub-network. These are PATHBLAST, NetMatch, QPath, QNet and GraphMatch. Such techniques are therefore applicable only to small problems. In detail, PATHBLAST (Kelley, 2003) runs in time $O(q!l)$, where l is the number of edges of the global alignment graph. NetMatch's time complexity (Ferro, 2007) is $\Theta(q!q)$, while QPath (Shlomi, 2006) runs in $\ln(n/\varepsilon) \cdot 2^{O(q+N_{ins})} Nm_{del}$, where ε is the probability that the algorithm does not find an optimal solution; therefore the running time is exponential in q, logarithmic in n, and linear in m. QNet's time complexity (Dost, 2007) is $\ln(l/\varepsilon) \cdot 2^{O(q+N_{ins})} m$ for tree queries and $2^{O(q)} n^{t+1}$ for bounded treewidth graph queries, where ε has the same meaning as explained above, and t is the maximum allowed treewidth of the query graph; so, the running time is polynomial in n (consider that $m \leq n^2$) and exponential in q and t. Finally, GraphMatch (Yang, 2007) runs in $O(2^q q^2)$, so its running time is exponential in q. It is worth pointing out that the exponential trend in q might not be as much problematic as an exponential trend in n, since in real applications q is expected to be relatively small compared to n.

Test Results (h)

The approaches developed specifically for querying PINs (Dost, 2007; Fionda, 2008; Kelley, 2003; Shlomi, 2006; Yang, 2007) were tested by their developers on different organisms, by downloading the data from several databases (see Table 2). All those approaches were evaluated on the networks downloaded from DIP (*Database of Interacting Proteins*) (Salwinski, 2004), though some of them also used other databases to obtain additional information or to perform evaluations on different data. For example, QPath (Shlomi, 2006) and QNet (Dost, 2007) used FlyGrid (the section of BioGRID containing the interaction data for the fly) (Stark, 2006), and Fionda also exploits the data downloaded from MINT (Molecular INTeraction database) (Chatr-aryamontri, 2006). Moreover, we note that the systems were evaluated on networks of different organisms (see Table 2). In detail, PATHBLAST (Kelley, 2003) was tested on the PIN of *S. cerevisiae* (yeast); QPath and QNet were evaluated using the networks of *S. cerevisiae*, *D. melanogaster* (fly), *H. sapiens* (human); PathMatch (Yang, 2007) was run on the networks of *S. cerevisiae*, *D. melanogaster*, *C. elegans* (worm), *H. pylori*, *E. coli* (the latter two are bacteria); for GraphMatch the networks of *S. cerevisiae*, *D. melanogaster* and *C. elegans* were used; Fionda was tested on the networks of *S. cerevisiae*, *D. melanogaster*, *C. elegans* and *H. sapiens*.

The quality of the results obtained by the different systems were evaluated from a biological point of view, manually or using the well known information stored in some databases, such as Swiss-Prot (Bairoch, 2004). The result returned from such tests are discussed below in detail.

PATHBLAST

The tests performed to validate PATHBLAST (Kelley, 2003) were executed on the PIN of *S. cerevisiae* downloaded from DIP. Specific yeast pathways of interest were used to query the yeast network to find pathways performing the same functions. In particular, the yeast MAPK pathway associated with the filamentation response, composed by a MAPK (Ste11), a MAPK kinase (Ste7) and a MAPK kinase kinase (Kss1), was chosen as first pathway (Figure 8). The algorithm identified two other well known MAPK pathways (Bck1 – Mkk1 – Stl2 and Ssk2 – Pbs2 – Hog1). Moreover, the authors of PATHBLAST used the system to search for new elements of the ubiquitin and ubiquitin-like conjugation machinery. In particular, the query pathways were two well characterized ubiquitin conjugation systems: the Spk1-Cdc53/cullin-F-box (SCF) complex (consisting of Cdc53, Cdc4, Cdc34 and Ubi4; Figure 9) and the anaphase-promoting complex (consisting of Apc11, Apc2, Pex4 and Cdc20), and a pathway based on SUMO, a ubiquitin-like protein (consisting of Ubc9, Aos1, Uba2). In these latter cases, the method was capable to align the queries with components of a paralogous Rub1-conjugating complex and propose new F-box and cullin-like proteins, namely respectively Ycr072c and Ylr106c, for the Rub1 ligation system.

QPath

Like PATHBLAST, QPath (Shlomi, 2006) was also tested by querying the PIN of *S. cerevisiae* with the yeast filamentous growth MAPK cascade, and it was able to identify the same two pathways found by PATHBLAST.

Figure 8. PATHBLAST. Query pathway in yeast and its hit results against the yeast network

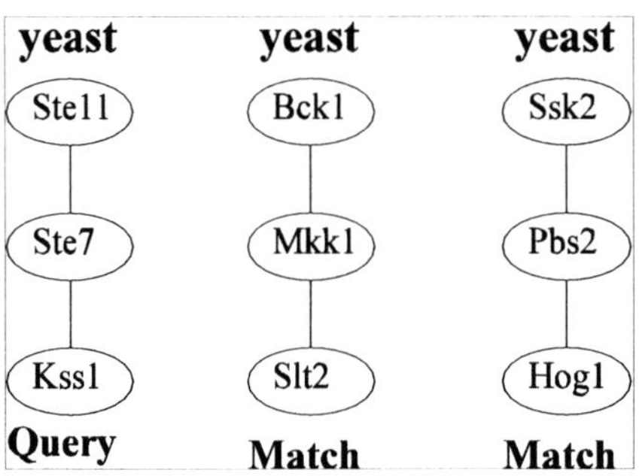

Figure 9. PATHBLAST. "Skp1-Cdc53/cullin-F-box (SCF) complex" query and its results in the yeast network.

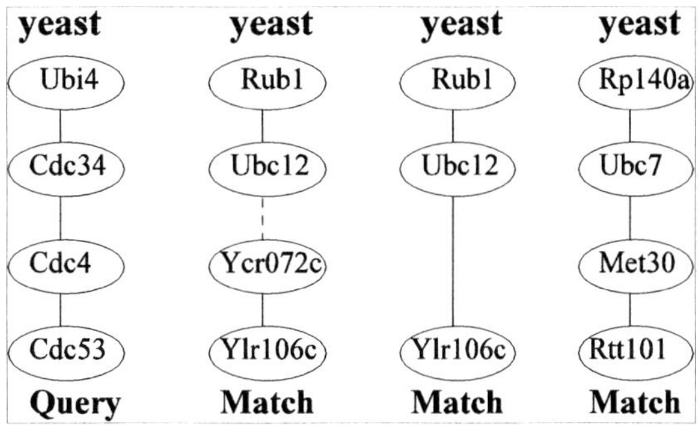

In addition, QPath was also used to query the PIN of *D. melanogaster* with 271 putative yeast pathways. It found at least one match in fly for the 63% of the 271 yeast queries, with up to three insertions and deletions. Moreover a third set of tests consisted in querying the fly network with well known signaling protein pathways from yeast and human. In particular, the authors of QPATH used a yeast ubiquitin-ligation pathway (Figure 10), a human MAPK cascade and a human Hedgehog signalling pathway. The interaction data were downloaded from DIP and FlyGrid.

QNet

Like QPath, QNet (Dost, 2007) was also tested using the PINs of *S. cerevisiae*, *D. melanogaster* and *H. sapiens* downloaded from DIP and FlyGrid. The method was first evaluated using some synthetic query trees to estimate its accuracy. Then, real data were considered. In particular, human and yeast MAPK pathways were used to query the fly network (Figure 11).

Moreover, the authors of QNet also queried the fly network using 94 yeast protein complexes downloaded from the MIPS database (Mewes, 2006). They considered only the hand curated complexes, which are more reliable than the ones obtained by high throughput methods. They used 94 complexes, each consisting of at least four proteins, and they tested the result correctness using the gene ontology annotations (The Gene Ontology Consortium, 2000), and illustrated the match in fly relative to the Cdc28p yeast complex query (Figure 12).

Figure 10. QPath. Yeast ubiquitin ligation pathway query and the result pathway from the fly network.

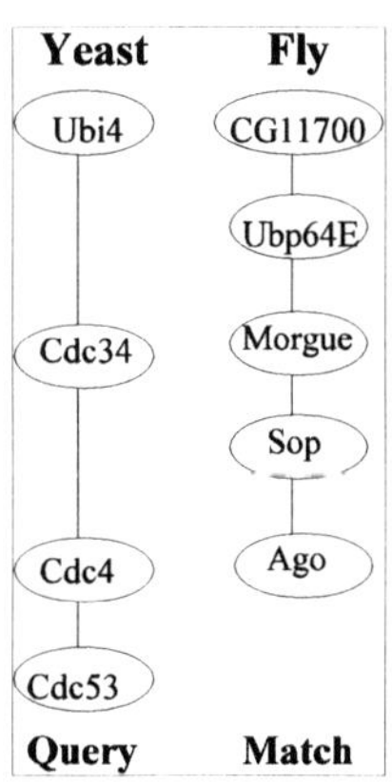

Figure 11. QNet. Human and yeast MAPK queries and their results from the fly network.

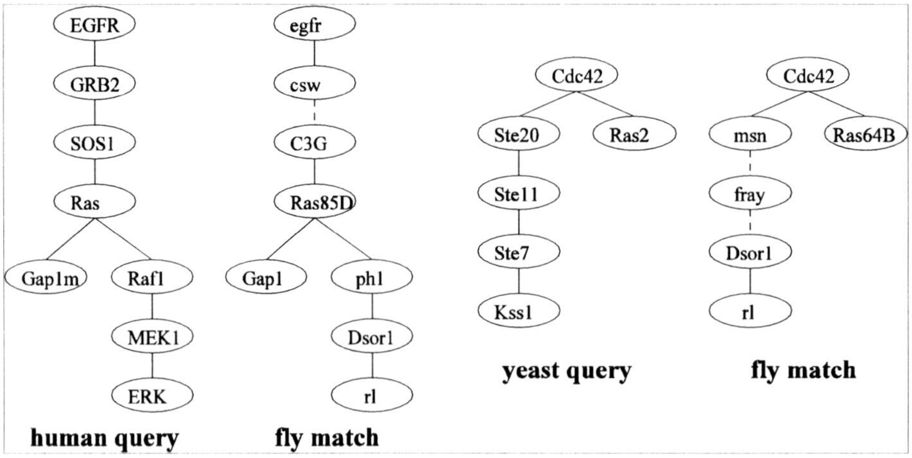

Figure 12. QNet. Cdc28p yeast complex query and its result in the fly network. White proteins represents the matches in fly, the grey ones represent node insertions.

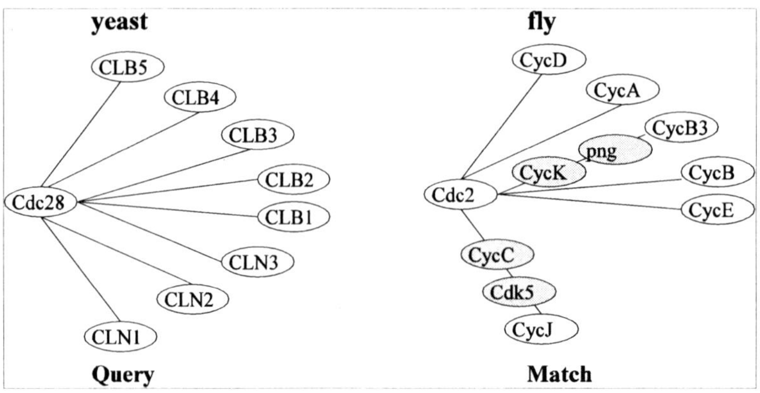

PathMatch

PathMatch (Yang, 2007) was evaluated on the *S. cerevisiae*, *D. melanogaster*, *C. elegans*, *H. pylori* and *E. coli* networks downloaded from the DIP database. The authors of PathMatch used the same yeast MAPK pathway that was used to test PATHBLAST (note that the protein Fus3 is very similar to the protein Kss1) to find pathways in the yeast network performing the same function (Figure 13). The results obtained by PathMatch are the same also obtained by PATHBLAST.

As a second example, PathMatch was tested using the longer yeast mating-pheromone response pathway for querying the networks of *D. malanogaster*, *C. elegans* , *H. pylori* and *E. coli*. This query pathway contains the MAP kinase cascade that appears in the previous query. The results were significant only for the networks of fly and worm (Figure 14), and it is consistent with the fact that the MAPK cascades are not known to occur in the other two bacteria organisms. PathMatch was also tested on metabolic networks.

Figure 13. PathMatch. MAPK pathway query in yeast and its hit results against the yeast network

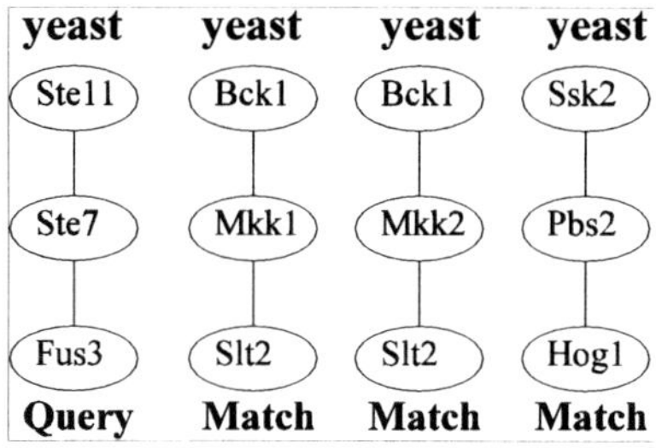

GraphMatch

GraphMatch (Yang, 2007) was tested on the PINs of *S. cerevisiae*, *D. melanogaster* and *C. elegans*, performing a graph query extracted from the yeast network against the fly and worm networks (Figure 15). The exploited query graph involve the MAPK kinase pathway, already discussed in connection with PathMatch. Moreover, we note that this graph is a related functional module from Spirin and Mirny (2003). Some of the correspondences are proteins with unknown functions. GraphMatch was also tested on metabolic networks.

Fionda

Fionda was tested on the PINs of *S. cerevisiae*, *D. melanogaster*, *C. elegans* and *H. sapiens* downloaded from the DIP and MINT databases. In particular, the same yeast query pathway employed to test PathMatch was used to

Figure 14. PathMatch. Yeast "mating-pheromone response pathway" used as query, and its results in the worm and fly networks.

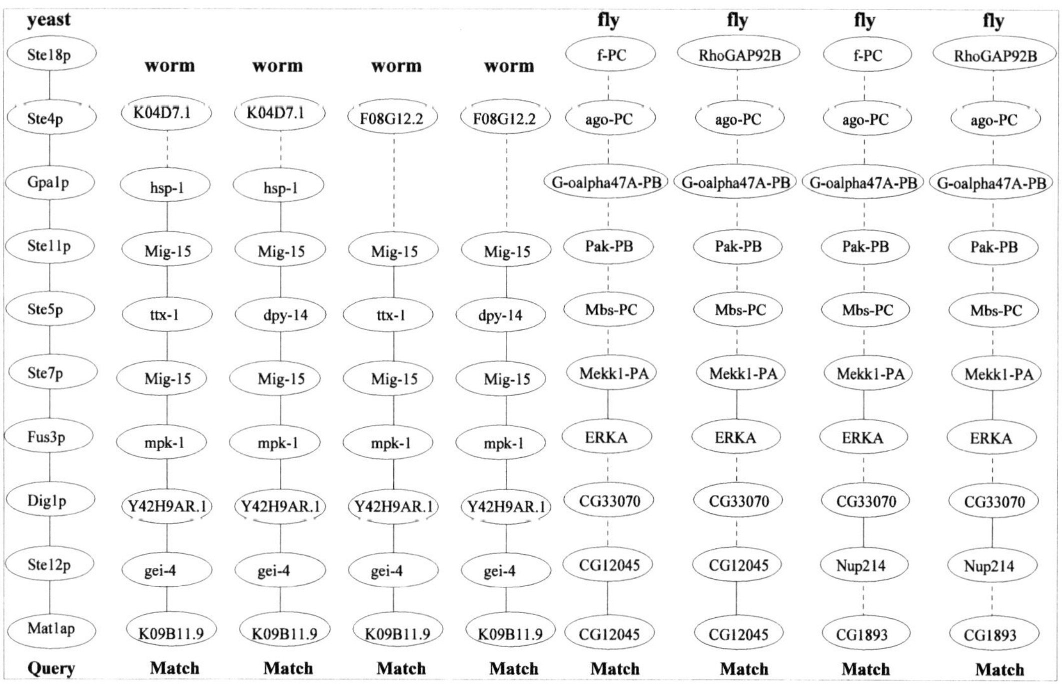

Figure 15. GraphMatch. A related functional module from yeast (Spirin, 2003) used as query and its results in the worm and fly networks

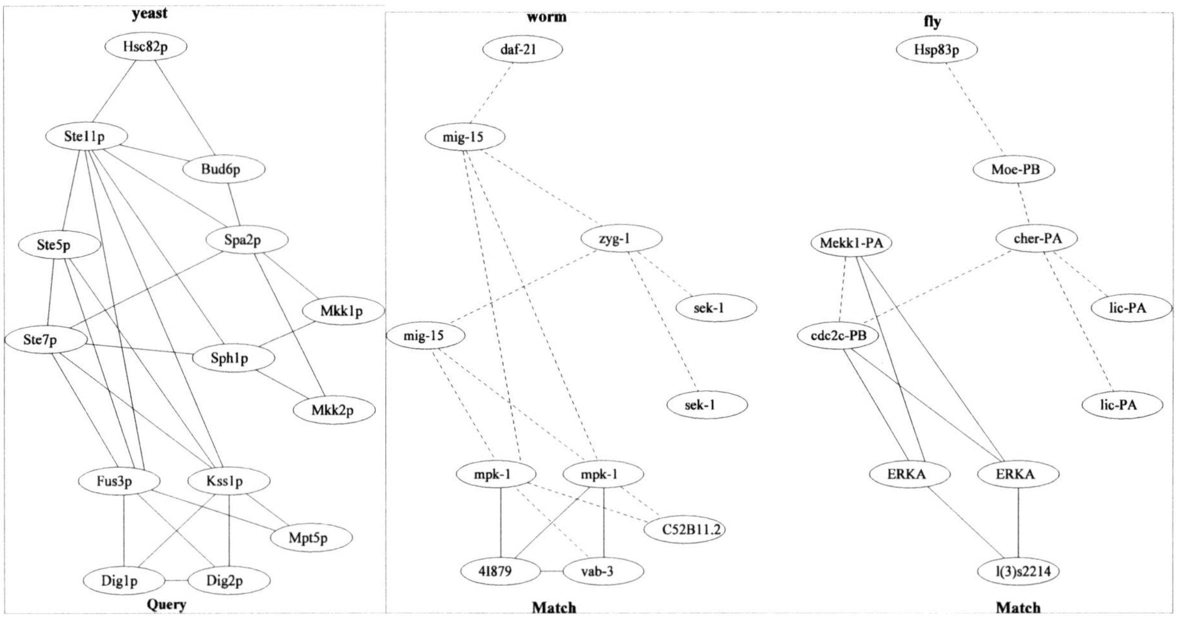

query the worm network (Figure 16) and the same yeast query graph employed to test GraphMatch was used to query the worm and fly ones. Moreover, two yeast query modules, shaped as graphs of general structure, were used to query the PIN of *H. sapiens* downloaded from the MINT database, where the first query module involves proteins that control cell-cycle transitions (Figure 17), while the second one concerns yeast act-in related proteins (Figure 18).

DISCUSSIONS AND FUTURE TRENDS

As implied by the previous descriptions, the approaches developed in the field of biological network querying are rather different from one another. There are methods that, due to the adoption of exact algorithms (e.g., GraphMatch (Yang, 2007)), or the handling of generic graph queries (such as NetMatch (Ferro, 2007)), result in rather high time complexities. Improvements in execution times are obtained by exploiting heuristic algorithms, like in Fionda (Fionda, 2008), by restricting query structure, as done in PathMatch (Yang, 2007), or by allowing only few types of approximations in the result sub-graphs, such as in MetaPathwayHunter (Pinter, 2005).

As far as the results obtained by running the developed tools are concerned, all of them appear able to find significant results. In particular, the tools tested querying the PIN of *S. cerevisiae* with the well known MAPK pathway (PATHBLAST (Kelley, 2003), QPath (Shlomi, 2006) and PathMatch (Yang, 2007)), all obtained the same results (Figure 6 and Figure 11), which are well known MAPK pathways. Moreover Fionda (Fionda, 2008) and PathMatch were evaluated and compared on the same yeast path query used to query the worm network, and both the methods matched the well known MAP kinases pair of proteins (Fus3 and mpk-1) and the pairs Ste7-Mig-15, Mat1a-K09B11.9, Ste4- F08G12.2, Ste5-ttx-1 and Dig1-Y42H9AR.1. Fionda was also compared to GraphMatch (Yang, 2007) using the same yeast sub-graph, a well characterized yeast signalling cascade that controls pheromone response (via Fus3) and pseudohyphal invasive growth pathway (via Kiss1) through the so-called MAPK pathway. This sub-graph was used to query the fly and worm networks. Both the approaches

Figure 16. Fionda. Yeast mating-pheromone response pathway used as query and its match to the worm network

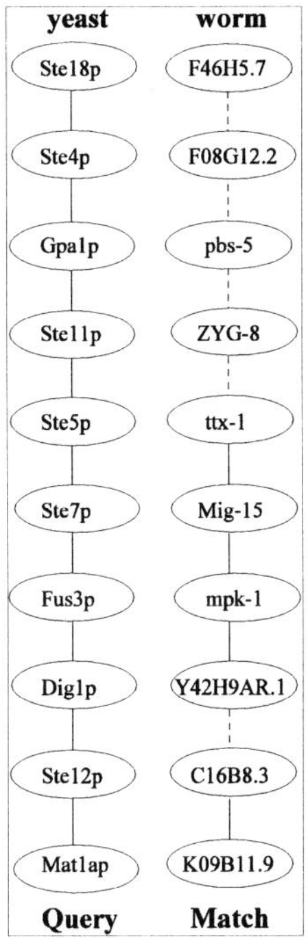

Figure 17. Fionda. Yeast query protein complex involving proteins that control cell-cycle transitions and its match to the human network

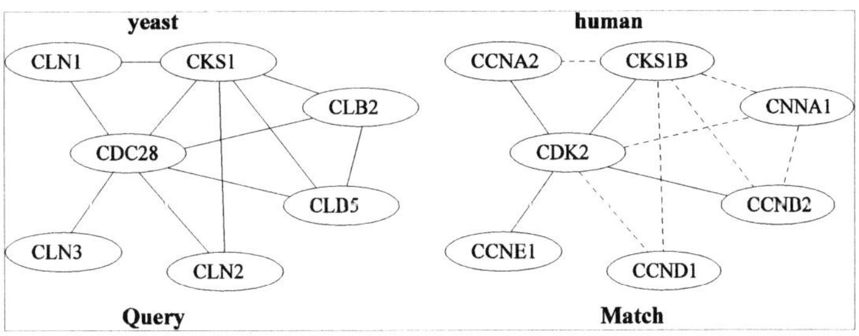

Figure 18. Fionda. Yeast query protein complex involving yeast act-in related proteins and its match to the human network

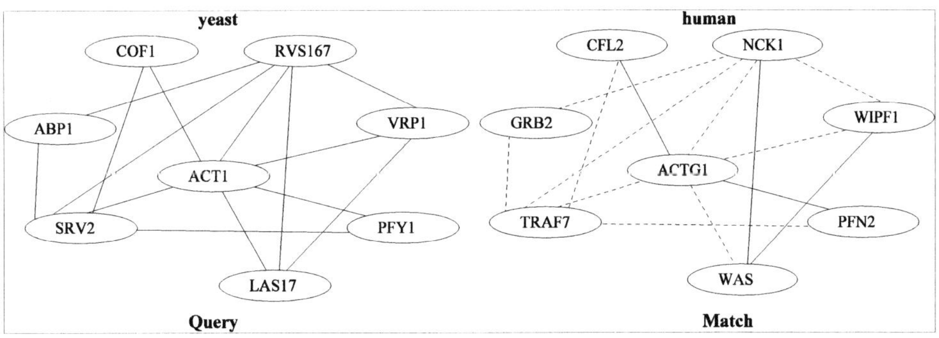

Table 1. Comparison summary table

	Query structure restrictions	Types of networks treated	Additional information	Exact vs approximate results	Types of approximation handled	Exact vs heuristic algorithm	Time complexity
PATHBLAST (Kelley, 2003)	Pathways	Tested on protein interaction networks	BLAST E-value	Approximate	Node insertions, node deletions, mismatches	Heuristic	$5q!O(l)$
MetaPathwayHunter (Pinter, 2005)	Trees in a forest	Tested on metabolic pathways	Functional classification	Approximate	Node insertions	Exact	$O(q^2 n / \log q + q \log n)$
QPath (Shlomi 2006)	Pathways	Tested on protein interaction networks	Interaction reliability, BLAST E-value	Approximate	Node insertions, node deletions	Heuristic	$\ln(n/e) \cdot 2^{O(q+N_{ins})} N_{del}$
GenoLink (Durand, 2006)	General graphs	General	None	Exact	None	Exact	Not evaluated
QNet (Dost, 2007)	Trees or graphs with bounded treewidth	Tested on protein interaction networks	Interaction reliability, BLAST E-value	Approximate	Node insertions, node deletions	Heuristic	$\ln(1/e) \cdot 2^{O(q+N_{ins})} m$ (tree) $2^{O(q)} n^{t+1}$ (bounded treewidth graph)
NetMatch (Ferro, 2007)	General graphs	General	None	Exact (but wildcards allowed in the query network)	None	Exact	$\Theta(q!q)$
SAGA (Tian, 2007)	General graphs	Tested on metabolic pathways	Functional classification	Approximate	Node insertions, node deletions, mismatches	Heuristic	Not evaluated
PathMatch (Yang, 2007)	Pathways	Tested both on protein interaction networks and metabolic pathways	BLAST E-value, Functional classification	Approximate	Node insertions, node deletions, mismatches	Heuristic	$O(m+n+k)$
GraphMatch (Yang, 2007)	General graphs	Tested both on protein interaction networks and metabolic pathways	BLAST E-value, Functional classification	Approximate	Node insertions, node deletions	Exact	$O(2^q q^2)$
MetaPAT (Wernicke, 2007)	General graphs	Tested on metabolic networks	Functional classification	Approximate	Node insertions, node deletions	Heuristic	Not evaluated
Fionda (Fionda, 2008)	General graphs	Tested on protein interaction networks	Interaction reliability, BLAST E-value	Approximate	Node insertion, node deletions	Heuristic	$O(MAXITERATION \cdot n^3)$

Table 2. Exploited organisms and data sets

	Analyzed organisms	Databases used to build protein interaction networks
PATHBLAST (Kelley, 2003)	S. cerevisiae (Yeast)	DIP
QPath (Shlomi, 2006)	S. cerevisiae, D. melanogaster (Fly), H. sapiens (Human)	DIP, FlyGRID
QNet (Dost, 2007)	S. cerevisiae, D. melanogaster, H. sapiens	DIP, FlyGRID
PathMatch (Yang, 2007)	S. cerevisiae, D. melanogaster, C. elegans (Worm), H. pilori (Bacteria), E. coli (Bacteria)	DIP
GraphMatch (Yang, 2007)	S. cerevisiae, D. melanogaster, C. elegans	DIP
Fionda (Fionda, 2008)	S. cerevisiae, D. melanogaster, C. elegans, H. sapiens	DIP, MINT

identified solutions involving several MAPK (sek-1, lic-PA, mpk-1 and ERKA for GraphMatch and mkk-4,pmk-4, mpk-1, jnk-1, ERKA and CG7717 for Fionda).

As the methods are equivalent with respect to the "biological" quality of the returned results, it is sensible to look at the application domain and complexity in order to find the best method to use. In this respect, for pathway queries, the optimal option seems to be PathMatch (Yang, 2007) which is able to deal with node mismatches, insertions and deletions, with the lowest time complexity among the developed systems. For general topology graph queries the most promising tools seem to be Fionda (Fionda, 2008) and GraphMatch (Yang, 2007). Fionda seems to be a promising method in as much as it exploits a heuristic technique that keeps the time complexity low while still considering both node insertions and deletion in the result sub-graph. GraphMatch is promising since, by exploiting an exact search algorithm, it guarantees to find the best solution according to the adopted scoring schema.

CONCLUDING REMARKS

In this chapter, a comparative survey of the methods developed to query protein interaction networks (PINs) has been carried out. In short, some tools have been developed and some of them are efficient and effective, but not yet sufficiently consistent with the requirements of any biologists. In fact, some of the tools do not yet exploit all the additional information (e.g. go terms or interaction reliability coefficients) that might improve the result quality, and they do not take into account all the possible biological diversities that might be permitted to obtain better results (structure approximations). Other tools, on the contrary, are more rigorous, for example by using exact algorithms. However, these tools have too much high time complexity and runtime to be able to handle large query sub-graphs.

Despite these limitations, the efforts within this research field have been increasing quickly in the last few years, and this area seems to be a promising research domain to improve significantly the knowledge of biological data and mechanisms at the basis of life processes.

REFERENCES

Alon, N., Yuster, R., & Zwick, U. (1995). Color-coding. *Journal of the ACM (JACM)*, *42*(4), 844-856.

Bairoch, A., Boeckmann, B., Ferro, S., & Gasteiger, E. (2004). Swiss-Prot: Juggling between evolution and stability. *Briefings in Bioinformatics*, *5*(1), 39-55.

Berg, J., Lässig, M., & Wagner, A. (2004). Structure and evolution of protein interaction networks: A statistical model for link dynamics and gene duplications. *BMC Evolutionary Biology*, *4*(51).

Chatr-aryamontri, A., Ceol, A., Montecchi Palazzi, L., Nardelli, G., Schneider, M. V., Castagnoli, L., & Cesarei, G. (2006). MINT: the Molecular INTeraction database. *Nucleic Acids Research*, *35*(Database issue), D572-D574.

Dost, B., Shlomi, T., Gupta, N., Ruppin, E., Bafna, V., & Sharan, R. (2007). QNet: A Tool for Querying Protein Interaction Networks. *Proceeding of the eleventh Annual International Conference Research in Computational Molecular Biology (RECOMB 2007)* (pp. 1-15). Springer Berlin: Heidelberg.

Durand, P., Labarre, L., Meil, A., Divol, J.-L., Vandenbrouck, Y., Viari, A., & Wojcik, J. (2006). GenoLink: a graph-based querying and browsing system for investigating the function of genes and proteins. *BMC Bioinformatics*, *7*(21).

Ferro, A., Giugno, R., Pigola, G., Pulvirenti, A., Skripin, D., Bader, G. D., & Shasha, D. (2007). NetMatch: a Cytoscape plugin for searching biological networks. *Bioinformatics*, *23*(7), 910-912.

Fionda, V., Palopoli, L., Panni, S., & Rombo, S. E. (2008). Protein-protein interaction network querying by a "focus and zoom" approach. *Proceeding of 2nd International Conference on Bioinformatics Research and Development (BIRD 08)*, (pp. 331-346).

Garey, M., & Johnson, D. (Ed.) (1979). *Computers and intractability: A guide to the theory of NP-completeness*. New York: Freeman.

Kalaev, M., Smoot, M., Ideker, T., & Sharan, R. (2008). NetworkBLAST: Comparative analysis of protein networks. *Bioinformatics*, *24*(4), 594-596.

Kanehisa, M., & Goto, S. (2000). KEGG: Kyoto Encyclopedia of Genes and Genomes. *Nucleic Acids Research*, *28*, 27-30.

Kelley, B. P., Sharan, R., Karp, R. M., Sittler, T., Root, D. E., Stockwell, B. R., & Ideker, T. (2003). Conserved pathways within bacteria and yeast as revealed by global protein network alignment. *Proceedings of the National Academy of Sciences of the United States of America*, *100*(20), 11394-11399.

Kelley, R., & Ideker, T. (2005). Systematic interpretation of genetic interactions using protein networks. *Nature Biotechnology*, *23*, 561-566.

Mewes, H. W., Frishman, D., Mayer, K. F. X., Münsterkötter, M., Noubibou, O., Pagel, P., Rattei T., Oesterheld, M., Ruepp, A., & Stümpflen, V. (2006). MIPS: Analysis and annotation of proteins from whole genomes in 2005. *Nucleic Acids Research*, *34*(Database Issue), D169–D172.

Pinter, R. Y., Rokhlenko, O., Yeger-Lotem, E., & Ziv-Ukelson, M. (2005). Alignment of metabolic pathways. *Bioinformatics*, *21*(16), 3401-3408.

Salwinski, L., Miller, C. S., Smith, A. J., Pettit, F. K., Bowie, J. U., & Eisenberg D. (2004). The Database of Interacting Proteins: 2004 update. *Nucleic Acids Research*, *32*(Database Issue), D449-D451.

Sharan, R., & Ideker, T. (2006). Modeling cellular machinery through biological network comparison. *Nature Biotechnology*, *24*(4), 427-433.

Sharan, R., Suthram, S., Kelley, R. M., Kuhn, T., McCuine, S., Uetz, P., Sittler, T., Karp, R. M., & Ideker, T. (2005). Conserved patterns of protein interaction in multiple species. *Proceedings of the National Academy of Sciences of the United States of America, 102*(6), 1974-1979.

Shlomi, T., Segal, D., Ruppin, E., & Sharan, R. (2006). QPath: a method for querying pathways in a protein-protein interaction network. *BMC Bioinformatics, 7*(199)

Singh, R., Xu, J., & Berger, B. (2007). Pairwise global alignment of protein interaction networks by matching neighbourhood topology. *Proceedings of the Eleventh Annual International Conference on Research in Computational Molecular Biology (RECOMB 2007), 4453*, 16-31.

Spirin, V., & Mirny, L. A. (2003). Protein Complexes and functional modules in molecular networks. *Proceedings of the National Academy of Sciences of the United States of America, 100*, 12123-12128.

Stark, C., Breitkreutz, B. J., Reguly, T., Boucher, L., Breitkreutz, A., & Tyers, M. (2006). BioGRID: A general repository for interaction datasets. *Nucleic Acids Research, 34*(Database Issue), D535-D539.

Tatusova, T. A., & Madden, T. L. (1999). Blast 2 sequences, a new tool for comparing protein and nucleotide sequences. *FEMS Microbiology Letters, 174*(2), 247-250.

The Gene Ontology Consortium. (2000). Gene Ontology: tool for the unification of biology. *Nature Genetics, 25*, 25-29.

Tian, Y., McEachin, R. C., Santos, C., States, D. J., & Patel, J. M. (2007). SAGA: A subgraph matching tool for biological graphs. *Bioinformatics, 23*(2), 232-239.

Wernicke, S., & Rasche, F. (2007). Simple and fast alignment of metabolic pathways by exploiting local diversity. *Bioinformatics, 23*(15), 1978-1985.

Yang, Q., & Sze, S.-H. (2007). Path Matching and Graph Matching in Biological Networks. *Journal of Computational Biology, 14*(1), 56-67.

Zhang, S., Zhang, X.-S., & Chen, L. (2008). Biomolecular network querying: a promising approach in systems biology. *BMC Systems Biology, 2*(5).

Zhang, L. V., King, O. D., Wong, S. L., Goldberg, D. S., Tong, A. H. Y., Lesage, G., Andrews, B., Bussey, H., Boone, C., & Roth, F. P. (2005). Motifs, themes and thematic maps of an integrated Saccharomyces cerevisiae interaction network. *Journal of Biology, 4*, 6

APPENDIX: DATA RESOURCE LIST

DIP (Database of Interacting Proteins). The DIP database (http://dip.doe-mbi.ucla.edu/) catalogues experimentally discovered interactions between proteins, combining information from a variety of sources to create a single, consistent set of protein-protein interactions. The data stored within the DIP database were curated both manually, by expert curators, and automatically, using computational approaches that exploit the knowledge about the protein-protein interaction networks extracted from the most reliable core subset of the DIP data. DIP is a member of the International Molecular Exchange Consortium (IMEx), a group of the major public providers of interaction data. The databases of IMEx are meant to work together to prevent duplications of effort, collecting data from non-overlapping sources and sharing the curated interaction data.

MINT (Molecular INTeraction database). MINT (http://mint.bio.uniroma2.it/mint/) focuses on experimentally verified protein-protein interactions extracted from the scientific literature by expert curators. The curated data can be analyzed from text or viewed graphically with the 'MINT Viewer'. MINT has signed the IMEx agreement and supports the Protein Standard Initiative (PSI) recommendation. The whole dataset can be freely accessed online in both interactive and batch modes through web-based interfaces and an FTP server.

MIPS Database. The Munich Information Centre for Protein Sequences (MIPS) (http://mips.gsf.de/), Neuherberg, Germany, provides resources related to genome information. Manually curated databases for several reference organisms are maintained. In particular, the MIPS protein–protein interaction resource (MPact) extends the single-protein view into a network perspective. The data model allows extending the interactions of proteins with other biochemical entities (e.g. RNA, drugs, etc.). Interacting objects can not only be associated with each other to represent, for example, complexes, but also with external information describing the corresponding experiments (e.g. yeast two-hybrid, co-immunoprecipitation or mass spectrometry data).

BioGRID (Biological General Repository for Interaction Datasets). The BioGRID (http://www.thebiogrid.org/) is a curated biological database of protein-protein interactions created in 2003 (originally simply referred to as the General Repository for Interaction Datasets - GRID). It provides protein–protein interaction data for all major species and is attempting to remove redundancy to create a single mapping of protein interactions. Users of BioGRID can search for their protein of interest and retrieve annotations, as well as physical and genetic interaction data. The reported interaction data are retrieved by the primary literature and curated by the BioGRID curators. BioGRID curators and contributors include experts in various model organism systems and associated databases. BioGRID is a member of IMEx.

KEGG (Kyoto Encyclopedia of Genes and Genomes). KEGG (http://www.genome.jp/kegg/) is a collection of online databases referring to genomes, enzymatic pathways, and biological chemicals. KEGG links information about molecular interaction networks, such as pathways and complexes (the Pathway Database), information about genes and proteins generated by genome projects (including the gene database) and information about biochemical compounds and reactions.

GO (Gene Ontology). The Gene Ontology (http://www.geneontology.org/) project, or GO, provides a controlled vocabulary to describe gene and gene product attributes in any organism. It can be broadly split into two parts. The first one is the ontology itself. There are three organizing principles in GO, each representing a key concept in Molecular Biology. These are: the *molecular function* of gene products, their role in multi-step *biological processes* and their localization in *cellular components*. The Gene Ontology is continuously updated, and new versions are made available on a monthly basis. The second part is the annotation set, characterizing gene products using terms from the ontology. The members of the GO Consortium submit their data, which are then made publicly available through the GO website.

Chapter XVIII
Module Finding Approaches for Protein Interaction Networks

Tero Aittokallio
University of Turku, Finland

ABSTRACT

This chapter provides an overview of the computational approaches developed for exploring the modular organization of protein interaction networks. A special emphasis is placed on the module finding tools implemented in three freely available software packages, VisANT, Cytoscape and MATISSE, as well as on their biomedical applications. The selected methods are presented in the broader context of module discovery options, ranging from approaches that rely merely on topological properties of the underlying network to those that take into account also other complementary data sources, such as the mRNA levels of the proteins. The author will also highlight some current limitations in the measured network data that should be understood when developing and applying module finding methodology, and discuss some key future trends and promising research directions with potential implications for clinical research.

INTRODUCTION

Recent advances in experimental technologies and computational methods have made it possible to measure and predict protein-protein interactions on a global scale (see Chapters I-VI and Shoemaker & Panchenko, 2007a, b). While the large-scale interaction datasets can provide an unprecedented glimpse into the cellular mechanisms underlying the behavior of various biological systems, the increasing sizes and densities of the protein interaction networks available today pose also many challenging computational problems. In particular, the inherent complexity of most biological processes and the large number of possible interactions involved can make it difficult to interpret and mine the interaction networks only by eye, even with the help of sophisticated visualization and layout tools available. Software packages that implement computational tools for more advanced network data mining can facilitate the explorative network analysis by identifying the key players and their interactions that contribute to the cellular processes of interest. This may allow e.g. to pinpoint errors in experimentally or computationally derived interaction links, identify proteins directly involved in the particular process, and to formulate hypotheses for follow-up experiments. It should be realized, however, that even if the rapidly developing complex network theory has successfully been applied to analysis of various social and technological networks, such as the Internet and computer chips, its impact on studying molecular interaction networks is still an emerging area of research and therefore all these methods should be considered as experimental. At the moment, the computational tools are best used together with network visualization and analysis software that enable interactive and fully-controlled mining of the complex protein interaction networks.

One of the most fundamental properties found in many biological networks is their modular organization (Hartwell et al., 1999). Consequently, the decomposition of large networks into a hierarchy of possible overlapping sub-networks (so-called *modules*) has become a principal analytical approach to deal with the complexity of large cellular networks (Barabási & Oltvai, 2004). In protein interaction networks, a functional module refers to a group of physically connected proteins that work together to carry out a specific cellular function in a particular spatio-temporal context. A large number of computational tools of increasing complexity have recently been developed for investigating the modular organization of interaction networks. These tools cannot only identify whether a given network is modular or not, but also detect the modules and their inter-relationships in the underlying network. By relating the found sub-networks with complementary functional genomics or proteomics data, such as gene expression profiles from genome-wide microarray experiments or protein abundance measurements from mass-spectrometry-based assays, it is also possible to identify a hierarchy of connected groups of components that show coherent expression patterns. Such functionally organized modules cannot only emphasize the biological meaning of the modules discovered, but also allow us to gradually focus on the active subsystems of particular interest (*active modules*), which can lead to concrete hypotheses about the regulatory mechanisms and pathways most important for the given process (Ideker et al., 2002). Moreover, these modules can subsequently be used in predictive modeling studies, with the aim of suggesting new biological hypotheses, such as unexplored new interactions or the function of individual components, or even distinguishing different biomedical phenotypes (*discriminative modules*).

While several computational approaches to discovering functional modules in protein interactions networks are available to molecular biologists, and new ones are constantly being developed by the methodologists, there is no general consensus on how to choose between the different strategies and definitions. The choice of the network mining approach is an important aspect which depends both on the network and questions under the analysis. Therefore, basic understanding of the graph-theoretic concepts and algorithms behind the software packages is required so that the specific questions can be addressed and biologically meaningful interpretations can be made. This chapter describes several computational strategies developed for investigating the modularity of protein interaction networks, together with their relative merits and potential pitfalls. Since this topic has witnessed great progress lately, only representative examples of the different approaches can be presented. Among the multitude of algorithms and methods available in the literature and Internet, I describe here those features and options available in three popular software tools, VisANT, Cytoscape and MATISSE, which support the discovery of functionally coherent network modules. These publicly available software were selected because they come with a convenient graphical user interface making the methods easily accessible also to the end-users without programming skills. A special emphasis is placed on four specific tools implemented in these software, which represent typical computational strategies in the wide spectrum of module finding options, ranging from approaches that are based on topological network properties, such as nodes with a particular connectivity level (e.g. VisANT) or sub-networks that are highly connected (MCODE), to more sophisticated approaches that take into account also other related data sources, such as significant differences or co-expression patterns in the mRNA levels of the proteins (jActiveModules and JACS, respectively).

BACKGROUND AND MOTIVATION

Many advanced network analysis software complement the visual network exploration with options that facilitate the characterization of topological and functional properties of the underlying network (comparison of the features available in the popular software tools can be found from Chapter XVI and Suderman & Hallett, 2007; Zhang et al., 2007; Cline et al., 2007). Typical topological properties include, e.g., the *degree distribution* for determining hub nodes, *clustering coefficient* for detecting dense neighborhoods, and *shortest paths* for finding

Figure 1. Basic graph-theoretic concepts. An example network graph G describing 18 binary interactions (edges) among 12 proteins (nodes), i.e. the cardinalities are $|V| = 12$ and $|E| = 18$. The ***neighborhood graph*** *$N_a = (V_a, E_a)$ of the node a is the set of nodes in G and their interactions that are directly connected to a; the left-hand circle in the figure identifies N_a. By definition, the node a itself and its own connections are excluded from the sets V_a and E_a, respectively. The* ***degree*** *d of a node is the number of its first neighbors, e.g. $d(a) = |V_a| = 5$, making it a hub protein in the network. The* ***path length*** *between two nodes is the number of edges on a shortest path between them; for instance, node b belongs to the 4-neighborhood of a. This particular toy network is said to be* ***connected,*** *because there exists a path connecting each pair of the nodes. A sub-network is called a* ***k-core*** *if the minimal degree among its nodes is k. The* ***density*** *D of a graph G is the fraction of edges it has out of all possible node pairs; formally, $D(G) = 2|E|/|V|(|V|-1)$. The* ***clustering coefficient*** *c of a node defines the proportion of its first neighbors that are connected to each other; for instance, since there is only one edge connecting the neighbors of a, $c(a) = D(N^a) = 2 \cdot 1/20$. In other words, clustering coefficient quantifies how close the connectivity of the neighborhood of a node is to a sub-network in which every node is connected to every other node (so-called clique). For instance, the neighborhood of b in the right-hand circle forms a fully connected sub-network, and therefore $c(b) = 1$.*

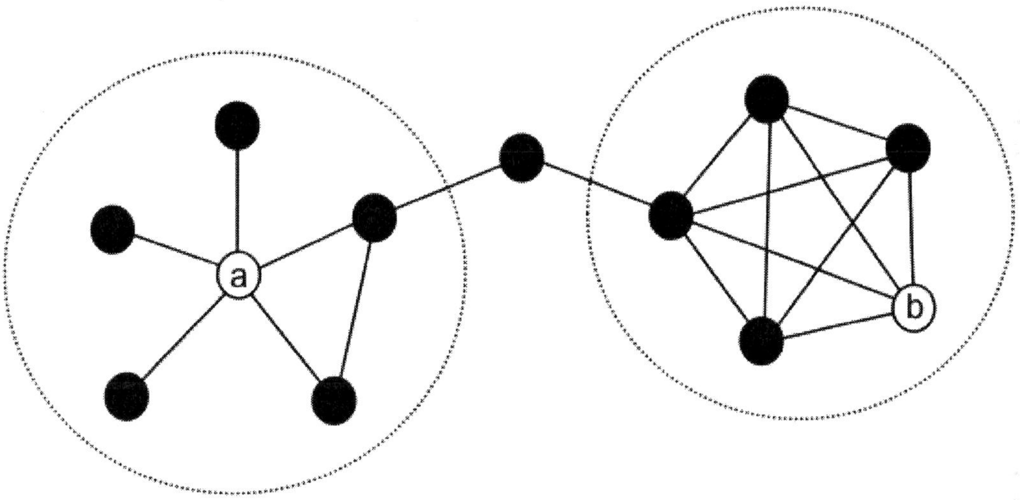

Box 1. Network nomenclature and biological counterparts

> *A* **graph** ***G = (V, E)*** *is a mathematical object that represents a protein interaction network. Conventionally, protein-protein interactions are modeled by undirected graphs* ***G****, in which* ***V*** *is a set of vertices (protein nodes), and* ***E*** *is a set of edges connecting pairs of nodes (binary interactions). The activity states of the proteins can be represented as node attributes (e.g. continuous weighting by the mRNA level of the gene encoding the particular protein under a specific condition), and the information how and when these biomolecules interact with each other as edge attributes (e.g. discrete labeling by the conditions under which the particular interaction is active).*
>
> *A* **module** *is a sub-network of G with specific properties. Typical topological properties include neighborhoods of highly connected genes (****hubs****), highly connected subnetworks (****clusters****), and highly-occuring subnetworks with particular connectivity patterns (****motifs****). Additionally, it can be required that the sub-networks show coherent expression changes (****active modules****), or allow distinguishing between distinct phenotypes (****disriminative modules****). In cell biology, such modules may correspond to functionally coherent groups of biomolecules, which act together in cellular mechanisms or processes, e.g., sub-units of protein complexes or biological pathways.*

n-neighborhoods of selected nodes; see Figure 1 and Box 1 for the basic graph-theoretic concepts and their biological counterparts. Such measures, either alone or when compared to those of the whole network or its randomized counterpart, can be used to identify topologically coherent sub-networks, which may encode functional modules – i.e. groups of protein nodes and their interactions that can be attributed to a particular biological function (Hartwell et al., 1999). Several computational definitions of the topological properties of specific sub-networks, together with their biological interpretations, have been proposed during the past few years (Barabási & Oltvai, 2004), many of which – e.g. clusters, complexes and motifs – are described in detail in Chapters VIII and IX. In this chapter, the focus is on those network properties that have widely been recognized in many protein interaction networks: (1) modules are highly connected sub-networks (i.e. intra-module connectivity is higher than inter-module connectivity), and (2) protein interaction networks have hierarchical modular organization (modules are not isolated but they can interact and overlap at multiple levels). The first criterion reflects the fact that certain proteins are involved in common elementary biological functions, while the second implies that the same proteins can be active in multiple cellular processes at different points of time or under different conditions. In addition to such topological properties, requirement of functionally coherent modules often involves also some type of complementary information on the protein nodes or their interactions that can facilitate revealing the multi-level functional organization of the protein interaction networks.

A useful source of complementary information on the activity of the protein nodes originates from the high-throughput transcriptomics and proteomics technologies. In particular, genome-wide mRNA or protein expression measurements using microarrays or mass-spectrometry, respectively, are increasingly being used to illuminate the molecular mechanisms involved in the control of cellular systems in various organisms. However, interpretation of the resulting lists of genes or proteins remains a labor-intensive and error-prone task, because listing the individual elements alone can provide only a limited insight into the multitude of biological processes the individual elements participate in. To facilitate distinguishing elements directly involved in a particular process from bystander elements, whose expression have been altered by secondary effects or technical artifacts, the expression data can be mapped on a protein interaction networks to provide better understanding of the overall organization of the system in its entirety, rather than investigating individual genes, proteins or pathways separately. Moreover, instead of using the conventional pathway enrichment analysis among pre-defined gene/protein sets, which are limited to well-studied systems and known pathways only, identification of network modules that are active in the particular experiment is better targeted at finding novel interactions, cross-talks between the established pathways, and eventually the biological meaning of each individual component. Therefore, analyzing experimental data in the context of protein interaction networks can, in general, help to determine system-level cellular mechanisms explaining the experimental observations. Functional modules extracted from the human interaction networks, in particular, that are becoming available at an increasing rate, can provide systematic strategies to unravel the molecular basis of complex human diseases, and therefore enable identification of novel molecular markers for pharmaceutical and/or diagnostics developments for many disorders (Ideker & Sharan, 2008).

MODULE DISCOVERY ON THE BASIS OF NETWORK TOPOLOGY ALONE

At the most general abstraction level, the protein interaction networks can be investigated and compared on the basis of their topology (i.e. the connection structure between the nodes). The representation of protein interaction networks as undirected graphs makes it possible to systematically investigate the structure-function relationships of these networks using well-understood graph-theoretical concepts and algorithms (Przulj et al., 2004). Global network measures, such as distributions of the node degrees, clustering coefficients, or path lengths, that can capture the overall network organization, have been proposed as topological correlates of the network evolution, function, stability or even dynamic responses (Yook et al., 2004; Albert, 2005). Alternatively, characterizing local interconnectivity patterns in terms of relative importance of individual nodes (centrality measures) or connecting paths (pathway redundancy), can provide more detailed functional information on the node neighborhoods and alternative pathways, and hence facilitate understanding of fundamental biological concepts, such as proteins' essentiality and robustness of a particular system under internal or external perturbations (Estrada, 2006; Aittokallio & Schwikowski, 2006). Since the topological parameters can be rather sensitive to the chosen experimental setup and data analysis protocol (Hakes et al., 2008), it is essential to utilize these network measures in software packages, which allow multi-resolution network display and mining, to confirm that the properties discovered are not due to experimental or computational artifacts only.

Characterizing Network Connectivity Patterns using VisANT

VisANT software package is a visual analysis tool for biological networks and pathways, which provides a number of useful tools for both exploring and displaying the topological properties of a given protein interaction network (Hu et al., 2004; Hu et al., 2008). For instance, the user can easily display a scatter-plot or log-linear fit of the degree distribution of the connections in a separate window, which is directly linked to the main network view so that nodes with a particular level of connectivity can be readily identified. Such plot allows for distinguishing highly-connected nodes (*hubs*) from weakly-connected nodes (*leaves*), whereas the degree exponent of the log-linear fit can be used to characterize the scale of the network (Barabási & Oltvai, 2004). Without entering to the debate whether or not protein interaction networks exhibit a scale-free topology, the degree distribution provides a quick summary of the overall connectivity structure and enables differentiation between distinct classes of network topologies (e.g. power-law type of distribution from a peaked one as seen in random network topologies). For identifying the modular organization of the network, VisANT provides similar plots for the distribution of the clustering coefficient over all of the nodes in the network, which can characterize the overall tendency of the nodes to form clusters or groups. Moreover, VisANT enables plotting of the average clustering coefficient of the nodes as a function of their degree k:

$$C(k) = \frac{1}{n_k} \sum_{d(v)=k} D(N_v),$$

where n_k is the number of nodes with degree k. The function $C(k)$ can be used to characterize the network's hierarchical structure, and it allows the user to identify local neighborhood of nodes at a particular connectivity and density level. VisANT software also enables evaluation of shortest path lengths between nodes, which can help to further expand the selected modules, and the occurrence of sub-networks with a pre-defined connection structure (network *motifs*). One way to evaluate the modular organization of the network with VisANT is to exclude part of nodes according to their degree value. Interactive experimenting with increasing degree cut-off thresholds, coupled with dynamic navigation of the reduced networks, can reveal densely connected areas when sufficiently many low-degree regions have been removed. The downside of this rather crude multi-level filtering process is that it may also destroy the inter-module connections. Once the user has selected the node groups, based on the topology statistics or motif configurations, or imported existing groupings from external databases or from own files, VisANT provides many options for displaying and manipulating these higher-level graph structures as meta-graphs, with the possibility to expand and collapse them iteratively to show reduced network views at multiple scales. Nested meta-nodes allow also one node to have multiple instances in the network and these instances are automatically tracked by the software.

Identifying Highly-Connected Sub-Networks using MCODE

Cytoscape software platform has a wide spectrum of tools and plug-ins for extracting the basic topological properties, such degree distributions, clustering coefficients, shortest paths, centrality statistics, and over-represented motifs, in addition to the advanced filtering and displaying options based on such attributes available in the core of the software package (Shannon et al., 2003; Cline et al., 2007). Cytoscape also includes specific plug-ins devoted for discovering modules – or putative complexes and pathways – in the imported networks. The molecular complex detection (MCODE) algorithm by Bader and Hogue (2003) was developed to enable an automated discovery of densely-connected sub-networks in large protein interaction networks. The requirement of high density originates from the observation that highly interconnected regions of protein interaction networks often correspond to known protein complexes and parts of biological pathways (Barabási & Oltvai, 2004). Like VisANT, MCODE considers local groups of nodes with a given connectivity and density level, but it can also expand these core modules automatically by recursively including nearby nodes that either can preserve or increase the sub-graph density. More formally, the operation of the module detection algorithm is organized through three stages. First, the node weighting is done on the basis of the density of the node's neighborhood. Instead of using the standard clustering coefficient, the algorithm uses so-called *core clustering coefficient* of a node a, which is the density of the largest k-core S_a in its neighborhood (including also a itself):

$$C_k(a) = D(S_a), \quad \text{where } S_a = \arg\max\{|S|: S \subseteq N_a \cup a, \forall v \in S: d(v) \geq k\}.$$

Figure 2. An example output from the MCODE. The selected sub-network corresponds to the 3rd highest ranked module in a sample interaction network of Cytoscape (galFiltered). The red intensity in the node color reflects its score, while the white nodes remain unscored and therefore unclustered. The shape of the nodes indicates their roles in the clustering process (seed node – square, core cluster member – circle and non-core cluster member – diamond). The default parameter values were used both in scoring and finding the core clusters. The Include Loops option controls whether or not also self-edges are used when calculating the neighborhood density. The Degree Cutoff parameter controls the minimum number of connections for a node to be scored. When the Haircut option is turned on, all the singly-connected nodes will be removed from the core cluster. The Fluff option enables expanding the cluster cores by one neighbor outwards. The Node Score Cutoff is the most influential parameter for the cluster size. New cluster members are added only if their node score deviates from the seed node's score by less than the set cutoff percentage. The Size Threshold slider on the right allows the user interactively shrink or expand the core clusters by adjusting the threshold below or above its original value (^). The Node Attribute Enumerator provides a summary of the protein attributes and their frequency in the module (here the GO Biological Process annotation is selected). The K-Core parameter enables filtering out modules that do not contain an inter-connected sub-network of at least k degrees. The Max. Depth parameter controls how far from the seed node the search for new members can extend (virtually unlimited by default). For a more detailed and up-to-date parameter information, consult the MCODE website (http:www.baderlab.org/Software/MCODE) and the original article (Bader and Hogue, 2003).

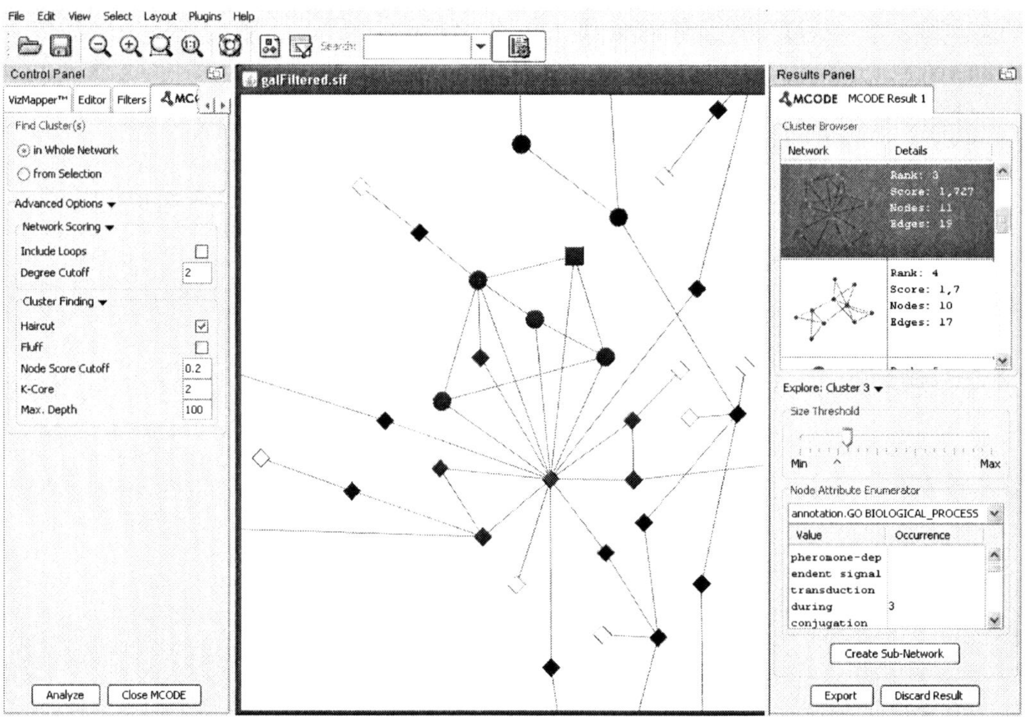

Depending on the chosen value of k, the definition of $C_k(a)$ can emphasize the hub nodes, while giving smaller weights to the less-connected nodes. The final weight attached to the node a is the product of its core clustering coefficient $C_k(a)$ and the size $|S_a|$ of the highest k-core in its neighborhood. In the second stage, greedy optimization algorithm is used, which starts from the seed nodes with highest weights, and then recursively traverses their neighborhoods by including those nodes into the module whose weights deviate from the seed node's weight by less than a user-defined cutoff value. This parameter controls the size and density of the resulting sub-networks. The third stage involves post-processing steps that can filter out or add nodes based on their connectivity. The output of the MCODE algorithm is a list of possible overlapping modules ranked according to their density and size values. The algorithm includes several user-adjustable parameters, with different impacts on the results. The fast running time makes it possible to experiment with different parameter combinations, and their fine-tuning can be carried out according to existing biological knowledge by starting the optimization algorithm from known seed nodes. A further advantage over many traditional network clustering algorithms is that MCODE assigns to the modules only those nodes above a given local threshold density, whereas the rest of the nodes remain unclustered (Figure 2).

MODULE DISCOVERY ON THE BASIS OF TOPOLOGY AND EXPRESSION

One potential pitfall of the module finding approaches that rely solely on the topology of the networks is that false negative and positive interactions can have severe effects on their results, regardless of whether one identifies protein hubs, clusters or network motifs. Since the current interaction networks are still limited both in accuracy and coverage, especially in more complex organisms like human, the topological structure alone may not always be sufficient for fully understanding of the complex cellular mechanisms and functions. To make the identification of functional modules more robust against experimental and modeling artifacts, a number of approaches have proposed the usage of protein interaction networks in conjunction with complementary information such as protein localization, expression or function (Tornow and Mewes, 2003; Lu et al., 2006; Lubovac et al., 2006; Cho et al., 2007). The presumption behind all such integrative module discovery approaches is that exploiting heterogeneous types of data sources in concert may be impacted less by the noise and biases associated with each individual data source if they were used separately. However, such integrative analysis also complicates the validation of the methods due to the overlap between complementary data sources and functional annotations, and thereby may increase the risk for circular reasoning. Here, I focus on those approaches only that combine global interaction networks with mRNA expression data, although any continuous type of state data on the nodes, such as protein abundance, could in principle be used instead.

Most computational approaches that use both expression and network datasets are motivated by those early experimental studies which demonstrated that genes with similar expression profiles are more likely to encode interacting proteins (Ge et al. 2001). In particular, it has been shown that many stable protein complexes show significant co-expression in their mRNA levels (Jansen et al., 2002). Inspired by these observations, several computational studies have used genome-wide mRNA expression data to characterize and predict protein interactions (Tirosh & Barkai, 2005; Sprinzak et al., 2006; Shoemaker & Panchenko 2007b), or used these complementary data resources together to reveal dynamic organization of modularity in protein interaction networks (Han et al., 2004). Most network visualization software tools also allow mapping gene expression data on the global interaction networks to facilitate their interpretation and integrated analysis. For instance, VisANT 3.0 enables the user to select node groups based on expression ranges or profile correlations, making it possible e.g. to discover new proteins with similar function to a given list of query proteins already known to have closely related function (Hu et al., 2007). Multiple experiments can be navigated using a sliding bar and the navigation process can be animated using color coding. However, VisANT currently lacks more automated module finding options based on both the networks connectivity and expression correlation structure.

Identifying Sub-Networks with Similar Expression Profiles using MATISSE

MATISSE software by Ulitsky & Shamir (2007) provides automated module detection tools that makes use of both the topology of interaction patterns as well as similarity sets of nodes inferred from expression data. In addition to several visualization and clustering options for network and expression datasets separately, it implements an efficient algorithm for identifying jointly active connected sub-networks (JACS) – i.e. connected sub-networks in the interaction data that show high internal similarity in their expression profiles (Ulitsky & Shamir, 2007). The algorithm finds such putative modules using statistical hypothesis testing framework, which evaluates the significance of the pairwise expression similarities S_{ij} between all those node pairs (i,j) that belong to a connected sub-network U (candidate modules). Pearson correlation between the expression patterns of nodes i and j is used as a default S_{ij}, but in principle any similarity measure could be used instead. More formally, let $\gamma_{ij}^{m} = \beta\ P_iP_j$ be the probability that the nodes i and j are in the same group (so-called module hypothesis), where β is a parameter reflecting the general probability of any pair to group together and P_i is a prior probability describing how likely is it that the node i is transcriptionally regulated. Similarly, let $\gamma_{ij}^{n} = p_m\ P_iP_j$ be the probability that the nodes i and j are in the same group under random selection of pairs (the null hypothesis), where p_m is the fraction of node pairs expected at random. The putative module U is then evaluated using the log-likelihood ratio score:

$$L(U) = \prod_{i,j \in U} \frac{\gamma_{ij}^{m} P(S_{ij} \mid M_{ij}) + (1 - \gamma_{ij}^{m}) P(S_{ij} \mid N_{ij})}{\gamma_{ij}^{n} P(S_{ij} \mid M_{ij}) + (1 - \gamma_{ij}^{n}) P(S_{ij} \mid N_{ij})}.$$

This formula stems from the assumption that the pairwise similarities S_{ij} are conditionally independent, and that they originate from a mixture of two Gaussian distributions, in which the two normal components $P(S_{ij} \mid M_{ij})$ and $P(S_{ij} \mid N_{ij})$ correspond to the node pairs (i, j) that have high or low expression similarity, respectively. These distributions, together with the prior probabilities under the two rivaling hypotheses, are automatically learned with the expectation-maximization algorithm. Using this scoring function, MATISSE identifies modules through a heuristic multi-stage optimization procedure. First, relatively small, high-scoring, connected node sets are detected as seeds. Second, a greedy iterative algorithm is used to optimize all the seeds simultaneously by considering the following four operations: (i) adding an unassigned node to a module, (ii) removing a node from a module, (iii) exchanging a node assignment between two modules, and (iv) merging two modules by taking their union. An operation is accepted if it improves the overall score calculated over all the modules and also preserves the connectivity of the modules. In the last stage, significance filtering of the resulting modules is performed by sampling random node sets of the same size and comparing their scores to those of the detected ones. While it assumes no prior knowledge on the number of modules (size constraints can be introduced), the optimization procedure forces the modules to be non-overlapping, which may hinder the assignment of proteins whose expression levels are a result of multiple cellular functions reflected in the experiment.

Identifying Sub-Networks with Coherent Expression Changes using jActiveModules

The usage of pairwise similarities, like Pearson correlation, in the identification of functional modules makes it possible to use various types of expression dataset, e.g., time-series or steady-state experiments. This comes with the potential limitation that such overall measures of co-expression may ignore the fact that a protein can be active in a particular process over a subset of experimentally measured conditions or time-points only – a phenomenon termed just-in-time assembly (De Lichtenberg et al., 2005). Experimental studies have indeed demonstrated that some components of protein complexes may show coherent expression changes under a few conditions only and those components often correspond to known sub-complexes with specific functions (Simonis et al. 2006). This motivates those computational approaches that search for connected sub-networks whose nodes show differential expression over a particular subset of conditions only. One such method by Ideker et al. (2002) is implemented in the Cytoscape software as a plug-in called jActiveModules. The algorithm requires as an input the condition-specific $p_{a,c}$-values representing the significance of differential expression of a node a under each condition c.

The method first converts the *p*-values to *z*-scores, using the inverse normal cumulative distribution function, that is, $z_{a,c} = P^{-1}(1 - p_{a,c})$, and then derives an aggregate score $S_c(A)$ for a sub-network *A* under the condition *c* by averaging the *z*-scores over the nodes in *A*. Using the simplifying assumption that these condition-specific scores are independent, the significance of the j th highest scoring condition $S_j(A)$ can be computed using the binomial order statistic:

$$p_j(A) = \sum_{h=j}^{m} \binom{m}{h} \left[1 - P(S_j(A))\right]^h \left[P(S_j(A))\right]^{m-h}.$$

The $p_j(A)$-value gives the probability that at least *j* of the *m* conditions have scores above $S_j(A)$, which reflects the activity of the sub-network *A* under conditions ranked 1 through *j*. After adjusting the scores for the rank and

Figure 3. Network modules involved in HIV-1 reactivation. An example of integrated analysis using both gene expression profiling of human T-cell lines infected with human immunodeficiency virus (HIV), as well as curated human-human and human-HIV protein interactions from HPRD and HIV-1 interaction database, respectively. The active sub-networks during different stages of the process were identified using a local search, with parameters depth=2 and max depth=2, in the jActiveModules plugin to Cytoscape software (Ideker et al. 2002). The significance of differential expression at different activation stages was assessed using p-values from the t-test. The color of the nodes corresponds to the mean expression change and the shape indicates the species (HIV – diamond and human - circle). The sub-network shown in the inset was active during the early stages of the reactivation. This connected sub-network included five hub-proteins (Rb1, Polr2a, Jun, HSPca and Pten), each of which had significant numbers of differentially expressed neighbors (p<0.01). As expected, the module contains HIV-1 proteins, such as p53 (TP53), which co-operates with Tat in the activation of HIV-1. Moreover, casein kinase 2 (Csnk2a1) is involved in the activation of the HIV-1 replication and it interacts with PTEN and HSPca, both of which are known to interact with Tat. In addition to the known players, the active sub-network can suggest novel functions relevant for the HIV pathogenesis (Bandyopadhyay et al., 2006). The figure was reproduced with permissions from the authors of the original work and the editors of the Proceedings of the Pacific Symposium on Biocomputing 2006.

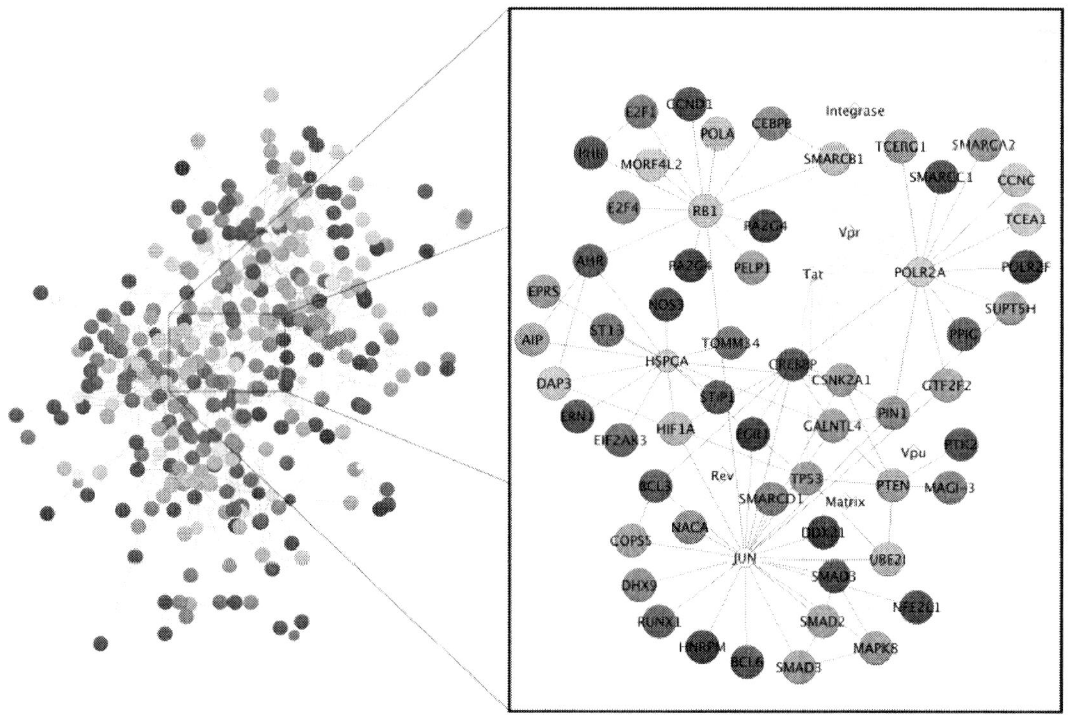

correcting for the background score distribution by sampling randomly node sets of similar size, independently of the underlying interactions, the scoring function can be use to rank both the sub-networks and the subsets of conditions. For identifying high-scoring connected sub-networks, jActiveModules enables using a stochastic optimization heuristic based on simulated annealing (Kirkpatrick et al., 1983). Briefly, in each iteration, a random node of the network is either removed or added to the module and the score for the resulting new sub-graph is calculated. If the new score is larger than the previous one, then the move is accepted; otherwise the move is accepted with a probability proportional to the current temperature, which decreases geometrically with the iteration number. Finally, the temperature is set to zero, meaning that the optimization becomes greedy, until the algorithm reaches the closest local maximum. The algorithm can be adjusted by allowing the user to define the number of modules searched simultaneously and by reducing the effect of hub nodes on the search results (Ideker et al. 2002). To output smaller modules for visualization, one can repeat the search within each original sub-network using the local search option (see Figure 3). Several modifications to the original scoring function and search algorithm have been developed since then (Sohler et al., 2004; Rajagopalan & Agarwal, 2005; Cabusora et al., 2005; Nacu 2007; Guo et al., 2007; Dittrich et al. 2008).

EVALUATION OF THE SIGNIFICANCE OF THE MODULES DISCOVERED

An increasing number of methodological papers have recently introduced new and improved computational methods for finding functional modules from protein interaction networks and other information sources; see e.g. Chapter IX and the recent review by Sharan et al (2007). Each of these papers concludes that the modules found by the particular method significantly overlap with known protein complexes (extracted from databases such as MIPS) and show significant homogeneity in their functional annotation (typically assessed using enriched GO categories). Given the known association of protein interaction patterns and expression profile similarity with the established protein complexes and functional classifications available, these results are not very surprising and cannot indisputably demonstrate the relative benefits gained by using the particular method, unless a set of reference methods has been applied to the same datasets. As a reference method, many methodology papers use random node sets of similar sizes to asses the statistical significance of the modules detected, and evaluate overlap with known functional classes to show functional homogeneity observed in these sets. However, it could be argued that such comparisons against random sub-networks, regardless whether or not the random model preserves the connectivity structure of the modules, should be considered more like a sanity check for the proposed method, and as such can provide only a little indication of the biological significance of the putative modules discovered. A potential caveat of using functional annotations either in module finding or in their validation is the limited accuracy and availability of annotations for some organisms, especially for human, and for some biological processes that are not yet well described in GO or other knowledge databases.

Evaluating the Methods in terms of Robustness Against Technical Factors

In the very systematic comparison by Brohée & van Helden (2006), the authors evaluated a number of topology-based module finding algorithms also in terms of their sensitivity to their user-adjustable parameters values and against different levels of false positive and negative links known to occur in the measured protein interaction networks. The authors found that the most robust method was the one based on the Markov clustering algorithm (Pereira-Leal 2004). Recently, Sharan et al. (2007) reviewed a number of computational approaches to protein interaction network-based function prediction and classified them into direct methods, which infer the functional role of a protein through its connections in the network, and module-assisted methods, which use the detected sub-networks for the annotation task. Using a small-scale comparison in a single protein interaction network from BioGRID, they observed that the straightforward neighbor-counting method of Schwikowski et al. (2000) had a higher specificity in predicting GO Biological Process annotations when compared to the MCODE-assisted method (Sharan et al., 2007). The authors attributed this difference in the function prediction performance to the fact that the current module discovery methods are biased towards those processes that impose highly-connected sub-networks. However, more comprehensive evaluation studies are critically needed that consider also other technical factors, e.g. how the module findings are affected by the different graph models used for representing the measured interaction patterns (Scholtens et al. 2005; Rungsarityotin et al., 2007), and what is the contribution of the particular experimental design, known sampling biases, or the varying experimental techniques for measuring the protein interactions (Pu et al., 2007; Gentleman & Huber, 2007).

Assessing the Biological Significance of the Modules Discovered

Beyond evaluating the modules in statistical or technical terms only, the true relevance of the computational approaches can be assessed with concrete biomedical applications by showing how the putative modules can provide e.g. added biological insights into the molecular processes being studied, or even novel medical information e.g. in terms of molecular markers for diseases. While the technical comparisons are important prerequisites for performing such real applications, their practical contribution may remain very limited from the biologist point of view. At the moment, there seems to be no systematic comparative studies among the different module discovery approaches that would demonstrate their relative benefits in more biological terms. The situation is very similar with the ongoing debate on the biological meaning of other topological properties, such as the scale-free property (Stumpf et al., 2005; Han et al., 2005; Khanin & Wit, 2006; Friedel & Zimmer, 2006a), or the concepts of 'date hubs' and 'party hubs' (Ekman et al., 2006; Batada et al., 2006; Bertin et al., 2007). The biological significance of these rather abstract network properties are still partly unclear and open to many opinions and speculations. This means that the biological conclusions drawn from the global topological measures should be viewed with some skepticism (Hakes et al., 2008). The same caution is also needed when interpreting the modular organization of the protein interaction networks. An additional dimension that may be needed when assessing the eventual biological relevance of the network hubs and modules comes from an evolutionary perspective (see Chapter X and Poyatos & Hurst, 2004; Batada et al., 2006; Wang & Zhang, 2007).

Elucidating the Biological Processes Implicated in Human Diseases

While the success of the module-based mining and interpretation of biological processes has originally been demonstrated in model organisms, such as yeast, functional modules have increasingly been identified also in more complex organisms, especially in the human interactome. Showing concrete implications, such as disease mechanisms or molecular marker, for clinically important phenotypes can be regarded as one important form of evaluation of the significance of the putative network modules. Similarly as the topology-based modules can be very sensitive to the artifacts in the interaction data, it has increasingly been recognized that disease biomarkers discovered solely from microarray or mass-spectrometry experiments may show very poor reproducibility due to several sources of experimental, technical and biological variation that can severely confound the actual phenotypic variability. Integrated computational analysis of expression datasets together with network modules has been shown to improve elucidating the underlying biological processes associated with complex disorders, like cancer phenotypes, as well as to provide more reproducible set of biomarkers having better coverage with known disease risk genes and improved sample classification accuracy as compared to the standard analysis of differential expression alone (Efroni et al., 2007; Chuang et al., 2007). Other promising applications of molecular networks to human disease include identifying new disease gene nodes and their network properties, as well as identifying disease-related protein complexes or sub-networks (Hu et al., 2007; Lage et al., 2007; Ideker & Sharan, 2008). Such integrated analysis can e.g. identify relevant components, which are not differentially expressed or may even not have been profiled under a given condition (e.g. JUN protein in Figure 3). Several additional approaches and illustrative examples how protein interaction networks can help prioritizing disease genes and understanding disease pathways are described in Chapter XIV.

LIMITATIONS, EXTENSIONS AND FUTURE RESEARCH DIRECTIONS

Perhaps the biggest challenges the current computational approaches are facing still originate from the limited coverage and accuracy of the interaction data available. Protein-protein interaction maps for yeast and human were estimated to contain only roughly 50% and 10% of all of the possible interactions in 2006 (Hart et al., 2006). There are various experimental techniques for identifying protein interactions, the major differences arising from their scale and ability to detect binary or complex interactions (Shoemaker & Panchenko, 2007a). However, all the techniques currently used in large-scale interactions screens are poor at detecting very transient interactions and those that depend on posttranslational modifications. Another technical complication comes from the fact that these techniques detect interactions that are biophysically possible but may not actually occur in the living cells. Moreover, the measured interactions are very much dependent upon the cell types studied and on the developmental stages or external conditions under which the measurements were performed. In particular, the completion of the human interactome will pose many experimental problems, not least because of its large estimated scale and relative sparseness (Stumpf et al. 2008). The context-dependency together with the large number of false positive and false negative interactions, that are bound to occur in the current human protein interaction datasets, also explains their small overlap (Ramírez et al., 2007; Futschik et al., 2007). These experimental limitations should be understood when developing and applying data mining methods. Computational strategies can also help to increase the coverage and accuracy of the interaction networks by predicting and prioritizing those components and their interactions, either observed or missing, that warrant further experimental testing (Qi et al., 2006; Qiu & Noble, 2008; Skrabanek et al., 2008). It can be expected that the development of data mining approaches to protein interaction networks will benefit greatly also from the ongoing community efforts to standardize the representation and storage of the interaction datasets (Orchard et al., 2007; Kerrien et al., 2007; Klipp et al., 2007), as well as from the integrated tools for accessing and querying the public interaction databases (Cerami et al., 2006; Strömbäck et al., 2006; Dhanapalan & Chen. 2007; Newman et al., 2008; Splendiani, 2008).

Albeit being error-prone and incomplete, the protein interaction networks mapped so far have already proved useful in understanding biological processes and have also triggered many biomedical applications with relevance to human health. The large-scale experimental datasets for human cell biology, in particular, that are constantly improving both in quality and coverage, are enabling computational methods to systematically address exciting biological questions, such as identifying the key players and their roles responsible for the multi-factorial behavior in complex diseases. An approach with a particular biomedical potential involves using protein interaction networks as priors when analyzing genome-wide gene or protein expression datasets; network modules can be used for distinguishing those genes and proteins directly involved in a particular disease process from the background variation originating from other biological factors or experimental noise. Besides such false positives detections, the use of connectivity patterns can help to decrease the number of false negatives as well, caused e.g. by those members of the module that are expressed at low or constant levels. The need of such integrated analysis is emphasized when the sample sizes are small and the phenotypes of interest involve also components with subtle expression changes. Accordingly, such network-based analysis of complex phenotypes is likely to be extremely useful in diseases like diabetes (Liu et al., 2007). Possible future extensions of the network-based identification of potential biomarkers include accounting for inter-individual variability in the disease responses: by aggregating the module-level expression changes between different subjects, it may be possible to provide multiple sets of molecular markers that correspond to different subsets of patients. Such modular approach to biomarker discovery should be robust to high measurement noise and intrinsic biological variation, thus enabling in the future tests to distinguish between different disease subtypes, progression stages, individual risk factors, or even individual response to different treatments.

To increase the coverage and accuracy of the protein interaction networks, the module searches can also be extended across multiple species (Dutkowski & Tiuryn, 2007; Kalaev et al., 2008). Such conserved sub-networks can suggest e.g. undescribed protein functions or interactions (Sharan et al., 2005), and when integrated with transcriptional interaction data, the cross-species discoveries can also provide insights into the evolutionary dynamics of protein complex regulation (Tan et al., 2007). Beyond different species, similar approaches could be extended to include also information on the varying external conditions, interaction types or time points, thus making the predictions more reliable by uncovering modules that are supported by interactions patterns of different types. An interesting future investigation would also be to assess how much adding complementary data sources can really contribute to the success of the module finding algorithms; in particular, the extent to which gene expression data can help to improve the pure topology-based methods and what is the optimal way of incorporating the high-throughput expression measurements into the module detection algorithms. An inherent limitation of the current methods that utilize both the network connectivity and expression levels, including JACS and jActiveModules algorithms, is that they cannot assign into the same module genes that show different behavior in their co-expression or differential expression patterns under the specific spatio-temporal context (e.g. just-in-time assembly). These limitations demonstrate that there is still room for many improvements in the module finding methodology, perhaps one of the most striking challenges being how to take the full advantage of the temporal information in the many time-series expression datasets available today, rather than treating the time points as independent conditions. By addressing these and related questions, we may in the future be able to more accurately distinguish between different types of network modules, with distinct biological roles, such as those corresponding to stable protein complexes and dynamic functional modules (Spirin & Mirny, 2003; Komurov & White, 2007).

Most computational approaches to module finding can be formulated as an optimization problem, which takes into account the networks topological properties and possible other data sources in their scoring function. Depending on the search strategy used, the solution can output either distinct or overlapping sub-networks. While greedy search is typically used because it is computationally inexpensive and provides reasonably good results in practice (Breitling et al., 2004; Nacu et al., 2008), more comprehensive search strategies have also been used (King et al., 2005; Rajagopalan & Agarwal, 2005). In the initialization phase, one can either use completely random seeds, high scoring nodes or proteins that are a part of the complex or pathway under investigation (Altaf-Ul-Amin et al., 2006; Maraziotis et al., 2007). The network information can be incorporated into the analysis of genomic datasets also using probabilistic frameworks, such as Markov random field models (Segal et al., 2003; Wei & Li, 2007) or spatially-correlated mixture models (Ulitsky & Shamir, 2007; Wei & Pan, 2008). Several network properties can be considered when decomposing the network into modules (Aittokallio & Schwikowski, 2006). An interesting approach frequently used when discovering community structures in complex social networks is the so-called edge-betweenness centrality (Girvan & Newman, 2002). It has been shown that protein clusters defined by removing the edges with the highest betweenness centralities tend to share similar functions (Dunn et al. 2005), and nodes with high betweenness centrality (e.g. the node connecting the two modules in Figure 1) are more likely to be essential proteins as compared to the those with high degree (Yu et al., 2007). This local node property can also be applied when searching modules in weighted graphs (Yoon et al., 2006), making it possible to weight the interactions using e.g. gene expression microarray measurements (Chen & Yuan, 2006). Regardless of the topological property used as an objective function in the optimization, however, it is important to take into account the limited sampling and the large proportion of false positive and negative interactions when estimating the particular network property (Stumpf et al., 2005; Friedel & Zimmer, 2006b; Scholtens et al., 2008).

CONCLUDING REMARKS

Graph-based models of protein interaction networks can combine heterogeneous system-level data sources and their integrative mining can provide valuable information about local and global characteristics of their modular and functional organization. The global structural properties try to characterize the networks as a whole, whereas local network analyses aim at discovering such individual interaction patterns that may carry significant information about their roles in specific cellular mechanisms. Module analyses seek to partition the complex networks into functionally organized hierarchy of interconnected groups that are involved in common cellular functions. While most current network modularization approaches are based solely on the topological properties, such as network hubs, clusters or motifs, it is likely that biologically relevant findings can only be made after contrasting the interaction patterns with other large-scale functional genomics or proteomics measurements. In particular, when mining human interaction networks, that are still limited both in their accuracy and coverage, integrated analysis of interaction networks together with complementary datasets is probably needed to identify robust functional modules, which may eventually prove valuable also in many medical applications. The limitations of the current protein interaction datasets should be taken into account when developing new data mining methods and when applying these methods to real biological problems. The future success of the module-based analysis is likely to be driven also by parallel improvements both in the experimental techniques for mapping global interaction networks as well as in the modeling frameworks for representing the existing and emerging types of interaction information.

ACKNOWLEDGMENT

The author thanks Jukka Hiissa for his technical assistance with the figures and Igor Ulitsky for the helpful discussions on the JACS procedure. The work was supported by the Academy of Finland (grant 120 569).

REFERENCES

Aittokallio, T., & Schwikowski, B. (2006). Graph-based methods for analysing networks in cell biology. *Briefings in Bioinformatics*, *7*(3), 243-255.

Albert, R. (2005). Scale-free networks in cell biology. *Journal of Cell Science*, *118*(Pt 21), 4947-4957.

Altaf-Ul-Amin, M., Shinbo, Y., Mihara, K., Kurokawa, K., & Kanaya, S. (2006). Development and implementation of an algorithm for detection of protein complexes in large interaction networks. *BMC Bioinformatics*, 2006, *7*, 207.

Bader, G. D., & Hogue, C. W. (2003). An automated method for finding molecular complexes in large protein interaction networks. *BMC Bioinformatics*, *4*(2).

Bandyopadhyay, S., Kelley, R., & Ideker, T. (2006). Discovering regulated networks during HIV-1 latency and reactivation. *Pacific Symposium on Biocomputing*, *11*, 354-366.

Barabási, A. L., & Oltvai, Z. N. (2004). Network biology: understanding the cell's functional organization. *Nature Reviews Genetics*, *5*(2), 101–113.

Batada, N. N., Hurst, L. D., & Tyers, M. (2006). Evolutionary and physiological importance of hub proteins. *PLoS Computational Biology*, *2*(7), e88.

Batada, N. N., Reguly, T., Breitkreutz, A., Boucher, L., Breitkreutz, B. J., Hurst, L. D., & Tyers, M. (2006). Stratus not altocumulus: a new view of the yeast protein interaction network. *PLoS Biology*, *4*(10), e317.

Bertin, N., Simonis, N., Dupuy, D., Cusick, M. E., Han, J. D., Fraser, H. B., Roth, F. P., & Vidal, M. (2007). Confirmation of organized modularity in the yeast interactome. *PLoS Biology*, *5*(6), e153.

Breitling, R., Amtmann, A., & Herzyk, P. (2004). Graph-based iterative group analysis enhances microarray interpretation. *BMC Bioinformatics*, *5*, 100.

Brohée, S., & van Helden, J. (2006). Evaluation of clustering algorithms for protein-protein interaction networks. *BMC Bioinformatics*, *7*, 488.

Cabusora, L., Sutton, E., Fulmer, A., & Forst, C. V. (2005). Differential network expression during drug and stress response. *Bioinformatics*, *21*(12), 2898-905.

Cerami, E. G., Bader, G. D., Gross, B. E., &Sander, C. (2006). cPath: open source software for collecting, storing, and querying biological pathways. *BMC Bioinformatics*, *7*, 497.

Chen, J., & Yuan, B. (2006). Detecting functional modules in the yeast protein-protein interaction network. *Bioinformatics*, *22*(18), 2283-2290.

Cho, Y. R., Hwang, W., Ramanathan, M., & Zhang, A. (2007). Semantic integration to identify overlapping functional modules in protein interaction networks. *BMC Bioinformatics*, *8*, 265.

Chuang, H. Y., Lee, E., Liu, Y. T., Lee, D., & Ideker, T. (2007). Network-based classification of breast cancer metastasis. *Molecular Systems Biology*, *3*,140.

Cline, M. S., Smoot, M., Cerami, E., Kuchinsky, A., Landys, N., Workman, C., Christmas, R., Avila-Campilo, I., Creech, M., Gross, B., Hanspers, K., Isserlin, R., Kelley, R., Killcoyne, S., Lotia, S., Maere, S., Morris, J., Ono, K., Pavlovic, V., Pico, A. R., Vailaya, A., Wang, P. L., Adler, A., Conklin, B. R., Hood, L., Kuiper, M., Sander, C., Schmulevich, I., Schwikowski, B., Warner, G. J., Ideker, T., & Bader, G. D. (2007). Integration of biological networks and gene expression data using Cytoscape. *Nature Protocols*, *2*(10), 2366-82.

Dhanapalan, L., & Chen, J. Y. (2007). A case study of integrating protein interaction data using semantic web technology. *International Journal of Bioinformatics Research and Applications*, *3*(3), 286-302.

de Lichtenberg, U., Jensen, L. J., Brunak, S., & Bork, P. (2005). Dynamic complex formation during the yeast cell cycle. *Science*, *307*(5710), 724-727.

Dittrich, M. T., Klau, G. W., Rosenwald, A., Dandekar, T., & Müller, T. (2008). Identifying functional modules in protein-protein interaction networks: an integrated exact approach. *Bioinformatics*, *24*(13), i223-31.

Dunn, R., Dudbridge, F., & Sanderson, C. M. (2005). The use of edge-betweenness clustering to investigate biological function in protein interaction networks. *BMC Bioinformatics*, *6*, 39.

Dutkowski, J., & Tiuryn, J. (2007). Identification of functional modules from conserved ancestral protein-protein interactions. *Bioinformatics*, *23*(13), i149-58.

Efroni, S., Schaefer, C. F., & Buetow, K. H. (2007). Identification of key processes underlying cancer phenotypes using biologic pathway analysis. *PLoS ONE*, *2*(5), e425.

Ekman, D., Light, S., Björklund, A. K., & Elofsson, A. (2006). What properties characterize the hub proteins of the protein-protein interaction network of Saccharomyces cerevisiae? *Genome Biology*, *7*(6), R45.

Estrada, E. (2006). Virtual identification of essential proteins within the protein interaction network of yeast. *Proteomics*, *6*(1), 35-40.

Friedel, C. C., & Zimmer, R. (2006a). Toward the complete interactome. *Nature Biotechnology*, *24*(6), 614-615.

Friedel, C. C., & Zimmer, R. (2006b). Inferring topology from clustering coefficients in protein-protein interaction networks. *BMC Bioinformatics*, *7*, 519.

Futschik, M. E., Chaurasia, G., & Herzel, H. (2007). Comparison of human protein-protein interaction maps. *Bioinformatics*, *23*(5), 605-611.

Ge, H., Liu, Z., Church, G. M., & Vidal, M. (2001). Correlation between transcriptome and interactome mapping data from Saccharomyces cerevisiae. *Nature Genetics*, *29*(4), 482-486.

Gentleman, R., & Huber, W. (2007). Making the most of high-throughput protein-interaction data. *Genome Biology*, *8*(10), 112.

Girvan, M., & Newman, M. E. (2002). Community structure in social and biological networks. *Proceedings of the National Academy of Sciences USA*, *99*(12), 7821-7826.

Guo, Z., Li, Y., Gong, X., Yao, C., Ma, W., Wang, D., Li, Y., Zhu, J., Zhang, M., Yang, D., & Wang, J. (2007). Edge-based scoring and searching method for identifying condition-responsive protein-protein interaction subnetwork. *Bioinformatics*, *23*(16), 2121-2128.

Hakes, L., Pinney, J. W., Robertson, D. L., & Lovell, S. C. (2008). Protein-protein interaction networks and biology -- what's the connection? *Nature Biotechnology*, *26*(1), 69-72.

Han, J. D., Bertin, N., Hao, T., Goldberg, D. S., Berriz, G. F., Zhang, L. V., Dupuy, D., Walhout , A. J., Cusick, M. E., Roth, F. P., & Vidal M. (2004). Evidence for dynamically organized modularity in the yeast protein-protein interaction network. *Nature*, *430*(6995), 88-93

Han, J. D., Dupuy, D, Bertin, N., Cusick, M. E., & Vidal, M. (2005). Effect of sampling on topology predictions of protein-protein interaction networks. *Nature Biotechnology*, *23*(7), 839-844.

Hart, G. T., Ramani, A. K., & Marcotte, E. M. (2006). How complete are current yeast and human protein-interaction networks? *Genome Biology*, *7*(11), 120.

Hartwell, L. H., Hopfield, J. J., Leibler, S., & Murray, A. W. (1999). From molecular to modular cell biology. *Nature*, *402*(6761 Suppl), C47–52.

Hu, P., Bader, G., Wigle, D. A., & Emili, A. (2007). Computational prediction of cancer-gene function. Nature Reviews Cancer, *7*(1), 23-34.

Hu, Z., Mellor, J., Wu, J., & DeLisi C. (2004). VisANT: an online visualization and analysis tool for biological interaction data. *BMC Bioinformatics*, *5*, 17.

Hu, Z., Ng, D. M., Yamada, T., Chen, C., Kawashima, S., Mellor, J., Linghu, B., Kanehisa, M., Stuart, J. M., & DeLisi, C. (2007). VisANT 3.0: new modules for pathway visualization, editing, prediction and construction. *Nucleic Acids Research*, *35*(Web Server issue), W625-32.

Hu, Z, Snitkin, E.S, & DeLisi, C. (2008). VisANT: an integrative framework for networks in systems biology. *Briefings in Bioinformatics*, *9*(4), 317-25.

Ideker, T., Ozier, O., Schwikowski, B., & Siegel, A.F. (2002). Discovering regulatory and signalling circuits in molecular interaction networks. *Bioinformatics*. *18*(Suppl 1), S233-240.

Ideker, T., & Sharan, R. (2008). Protein networks in disease. *Genome Research*, *18*(4), 644-652.

Jansen, R., Greenbaum, D., & Gerstein, M. (2002). Relating whole-genome expression data with protein-protein interactions. *Genome Research*, *12*(1), 37-46.

Kalaev, M., Smoot, M., Ideker, T., & Sharan, R. (2008). NetworkBLAST: comparative analysis of protein networks. *Bioinformatics*, *24*(4), 594-596.

Kerrien, S., Orchard, S., Montecchi-Palazzi, L., Aranda, B., Quinn, A. F., Vinod, N., Bader, G. D., Xenarios, I., Wojcik, J., Sherman, D., Tyers, M., Salama, J. J., Moore, S., Ceol, A., Chatr-Aryamontri, A., Oesterheld, M., Stümpflen, V., Salwinski, L., Nerothin, J., Cerami, E., Cusick, M. E., Vidal, M., Gilson, M., Armstrong, J., Woollard, P., Hogue, C., Eisenberg, D., Cesareni, G., Apweiler, R., & Hermjakob, H. (2007). Broadening the horizon - level 2.5 of the HUPO-PSI format for molecular interactions. *BMC Biology*, *5*, 44.

Khanin, R., & Wit, E. (2006). How scale-free are biological networks. *Journal of Computational Biology*, *13*(3), 810-818.

King, A. D., Przulj, N., & Jurisica, I. (2004). Protein complex prediction via cost-based clustering. *Bioinformatics*, *20*(17), 3013-2020.

Kirkpatrick, S., Gelatt, C. D., & Vecchi, M. P. (1983). Optimization by simulated annealing. *Science*, *220*(4598), 671-680.

Klipp, E., Liebermeister, W., Helbig, A., Kowald, A., & Schaber, J. (2007). Systems biology standards - the community speaks. *Nature Biotechnology*, *25*(4), 390-391.

Komurov, K., & White, M. (2007). Revealing static and dynamic modular architecture of the eukaryotic protein interaction network. *Molecular Systems Biology*, *3*, 110.

Lage, K., Karlberg, E. O., Størling, Z. M., Olason, P. I., Pedersen, A. G., Rigina, O., Hinsby, A. M., Tümer, Z., Pociot, F., Tommerup, N., Moreau, Y., & Brunak, S. (2007). A human phenome-interactome network of protein complexes implicated in genetic disorders. *Nature Biotechnology*, *25*(3), 309-16.

Liu, M., Liberzon, A., Kong, S. W., Lai, W. R., Park, P. J., Kohane, I. S., & Kasif, S. (2007). Network-based analysis of affected biological processes in type 2 diabetes models. *PLoS Genetics*, *3*(6), e96.

Lu, H., Shi, B., Wu, G., Zhang, Y., Zhu, X., Zhang, Z., Liu, C., Zhao, Y., Wu, T., Wang, J., & Chen, R. (2006). Integrated analysis of multiple data sources reveals modular structure of biological networks. *Biochemical and Biophysical Research Communications*, *345*(1), 302-309.

Lubovac, Z., Gamalielsson, J., & Olsson, B. (2006). Combining functional and topological properties to identify core modules in protein interaction networks. *Proteins*, *64*(4), 948-959.

Maraziotis, I. A., Dimitrakopoulou, K., & Bezerianos, A. (2007). Growing functional modules from a seed protein via integration of protein interaction and gene expression data. *BMC Bioinformatics*, *8*, 408.

Nacu, S., Critchley-Thorne, R., Lee, P., & Holmes, S. (2007). Gene expression network analysis and applications to immunology. *Bioinformatics*, *23*(7), 850-858.

Newman, A., Hunter, J., Li, Y. F., Bouton, C., & Davis, M. (2008). BioMANTA ontology: the integration of protein-protein interaction data. In *Interdisciplinary Ontology Conference*, Tokyo, Japan, 26-27 February, 2008.

Orchard, S., Salwinski, L., Kerrien, S., Montecchi-Palazzi, L., Oesterheld, M., Stümpflen, V., Ceol, A., Chatr-aryamontri, A., Armstrong, J., Woollard, P., Salama, J. J., Moore, S., Wojcik, J., Bader, G. D., Vidal, M., Cusick, M. E., Gerstein, M., Gavin, A. C., Superti-Furga, G., Greenblatt, J., Bader, J., Uetz, P., Tyers, M., Legrain, P., Fields, S., Mulder, N., Gilson, M., Niepmann, M., Burgoon, L., De Las Rivas, J., Prieto, C., Perreau, V. M., Hogue, C., Mewes, H. W., Apweiler, R., Xenarios, I., Eisenberg, D., Cesareni, G., & Hermjakob, H. (2007). The minimum information required for reporting a molecular interaction experiment (MIMIx). *Nature Biotechnology*, *25*(8), 894-898.

Pereira-Leal, J. B., Enright, A. J., & Ouzounis, C. A. (2004). Detection of functional modules from protein interaction networks. *Proteins*, *54*(1), 49-57.

Poyatos, J. F., & Hurst, L. D. (2004). How biologically relevant are interaction-based modules in protein networks? *Genome Biology*, *5*(11), R93.

Przulj, N., Wigle, D. A., & Jurisica, I. (2004). Functional topology in a network of protein interactions. *Bioinformatics*, *20*(3), 340-348.

Pu, S., Vlasblom, J., Emili, A., Greenblatt, J., & Wodak, S. J. (2007). Identifying functional modules in the physical interactome of Saccharomyces cerevisiae. *Proteomics*, *7*(6), 944-960.

Qi, Y., Bar-Joseph, Z., & Klein-Seetharaman, J. (2006). Evaluation of different biological data and computational classification methods for use in protein interaction prediction. *Proteins*, *63*(3), 490-500.

Qiu, J., & Noble, W. S. Predicting co-complexed protein pairs from heterogeneous data. *PLoS Computational Biology*, *4*(4), e1000054.

Rajagopalan, D., & Agarwal, P. (2005). Inferring pathways from gene lists using a literature-derived network of biological relationships. *Bioinformatics*, *21*(6), 788-793.

Ramírez, F., Schlicker, A., Assenov, Y., Lengauer, T., & Albrecht, M. (2007). Computational analysis of human protein interaction networks. *Proteomics*, *7*(15), 2541-2552.

Rungsarityotin, W., Krause, R., Schödl, A., & Schliep, A. (2007). Identifying protein complexes directly from high-throughput TAP data with Markov random fields. *BMC Bioinformatics*, *8*, 482.

Scholtens, D., Chiang, T., Huber, W., & Gentleman, R. (2008). Estimating node degree in bait-prey graphs. *Bioinformatics*, *24*(2), 218-224.

Scholtens, D., Vidal, M., & Gentleman, R. (2005). Local modeling of global interactome networks. *Bioinformatics, 21*(17), 3548-3557.

Schwikowski, B., Uetz, P., & Fields, S. (2000). A network of protein-protein interactions in yeast. *Nature Biotechnology, 18*(12), 1257-1261.

Segal, E., Wang, H., & Koller, D. (2003). Discovering molecular pathways from protein interaction and gene expression data. *Bioinformatics, 19* (Suppl 1), i264-i271.

Shannon, P., Markiel, A., Ozier, O., Baliga, N. S., Wang, J. T., Ramage, D., Amin, N., Schwikowski, B., & Ideker, T. (2003). Cytoscape: a software environment for integrated models of biomolecular interaction networks. *Genome Research, 13*(11), 2498-2504.

Sharan, R., Suthram, S., Kelley, R. M., Kuhn, T., McCuine, S., Uetz, P., Sittler, T., Karp, R. M., & Ideker, T. (2005). Conserved patterns of protein interaction in multiple species. *Proceedings of the National Academy of Sciences USA, 102*(6), 1974-1979.

Sharan, R., Ulitsky, I., & Shamir, R. (2007). Network-based prediction of protein function. *Molecular Systems Biology, 3*, 88.

Shoemaker, B. A., & Panchenko, A. R. (2007a). Deciphering protein-protein interactions. Part I. Experimental techniques and databases. *PLoS Computational Biology, 3*(3), e42.

Shoemaker, B. A., & Panchenko, A. R. (2007b). Deciphering protein-protein interactions. Part II. Computational methods to predict protein and domain interaction partners. *PLoS Computational Biology, 3*(4), e43.

Simonis, N., Gonze, D., Orsi, C., van Helden, J., & Wodak, S. J. (2006). Modularity of the transcriptional response of protein complexes in yeast. *Journal of Molecular Biology, 363*(2), 589-610.

Skrabanek, L., Saini, H. K., Bader, G. D., & Enright, A. J. (2008). Computational prediction of protein-protein interactions. *Molecular Biotechnology, 38*(1), 1-17.

Sohler, F., Hanisch, D., & Zimmer, R. (2004). New methods for joint analysis of biological networks and expression data. *Bioinformatics, 20*(10), 1517-1521.

Splendiani, A. (2008). RDFScape: semantic Web meets systems biology. *BMC Bioinformatics, 9*(Suppl 4), S6.

Sprinzak, E., Altuvia, Y., & Margalit, H. (2006). Characterization and prediction of protein-protein interactions within and between complexes. *Proceedings of the National Academy of Sciences USA, 103*(40), 14718-14723.

Spirin, V, & Mirny, L. A. (2003). Protein complexes and functional modules in molecular networks. *Proceedings of the National Academy of Sciences USA, 100*(21), 12123-12128.

Strömbäck L, Jakoniene V, Tan H, & Lambrix P. (2006). Representing, storing and accessing molecular interaction data: a review of models and tools. *Briefings in Bioinformatics, 7*(4), 331-338.

Stumpf, M. P, Thorne, T., de Silva, E., Stewart, R., An, H. J., Lappe, M., & Wiuf, C. (2008). Estimating the size of the human interactome. *Proceedings of the National Academy of Sciences USA, 105*(19), 6959-6964.

Stumpf, M. P., Wiuf, C., & May, R. M. (2005). Subnets of scale-free networks are not scale-free: sampling properties of networks. *Proceedings of the National Academy of Sciences USA, 102*(12), 4221-4224.

Suderman, M., & Hallett, M. (2007). Tools for visually exploring biological networks. *Bioinformatics, 23*(20), 2651-2659.

Tan, K., Shlomi, T., Feizi, H., Ideker, T., & Sharan, R. (2007). Transcriptional regulation of protein complexes within and across species. *Proceedings of the National Academy of Sciences USA, 104*(4), 1283-1288.

Tirosh, I., & Barkai, N. (2005). Computational verification of protein-protein interactions by orthologous co-expression. *BMC Bioinformatics*, *6*, 40.

Tornow, S., & Mewes, H. W. (2003). Functional modules by relating protein interaction networks and gene expression. *Nucleic Acids Research*, *31*(21), 6283-6289.

Ulitsky, I., & Shamir, R. (2007). Identification of functional modules using network topology and high-throughput data. *BMC Systems Biology*, *1*, 8.

Wang, Z., & Zhang, J. (2007). In search of the biological significance of modular structures in protein networks. *PLoS Computational Biology*, *3*(6), e107.

Wei, P., & Pan, W. (2008). Incorporating gene networks into statistical tests for genomic data via a spatially correlated mixture model. *Bioinformatics*, *24*(3), 404-411.

Wei, Z., & Li, H. (2007). A Markov random field model for network-based analysis of genomic data. *Bioinformatics*, *23*(12), 1537-1544.

Yook, S. H., Oltvai, Z. N., & Barabási, A. L. (2004). Functional and topological characterization of protein interaction networks. *Proteomics*, *4*(4), 928-942.

Yoon, J., Blumer, A., & Lee, K. (2006). An algorithm for modularity analysis of directed and weighted biological networks based on edge-betweenness centrality. *Bioinformatics*, *22*(24), 3106-3108.

Yu, H., Kim, P. M., Sprecher, E., Trifonov, V., & Gerstein, M. (2007). The importance of bottlenecks in protein networks: correlation with gene essentiality and expression dynamics. *PLoS Computational Biology*, *3*(4), e59.

Zhang, S., Jin, G., Zhang, X. S., & Chen, L. (2007). Discovering functions and revealing mechanisms at molecular level from biological networks. *Proteomics*, *7*(16), 2856-2869.

Compilation of References

Adamcsek, B., Palla, G., Farkas, I. J., Dernyi, I., & Vicsek, T. (2006). Cfinder: locating cliques and overlapping modules in biological networks. *Bioinformatics, 22*(8), 1021-1023.

Adie, E. A., Adams, R. R., Evans, K. L., Porteous, D. J., & Pickard, B. S. (2006). SUSPECTS: Enabling fast and effective prioritization of positional candidates. *Bioinformatics, 22*(6), 773-774.

Aebersold, R., & Mann, M. (2003). Mass spectrometry-based proteomics. *Nature, 422*(6928), 198-207.

Aerts, S., Lambrechts, D., Maity, S., Van Loo, P., Coessens, B., De Smet, F., et al. (2006). Gene prioritization through genomic data fusion. *Nat Biotechnol, 24*(5), 537-544.

Agrafioti, I., Swire, J., Abbott, J., Huntley, D., Butcher, S., & Stumpf, M. P. (2005). Comparative analysis of the saccharomyces cerevisiae and caenorhabditis elegans protein interaction networks. *BMC Evol Biol, 5*(1), 23.

Agrawal, R., & Srikant, R. (1994). Fast algorithms for mining association rules in large databases. *Proc. VLDB 1994* (pp. 487–499).

Aho, A. V., Hopcroft, J. E., & Ullman, J. D. (1974). *The design and analysis of computer algorithms.* Reading, USA: Addison-Wesley Publishing Co.

Aitken, S., Korf, R., Webber, B., & Bard, J. (2005). COBrA: A bio-ontology editor. *Bioinformatics, 21*(6), 825-826.

Aittokallio, T., & Schwikowski, B. (2006). Graph-based methods for analysing networks in cell biology. *Briefings in Bioinformatics*, *7*(3), 243-255.

Aittokallio, T., & Schwikowski, B. (2006). Graph-based methods for analysing networks in cell biology. *Briefing in Bionformatics*, *7*(3), 243-255.

Albert, I., & Albert, R. (2004). Conserved network motifs allow protein-protein interaction prediction. *Bioinformatics*, *20*(18):3346-52.

Albert, R. (2005). Scale-free networks in cell biology. *Journal of Cell Science*, *118*(Pt 21), 4947-4957.

Albert, R., & Baraba´si, A-L. (2002). Statistical mechanics of complex networks. *Reviews of Mordern Physics, 74 (1),* 47-96.

Alberts, B. (1998). The cell as a collection of protein machines: preparing the next generation of molecular biologists. *Cell, 92*(3), 291-294.

Alfarano, C., Andrade, C. E., Anthony, K., Bahroos, N., Bajec, M., Bantoft, K., et al. (2005). The Bio-molecular Interaction Network Database and related tools 2005 update. *Nucleic Acids Research. 33*(Database Issue), D418-24.

Alfarano, C., Andrade, C. E., Anthony, K., Bahroos, N., Bajec, M., Bantoft, K., et al. (2005). The Biomolecular Interaction Network Database and related tools 2005 update. *Nucleic Acids Res, 33*(Database issue), D418-424.

Alfarano, C., Andrade, C. E., Anthony, K., Bahroos, N., Bajec, M., Bantoft, K., et al. (2005). The Biomolecular Interaction Network Database and related tools 2005 update. *Nucleic Acids Res, 33*(Database issue), D418-424.

Alon, N., Yuster, R., & Zwick, U. (1995). Color-coding. *Journal of the ACM (JACM)*, *42*(4), 844-856.

Alon, U. (2007, June). Network motifs: theory and experimental approaches. *Nature Reviews Genetics*, *8*(6), 450–461.

Aloy, P. (2007). Shaping the future of interactome networks. *Genome Biol, 8*(10), 316.

Aloy, P., & Russell, R. B. (2002). Potential artefacts in protein-interaction networks. *FEBS Lett, 530*(1-3), 253-254.

Aloy, P., Bottcher, B., Ceulemans, H., Leutwein, C., Mellwig, C., Fischer, S., et al. (2004). Structure-based assembly of protein complexes in yeast. *Science, 303*(5666), 2026-2029.

Al-Shahrour *et al.* (2006). BABELOMICS: a systems biology perspectve in the functional annotation of genome-scale experiments. *Nucleic Acids Research, 34*(Web Server Issue), W472-W476

Al-Shahrour, F., Carbonell, J., Minguez, P., Goetz, S., Conesa, A., Tárraga, J., Medina, I., Alloza, E., Montaner, D., & Dopazo, J. (2008). Babelomics: Advanced functional profiling of transcriptomics, proteomics and genomics experiments. *Nucleic Acid Research, 36*, 341-346.

Al-Shahrour, F., Díaz-Uriarte, R., & Dopazo, J. (2004). FatiGO: A web tool for finding significant associations of Gene Ontology terms with groups of genes. *Bioinformatics, 20*, 578-580

Al-Shahrour, F., Minguez, P., Tárraga, J., Medina, I., Alloza, E., Montaner, D., & Dopazo, J. (2007). FatiGO+: a functional profiling tool for genomic data. Integration of functional annotation, regulatory motifs and interpretation data with microarray experiments. *Nucleic Acids Research, 35*(Web Server issue), W91-96

Altaf-Ul-Amin, M., Shinbo, Y., Mihara, K., Kurokawa, K., & Kanaya, S. (2006). Development and implementation of an algorithm for detection of protein complexes in large interaction networks. *BMC Bioinformatics*, 2006, *7*, 207.

Altman, D. G., & Bland, J. M. (1994). Diagnostic tests. 1: Sensitivity and specificity. *BMJ*, *308*(6943):1552.

Altschul, S. F., Gish, W., et al. (1990). Basic local alignment search tool. *J Mol Biol,, 215*(3), 403-10.

Altschul, S. F., Madden, T. L., et al. (1997). Gapped BLAST and PSI-BLAST: A new generation of protein database search programs. *Nucleic Acids Res, 25*(17), 3389-402.

Altschul, S. F., Madden, T. L., Schaffer, A. A., Zhang, J., Zhang, Z., Miller, W., & Lipman, D. J. (1997). Gapped BLAST and PSI-BLAST: A new generation of protein database search programs. *Nucleic Acids Res, 25*, 3389-3402.

Antikainen, J., Kupannen, V., Lähteenmäk, K., & Korhonen, T. K. (2007). pH-Dependent Association of Enolase and Glyceraldehyde-3-Phosphate Dehydrogenase of Lactobacillus crispatus with the Cell Wall and Lipoteichoic Acids. *J Bacteriol., 189*(12), 4539-4543.

Aragues R., Jaeggi D., & Oliva, B. (2006). PIANA: Protein Interactions and Network Analysis. *Bioinformatics, 22*(8), 1015-7.

Aravind, L., Iyer, L. M., Wellems, T. E., & Miller, L. H. (2003). Plasmodium Biology: Genomic Gleanings. *Cell, 115*, 771-785.

Armon, A., Graur, D., & Ben-Tal, N. (2001). ConSurf: an algorithmic tool for the identification of functional regions in proteins by surface mapping of phylogenetic information. *J Mol Biol, 307*, 447-463.

Arnau,V., Mars, S., & Marìn, I. (2004). Iterative cluster analysis of protein interaction data. *Bioinformatics, 21*(3), 364–378.

Artzy-Randrup, Y., Fleishman, S. J., Ben-Tal, N., & Stone, L. (2004, August). Comment on 'Network motifs: Simple building blocks of complex networks' and 'Superfamilies of evolved and designed networks'. *Science, 305*(5687), 1107c.

Asai, T., Abe, K., Kawasoe, S., Arimura, H., Sakamoto, H., & Arikawa, S. (2002). Efficient substructure discovery from large semi-structured data. *Proc 2nd SIAM Data Mining Conference (SDM)* (pp. 158–174).

Asburner, S. et al. (2000). Gene ontology: Tool for the unification of biology. The gene ontology consortium. *Nature Genetics*, *25*, 25-29.

Ashburner, M. et. al. (2000). Gene ontology: tool for the unification of biology. The Gene Ontology Consortium. *Nature Genetics, 25*(1), 25-29.

Ashburner, M., Ball, C. A., Blake, J. A., Botstein, D., Butler, H., Cherry, J. M., et al. (2000). Gene ontology: tool for the unification of biology. *Nature Genetics*, *25*, 25-9.

Ashburner, M., Ball, C. A., Blake, J. A., Botstein, D., Butler, H., Cherry, J. M., Davis, A. P., Dolinski, K., Dwight, S. S., Eppig, J. T. et al. (2000). Gene ontology: tool for the unification of biology. *The Gene Ontology Consortium. Nat Genet, 25*, 25-29.

Ashburner, M., Ball, C. A., et al. (2000). Gene ontology: tool for the unification of biology. The Gene Ontology Consortium. *Nat Genet*, *25*(1), 25-9.

Assenov, Y., Ramírez, F., Schelhorn, S. E., Lengauer, T., & Albrecht, M. (2008). Computing topological parameters of biological networks. *Bioinformatics, 24*(2), 282-284.

Asthana, S., King, O. D., Gibbons, F. D., & Roth, F. P. (2004). Predicting protein complex membership using probabilistic network reliability. *Genome Res, 14*(6), 1170-1175.

Asur, S., Ucar, D., & Parthasarathy, S. (2007). An ensemble framework for clustering protein-protein interaction networks. *Bioinformatics, 23*, i29-i40.

Attwood, T. K., P. Bradley, et al. (2003). PRINTS and its automatic supplement, prePRINTS." Nucleic Acids Res 31(1): 400-2.

Ausiello, G., Gherardini, P. F., Marcatili, P., Tramontano, A., Via, A., & Helmer-Citterich, M. (2008). FunClust: a web server for the identification of structural motifs in a set of non-homologous protein structures. *BMC Bioinformatics, 9 Suppl 2*, S2.

Ausiello, G., Zanzoni, A., Peluso, D., Via, A., & Helmer-Citterich, M. (2005). pdbFun: Mass selection and fast comparison of annotated PDB residues. *Nucleic Acids Res, 33*(Web Server issue), W133-137.

Badano, J. L., & Katsanis, N. (2002). Beyond Mendel: an evolving view of human genetic disease transmission. *Nat. Rev. Genet., 3*, 779–789.

Bader, D. A., & Madduri, K. (2007). A Graph-Theoretic Analysis of the Human Protein-Interaction Network Using Multi-core Parallel Algorithms. *In Proceedings Parallel and Distributed Processing Symposium* (pp. 1-8).

Bader, G. D., & Hogue, C. W. (2000). BIND--a data specification for storing and describing biomolecular interactions, molecular complexes and pathways. Bioinformatics, *16*(5), 465-77.

Bader, G. D., & Hogue, C. W. (2002). Analyzing yeast protein-protein interaction data obtained from different sources. *Nat Biotechnol., 20*(10), 991-7.

Bader, G. D., & Hogue, C. W. (2003). An automated method for finding molecular complexes in large protein interaction networks. *BMC Bioinformatics*, *4*(2).

Bader, G. D., & Hogue, C. W. (2003). An automated method for finding molecular complexes in large protein interaction networks. *BMC Bioinformatics, 4*(2), doi:10.1186/1471-2105-4-2.

Bader, G. D., & Hogue, C. W. (2003). An automated method for finding molecular complexes in large protein interaction networks. *BMC Bioinformatics, 4*(2).

Bader, G. D., Betel, D., & Hogue, C. W. (2003). BIND: The Biomolecular Interaction Network Database. *Nucleic Acids Res, 31*(1), 248-250.

Bader, G. D., Betel, D., & Hogue, C. W. (2003). BIND: the Biomolecular Interaction Network Database. *Nucleic Acids Res*, *31*(1), 248-50.

Bader, G. D., Betel, D., & Hogue, C. W. V. (2003). BIND: The Biomolecular Interaction Network Database. *Nucleic Acids Research, 31*(1).

Bader, G. D., Cary, M. P., & Sander, C. (2006). Pathguide: A pathway resource list. *Nucleic Acids Research, 34*(Database issue), D504-6.

Bader, G., & Hogue, H. (2003). An automated method for finding molecular complexes in large protein-protein interaction networks. *BMC Bioinformatics*, *5*(2).

Bader, G.D., Donaldson, I., Wolting, C., Ouellette, B. F., Pawson, T., & Hogue, C. W. (2001). BIND--The Biomolecular Interaction Network Database. *Nucleic Acids Res*, *29*(1), 242-5.

Bader, J. S. et al. (2004). Gaining confidence in high-throughput protein interaction networks. *Nature Biotechnology, 22*(1), 78-84.

Bader, J. S., Chaudhuri, A. et al. (2004). Gaining confidence in high-throughput protein interaction networks. *Nat Biotechnol*, *22*(1), 78-85.

Bader, J. S., Chaudhuri, A., Rothberg, J. M., & Chant, J. (2004). Gaining confidence in high-throughput protein interaction networks. *Nat. Biotechnol., 22*, 78–85.

Bader, J. S., Chaudhuri, A., Rothberg, J. M., & Chant, J. (2004). Gaining confidence in high-throughput protein interaction networks. *Nature Biotechnology*, *22*, 78-85.

Bader, J. S., Chaudhuri, A., Rothberg, J. M., & Chant, J. (2004). Gaining confidence in high-throughput protein interaction networks. *Nat Biotechnol, 22*(1), 78-85.

Bahl, A., Brunk, B., Crabtree, J., Fraunholz, M., Gajria, B., Grant, G., et al. (2003). PlasmoDB: the Plasmodium genome resource. A database integrating experimental and computational data. *Nucleic Acids Res, 31*, 212-215.

Bailey, T. L., Williams, N., Misleh, C., & Li, W. W. (2006). Meme: discovering and analyzing dna and protein sequence motifs. *Nucleic Acids Res, 34*, W369–W373.

Bairoch, A. (1993). The PROSITE dictionary of sites and patterns in proteins, its current status. *Nucleic Acids Res*, *21*(13), 3097-103.

Bairoch, A., & Apweiler, R. (1996) The swiss-prot protein sequence data bank and its new supplement trembl. *Nucleic Acids Research*, *24*(1), 21–25.

Bairoch, A., Boeckmann, B., Ferro, S., & Gasteiger, E. (2004). Swiss-Prot: Juggling between evolution and stability. *Briefings in Bioinformatics*, *5*(1), 39-55.

Bairoch, A., R. Apweiler, et al. (2005). The Universal Protein Resource (UniProt). *Nucleic Acids Res*, *33*(Database issue), D154-9.

Baitaluk, M., Qian, X., Godbole, S., Raval, A., Ray, A., & Gupta, A. (2006). PathSys: integrating molecular interaction graphs for systems biology. *BMC Bioinformatics, 7,* 55.

Baitaluk, M., Sedova, M., Ray, A., & Gupta, A. (2006). BiologicalNetworks: visualization and analysis tool for systems biology. *Nucleic Acids Research*, *34*(Webserver Issue), W466-71.

Balla, S., et. al. (2006). Minimotif Miner: a tool for investigating protein function. *Nature Methods, 3*(3), 175-177.

Bandyopadhyay, S., Kelley, R., & Ideker, T. (2006). Discovering regulated networks during HIV-1 latency and reactivation. *Pacific Symposium on Biocomputing*, *11*, 354-366.

Bandyopadhyay, S., Sharan, R. et al. (2006). Systematic identification of functional orthologs based on protein network comparison. Genome Res, *16*(3), 428-35.

Bansal, M., Belcastro, V., et al. (2007). How to infer gene networks from expression profiles. *Molecular Sytems Biology, 3,* 1-10.

Barabasi, A. L. (2007). Network medicine--from obesity to the "diseasome". *N Engl J Med, 357*(4), 404-407.

Barabási, A. L., & Albert, R. (1999). Emergence of scaling in random networks *Science*, *286*(5439), 509-12.

Barabasi, A. L., & Bonabeu, E. (2003). Scale-Free Networks. *Scientific American 288*(5), 60-9.

Barabasi, A. L., & Oltvai, Z. N. (2004). Network Biology: Understanding the cell's functional organization. *Nature Reviews, 5*(2), 101-13.

Barabasi, A. L., & Oltvai, Z. N. (2004). Network biology: Understanding the cell's functional organization. *Nat Rev Genet, 5*(2), 101-113.

Barabási, A. L., & Oltvai, Z. N. (2004). Network biology: understanding the cell's functional organization. *Nature Reviews Genetics*, *5*(2), 101–113.

Barabási, A. L., & Oltvai, Z. N. (2004). Network biology: understanding the cell's functional organization. *Nature Reviews Genetics, 5*, 101-113.

Barabási, A.-L., & Albert, R. (1999, October). Emergence of scaling in random networks. *Science, 286*(5439), 509–512.

Barabási, A-L., & Oltvai, Z. N. (2004). Network biology: understanding the cells' functional organization. *Nature Reviews Genetics, 5*, 101-113.

Barabási, A-L., Albert, R., & Jeong, H. (1999). Mean-field theory for scale-free random networks. *Physica A, 272*, 174-187.

Barash, Y., Kaplan, T., Friedman, N., & Elidan, G. (2003). *Modeling dependencies in protein-DNA binding sites.* Paper presented at the the 7th International Conference on Research in Computational Molecular Biology (RECOMB), San Diego.

Barbosa, M. S., Báo, S. N., Andreotti, P. F., Faria, F. P. d., Felipe, M. S. S., Feitosa, L. d. S., et al. (2006). Glyceraldehyde-3-Phosphate Dehydrogenase of Paracoccidioides brasiliensis Is a Cell Surface Protein Involved in Fungal Adhesion to Extracellular Matrix Proteins and Interaction with Cells. *Infection and Immunity, 74*(1), 382-389.

Bar-Joseph, Z., Gerber, G. K., Lee, T. I., Rinaldi, N. J., Yoo, J. Y., Robert, F., Gordon, D. B., Fraenkel, E., Jaakkola, T. S., Young, R. A., & Gifford, D. K. (2003). Computational discovery of gene modules and regulatory networks. *Nat Biotechnol., 21*(11), 1337-42.

Barrett, T., Troup, D. B., Wilhite, S. E., Ledoux, P., Rudnev, D., Evangelista, C., et al. (2007). NCBI GEO: Mining tens of millions of expression profiles--database and tools update. *Nucleic Acids Res, 35*(Database issue), D760-765.

Batada, N. N., Hurst, L. D., & Tyers, M. (2006). Evolutionary and Physiological Importance of Hub Proteins. *Plos Computational Biology, 2*(7), e88.

Batada, N. N., Reguly, T., Breitkreutz, A., Boucher, L., Breitkreutz, B. J., Hurst, L. D., & Tyers, M. (2006). Stratus not altocumulus: a new view of the yeast protein interaction network. *PLoS Biology, 4*(10), e317.

Bateman, A., Coin, L., Durbin, R., Finn, R. D., Hollich, V., Griffiths-Jones, S., Khanna, A., Marshall, M., Moxon, S., Sonnhammer, E. L., Studholme, D. J., Yeats, C., Eddy, S. R. (2004). The Pfam protein families database. *Nucleic Acids Res., 32*(Database issue), D138-41.

Baudin, A., Ozier-Kalogeropoulos, O., Denouel, A., Lacroute, F., & Cullin, C. (1993). A simple and efficient method for direct gene deletion in *Saccharomyces cerevisiae. Nucleic Acids Res, 21*(14), 3329-30.

Baudot, A., Jacq, B., & Brun, C. (2004). A scale of functional divergence for yeast duplicated genes revealed from analysis of the protein-protein interaction network. *Genome Biol, 5*(10), R76.

Beissbarth, T., & Speed, T. P. (2004). GOstat: find statistically overrepresented Gene Ontologies within a group of genes. *Bioinformatics, 20*(9), 1464-1465.

Beltrao, P., & Serrano, L. (2007). Specificity and evolvability in eukaryotic protein interaction networks. *PLoS Comput Biol, 3*(2), e25.

Ben Mamoun, C., Gluzman, I. Y., Hott, C., MacMillan, S. K., Amarakone, A. S., Anderson, D. L., et al. (2001). Co-ordinated programme of gene expression during asexual intraerythrocytic development of the human malaria parasite Plasmodium falciparum revealed by microarray analysis. *Mol Microbiol., 39*, 26-36.

Benfey, P. N., & Mitchell-Olds, T. (2008). From genotype to phenotype: Systems biology meets natural variation. *Science, 320*(5875), 495-497.

Ben-Gal, I., Shani, A., Gohr, A., Grau, J., Arviv, S., Shmilovici, A., et al. (2005). Identification of transcription factor binding sites with variable-order Bayesian networks. *Bioinformatics, 21*(11), 2657-2666.

Ben-Hur, A., & Noble, W. S. (2005). Kernel methods for predicting protein-protein interactions. *Bioinformatics, 21, Suppl 1*, i38-46.

Ben-Hur, A., & Noble, W. S. (2006). Choosing negative examples for the prediction of protein-protein interactions. *BMC Bioinformatics, 7 Suppl 1*, S2.

Benson, D. A., Karsch-Mizrachi, I. et al. (2007). GenBank. *Nucleic Acids Res*, *35*(Database issue): D21-5.

Berg, J., Lassig, M., & Wagner, A. (2004). Structure and evolution of protein interaction networks: A statistical model for link dynamics and gene duplications. *BMC Evol Biol*, *4*(1), 51.

Berg, J., Lässig, M., & Wagner, A. (2004). Structure and evolution of protein interaction networks: A statistical model for link dynamics and gene duplications. *BMC Evolutionary Biology*, *4*(51).

Berg, J., Lässig, M., & Wagner, A. (2004). Structure and evolution of protein interaction networks: a statistical model for link dynamics and gene duplications. *BMC Evolutionary Biology*, *4*, 51.

Berger, S. I., Posner, J. M., & Ma'ayan, A. (2007). Genes2Networks: Connecting lists of gene symbols using mammalian protein interactions databases. *BMC Bioinformatics*, *8*, 372.

Berggård, T., Linse, S., & James, P. (2007). Methods for the detection and analysis of protein-protein interactions. *Proteomics*, *7*(16), 2833-2842.

Berman, H. M., et. al. (2000). The Protein Data Bank, *Nucleic Acids Research*, *28*(1), 235-242.

Berman, H. M., J. Westbrook, et al. (2000). The Protein Data Bank. *Nucleic Acids Res*, *28*(1), 235-42.

Bertin, N., Simonis, N., Dupuy, D., Cusick, M. E., Han, J. D., Fraser, H. B., Roth, F. P., & Vidal, M. (2007). Confirmation of organized modularity in the yeast interactome. *PLoS Biology*, *5*(6), e153.

Betel, D., et. al. (2007). Structure-templated predictions of novel protein interactions from sequence information. *PloS Computational Biology*, *3*(9), 1783-1789.

Birney, E., Andrews, D., Caccamo, M., Chen, Y., Clarke, L., Coates, G., et al. (2006). Ensembl 2006. *Nucl. Acids Res.*, *34*(suppl_1), D556-561.

Blatt, M., Wiseman, S., & Domany, E. (1996). Superparamagnetic clustering of data. *Phisical Review Letters*, *76*(18), 3251-3254.

Blekas, K., Fotiadis, D. I., & Likas, A. (2003). Greedy mixture learning for multiple motif discovery in biological sequences. *Bioinformatics*, *19*, 607–617.

Block, P., Paern, J., Hullermeier, E., Sanschagrin, P., Sotriffer, C. A., & Klebe, G. (2006). Physicochemical descriptors to discriminate protein-protein interactions in permanent and transient complexes selected by means of machine learning algorithms. *Proteins*, *65*, 607-622.

Bloom, J. D., & Adami, C. (2003). Apparent dependence of protein evolutionary rate on number of interactions is linked to biases in protein-protein interactions data sets. *BMC Evol Biol*, *3*, 21.

Bodenreider, O. (2004). The Unified Medical Language System (UMLS): Integrating biomedical terminology. *Nucleic Acids Res*, *32*(Database issue), D267-270.

Boeckmann, B., Bairoch, A. et al. (2003). The SWISS-PROT protein knowledgebase and its supplement TrEMBL in 2003. *Nucleic Acids Res*, *31*(1), 365-70.

Boeckmann, B., et. al. (2003). The SWISS-PROT protein knowledgebase and its supplement TrEMBL in 2003. *Nucleic Acids Research, Database issue 31*, D365-D370.

Bogan, A. A., & Thorn, K. S. (1998). Anatomy of hot spots in protein interfaces. *J Mol Biol*, *280*, 1-9.

Bollobás, B. (2001). *Random graphs* (Second ed., Vol. 73). Cambridge, England: Cambridge University Press.

Bonacich, P. (1987). Power and Centrality: A Family of Measures. *The American Journal of Sociology*, *92*(5), 1170-1182.

Boone, C., Bussey, H., & Andrews, B. J. (2007). Exploring genetic interactions and networks with yeast. *Nat Rev Genet*, *8*(6), 437-449.

Borgwardt, K. M., Ong, C. S., Schoenauer, S., Vishwanathan, S. V. N., Smola, A. J., & Kriegel, H.-P. (2006). Protein function prediction via graph kernels. *Bioinformatics*, *21*(suppl. 1), i47–i56.

Botstein, D., & Risch, N. (2003). Discovering genotypes underlying human phenotypes: past successes for mendelian disease, future approaches for complex disease. *Nat Genet, 33 Suppl*, 228-237.

Bouwmeester, T., Bauch, A., Ruffner, H., Angrand, P. O., Bergamini, G., Croughton, K., et al. (2004). A physical and functional map of the human TNF-alpha/NF-kappa B signal transduction pathway. *Nat Cell Biol, 6*(2), 97-105.

Boyle, E. I., Weng, S., Gollub, J., Jin, H., Botstein, D., Cherry, J. M., & Sherlock, G. (2004). GO TermFinder--open source software for accessing Gene Ontology information and finding significantly enriched Gene Ontology terms associated with a list of genes. *Bioinformatics 20*(18), 3710-5.

Bradford, J. R., & Westhead, D. R. (2005). Improved prediction of protein-protein binding sites using a support vector machines approach. *Bioinformatics, 21*, 1487-1494.

Breiman, L. (1996). Bagging predictors. *Machine Learning, 24*(2), 123-40.

Breiman, L. (2001). Random Forests. *Machine Learning, 45*(1), 5-32.

Breiman, L., Freidman, J. H., Olshen, R. A., & Stone, C. J. (1984). *Classification and Regression Trees*. Wadsworth.

Breitkreutz, B. J., Stark, C. et al. (2002). The GRID: The General Repository for Interaction Datasets. *Genome Biol, 3*(12), PREPRINT0013.

Breitkreutz, B. J., Stark, C., & Tyers, M. (2002). Osprey: A network visualization system. *Genome Biol, 3*(12), PREPRINT0012.

Breitkreutz, B. J., Stark, C., Reguly, T., Boucher, L., Breitkreutz, A., Livstone, M., Oughtred, R., Lackner, D. H., Bähler, J., Wood, V., Dolinski, K., & Tyers, M. (2008). The BioGRID Interaction Database: 2008 update. *Nucleic Acids Research, 36*(Database Issue), D637-40.

Breitkreutz, B., et al. (2008). The BioGRID Interaction Database: 2008 update. *Nucleic Acids Research, 36*(Database issue)D637-D640.

Breitkreutz, B., Stark, C., & Tyers, M. (2003). Osprey: A network visualization system. *Genome Biology, 4*, R22.

Breitling, R., Amtmann, A., & Herzyk, P. (2004). Graph-based iterative group analysis enhances microarray interpretation. *BMC Bioinformatics, 5*, 100.

Breman, J., Egan, A., & Keusch, G. (2001). The intolerable burden of malaria: a new look at the numbers. *Am J Trop Med Hyg., 64*, 4-7.

Brohée, S., & van Helden, J. (2006). Evaluation of clustering algorithms for protein-protein interaction networks. *BMC Bioinformatics, 7*, 488.

Brown, K. R., & Jurisica, I. (2007). Unequal evolutionary conservation of human protein interactions in interologous networks. *Genome Biol, 8*(5), R95.

Bru, C., Courcelle, E., Carrère, S., Beausse, Y., Dalmar, S., & Kahn, D. (2005). The ProDom database of protein domain families: more emphasis on 3D. *Nucleic Acids Research, 33*, D212-D215.

Brun, C., Chevenet, F., et al. (2003). Functional classification of proteins for the prediction of cellular function from a protein-protein interaction network. *Genome Biol, 5*(1), R6.

Brun, C., Herrmann, C., & Guenoche, A. (2004) Clustering proteins from interaction networks for the prediction of cellular functions. *BMC Bioinformatics, 5*.

Brunner, H. G., & van Driel, M. A. (2004). From syndrome families to functional genomics. *Nat. Rev. Genet., 5,* 545–551.

Bu, D., Zhao, Y., Cai, L., Xue, H., Zhu, X., Lu, H., Zhang, J., Sun, S., Ling, L., Zhang, N., Li, G., & Chen, R. (2003). Topological structure analysis of the protein-protein interaction network in budding yeast. *Nucleic Acids Res, 31*(9), 2443-50.

Buhler, J., & Tompa, M. (2002). Finding motifs using random projections. *Journal of Computational Biology, 9*(2), 225-242.

Burgoyne, N. J., & Jackson, R. M. (2006). Predicting protein interaction sites: binding hot-spots in protein-protein and protein-ligand interfaces. *Bioinformatics, 22*, 1335-1342.

Burz, D. S., Dutta, K., Cowburn, D., & Shekhtman, A. (2006a). Mapping structural interactions using in-cell NMR spectroscopy (STINT-NMR). *Nat Methods, 3*(2), 91-93.

Burz, D. S., Dutta, K., Cowburn, D., & Shekhtman, A. (2006b). In-cell NMR for protein-protein interactions (STINT-NMR). *Nat Protoc, 1*(1), 146-152.

Butland, G., Peregrin-Alvarez, J. M., Li, J., Yang, W., Yang, X., Canadien, V., et al. (2005). Interaction network containing conserved and essential protein complexes in escherichia coli. *Nature, 433*(7025), 531-537.

Cabusora, L., Sutton, E., Fulmer, A., & Forst, C. V. (2005). Differential network expression during drug and stress response. *Bioinformatics, 21*(12), 2898-905.

Caffrey, D. R., Somaroo, S., Hughes, J. D., Mintseris, J., & Huang, E. S. (2004). Are protein-protein interfaces more conserved in sequence than the rest of the protein surface? *Protein Sci, 13*, 190-202.

Camargo A., & Azuaje, F. (2007). Linking Gene Expression and Functional Network Data in Human Heart Failure. *PloS One, 12*, e1347.

Capra, J. A., & Singh, M. (2007). Predicting functionally important residues from sequence conservation. *Bioinformatics, 23*, 1875.

Caspi, R., Foerster, H., Fulcher, C., Kaipa, P., Krummenacker, M., Latendresse, M., et al. (2008). The MetaCyc Database of metabolic pathways and enzymes and the BioCyc collection of Pathway/Genome Databases. *Nucleic Acids Res., 36*, D623-631.

Ceol, A., Chatr-Aryamontri A., Liacata L., & Cesareni G. (2008). Linking entries in protein interaction database to structured text: The FEBS Letters experiment. *FEBS Lettwers, 585*, 1171-1177.

Ceol, A., et. al. (2007). DOMINO: A database of domain peptide interactions. *Nucleic Acids Research, Database issue 35,* D557-D560.

Cerami, E. G., Bader, G. D., Gross, B. E., &Sander, C. (2006). cPath: open source software for collecting, storing, and querying biological pathways. *BMC Bioinformatics, 7*, 497.

Cesareni, G., Ceol, A., Gavrila, C., Palazzi, L. M., Persico, M., & Schneider, M. V. (2005). Comparative interactomics. *FEBS Lett, 579*(8), 1828-1833.

Chatr-aryamontri, A., Ceol, A., Montecchi Palazzi, L., Nardelli, G., Schneider, M. V., Castagnoli, L., & Cesarei, G. (2006). MINT: the Molecular INTeraction database. *Nucleic Acids Research, 35*(Database issue), D572-D574.

Chatr-aryamontri, A., Ceol, A., Palazzi, L. M., Nardelli, G., Schneider, M. V., Castagnoli, L., & Cesareni, G. (2007). MINT: The Molecular INTeraction database. *Nucleic Acids Res, 35*(Database issue): D572–D574.

Chen, J., & Yuan, B. (2006). Detecting Functional Modules in the yeast protein-protein interaction network. *Bioinformatics, 22*(18), 2283-2290.

Chen, J., & Yuan, B. (2006). Detecting functional modules in the yeast protein-protein interaction network. *Bioinformatics, 22*(18), 2283-2290.

Chen, J., Chua, H. N., Hsu, W., Lee, M. L., Ng, S. K., Saito, R., et al. (2006). Increasing confidence of protein-protein interactomes. *Genome Inform, 17*(2), 284-297.

Chen, J., Hsu, W., et al. (2006). Increasing confidence of protein interactomes using network topological metrics. *Bioinformatics, 22*(16), 1998-2004.

Chen, J., Hsu, W., et al. (2007). Labeling network motifs in protein interactomes for protein function prediction. *Proceedings of the IEEE 23rd International Conference on Data Engineering.*

Chen, J., Hsu, W., Lee, M. L. & Ng, S.-K. (2007). Labeling network motifs in protein interactomes for protein function prediction. In *IEEE 23rd International Conference on Data Engineering* (pp. 546-555).

Chen, J., Hsu, W., Lee, M. L., & Ng, S. K. (2005). Discovering reliable protein interactions from high-throughput experimental data using network topology. *Artif Intell Med, 35*(1-2), 37-47.

Chen, J., Hsu, W., Lee, M. L., & Ng, S. K. (2006). Increasing confidence of protein interactomes using network topological metrics. *Bioinformatics, 22*(16), 1998-2004.

Chen, J., Hsu, W., Lee, M. L., & Ng, S.-K. (2006). NeMoFinder: dissecting genome-wide protein-protein interactions with meso-scale network motifs. In *Proc. 12th ACM International Conference on Knowledge Discovery and Data Mining (SIGKDD)* (pp. 106–115). New York, USA: ACM Press.

Chen, J., Hsu, W., Lee, M. L., & Ng, S.-K. (2007, April). Labeling network motifs in protein interactomes for protein function prediction. In *Proc. IEEE 23rd International Conference on Data Engineering* (pp. 546–555). Washington, DC, USA: IEEE Computer Society Press.

Chen, L., Wu, L-Y., Wang, Y., & Zhang, X-S. (2006). Inferring protein interactions from experimental data by association probabilistic method. *PROTEINS: Structure, Function, and Bioinformatics*, *62*, 833-827.

Chen, R., Mintseris, J., Janin, J., & Weng, Z. (2003). A protein-protein docking benchmark. *Proteins, 52*, 88-91.

Chen, S., Li, B., Grundke-Iqbal, I., & Iqbal, K. (2008). I1PP2A affects tau phosphorylation via association with the catalytic subunit of protein phosphatase 2A. *J Biol Chem, 283*(16), 10513-10521.

Chen, X. W., & Liu, M. (2005). Prediction of protein-protein interactions using random decision forest framework. *Bioinformatics*, *21*(24), 4394-4400.

Chen, X-W., & Liu, M. (2006). Domain-based predictive models for protein-protein interaction predictions. *EURASIP Journal on Applied Signal Processing, 2006*, 32767.

Chen, Y., & Xu, D. (2004). Global protein function annotation through mining genome-scale data in yeast Saccharomyces cerevisiae. *Nucleic Acids Res*, *32*(21), 6414-24.

Cherry, J. M., Adler, C. et al. (1998). SGD: Saccharomyces Genome Database. *Nucleic Acids Res*, *26*(1), 73-9.

Chiang, T., Scholtens, D., Sarkar, D., Gentleman, R., & Huber, W. (2007). Coverage and error models of protein-protein interaction data by directed graph analysis. *Genome Biol, 8*(9), R186.

Chiaraluce, R., Florio, R. et al. (2007). Tertiary structure in 7.9 M guanidinium chloride--the role of Glu53 and Asp287 in Pyrococcus furiosus endo-beta-1,3-glucanase. *Febs J, 274*(23), 6167-79.

Cho, R. J., Campbell, M. J., Winzeler, E. A., Steinmetz, L., Conway, A., Wodicka, L., et al. (1998). A genome-wide transcriptional analysis of the mitotic cell cycle. *Mol Cell, 2*(1), 65-73.

Cho, S. et al. (2003). Protein-protein Interaction Networks: from Interactions to Networks. *Journal of Biochemistry and Molecular Biology, 37*, 1, 45-52.

Cho, Y. R., Hwang, W., Ramanathan, M., & Zhang, A. (2007). Semantic integration to identify overlapping functional modules in protein interaction networks. *BMC Bioinformatics*, *8*, 265.

Cho, Y.-R., Hwang, W., & Zhang, A. (2006). Identification of Overlapping Functional Modules in Protein interaction networks: Information Flow-based Approach, in Proceedings of the *Sixth International Conference on Data Mining-Workshops (ICDMW'06)*.

Cho, Y.-R., Hwang, W., Ramanathan, M., & Zhang, A. (2007). Semantic integration to identify overlapping functional modules in protein interaction networks. *BMC Bioinformatics*, 8:265.

Christie, K. R., Weng, S., Balakrishnan, R., Costanzo, M. C., Dolinski, K., Dwight, S. S., et al. (2004). Saccharomyces Genome Database (SGD) provides tools to identify and analyze sequences from *Saccharomyces cerevisiae* and related sequences from other organisms. *Nucleic Acids Res.*, *32*(Database issue), D311-4.

Chua, H. N., & Wong, L. (2008). Increasing the reliability of protein interactomes. *Drug Discov Today.*

Chua, H. N., Ning, K., Sung, W. K., Leong, H. W., & Wong, L. (2007). Using indirect protein-protein interactions for protein complex prediction. In Proceedings of

IEEE Computational Systems Bioinformatics Conference (CSB07) (pp. 97-109).

Chua, H. N., Sung, W. K. et al. (2006). Exploiting indirect neighbours and topological weight to predict protein function from protein-protein interactions. *Bioinformatics, 22*(13), 1623-30.

Chua, H. N., Sung, W. K. et al. (2007). Using indirect protein interactions for the prediction of Gene Ontology functions. *Bioinformatics, 8*(Suppl 4), S8.

Chua, H. N., Sung, W.-K. et al. (2007). An efficient strategy for extensive integration of diverse biological data for protein function prediction. *Bioinformatics: btm520.*

Chuang, H. Y., Lee, E., Liu, Y. T., Lee, D., & Ideker, T. (2007). Network-based classification of breast cancer metastasis. *Molecular Systems Biology, 3*,140.

Chuang, H., Lee, E., Liu, Y., Lee, D., & Ideker, T. (2007). Network-based classification of breast cancer metastasis. *Molecular Systems Biology, 3*, 140

Chuang, H-Y., Lee, E., Liu, Y-T., Lee, D., & Ideker, T. (2007) Network-based classification of breast cancer metastasis. *Molecular Systems Biology, 3*, 140.

Ciriello, G., & Guerra, C. (2008). A review on models and algorithms for motif discovery in protein-protein interaction networks. *Briefings in Functional Genomics and Proteomics.*

Cline, M. S., Smoot, M., Cerami, E., Kuchinsky, A., Landys, N., Workman, C., et al. (2007). Integration of biological networks and gene expression data using Cytoscape. *Nature Protocols, 2*(10), 2366-82.

Collins, S. R., Kemmeren, P., Zhao, X. C., Greenblatt, J. F., Spencer, F., Holstege, F. C., et al. (2007). Toward a comprehensive atlas of the physical interactome of Saccharomyces cerevisiae. *Mol Cell Proteomics, 6*(3), 439-450.

Commichau, F. M., Forchhammer, K., & Stulke, J. (2006). Regulatory links between carbon and nitrogen metabolism. *Curr Opin Microbiol, 9*(2), 167-172.

Cong, G., Tan, K.-L., Tung, A., & Xu, X. (2005). Mining top-k covering rule groups for gene expression data. *Proceedings of the 2005 ACM SIGMOD International Conference on Management of Data* (pp. 670–681).

Cormen, T. H., Leiserson, C. E., Rivest, R. L., & Stein, C. (2001). *Introduction to algorithms* (Second ed.). Cambridge, USA: The MIT Press.

Cortes, C., & Vapnik, V. (1995). Support-Vector Networks. *Machine Learning, 20*(3), 273-297.

Cote, R. G., Jones, P., Apweiler, R., & Hermjakob, H. (2006). The Ontology Lookup Service, a lightweight cross-platform tool for controlled vocabulary queries. *BMC Bioinformatics, 7*, 97.

Cover, T., & Hart, P. (1967). Nearest neighbor pattern classification. *IEEE Transactions on Information Theory, 13*(1), 21-27.

Creighton, C., & Hanash, S. (2003). Mining gene expression databases for association rules. *Bioinformatics, 19*, 79–86.

Cristianini, N., & Shawe-Taylor, J. (2000). *An Introduction to Support Vector Machines.* Cambridge University.

Cui, J., Li, P., Li, G., Xu, F., Zhao, C., Li, Y., Yang, Z., Wang, G., Yu, Q., Li, Y., & Shi, T. (2008) AtPID: *Arabidopsis thaliana* protein interactome database--an integrative platform for plant systems biology. *Nucleic Acids Research, 36*(Database issue), D999-1008.

D'Haeseleer, P., & Church, G. M. (2004). Estimating and improving protein interaction error rates. *Proc IEEE Comput Syst Bioinform Conf,* (pp. 216-223).

Dalkilic, M. M., Costello, J. C., Clark, W. T., & Radivojac, P. (2008). From protein-disease associations to disease informatics. *Front Biosci, 13*, 3391-3407.

Dandekar, T., Snel, B. et al. (1998). Conservation of gene order: a fingerprint of proteins that physically interact. *Trends Biochem Sci, 23*(9), 324-8.

Davis, F. P., Barkan, D. T., Eswar, N., McKerrow, J. H., & Sali, A. (2007). Host pathogen protein interactions predicted by comparative modeling. *Protein Sci, 16*(12), 2585-2596.

Dayhoff, M. O., Barker, W. C., & Hunt, L. T. (1983). Establishing homologies in protein sequences. *Methods Enzymol., 91*, 524-545.

De Bie, T., Tranchevent, L. C., van Oeffelen, L. M., & Moreau, Y. (2007). Kernel-based data fusion for gene prioritization. *Bioinformatics, 23*(13), i125-132.

de Chiara, C., Menon, R. P., & Pastore, A. (2008). Structural bases for recognition of Anp32/LANP proteins. *FEBS J, 275*(10), 2548-2560.

de Lichtenberg, U., Jensen, L. J., Brunak, S., & Bork, P. (2005). Dynamic complex formation during the yeast cell cycle. *Science, 307*(5710), 724-727.

de Vries, S. J., van Dijk, A. D., & Bonvin, A. M. (2006). WHISCY: what information does surface conservation yield? Application to data-driven docking. *Proteins, 63*, 479-489.

Deane, C. M., Salwinski, L. et al. (2002). Protein interactions: two methods for assessment of the reliability of high throughput observations. *Mol Cell Proteomics, 1*(5), 349-56.

Deane, C. M., Salwinski, L., Xenarios, I., & Eisenberg, D. (2002). Protein interactions: Two methods for assessment of the reliability of high throughput observations. *Mol Cell Proteomics, 1*(5), 349-356.

Delgado, M. L., O'Connor, J. E., Azorín, I., Renau-Piqueras, J., Gil, M. L., & Gozalbo, D. (2001). The glyceraldehyde-3-phosphate dehydrogenase polypeptides encoded by the Saccharomyces cerevisiae TDH1, TDH2 and TDH3 genes are also cell wall proteins. *Microbiology, 147*, 411-417.

DeLuna, A., Vetsigian, K., Shoresh, N., Hegreness, M., Colon-Gonzalez, M., Chao, S., et al. (2008). Exposing the fitness contribution of duplicated genes. *Nat Genet, 40*(5), 676-681.

Demirel, M. C., & Cherny, D. (2005). Clustering and diversity of fluctuations for proteins. *Nanomedicine: Nanotechnology, Biology, and Medicine, 1*, 41-46.

Dempster, A., Laird, N., & Rubin, D. (1977). Maximum Likelihood from Incomplete Data via the EM Algorithm. *Journal of the Royal Statistical Society, 39*(B), 1-38.

Deng, M., Chen, T. et al. (2004). An integrated probabilistic model for functional prediction of proteins. *J Comput Biol, 11*(2-3), 463-75.

Deng, M., Mehta, S., et al. (2002). Inferring domain-domain interactions from protein-protein interactions. *Genome, 12*(10), 1540-8.

Deng, M., Sun, F., & Chen, T. (2003) Assessment of the reliability of protein protein interactions and protein function prediction. In *Pac. Symp. Biocomput. (PSB2003)* (pp. 140-151).

Deng, M., Zhang, K., et al. (2003). Prediction of protein function using protein-protein interaction data. *J Comput Biol, 10*(6), 947-60.

Deng, M., Zhang, K., Mehta, S., Chen, T., & Sun, F. (2003) Prediction of protein function using protein-protein interaction data. *Journal of Computational Biology, 10*(6), 947-960.

Dennis Jr., G., Sherman, B. T., Hosack, D. A., Yang, J., Gao, W., Lane, H. C., &Lempicki, R. A. (n/d). DAVID: Database for Annotation, Visualization, and Integrated Discovery. *Genome Biology, 2003, 4*(5):P3.

Dennis, G., Jr., Sherman, B. T., Hosack, D. A., Yang, J., Gao, W., Lane, H. C., et al. (2003). DAVID: Database for Annotation, Visualization, and Integrated Discovery. *Genome Biol, 4*(5), P3.

Derenyi, I., Palla, G., & Vicsek, T. (2005). Clique percolation in random networks. *Physical Review Letters, 94*, 160202.

DeRisi, J. L., Iyer, V. R., & Brown, P. O. (1997). Exploring the metabolic and genetic control of gene expression on a genomic scale. *Science, 278*, 680-686.

Dhanapalan, L., & Chen, J. Y. (2007). A case study of integrating protein interaction data using semantic web technology. *International Journal of Bioinformatics Research and Applications, 3*(3), 286-302.

Dittrich, M. T., Klau, G. W., Rosenwald, A., Dandekar, T., & Müller, T. (2008). Identifying functional modules in protein-protein interaction networks: an integrated exact approach. *Bioinformatics, 24*(13), i223-31.

Dobson, S., May, T., Berriman, M., Del, V. C., Fairlamb, A., Chakrabarti, D., et al. (1999). Characterization of protein Ser/Thr phosphatases of the malaria parasite, Plasmodium falciparum: inhibition of the parasitic calcineurin by cyclophilin-cyclosporin complex. *Mol Biochem Parasitol., 99*(2), 167-181.

Dominguez, C., Boelens, R., & Bonvin, A. M. (2003). HADDOCK: A protein-protein docking approach based on biochemical or biophysical information. *J Am Chem Soc, 125*, 1731-1737.

Dopazo, J. (2006). Functional interpretation of microarray experiments. *Omics, 10*, 398-410.

Doray, B., & Kornfiel, S. (2001). Gamma subunit of the AP-1 adaptor complex binds clathrin: implications for cooperative binding in coated vesicle assembly. *Molecular Biology of the Cell, 12*(7), 1925-1935.

Dost, B., Shlomi, T., Gupta, N., Ruppin, E., Bafna, V., & Sharan, R. (2007). QNet: A Tool for Querying Protein Interaction Networks. *Proceeding of the eleventh Annual International Conference Research in Computational Molecular Biology (RECOMB 2007)* (pp. 1-15). Springer Berlin: Heidelberg.

Dosztanyi, Z., Chen, J., Dunker, A. K., Simon, I., & Tompa, P. (2006). Disorder and sequence repeats in hub proteins and their implications for network evolution. *J Proteome Res, 5*, 2985-2995.

Doyle, M. L. (2001). Titration microcalorimetry. *Curr Protoc Protein Sci, Chapter 20*, Unit20 24.

Drakas, R., Prisco, M., & Baserga, R. (2005). A modified tandem affinity purification tag technique for the purification of protein complexes in mammalian cells. *Proteomics 5*(1), 132-137.

Drees, B. L., Sundin, B., & et al. (2001) A protein interaction map for cell polarity development. *Journal of Cellular Biology, 154*, 549-571.

Drewes, G., & Bouwmeester, T. (2003). Global approaches to protein-protein interactions. *Current Opinion in Cell Biology, 15*, 199-205.

Drogrusoz, U., *et al.* (2006). PATIKAweb: A Web interface for analyzing biological pathways through advanced querying and visualization. *Bioinformatics, 22*(3), 374-375.

Dunker, A. K., Cortese, M. S., Romero, P., Iakoucheva, L. M., & Uversky, V. N. (2005). Flexible nets. The roles of intrinsic disorder in protein interaction networks. *FEBS J, 272*, 5129-5148.

Dunn, R., Dudbridge, F., & Sanderson, C. M. (2005). The use of edge-betweenness clustering to investigate biological function in protein interaction networks. *BMC Bioinformatics, 6*, 39.

Dunn, R., Dudbridge, F., & Sanderson, C. M. (2005). The use of edge-betweenness Clustering to Investigate Biological Function in Protein Interaction Networks. *BMC Bioinformatics, 6*, 39.

Durand, P., Labarre, L., Meil, A., Divol, J.-L., Vandenbrouck, Y., Viari, A., & Wojcik, J. (2006). GenoLink: a graph-based querying and browsing system for investigating the function of genes and proteins. *BMC Bioinformatics, 7*(21).

Dutkowski, J., & Tiuryn, J. (2007). Identification of functional modules from conserved ancestral protein-protein interactions. *Bioinformatics, 23*(13), i149-58.

Eddy, S. R. (1998). Profile Hidden Markov Models. *Bioinformatics, 14*, 755-763.

Efroni, S., Schaefer, C. F., & Buetow, K. H. (2007). Identification of key processes underlying cancer phenotypes using biologic pathway analysis. *PLoS ONE, 2*(5), e425.

Eisen, M. B., Spellman, P. T., Brown, P. O., & Botstein, D. (1998). Cluster analysis and display of genome-wide expression patterns. *Proc Natl Acad Sci U S A, 95*(25), 14863-14868.

Eisen, M. B., Spellman, P. T., et al. (1998). Cluster analysis and display of genome-wide expression patterns. *Proc Natl Acad Sci U S A, 95*(25), 14863-8.

Eisenberg, D., & McLachlan, A. D. (1986). Solvation energy in protein folding and binding. *Nature, 319*, 199-203.

Eisenberg, D., Marcotte, E. M., Xenarios, I., & Yeates, T. O. (2000). Protein function in the post-genomic era. *Nature, 405*(6788), 823-6.

Eisenberg, E., & Levanon, E. Y. (2003). Preferential attachment in the protein network evolution. *Phys Rev Lett, 91*(13), 138701.

Ekman, D., Light, S., Bjorklund, A. K., & Elofsson, A. (2006). What properties characterize the hub proteins of the protein-protein interaction network of Saccharomyces cerevisiae? *Genome Biol, 7*(6), R45.

Ekman, D., Light, S., Björklund, A. K., & Elofsson, A. (2006). What properties characterize the hub proteins of the protein-protein interaction network of Saccharomyces cerevisiae? *Genome Biology, 7*(6), R45.

Ekman, D., Light, S., Bjorklund, A. K., & Elofsson, A. 2006. What properties characterize the hub proteins of the protein-protein interaction network of Saccharomyces cerevisiae? *Genome Biology, 2007 8, R192* 7:R45.

Enright, A. J., Dongen, S. V., & Ouzounis, C.A. (2002). An efficient algorithm for large-scale detection of protein families. *Nucleic Acids Res, 30*(7), 1575-84.

Enright, A. J., & Ouzounis, C. A. (2001). Functional associations of proteins in entire genomes by means of exhaustive detection of gene fusions. *Genome Biol, 2*(9), RESEARCH0034.

Enright, A. J., Iliopoulos, I., et al. (1999). Protein interaction maps for complete genomes based on gene fusion events. *Nature, 402*(6757), 86-90.

Estrada, E. (2006). Virtual identification of essential proteins within the protein interaction network of yeast. *Proteomics, 6*(1), 35-40.

Estrada, E., & Rodríguez-Velázquez, J. A. (2005). Subgraph centrality in complex networks. *Phys Rev E Stat Nonlin Soft Matter Phys, 71*(5 Pt 2).

Everett, M. G., & Borgatti, S. P. (1999). The centrality of groups and classes. *Journal of Mathematical Sociology, 23*(3), 181-201.

Evlampiev, K., & Isambert, H. (2007). Modeling protein network evolution under genome duplication and domain shuffling. *BMC Systems Biology, 1*, 49.

Falk, R., *et al.* (2007). Approaches for systematic proteome exploration. *Biomolecular Engineering, 24*, 155-168.

Farutin, V., Robinson, K., et al. (2006). Edge-Count Probabilities for the Identification of Local Protein Communities and their Organization. *Proteins: Structure, Function, and Bioinformatics, 62*, 800-818.

Fawcett, T. (2004). *ROC Graphs: Notes and Practical Considerations for Researchers*. Technical report, Palo Alto, USA: HP Laboratories.

FCSB 2008. Retrieved August 15, 2008, from http://www.systems-biology.org/~myukiko/FCSB2008/doku.php.

Fernandez-Recio, J., Totrov, M., Skorodumov, C., & Abagyan, R. 2005. Optimal docking area: a new method for predicting protein-protein interaction sites. *Proteins, **58***, 134-143.

Ferre, S., & King, R. D. (2006). Finding motifs in protein secondary structure for use in function prediction. *J Comput Biol, 13*(3), 719-31.

Ferro, A., Giugno, R., Pigola, G., Pulvirenti, A., Skripin, D., Bader, G. D., & Shasha, D. (2007). NetMatch: a Cytoscape plugin for searching biological networks. *Bioinformatics, 23*(7), 910-912.

Fiedler, M., & Borgelt, C. (2007). Support computation for mining frequent subgraphs in a single graph. *Mining and Learning with Graphs (MLG).*

Fields, S., & Song, O. (1989). A novel genetic system to detect protein-protein interactions. *Nature, 340*(6230), 245-246.

Finn, R. D., et al. (2006). Pfam: clans, web tools and services. *Nucleic Acids Research, 34*, D247-D251.

Finn, R. D., et. al. (2005). iPfam: visualization of protein-protein interactions in PDB at domain and amino acid resolutions. *Bioinformatics, 21*(3), 410-412.

Finn, R. D., et. al. (2006). Pfam: clans, web tools and services. *Nucleic Acids Research, Database issue 34*, D247-D251.

Finn, R. D., Marshall, M., & Bateman, A. (2005). iPfam: visualization of protein-protein interactions in PDB at domain and amino acid resolutions. *Bioinformatics, 21*(3), 410-2.

Finn, R. D., Tate, J., et al. (2008). The Pfam protein families database. *Nucleic Acids Res, 36*(Database issue), D281-8.

Fionda, V., Palopoli, L., et al. (2007). GRAPPIN: Bipartite GRAph Based Protein-Protein Interaction Network Similarity Search. *2007 IEEE International Conference on Bioinformatics and Biomedicine (BIBM 2007).*

Fionda, V., Palopoli, L., Panni, S., & Rombo, S. E. (2008). Protein-protein interaction network querying by a "focus and zoom" approach. *Proceeding of 2nd International Conference on Bioinformatics Research and Development (BIRD 08)*, (pp. 331-346).

Fionda, V., Palopoli, L., Panni, S., & Rombo, S. E. (2008). Protein-protein interaction network querying by a "focus and zoom" approach. In Proceedings of *BioInformatics Research and Development (BIRD2008)*, To appear.

Fionda, V., Panni, S., Palopoli, S., & Rombo, S. E. (2007). Bi-GRAPPIN: Bipartite graph based protein-protein interaction networks similarity search. In *the IEEE International Conference on Bioinformatics and Biomedicine (BIBM)* (pp. 355-61).

Florens, L., Washburn, M., Raine, J., Anthony, R., Grainger, M., Haynes, J., et al. (2002). A proteomic view of the Plasmodium falciparum life cycle. *Nature, 419*(6906), 520-526.

Formstecher, E., Aresta, S., Collura, V., Hamburger, A., Meil, A., Trehin, A., et al. (2005). Protein interaction mapping: A drosophila case study. *Genome Res, 15*(3), 376-384.

Formstecher, E., *et al.* (2005). Protein interaction mapping: A Drosophila case study. Genome Research, *15*, 376-384.

Fortin, S. (1996). *The graph isomorphism problem* (Tech. Rep. No. 96-20). University of Alberta: Dept. of Computer Science.

Frank, E., Wang, Y., Holmes, G., & Witten, I. H. (1998). Using model trees for classification. *Machine Learning, 32*(1), 63-76.

Franke, L., van Bakel, H., Fokkens, L., de Jong, E. D., Egmont-Petersen, M., & Wijmenga, C. (2006). Reconstruction of a functional human gene network, with an application for prioritizing positional candidate genes. *Am J Hum Genet, 78*(6), 1011-1025.

Fraser, H. B., Hirsh, A. E., Steinmetz, L. M., Scharfe, C., & Feldman, M. W. (2002). Evolutionary rate in the protein interaction network. *Science, 296*(5568), 750-752.

Fraser, H. B., Hirsh, A. E., Steinmetz, L. M., Scharfe, C., & Feldman, M. W. (2002). Evolutionary rate in the protein interaction network. *Science, 296*(5568), 750-752.

Fraser, H. B., Wall, D. P., & Hirsh, A. E. (2003). A simple dependence between protein evolution rate and the number of protein-protein interactions. *BMC Evol Biol, 3*(1), 11.

Frazier, M. E., Johnson, G. M., et al. (2003). Realizing the potential of the genome revolution: the genomes to life program. *Science, 300*(5617), 290-3.

Freimer, N., & Sabatti, C. (2003). The human phenome project. *Nat Genet, 34*(1), 15-21. Giot, L., Bader, J. S., Brouwer, C., Chaudhuri, A., Kuang, B., Li, Y., et al. (2003). A protein interaction map of Drosophila melanogaster. *Science, 302*(5651), 1727-1736.

Friedel, C. C., & Zimmer, R. (2006). Toward the complete interactome. *Nature Biotechnology*, *24*(6), 614-615.

Friedel, C. C., & Zimmer, R. (2006). Inferring topology from clustering coefficients in protein-protein interaction networks. *BMC Bioinformatics*, *7*, 519.

Furney, S. J., Alba, M. M., & Lopez-Bigas, N. (2006). Differences in the evolutionary history of disease genes affected by dominant or recessive mutations. *BMC Genomics, 7*, 165.

Futschik, M. E., Chaurasia, G., & Herzel, H. (2007). Comparison of human protein-protein interaction maps. *Bioinformatics, 23*(5), 605-611.

G.Siek, J., Lee, L.-Q., & Lumsdaine, A. (2002). *The Boost graph library: User guide and reference manual*. New York, USA: Addison-Wesley Publishing Co.

Gabow, A., Leach, S., et al. (2008). Improving protein function prediction methods with integrated literature data. *BMC Bioinformatics, 9*(1), 198.

Gandhi, T. K., Zhong, J., Mathivanan, S., Karthick, L., Chandrika, K. N., Mohan, S. S., et al. (2006). Analysis of the human protein interactome and comparison with yeast, worm and fly interaction datasets. *Nat Genet, 38*(3), 285-293.

Gardina, P. J., Clark, T. A., Shimada, B., Staples, M. K., Yang, Q., Veitch, J., et al. (2006). Alternative splicing and differential gene expression in colon cancer detected by a whole genome exon array. *BMC Genomics, 7*, 325.

Garey, M., & Johnson, D. (Ed.) (1979). *Computers and intractability: A guide to the theory of NP-completeness*. New York: Freeman.

Gascuel, O. (1997). BIONJ: an improved version of the NJ algorithm based on a simple model of sequence data. *Mol Biol Evol, 14*(7), 685-95.

Gasteiger, J., & Engel, T. (2003). *Chemoinformatics: A textbook*. Wiley-VCH.

Gavin, A. C. *et al.* (2002). Functional organization of the yeast proteome by systematic analysis of protein complexes. *Nature, 415*, 141-147

Gavin, A. C., Aloy, P., Grandi, P., Krause, R., Boesche, M., Marzioch, M., et al. (2006). Proteome survey reveals modularity of the yeast cell machinery. *Nature, 440*(7084), 631-636.

Gavin, A. C., Aloy, P., Grandi, P., Krause, R., Boesche, M., Marzioch, M., et al. (2006). Proteome survey reveals modularity of the yeast cell machinery. *Nature, 440*(7084), 631-636.

Gavin, A. C., Bösche, M., Krause, R., Grandi, P., Marzioch, M., Bauer, A., et al. (2002). Functional organization of the yeast proteome by systematic analysis of protein complexes. *Nature, 415*(6868), 141-7.

Gavin, A.L., et. al. (2006). Proteome survey reveals modularity of the yeast cell machinery. *Nature, 440*, 631-636.

Ge, H., *et al.* (2001). Correlation between transcriptome and interactome mapping data from Saccharomyces cerevisiae. *Nature Genetics, 29*, 482-486

Ge, H., Liu, Z., Church, G. M., & Vidal, M. (2001). Correlation between transcriptome and interactome mapping data from Saccharomyces cerevisiae. *Nature Genetics, 29*(4), 482-486.

Geman, S., & Geman, D. (1990). Stochastic relaxation, Gibbs distributions, and the Bayesian restoration of images. *Readings in uncertain reasoning* (pp. 452-472). Morgan Kaufmann Publishers Inc.

Gentleman, R., & Huber, W. (2007). Making the most of high-throughput protein-interaction data. *Genome Biology, 8*(10), 112.

Ghaemmaghami, S., Huh, W. K., Bower, K., Howson, R. W., Belle, A., Dephoure, N., O'Shea, E.K., Weissman, J. S. (2003). Global analysis of protein expression in yeast. *Nature, 425*(6959), 737-41.

Ghazalpour, A., et al. (2006). Integrating Genetic and Network Analysis to Characterize Genes Related to Mouse Weight. *Plos Genetics, 2*(8), e130.

Ghosh, N., Patel, N., Jiang, K., Watson, J. E., Cheng, J., Chalfant, C. E., et al. (2007). Ceramide-activated protein phosphatase involvement in insulin resistance via Akt, serine/arginine-rich protein 40, and ribonucleic acid splicing in L6 skeletal muscle cells. *Endocrinology, 148*(3), 1359-1366.

Giaever, G., Chu, A. M., Ni, L., Connelly, C., Riles, L., Veronneau, S., et al. (2002). Functional profiling of the *Saccharomyces cerevisiae* genome. *Nature, 418*(6896), 387-91.

Gietz, R. D., Triggs-Raine, B., et al. (1997). Identification of proteins that interact with a protein of interest:

Applications of the yeast two-hybrid system. *Molecular and Cellular Biochemistry, 172*(1), 67-79.

Gil-Navarro, I., Gil, M., Casanova, M., O'Connor, J., Martinez, J., & Gozalbo, D. (1997). The glycolytic enzyme glyceraldehyde-3-phosphate dehydrogenase of Candida albicans is a surface antigen. *J Bacteriol., 179*(16), 4992-4999.

Gingras, A. C., Gstaiger, M., Raught, B., & Aebersold, R. (2005). Analysis of protein complexes using mass spectrometry. *Nat Rev Mol Cell Biol., 8*(8), 645-654.

Giot, L., Bader, J. S., Brouwer, C., Chaudhuri, A., Kuang, B., Li, Y., et al. (2003). A protein interaction map of drosophila melanogaster. *Science, 302*(5651), 1727-1736.

Giot, L., Bader, J. S., Brouwer, C., Chaudhuri, A., Kuang, B., Li, Y., et al. (2003). A protein interaction map of Drosophila melanogaster. *Science, 302*, 1727-1736.

Girvan, M., & Newman, M. E. (2002). Community structure in social and biological networks. *Proceedings of the National Academy of Sciences USA, 99*(12), 7821-7826.

Girvan, M., & Newman, M. E. J. (2002). Community structure in socila and biological networks. *Proc. Natl. Acad. Sci., 99*(12), 7821-7826.

Glaser, F., Pupko, T., Paz, I., Bell, R. E., Bechor-Shental, D., Martz, E., & Ben-Tal, N. 2003. ConSurf: identification of functional regions in proteins by surface-mapping of phylogenetic information. *Bioinformatics, 19*, 163-164.

Glaser, F., Rosenberg, Y., Kessel, A., Pupko, T., and Ben-Tal, N. (2005). The ConSurf-HSSP database: The mapping of evolutionary conservation among homologs onto PDB structures. *Proteins, 58*, 610-617.

Goebel, U., Sander, C., Schneider, R., & Valencia, A. (1994). Correlated mutations and residue contacts in proteins. *Proteins, 18*, 309-317.

Goehler, H., Lalowski, M., Stelzl, U., Waelter, S., Stroedicke, M., Worm, U., et al. (2004). A protein interaction network links GIT1, an enhancer of huntingtin aggregation, to Huntington's disease. *Mol Cell, 15*(6), 853-865.

Goh, K. I., Cusick, M. E., Valle, D., Childs, B., Vidal, M., & Barabasi, A. L. (2007). The human disease network. *Proc Natl Acad Sci U S A, 104*(21), 8685-8690.

Golemis, E. A., Serebriiskii, I., & Law, S. (1999). The yeast two-hybrid system: criteria for detecting physiologically significant protein-protein interactions. *Curr Issues Mol Biol., 1*(1-2), 31-45.

Golemis, G. A., Serebriiskii, I., Finley Jr., R. L., Kolonin, M. G., Gyuris, J., & Brent R. (2008). Interaction trap/two-hybrid system to identify interacting proteins. In F.M. Ausubel, R. Brent, R. E. Kingston, D. D. Moore, J. G. Seidman, J. A. Smith, et al. *Current Protocols in Molecular Biology* (pp. 20.1.1-20.1.35). New York: John Wiley & Sons, Inc.

Gomez, S. M., & Rzhetsky, A. (2002). Towards the prediction of complete protein--protein interaction networks. *Pac Symp Biocomput*, 413-424.

Gotfredsen, M., & Gerdes, K. (1998). The *Escherichia coli* relBE genes belong to a new toxin-antitoxin gene family. *Mol Microbiol, 29*(4), 1065-76.

Grahne, G., & Zhu, J. (2003). Efficiently using prefix-trees in mining frequent itemsets. *Proceedings of the Workshop on Frequent Itemset Mining Implementations 2003.*

Graur, D., & Li, W. H. (2000). Fundamentals of molecular evolution (second edition). *Sunderland, Massachusetts: Sinauer.*

Greenman, C. *et. al.* (2007). Patterns of somatic mutation in human cancer genome. *Nature, 446*, 153-158.

Grigoriev, A. (2001). A relationship between gene expression and protein interactions on the proteome scale: analysis of the bacteriophage T7 and the yeast Saccharomyces cerevisiae. *Nucleic Acids Res, 29*(17), 3513-3519.

Gross, J. L., & Yellen, J. (2005). *Graph theory and its applications (Discrete mathematics and its applications)* (Second ed.). Boca Raton, USA: Chapman & Hall/CRC.

Grossmann, S., Bauer, S., Robinson, P. N., & Vingron, M. (2007). Improved detection of overrepresentation of

Gene-Ontology annotations with parent child analysis. *Bioinformatics, 23*(22), 3024-3031.

Guan, Y., Dunham, M. J., & Troyanskaya, O. G. (2007). Functional analysis of gene duplications in saccharomyces cerevisiae. *Genetics, 175*(2), 933-943.

Güldener, U., Münsterkötter, M., Oesterheld, M., Pagel, P., Ruepp, A., Mewes, H. W., Stümpflen, V. (2006). MPact: the MIPS protein interaction resource on yeast. *Nucleic Acids Res, 34*(Database issue), D436-41.

Guo, J., Wu, X., Zhang, D. Y., & Lin, K. (2008). Genome-wide inference of protein interaction sites: lessons from the yeast high-quality negative protein-protein interaction dataset. *Nucleic Acids Res, 36*(6), 2002-2011.

Guo, Y., Yu, L., Wen, Z., & Li, M. (2008). Using support vector machine combined with auto covariance to predict protein-protein interactions from protein sequences. *Nucleic Acids Res, 36*(9), 3025-3030.

Guo, Z., Li, Y., Gong, X., Yao, C., Ma, W., Wang, D., Li, Y., Zhu, J., Zhang, M., Yang, D., & Wang, J. (2007). Edge-based scoring and searching method for identifying condition-responsive protein-protein interaction sub-network. *Bioinformatics, 23*(16), 2121-2128.

Hagopian, J. C., Ma, C. T., Meade, B. R., Albuquerque, C. P., Ngo, J. C., Ghosh, G., et al. (2008). Adaptable Molecular Interactions Guide Phosphorylation of the SR Protein ASF/SF2 by SRPK1. *J Mol Bio.*

Hahn, M. W., & Kern, A. D. (2005). Comparative genomics of centrality and essentiality in three eukaryotic protein-interaction networks. *Mol Biol Evol, 22*(4), 803-806.

Hahn, M. W., Conant, G. C., & Wagner, A. (2004). Molecular evolution in large genetic networks: Does connectivity equal constraint? *J Mol Evol, 58*(2), 203-211.

Hakes, L., Pinney, J. W., Robertson, D. L., & Lovell, S. C. (2008). Protein-protein interaction networks and biology -- what's the connection? *Nature Biotechnology, 26*(1), 69-72.

Halperin, I., Wolfson, H., & Nussinov, R. (2006). Correlated mutations: advances and limitations. A study on fusion proteins and on the Cohesin-Dockerin families. *Proteins 63*, 832-845.

Hamada, M., Tsuda, K., Kudo, T., Kin, T., & Asai, K. (2006). Mining frequent stem patterns from unaligned RNA sequences. *Bioinformatics, 22,* 2480–2487.

Hamosh, A., et. al. (2005). Online Mendelian Inheritance in Man (OMIM), a knowledge base of human genes and genetic disorders. *Nucleic Acids Research. Database issue 33*, D514-517.

Hamosh, A., Scott, A. F., Amberger, J. S., Bocchini, C. A., & McKusick, V. A. (2005). Online Mendelian Inheritance in Man (OMIM), a knowledgebase of human genes and genetic disorders. *Nucleic Acids Res, 33*(Database issue), D514-517.

Han, J. D. (2008). Understanding biological functions through molecular networks. *Cell Res, 18*(2), 224-237.

Han, J. D., Bertin, N., Hao, T., Goldberg, D. S., Berriz, G. F., Zhang, L. V., Dupuy, D., Walhout , A. J., Cusick, M. E., Roth, F. P., & Vidal M. (2004). Evidence for dynamically organized modularity in the yeast protein-protein interaction network. *Nature, 430*(6995), 88-93

Han, J. D., Dupuy, D, Bertin, N., Cusick, M. E., & Vidal, M. (2005). Effect of sampling on topology predictions of protein-protein interaction networks. *Nature Biotechnology, 23*(7), 839-844.

Han, J., & Kamber, M. (2000). *Data mining: Concepts and techniques*. Morgan Kaufmann.

Han, J-D.J. (2008). Understanding biological functions through molecular networks. *Cell Research, 18*, 224–237.

Han, J-D.J., Bertin, H., Hao, T., Goldberg, D. S., Berriz, G. F., Zhang, L. V. Dupuy, D., Walhout, A. J. M., Cusick, M. E., Roth, F. P. & Vidal, M. (2004). Evidence for dynamically organized modularity in the yeast protein–protein interaction network. *Nature, 430*, 88-93.

Han, J-DJ., *et al.* (2005). ffect of sampling on topology predictions of protein-protein interaction networks. *Nature Biotechnology, 23*, 839-844.

Harary, F., & Palmer, E. M. (1973). *Graphical enumeration.* New York, USA: Academic Press.

Harbison, C. T., Gordon, D. B., Lee, T. I., Rinaldi, N. J., Macisaac, K. D., Danford, T. W., et al. (2004). Transcriptional regulatory code of a eukaryotic genome. *Nature, 431*(7004), 99-104.

Harris, S. J., & Winzor, D. J. (1990). Interactions of glycolyltic enzymes with erythrocyte membrane. *Biochemica et Biophysica Acta, 1038*, 306-314.

Hart, G. T., Lee, I., & Marcotte, E. R. (2007). A high-accuracy consensus map of yeast protein complexes reveals modular nature of gene essentiality. *BMC Bioinformatics, 8*, 236.

Hart, G. T., Ramani, A. K., & Marcotte, E. M. (2006). How complete are current yeast and human protein-interaction networks? *Genome Biology, 7*(11), 120.

Hartuv, E., & Shamir, R. A. (2000). Clustering Algorithm based Graph Connectivity. *Information Processing Letters, 76*, 175-181.

Hartwell L. H., Hopfield J. J., Leibler S., & Murray A. W. (1999). From molecular to modular cell biology. *Nature, 402*, C47-C52.

Hawkins, T., & Kihara, D. (2007). Function prediction of uncharacterized proteins. *J Bioinform Comput Biol, 5*(1), 1-30.

Hawkins, T., Luban, S., et al. (2006). Enhanced automated function prediction using distantly related sequences and contextual association by PFP. *Protein Sci, 15*(6), 1550-6.

Hayashida, M., Ueda, N., & Akutsu, T. (2003). Inferring strengths of protein-protein interactions from experimental data using linear programming. *Bioinformatics, 19*, ii58-ii65.

Hayashida, M., Ueda, N., & Akutsu, T. (2004). A simple method for inferring strengths of protein-protein interactions. *Genome Informatics, 15-1*, 56-68.

He, X., & Zhang, J. (2005). Rapid subfunctionalization accompanied by prolonged and substantial neofunctionalization in duplicate gene evolution. *Genetics, 169*(2), 1157-1164.

He, X., & Zhang, J. (2006). Why do hubs tend to be essential in protein networks? *Plos Genetics, 2*(6), e88.

Henschel, A., et. al. (2006). Using structural motif descriptors for sequence-based binding site prediction. *BMC Bioinformatics, 8*(Suppl 4), S5.

Hermjakob, H., et al. (2004). The HUPO PSI's Molecular Interaction format-a community standard for the representation of protein interaction data. *Nature Biotechnology, 22*(2), 177-183.

Hermjakob, H., Montecchi-Palazzi, L., Bader, G., Wojcik, J., Salwinski, L., Ceol, A., et al. (2004). The HUPO PSI's molecular interaction format--a community standard for the representation of protein interaction data. *Nat Biotechnol, 22*(2), 177-183.

Hermjakob, H., Montecchi-Palazzi, L., Lewington, C., Mudali, S., Kerrien, S., Orchard, S., et al. (2004). IntAct: an open source molecular interaction database. *Nucleic Acids Res, 32*(Database issue), D452-455.

Hernandez, P., et al. (2007). Evidence for system-level molecular mechanisms of tumorgenesis. *BMC Genomics, 8*, 186

Hernandez-Toro, J., Prieto C., & De Las Rivas, J. (2007). APID2NET: unified interactome graphic analyzer. *Bioinformatics, 23*(18), 2495-2497

Higurashi, M., Ishida, T., & Kinoshita, K. (2008). Identification of transient hub proteins and the possible structural basis for their multiple interactions. *Protein Sci, 17*, 72-78.

Hinrichs, A. S., Karolchik, D., Baertsch, R., Barber, G. P., Bejerano, G., Clawson, H., et al. (2006). The UCSC Genome Browser Database: update 2006. *Nucl. Acids Res., 34*(suppl_1), D590-598.

Hishigaki, H., Nakai, K., et al. (2001). Assessment of prediction accuracy of protein function from protein--protein interaction data. *Yeast, 18*(6), 523-31.

Ho, Y., Gruhler, A., Heilbut, A., Bader, G. D., Moore, L., Adams, S. L., et al. (2002). Systematic identification of protein complexes in *Saccharomyces cerevisiae* by mass spectrometry. *Nature*, *415*(6868), 180-3.

Hofmann, T. (2001). Unsupervised learning by probabilistic latent semantic analysis. *Machine Learning, 42*, 177-196.

Hoheisel, J. D. (2006). Microarray technology: beyond transcript profiling and genotype analysis. *Nat Rev Genet, 7*(3), 200-210.

Hsu, W. H., King, A. L., Paradesi, M. S. R., Pydimarri, T., & Weninger, T. (2006). Collaborative and Structural Recommendation of Friends using Weblog-based Social Network Analysis. *Proc. of Computational Approaches to Analyzing Weblogs* - AAAI 2006 Technical Report SS-06-03, (pp. 55-60).

Hu, P., Bader, G., et al. (2007). Computational prediction of cancer-gene function. *Nat Rev Cancer, 7*(1), 23-34.

Hu, P., Bader, G., Wigle, D. A., & Emili, A. (2007). Computational prediction of cancer-gene function. Nature Reviews Cancer, 7(1), 23-34.

Hu, Z, Snitkin, E.S, & DeLisi, C. (2008). VisANT: an integrative framework for networks in systems biology. *Briefings in Bioinformatics*, *9*(4), 317-25.

Hu, Z. et al. (2007). VisANT 3.0: new modules for pathway visualization, editing, prediction and construction. *Nucleic Acids Research*, (pp. 1-8).

Hu, Z., Mellor, J., Wu, J., & DeLisi C. (2004). VisANT: an online visualization and analysis tool for biological interaction data. *BMC Bioinformatics*, *5*, 17.

Hu, Z., Mellor, J., Wu, J., Kanehisa, M., Stuart, J., & DeLisi, C. (2007). Towards zoomable multidimensional maps of the cell. *Nature Biotechnology, 25*, 547-554.

Hu, Z., Mellor, J., Wu, J., Yamada, T., Holloway, D., & Delisi, C. (2005). VisANT: data-integrating visual framework for biological networks and modules. *Nucleic Acids Res, 33*(Web Server issue), W352-357.

Hu, Z., Ng, D. M., Yamada, T., Chen, C., Kawashima, S., Mellor, J., Linghu, B., Kanehisa, M., Stuart, J. M., & DeLisi, C. (2007). VisANT 3.0: New modules for pathway visualisation, prediction and construction. *Nucleic Acids Research, 35* (Webserver issue), W625-W632.

Huan, J., Wang, W., & Prins, J. (2003). Efficient mining of frequent subgraphs in the presence of isomorphism. In *Proc. 3rd IEEE international conference on data mining* (pp. 549–552). Washington, DC, USA: IEEE Computer Society Press.

Huan, J., Wang, W., Prins, J., & Yang, J. (2004). SPIN: Mining maximal frequent subgraphs from graph databases. In *Proc. 10th ACM international conference on knowledge discovery and data mining (SIGKDD)* (pp. 581–586). New York, USA: ACM Press.

Huang, B., & Schroeder, M. (2008). Using protein binding site prediction to improve protein docking. *Gene, 422*, 14-21.

Huang, H., et. al. (2007). Defining the specificity space of the human Src-homology 2 domain. *Molecular & Cellular Proteomics, 7*(4), 768-784.

Huang, H., Jedynak, B. M., & Bader, J. S. (2007). Where have all the interactions gone? Estimating the coverage of two-hybrid protein interaction maps. *PLoS Comput Biol, 3*(11), e214.

Huang, J. Y., & Brutlag, D. L. (2001). The EMOTIF database. *Nucleic Acids Res, 29*(1), 202-4.

Huang, T. W., Lin, C. Y., & Kao, C. Y. (2007). Reconstruction of human protein interolog network using evolutionary conserved network. *BMC Bioinformatics, 8*, 152.

Hughes, T. R., Marton, M. J., et al. (2000). Functional discovery via a compendium of expression profiles. *Cell, 102*(1), 109-26.

Huh, W. K., Falvo, J. V., Gerke, L. C., Carroll, A. S., Howson, R. W., Weissman, J. S., O'Shea, E. K. (2003). Global analysis of protein localization in budding yeast. *Nature*, *425*(6959), 686-91.

Hulo, N., et. al. (2008). *The 20 years of PROSITE. Nucleic Acids Research, Database issue 36*, D245-D249.

Human Interactome Project. (2008). Retrieved August 15, 2008, from http://interactome.dfci.harvard.edu/H_sapiens/

Hurst, L. D., & Smith, N. G. (1999). Do essential genes evolve slowly? *Curr Biol 9*(14), 747-50.

Hurst, L. D., Pal, C., & Lercher, M. J. (2004). The evolutionary dynamics of eukaryotic gene order. *Nat Rev Genet, 5*(4), 299-310.

Huynen, M. A., Snel, B., & Bork, P. (2001). Inversions and the dynamics of eukaryotic gene order. *Trends Genet, 17*(6), 304-306.

Huynen, M., Snel, B., et al. (2000). Predicting protein function by genomic context: quantitative evaluation and qualitative inferences. *Genome Res, 10*(8), 1204-10.

Hwang, H., Pierce, B., Mintseris, J., Janin, J., & Weng, Z. (2008). Protein-protein docking benchmark version 3.0. *Proteins.*

Hwang, W., Cho, Y.-R., Zhang, A., & Ramanathan, M. (2006). A novel functional module detection algorithm for protein-protein interaction networks. *Algorithms for Molecular Biology, 1*(24).

Hwang, W., Kim, T., Cho, Y.-R., Zhang, A., & Ramanathan, M. (2007). SIGN: reliable protein interaction identification by integrating the Similarity In GO and the similarity in protein interaction Networks. In Proceedings of the *7th IEEE International Conference on Bioinformatics and Bioengineering (BIBE'2007)*, (pp. 1384-1388).

Ideker, T. et al. (2001). Integrated Genomic and Proteomic Analyses of a Systematically Perturbed Metabolic Network. *Science, 292*, 929-933

Ideker, T., & Sharan, R. (2008). Protein networks in disease. *Genome Research, 18*(4), 644-652.

Ideker, T., Ozier, O., Schwikowski, B., & Siegel, A. F. (2002). Discovering regulatory and signalling circuits in molecular interaction networks. *Bioinformatics, 18*(Supplement 1), S233–S240.

Ideker, T., Ozier, O., Schwikowski, B., & Siegel, A.F. (2002). Discovering regulatory and signalling circuits in molecular interaction networks. *Bioinformatics. 18*(Suppl 1), S233-240.

Inokuchi, A. (2005). Mining generalized substructures from a set of labeled graphs. *Proceedings of the 4th IEEE Internatinal Conference on Data Mining* (pp. 415–418). IEEE Computer Society.

Iragne, F., Nikolski, M., Mathieu, B., Auber, D., & Sherman, D. (2005). ProViz: Protein interaction visualization and exploration. *Bioinformatics, 21*(2), 272-4.

Islinger, M., Lüers, G.H., Li, K.W., Loos, M.,& Völkl, A. (2007). Rat liver peroxisomes after fibrate treatment. A survey using quantitative mass spectrometry. *J Biol Chem. 282*(32), 23055-23069.

Ispolatov, I., Krapivsky, P. L., & Yuryev, A. (2005). Duplication-divergence model of protein interaction network. *Physical Review E, 71*, 061911.

Ispolatov, I., Yuryev, A., Mazo, I., & Maslov, S. (2005). Binding properties and evolution of homodimers in protein-protein interaction networks. *Nucleic Acids Res, 33*(11), 3629-3635.

Ito, T. et al. (2001). A comprehensive two-hybrid analysis to explore the yeast protein interactome. *Proc. Natl. Acad. Sci. 98*, 4569-4574.

Ito, T., Chiba, T., Ozawa, R., Yoshida, M., Hattori, M., & Sakaki, Y. (2001) A comprehensive two-hybrid analysis to explore the yeast protein interactome. *Proceedings of the National Academy of Sciences, 98*(8), 4277-8.

Ito, T., Chiba, T., Ozawa, R., Yoshida, M., Hattori, M., & Sakaki, Y. (2001). A comprehensive two-hybrid analysis to explore the yeast protein interactome. *Proc Natl Acad Sci U S A, 98*(8), 4569-4574.

Ito, T., Chiba, T., Ozawa, R., Yoshida, M., Hattori, M., & Sakaki, Y. (2001). A comprehensive twohybrid analysis to explore the yeast protein interactome. *Proceedings of the National Academy of Sciences of the USA, 98*(8), 4569-4574.

Ito, T., Chiba, T., Ozawa, R., Yoshida, M., Hattori, M., & Sakaki, Y. (2001). A comprehensive two-hybrid analysis to explore the yeast protein interactome. *Proc Natl Acad Sci U S A, 98*(8), 4569-4574.

Ito, T., Tashiro, K., Muta, S., Ozawa, R., Chiba, T., Nishizawa, M., et al. (2000). Toward a protein-protein interaction map of the budding yeast: A comprehensive system to examine two-hybrid interactions in all possible combinations between the yeast proteins. *Proc Natl Acad Sci U S A, 97*(3), 1143-1147.

Itzhaki, Z., Akiva, E., Altuvia, Y., & Margalit, H. (2006). Evolutionary conservation of domain-domain interactions. *Genome Biol, 7*(12), R125.

Itzhaki, Z., Akiva, E., Altuvia, Y., & Margalit, H. (2006). Evolutionary conservation of domain-domain interactions. *Genome Biology, 7*, R125.

Jaimovich, A., Elidan, G., Margalit, H., & Friedman, N. (2006). Towards an integrated protein-protein interaction network: A relational Markov network approach. *J Comput Biol, 13*(2), 145-164.

Jain., R. D. A. (1988). *Algorithms for Clustering Data.* Prentice Hall.

Janin, J., Rodier, F., Chakrabarti, P., & Bahadur, R. P. (2007). Macromolecular recognition in the Protein Data Bank. *Acta Crystallogr D Biol Crystallogr, 63*, 1-8.

Jannière, L., Canceill, D., Suski, C., Kanga, S., Dalmais, B., & Lestini, R. (2007). Genetic evidence for a link between glycolysis and DNA replication. *PLoS ONE 2*, e447.

Jansen, R. et al. (2002). Relating whole-genome expression data with protein-protein inetractions. *Genome Research, 12*, 37-46

Jansen, R., Greenbaum, D., & Gerstein, M. (2002). Relating whole-genome expression data with protein-protein interactions. *Genome Research, 12*(1), 37-46.

Jansen, R., Yu, H., Greenbaum, D., Kluger, Y., Krogan, N. J., Chung, S., et al. (2003). A Bayesian networks approach for predicting protein-protein interactions from genomic data. *Science, 302*(5644), 449-453.

Jansen, R., Yu, H., Greenbaum, D., Kluger, Y., Krogan, N. J., Chung, S., et al. (2003). A Bayesian networks approach for predicting protein-protein interactions from genomic data. *Science, 302*(5644), 449-453.

Jeong, H., Mason, S. P., Barabasi, A. L., & Oltvai, Z. N. (2001). Lethality and centrality in protein networks. *Nature, 411*(6833), 41-2.

Jeong, H., Oltvai, Z. N., & Barabási, A. L. (2003). Prediction of Protein Essentiality Based on Genomic Data. *Complexus, 1*(1), 19-28.

Johsson, P. F., & Bates, P. A. (2006). Global topological features of cancer proteins in the human interactome. *Bioinformatics, 22*(18), 2291-2297.

Jones, D. (2008). All systems go. *Nat Rev Drug Discov, 7*(4), 278-279.

Jones, S. & Thornton, J. M. (1997). Prediction of protein-protein interaction sites using patch analysis. *J Mol Biol, 272*, 133-143.

Jones, S., & Thornton, J. M. (1996). Principles of protein-protein interactions. *Proc Natl Acad Sci USA, 93*, 13-20.

Jones, S., & Thornton, J. M. (1997). Analysis of protein-protein interaction sites using surface patches. *J Mol Biol, 272*, 121-132.

Jones, S., & Thornton, J. M., (1996). Principles of protein-protein interactions. *Proceedings of the National Academy of Sciences U S A., 93*(1), 13-20.

Jordan, I. K., Wolf, Y. I., & Koonin, E. V. (2003). No simple dependence between protein evolution rate and the number of protein-protein interactions: Only the most prolific interactors tend to evolve slowly. *BMC Evol Biol, 3*, 1.

Joshi-Tope, G., Gillespie, M., Vastrik, I., D'Eustachio, P., Schmidt, E., de Bono, B., Jassal, B., Gopinath, G. R., Wu, G. R., Matthews, L., Lewis, S., Birney, E., Stein L. (2005). Reactome: a knowledgebase of biological pathways. *Nucleic Acids Research, 33*(Database issue), D428-32.

Jothi, R., Cherukuri, P. F., Tasneem, A., & Przytycka, T. M. (2006). Co-evolutionary analysis of domains in interacting proteins reveals insights into domain-domain interactions mediating protein-protein interactions. *J Mol Biol, 362*(4), 861-875.

Joy, M. P., Brock, A., Ingber, D. E., & Huang, S. (2005). High-Betweenness Porteins in the Yeast Protein Interaction Network. *Journal of Biomedicine and Biotechnology, 2*, 96-103

Joy, M. P., Brock, A., Ingber, D. E., & Huang, S. (2005). High-betweenness proteins in the yeast protein interaction network. *J Biomed Biotechnol, 2005*(2), 96-103.

Kafri, R., Dahan, O., Levy, J., & Pilpel, Y. (2008). Preferential protection of protein interaction network hubs in yeast: Evolved functionality of genetic redundancy. *Proc Natl Acad Sci U S A, 105*(4), 1243-1248.

Kalaev, M., Smoot, M., Ideker, T., & Sharan, R. (2008). NetworkBLAST: comparative analysis of protein networks. *Bioinformatics, 24*(4), 594-596.

Kaltenbach, L. S., Romero, E., Becklin, R. R., Chettier, R., Bell, R., Phansalkar, A., et al. (2007). Huntingtin interacting proteins are genetic modifiers of neurodegeneration. *PLoS Genet, 3*(5), e82.

Kamada, T., & Kawai, S. (1989). An algorithm for drawing general undirected graphs. *Information Processing Letters, 31*(1), 7-15

Kanehisa, M., & Goto, S. (2000). KEGG: kyoto encyclopedia of genes and genomes. *Nucleic Acids Res, 28*(1), 27-30.

Kanehisa, M., & Goto, S. (2000). KEGG: Kyoto Encyclopedia of Genes and Genomes. *Nucleic Acids Res, 28*, 27-30.

Kanehisa, M., Araki, M., Goto, S., Hattori, M., Hirakawa, M., Itoh, M., Katayama, T., Kawashima, S., Okuda, S., Tokimatsu, T., & Yamanishi, Y. (2008). KEGG for linking genomes to life and the environment. *Nucleic Acids Research, 36*(Database issue), D480-4.

Kanehisa, M., Goto, S., Kawashima, S., Okuno, Y., & Hattori, M. (2004). The KEGG resource for deciphering the genome. *Nucleic Acids Res., 32*, D277-D280.

Kann, M. G. (2007). Protein interactions and disease: computational approaches to uncover the etiology of diseases. *Brief Bioinform, 8*(5), 333-346.

Karaoz, U., Murali, T. M., et al. (2004). Whole-genome annotation by using evidence integration in functional-linkage networks. *Proc Natl Acad Sci U S A, 101*(9), 2888-93.

Karev, G. P., Wolf, Y. I, Rzhetsky, A. Y., Berezovskaya, F. S., & Koonon, E. V. (2002). Birth and death of protein domains: A simple model of evolution explains power law behavior. *BMC Evolutionary Biology, 2*, 18.

Kashtan, N., Itzkovitz, S., Milo, R., & Alon, U. (2004). Efficient sampling algorithm for estimating subgraph concentrations and detecting network motifs. *Bioinformatics, 20*(11), 1746–1758.

Kazius, J., Nijssen, S., Kok, J., & Ijzerman, T. B. A. (2006). Substructure mining using elaborate chemical representation. *J. Chem. Inf. Model., 46*, 597–605.

KEGG PATHWAY: Wnt signaling pathway – Homo sapiens (human). (2008). Retrieved August 15, 2008, from http://www.genome.ad.jp/kegg/pathway/hsa/hsa04310.html

Kelley, B. P., Sharan, R., Karp, R. M., Sittler, T., Root, D. E., Stockwell, B. R., & Ideker, T. (2003). Conserved pathways within bacteria and yeast as revealed by global protein network alignment. *Proceedings of the National Academy of Sciences of the United States of America, 100*(20), 11394-11399.

Kelley, B. P., Yuan, B., Lewitter, F., Sharan, R., Stockwell, B. R., & Ideker, T. (2004). PathBLAST: a tool for alignment of protein interaction networks. *Nucleic Acids Res, 32*(Web Server issue), W83-8.

Kelley, R., & Ideker, T. (2005). Systematic interpretation of genetic interactions using protein networks. *Nature Biotechnology, 23*, 561-566.

Kelley, R., & Ideker, T. (2005). Systematic interpretation of genetic interactions using protein networks. *Nat Biotechnol, 23*(5), 561-566.

Kellis, M., Birren, B. W., & Lander, E. S. (2004). Proof and evolutionary analysis of ancient genome duplication in the yeast saccharomyces cerevisiae. *Nature, 428*(6983), 617-624.

Kemp, H. A., & Sprague, G. F. (2003). Far3 and five interacting proteins prevent premature recovery from pheromone arrest in the budding yeast *Saccharomyces cerevisiae. Molecular and Cellular Biology, 5*(23), 1750-63.

Kerppola, T. K. (2006). Complementary methods for studies of protein interactions in living cells. *Nat Methods, 3*(12), 969-971.

Kerrien, S. et al. (2006). IntAct – Open Source Resource for Molecular Interaction Data. *Nucleic Acids Research, 2006.*

Kerrien, S., Alam-Faruque, Y., Aranda, B., Bancarz, I., Bridge, A., Derow, C., et al. (2007). IntAct - open-source resource for molecular interaction data. *Nucleic Acids Research, 35*(Database Issue). D561-5.

Kerrien, S., Alam-Faruque, Y., Aranda, B., Bancarz, I., Bridge, A., Derow, C., et al. (2007). IntAct--open source resource for molecular interaction data. *Nucleic Acids Res, 35*(Database issue), D561-565.

Kerrien, S., Orchard, S., Montecchi-Palazzi, L., Aranda, B., Quinn, A. F., Vinod, N., et al. (2007). Broadening the horizon - level 2.5 of the HUPO-PSI format for molecular interactions. *BMC Biology, 5*, 44.

Kersey, P. J., Duarte, J., et al. (2004). The International Protein Index: an integrated database for proteomics experiments. *Proteomics, 4*(7), 1985-8.

Khan, S., Situ, G., et al. (2003). GoFigure: automated Gene Ontology annotation. *Bioinformatics, 19*(18), 2484-5.

Khanin, R., & Wit, E. (2006). How scale-free are biological networks. *Journal of Computational Biology, 13*(3), 810-818.

Kiemer, L., & Cesareni, G. (2007). Comparative interactomics: Comparing apples and pears? *Trends Biotechnol, 25*(10), 448-454.

Kihara, D., & Skolnick, J. (2004). Microbial genomes have over 72% structure assignment by the threading algorithm PROSPECTOR_Q. *Proteins, 55*, 464-473.

Kihara, D., Lu, H., Kolinski, A., & Skolnick, J. (2001). TOUCHSTONE: An ab initio protein structure prediction method that uses threading-based tertiary restraints. *Proc. Natl. Acad. Sci U. S. A, 98*, 10125-10130.

Kihara, D., Zhang, Y., Lu, H., Kolinski, A., & Skolnick, J. (2002). Ab initio protein structure prediction on a genomic scale: application to the Mycoplasma genitalium genome. *Proc Natl Acad Sci USA, 99*, 5993-5998.

Kim, I., Liu, Y., & Zhao, H. (2007). Bayesian methods for predicting interacting protein pairs using domain information. *Biometrics, 63*, 824-833.

Kim, J., & Dang, C. V. (2005). Multifaceted roles of glycolytic enzymes. *Trends Biochem Sci, 30*, 142–150.

Kim, P. M., Lu, L. J., Xia, Y., & Gerstein, M. B. (2006). Relating three-dimensional structures to protein networks provides evolutionary insights. *Science, 314*, 1938-1941.

Kim, W. K., Park, J., & Suh, J. K. (2002). Large scale statistical prediction of protein-protein interaction by potentially interacting domain (PID) pair. *Genome Informatics, 13*, 42-40.

Kimura, M., & Ota, T. (1974). On some principles governing molecular evolution. *Proc Natl Acad Sci U S A., 71*, 2848-52.

King, A. D., Przulj, N., & Jurisica, I. (2004). Protein complex prediction via cost-based clustering. *Bioinformatics, 20*(17), 3013-2020.

King, A. D., Przulj, N., & Jurisica, I. (2004). Protein complex prediction via cost-based clustering. *Bioinformatics, 20*(17), 3013–3020.

Kirac, M., & Ozsoyoglu, G. (2008). Protein Function

Prediction Based on Patterns in Biological Networks. *Research in Computational Molecular Biology,* (pp. 197-213).

Kirkpatrick, S., Gelatt, C. D., & Vecchi, M. P. (1983). Optimization by simulated annealing. *Science, 220*(4598), 671-680.

Kitano, H. (2002). Computational systems biology. *Nature, 420*(6912), 206-10.

Klipp, E., Liebermeister, W., Helbig, A., Kowald, A., & Schaber, J. (2007). Systems biology standards - the community speaks. *Nature Biotechnology, 25*(4), 390-391.

Kobayashi, K., Ehrlich, S. D., Albertini, A., Amati, G., Andersen, K. K., Arnaud, M., et al. (2003). Essential *Bacillus subtilis* genes. *Proc Natl Acad Sci U S A, 100*(8), 4678-83.

Kola, I. & Landis, J. (2004). Can the pharmaceutical industry reduce attrition rates? *Nat Rev Drug Discov 3*(8), 711-716.

Komander, D., & Barford, D. (2008). Structure of the A20 OTU domain and mechanistic insights into deubiquitination. *Biochem J, 409*(1), 77-85.

Komurov, K., & White, M. (2007). Revealing static and dynamic modular architecture of the eukaryotic protein interaction network. *Molecular Systems Biology, 3*, 110.

Kondrashov, F. A., Ogurtsov, A. Y., & Kondrashov, A. S. (2004). Bioinformatical assay of human gene morbidity. Nucleic Acids Res, *32*(5), 1731-7.

Krishnadev, O., & Srinivasan, N. (2008). A data integration approach to predict host-pathogen protein-protein interactions: application to recognize protein interactions between human and a malarial parasite. *In Silico Biology, 8*, 0020.

Krogan, N. J., Cagney, G. et al. (2006) Global landscape of protein complexes in the yeast saccharomyces cerevisiae. *Nature, 440*(7084), 637–643.

Krogan, N. J., Cagney, G., Yu, H., Zhong, G., Guo, X., Ignatchenko, A., et al. (2006). Global landscape of protein complexes in the yeast saccharomyces cerevisiae. *Nature, 440*(7084), 637-643.

Krogan, N. J., Cagney, G., Yu, H., Zhong, G., Guo, X., Ignatchenko, A., et al. (2006). Global landscape of protein complexes in the yeast Saccharomyces cerevisiae. *Nature, 440*(7084), 637-643.

Krylov, D. M., Wolf, Y. I., Rogozin, I. B., & Koonin, E. V. (2003). Gene loss, protein sequence divergence, gene dispensability, expression level, and interactivity are correlated in eukaryotic evolution. *Genome Res, 13*(10), 2229-2235.

Kufareva, I., Budagyan, L., Raush, E., Totrov, M., & Abagyan, R. (2007). PIER: Protein interface recognition for structural proteomics. *Proteins, 67*, 400-417.

Kulikova, T., Akhtar, R., et al. (2007). EMBL Nucleotide Sequence Database in 2006. *Nucleic Acids Res, 35*(Database issue), D16-20.

Kumar, R., Musiyenko, A., & Barik, S. (2005). Plasmodium falciparum calcineurin and its association with heat shock protein 90: mechanisms for the antimalarial activity of cyclosporin A and synergism with geldanamycin. *Mol Biochem Parasitol., 141*(1).

Kumar, R., Musiyenko, A., Oldenburg, A., Adams, B., & BMC, B. S. (2004). Post-translational generation of constitutively active cores from larger phosphatases in the malaria parasite, Plasmodium falciparum: implications for proteomics. *Mol Biol., 5*(6).

Kunin, V., Pereira-Leal, J. B., & Ouzounis, C. A. (2004). Functional evolution of the yeast protein interaction network. *Mol Biol Evol, 21*(7), 1171-1176.

Kurland, I. J., & Pilkis, S. J. (1995). Covalent control of 6-phosphofructo-2-kinase/bifunctional enzyme fructose-2,6-bisphosphatase: Insights into autoregulation of a bifunctional enzyme. *Prorein Sci., 4*, 1023-1027.

Kurochkin, I. V., Mizuno, Y., Konagaya, A., Sakaki, Y., Schönbach, C., & Okazaki, Y. (2007). Novel peroxisomal protease Tysnd1 processes PTS1- and PTS2-containing enzymes involved in beta-oxidation of fatty acids. *EMBO J, 26*(3), 835-845.

Kyte, J., & Doolittle, R. F. (1982). A simple method for displaying the hydropathic character of a protein. *J Mol Biol, 157*, 105-132.

LaCount, D. J., Vignali, M., Chettier, R., Phansalkar, A., Bell, R., Hesselberth, J. R., et al. (2005). A protein interaction network of the malaria parasite plasmodium falciparum. *Nature, 438*(7064), 103-107.

Lacroix, V., Fernandes, C. G., & Sagot, M.-F. (2006). Motif search in graphs: Application to metabolic networks. *IEEE/ACM Transactions on Computational Biology and Bioinformatics, 3*(4), 360–368.

Lage, K., Karlberg, E. O., Storling, Z. M., Olason, P. I., Pedersen, A. G., Rigina, O., et al. (2007). A human phenome-interactome network of protein complexes implicated in genetic disorders. *Nat Biotechnol, 25*(3), 309-316.

Lage, K., Karlberg, E. O., Størling, Z. M., Olason, P. I., Pedersen, A. G., Rigina, O., Hinsby, A. M., Tümer, Z., Pociot, F., Tommerup, N., Moreau, Y., & Brunak, S. (2007). A human phenome-interactome network of protein complexes implicated in genetic disorders. *Nature Biotechnology, 25*(3), 309-16.

Lalonde, S., Ehrhardt, D. W., Loque, D., Chen, J., Rhee, S. Y., & Frommer, W. B. (2008). Molecular and cellular approaches for the detection of protein-protein interactions: latest techniques and current limitations. *Plant J, 53*(4), 610-635.

Lan, J., & Steiner, R. F. (1991). The interaction of troponin C with phosphofructokinase comparision with calmodulin. *Biochem. J., 274*, 445-451.

Lanckriet, G. R., Deng, M., et al. (2004). Kernel-based data fusion and its application to protein function prediction in yeast. *Pac Symp Biocomput*, (pp. 300-11).

Landau, M., Mayrose, I., Rosenberg, Y., Glaser, F., Martz, E., Pupko, T., & Ben-Tal, N. (2005). ConSurf 2005: The projection of evolutionary conservation scores of residues on protein structures. *Nucleic Acids Res, 33*, W299-W302.

Laskowski, R. A., Watson, J. D., et al. (2005). ProFunc: a server for predicting protein function from 3D structure. *Nucleic Acids Res, 33*(Web Server issue), W89-93.

Laskowski, R. A., Watson, J. D., et al. (2005). Protein function prediction using local 3D templates. *J Mol Biol, 351*(3), 614-26.

Lasonder, E., Ishihama, Y., Andersen, J., Vermunt, A., Pain, A., Sauerwein, R., et al. (2002). Analysis of the Plasmodium falciparum proteome by high-accuracy mass spectrometry. *Nature, 419*(6906), 537-542.

Leach, S., Gabow, A., Hunter, L., & Goldberg, D. S. (2007). Assessing and combining reliability of protein interaction sources. *Pac Symp Biocomput*, (pp. 433-444).

Lécuyer, E. et. al. (2007). Global analysis of mRNA localization reveals a prominent role in organizing cellular architecture and function. *Cell, 131*(1), 174-187.

Lee et al. (2007). A single gene network accurately predicts phenotypic effects of gene perturbation in Caenorhabditis elegans. *Nature Genetics, 40*(2), 181-188

Lee, H., Deng, M., Sun, F., & Chen, T. (2006). An integrated approach to the prediction of domain-domain interactions. *BMC Bioinformatics, 7*, 269-284.

Lee, I., Date, S. V., Adai, A. T., & Marcotte, E. M. (2004). A probabilistic functional network of yeast genes. *Science, 306*(5701), 1555-1558.

Lee, I., Li, Z., & Marcotte, E. M. (2007). An improved, bias-reduced probabilistic functional gene network of baker's yeast, Saccharomyces cerevisiae. *PLoS ONE, 2*(10), e988.

Legrain, P., Wojcik, J., & Gauthier, J. M. (2001). Protein--protein interaction maps: a lead towards cellular functions. *Trends Genet, 17*(6), 346-352.

Lercher, M. J., Urrutia, A. O., & Hurst, L. D. (2002). Clustering of housekeeping genes provides a unified model of gene order in the human genome. *Nat Genet, 31*(2), 180-183.

Leskovec, J., & Faloutsos, C. (2006). Sampling from large graphs. In *Proc. 12th ACM International Conference on*

Knowledge Discovery and Data Mining (SIGKDD) (pp. 631–636). New York, USA: ACM Press.

Letovsky, S., & Kasif, S. (2003). Predicting protein function from protein/protein interaction data: a probabilistic approach. *Bioinformatics, 19*(Suppl 1)i197-204.

Leung, H.C., et. al. (2007). Finding linear motif pairs from protein interaction networks: a probabilistic approach. *Computational Systems Bioinformatics Conference, 6*, 111-119.

Levy, E. D., & Pereira-Leal, J. B. (2008). Evolution and dynamics of protein interactions and networks. *Curr Opin Struct Biol, 18*(3), 349-357.

Li, B., Turuvekere, S., Agrawal, M., La, D., Ramani, K., & Kihara, D. (2008). Characterization of local geometry of protein surfaces with the visibility criterion. *Proteins, 71*, 670-683.

Li, H. et. al. (2004). Discovery of binding motif pairs from protein complex structural data and protein interaction sequence data. *Pacific Symposium on Biocomputing*, (pp. 312-323).

Li, H. et. al. (2006). Discovering motif pairs at interaction sites from protein sequences on a proteome-wide scale. *Bioinformatics, 22*(8), 989-996.

Li, H., & Li, J. (2005). Discovery of stable and significant binding motif pairs from PDB complexes and protein interaction datasets. *Bioinformatics, 21*(3), 314-324.

Li, M., Makkinje, A., & Damuni, Z. (1996). Molecular identification of I1PP2A, a novel potent heat-stable inhibitor protein of protein phosphatase 2A. *Biochemistry, 35*(22), 6998-7002.

Li, S. (2004). A map of the interactome network of the metazoan C. elegans. *Science, 303,* 540-543

Li, S. Z. (1995). *Markov Random Field Modeling in Computer Vision*. Springer-Verlag New York, Inc.

Li, S., Armstrong, C. M., Bertin, N., Ge, H., Milstein, S., Boxem, M., & Vidalain, P. O. (2004). A map of the interactome network of the metazoan C.elegans. *Science, 303*, 540-543.

Li, S., Armstrong, C. M., Bertin, N., Ge, H., Milstein, S., Boxem, M., et al. (2004). A map of the interactome network of the metazoan c. Elegans. *Science, 303*(5657), 540-543.

Li, X.-L, Foo, C. S., & Ng, S.-K. (2007) Discovering protein complexes in dense reliable neighbourhoods of protein interaction networks. In *Proceedings of IEEE Computational Systems Bioinformatics Conference (CSB07)* (pp. 157-168).

Li, X.-L., Tan, S.-H., & Ng, S.-K. (2005). Interaction Graph Mining for Protein Complexes Using Local Clique Merging. *Genome Informatics,* (16), 260.

Li, X-L., Tan, S-H., & Ng, S-K. (2006). Improving domain-based protein interaction prediction using biologically-significant negative dataset. *International Journal of Data Mining and Bioinformatics, 1*, 138-149.

Liang, S., Zhang, C., Liu, S., & Zhou, Y. (2006). Protein binding site prediction using an empirical scoring function. *Nucleic Acids Res, 34*, 3698-3707.

Liang, S., Zhang, J., Zhang, S., & Guo, H. (2004). Prediction of the interaction site on the surface of an isolated protein structure by analysis of side chain energy scores. *Proteins 57*, 548-557.

Liang, Z., Xu, M., Teng, M., & Niu, L. (2006). Comparison of protein interaction networks reveals species conservation and divergence. *BMC Bioinformatics, 7*, 457.

Liao, B. Y., & Zhang, J. (2008). Null mutations in human and mouse orthologs frequently result in different phenotypes. *Proc Natl Acad Sci U S A, 105*(19), 6987-6992.

Liao, B. Y., Scott, N. M., & Zhang, J. (2006). Impacts of gene essentiality, expression pattern, and gene compactness on the evolutionary rate of mammalian proteins. *Mol Biol Evol, 23*(11), 2072-80.

Licamele, K., & Getoor, L. (2006). Predicting Protein-Protein Interactions Using Relational Features. *Proc. of ICML Workshop on Statistical Network Analysis.*

Lichtarge, O., Bourne, H. R., & Cohen, F. E. (1996). An evolutionary trace method defines binding surfaces common to protein families. *J Mol Biol, 257*, 342-358.

Lichtarge, O., Bourne, H. R., & Cohen, F. E. (1996). Evolutionarily conserved Galphabetagamma binding surfaces support a model of the G protein-receptor complex. *Proc Natl Acad Sci USA, 93*, 7507-7511.

Lim, J., Hao, T., Shaw, C., Patel, A. J., Szabo, G., Rual, J. F., et al. (2006). A protein-protein interaction network for human inherited ataxias and disorders of Purkinje cell degeneration. *Cell, 125*(4), 801-814.

Lin, C., Cho, Y., Hwang, W., Pei, P., & Zhang, A. (2006) Clustering Methods in Protein-Protein Interaction Network. *Knowledge Discovery in Bioinformatics: Techniques, Methods and Application.* John Wiley & Sons, Inc.

Lin, N., Wu, B., Jansen, R., Gerstein, M., & Zhao, H. (2004). Information assessment on predicting protein-protein interactions. *BMC Bioinformatics, 5*, 154.

Liolios, K., Mavromatis, K., et al. (2008). The Genomes On Line Database (GOLD) in 2007: status of genomic and metagenomic projects and their associated metadata. *Nucleic Acids Res, 36*(Database issue), D475-9.

Litzkow, M., Livny, M., & Mutka, M. W. (1988). Condor - A Hunter of Idle Workstations. *Proceedings of the 8th IEEE Distributed Computing Conference*, (San Jose, CA) (pp. 104-111).

Liu, B., Hsu, W., & Ma, Y. (1998). Integrating classification and association rule mining. *Proceedings of the Fourth International Conference on Knowledge Discovery and Data Mining (KDD-98)* (pp. 80–86).

Liu, H., & Motoda, H. (1998). *Feature Selection for Knowledge Discovery and Data Mining.* Kluwer.

Liu, M., Liberzon, A., Kong, S. W., Lai, W. R., Park, P. J., Kohane, I. S., & Kasif, S. (2007). Network-based analysis of affected biological processes in type 2 diabetes models. *PLoS Genetics, 3*(6), e96.

Lo Conte, L., Chothia, C., & Janin, J. (1999). The atomic structure of protein-protein recognition sites. *J Mol Biol, 285*, 2177-2198.

Loscalzo, J., Kohane, I., & Barabasi, A. L. (2007). Human disease classification in the postgenomic era: a complex systems approach to human pathobiology. *Mol Syst Biol, 3*, 124.

Lowe, H. J., & Barnett, G. O. (1994). Understanding and using the medical subject headings (MeSH) vocabulary to perform literature searches. *Jama, 271*(14), 1103-1108.

Lu, H., Shi, B., Wu, G., Zhang, Y., Zhu, X., Zhang, Z., Liu, C., Zhao, Y., Wu, T., Wang, J., & Chen, R. (2006). Integrated analysis of multiple data sources reveals modular structure of biological networks. *Biochemical and Biophysical Research Communications, 345*(1), 302-309.

Lu, L. J., Xia, Y., Paccanaro, A., Yu, H., & Gerstein, M. (2005). Assessing the limits of genomic data integration for predicting protein networks. *Genome Res, 15*(7), 945-953.

Lubovac, Z., Gamalielsson, J., & Olsson, B. (2006). Combining functional and topological properties to identify core modules in protein interaction networks. *Proteins, 64*(4), 948-959.

Luo, F. et al. (2006). Modular organization of protein interaction networks. *Bioinformatics, 23*(2), 207-214 .

Luo, W., & Lin, S. C. (2004). Axin: a master scaffold for multiple signaling pathways. *Neurosignals, 13*(3), 99-113.

Luscombe, N.M., Babu, M.M. et al. (2004). Genomic analysis of regulatory network dynamics reveal large topological changes. *Nature, 431*, 308-312.

MacBeath, G., & Schreiber, S. L. (2000). Printing proteins as microarrays for high-throughput function determination. *Science*; *289*(5485), 1760-3.

Mackay, J. P., Sunde, M., Lowry, J. A., Crossley, M., & Matthews, J. M. (2007). Protein interactions: is seeing believing? *Trends Biochem Sci, 32*(12), 530-531.

Madeira, S. C., & Oliveira, A. L. (2004). Biclustering algorithms for biological data analysis: A survey. *IEEE Transactions on Computational Biology and Bioinformatics, 1*(1), 24-45.

Maere, S., Heymans, K., & Kuiper, M. (2005). BiNGO: a Cytoscape plugin to assess overrepresentation of Gene Ontology categories in biological networks. *Bioinformatics. 21.* 3448-3449.

Maere, S., Heymans, K., & Kuiper, M. (2005). BiNGO: a Cytoscape plugin to assess overrepresentation of gene ontology categories in biological networks. *Bioinformatics, 21*(16), 3448-9.

Mahdavi, M. A., & Lin, Y. H. (2007). False positive reduction in protein-protein interaction predictions using gene ontology annotations. *BMC Bioinformatics, 8*, 262.

Makino, T., & Gojobori, T. (2006). The evolutionary rate of a protein is influenced by features of the interacting partners. *Mol Biol Evol, 23*(4), 784-789.

Makino, T., & McLysaght, A. (2008). Interacting gene clusters and the evolution of the vertebrate immune system. *Mol Biol Evol, 25*(9), 1855-1862.

Makino, T., Suzuki, Y., & Gojobori, T. (2006). Differential evolutionary rates of duplicated genes in protein interaction network. *Gene, 385*, 57-63.

Mamitsuka, H. (2005). Essential latent knowledge for protein interactions: Analysis by an unsupervised learning approach. *IEEE/ACM Transactions on Computational Biology and Bioinformatics, 2*(2), 119-130.

Maraziotis, I. A., Dimitrakopoulou, K., & Bezerianos, A. (2007). Growing functional modules from a seed protein via integration of protein interaction and gene expression data. *BMC Bioinformatics, 8*, 408.

Marcatili, P., Bussotti, G., & Tramontano, A. (2008). The MoVIN server for the analysis of protein interaction networks. *BMC Bioinformatics, 9 Suppl 2*, S11.

Marcotte, E. M., Pellegrini, M., et al. (1999). Detecting protein function and protein-protein interactions from genome sequences. *Science, 285*(5428), 751-3.

Martin, D. M., Berriman, M., et al. (2004). GOtcha: a new method for prediction of protein function assessed by the annotation of seven genomes. *BMC Bioinformatics, 5*, 178.

Maslov, S., & Sneppen, K. (2002). Specificity and stability in topology of protein networks. *Science, 296*(5569), 910-3.

Mathivanan, S. et al. (2006). An evaluation of human protein-protein inetraction data in the public domain. *BMC Bioinformatics, 7*(5), S19.

Matthews, L. R., Vaglio, P., Reboul, J., Ge, H., Davis, B. P., Garrels, J., et al. (2001). Identification of potential interaction networks using sequence-based searches for conserved protein-protein interactions or "interologs". *Genome Res, 11*(12), 2120-2126.

Mattingly, C. J., Rosenstein, M. C., Colby, G. T., Forrest, J. N., Jr., & Boyer, J. L. (2006). The Comparative Toxicogenomics Database (CTD): a resource for comparative toxicological studies. *J Exp Zoolog A Comp Exp Biol, 305*(9), 689-692.

Matys, V., Fricke, E., Geffers, R., Gossling, E., Haubrock, M., Hehl, R., et al. (2003). TRANSFAC: transcriptional regulation, from patterns to profiles. *Nucleic Acids Res, 31*(1), 374-378.

McCarroll, S. A., & Altshuler, D. M. (2007). Copy-number variation and association studies of human disease. *Nat Genet, 39*(7 Suppl), S37-42.

McCaskill, J. S. (1990). The equilibrium partition function and base pair binding probabilities for RNA secondary structure. *Biopolymers, 29*, 1105 1119.

McKay, B. D. (1981). Practical graph isomorphism. *Congressus Numerantium*, (pp. 45–87).

McNabb, D. S., & Pinto, I. (2005). Assembly of the Hap2p/Hap3p/Hap4p/Hap5p-DNA complex in Saccharomyces cerevisiae. *Eukaryot Cell, 4*(11), 1829-1839.

Mellor, J. C., Yanai, I., Clodfelter, K. H., Mintseris, J., & DeLisi, C. (2002). Predictome: A database of putative functional links between proteins. *Nucleic Acids Research, 30*(1), 306-9.

Mewes, H. W., et. al. (2008). MIPS: analysis and annotation of genome information in 2007. *Nucleic Acids Research, Database issue 36*, D196-D201.

Mewes, H. W., Frishman, D., Güldener, U., Mannhaupt, G., Mayer, K., Mokrejs, M., Morgenstern, B., Münsterkötter, M., Rudd, S., & Weil, B. (2002). MIPS: A database for genomes and protein sequences. *Nucleic Acids Res, 30*(1), 31-4.

Mewes, H. W., Frishman, D., Mayer, K. F. X., Münsterkötter, M., Noubibou, O., Pagel, P., Rattei T., Oesterheld, M., Ruepp, A., & Stümpflen, V. (2006). MIPS: Analysis and annotation of proteins from whole genomes in 2005. *Nucleic Acids Research, 34*(Database Issue), D169–D172.

Mewes, H. W., Heumann, K., Kaps, A., Mayer, K., Pfeiffer, F., Stocker, S., & Frishman, D. (1999). MIPS: A database for genomes and protein sequences. *Nucleic Acids Res, 27,* 44–48.

Middendorf, M., Ziv, E., & Wiggins, C. H. (2005). Inferring network mechanisms: the drosophila melanogaster protein interaction network. *Proc Natl Acad Sci U S A, 102*, 3192–3197.

Miernyk, J. A., & Thelen, J. J. (2008). Biochemical approaches for discovering protein-protein interactions. *Plant J, 53*(4), 597-609.

Mihalek, I., Res, I., & Lichtarge, O. (2007). On itinerant water molecules and detectability of protein-protein interfaces through comparative analysis of homologues. *J Mol Biol, 369*, 584-595.

Mihalek, I., Res, I., Yao, H., & Lichtarge, O. (2003). Combining inference from evolution and geometric probability in protein structure evaluation. *J Mol Biol, 331*, 263-279.

Mika, S., & Rost, B. (2006). Protein-protein interactions more conserved within species than across species. *PLoS Comput Biol, 2*(7), e79.

Miller, J. A., Oldham, M. C., & Geschwind, D. H. (2008). A systems level analysis of transcriptional changes in Alzheimer's disease and normal aging. *J Neurosci, 28*(6), 1410-1420.

Miller, J. P., Lo, R. S., Ben-Hur, A., Desmarais, C., Stagljar, I., Stafford Noble, W., & Fields, S. (2005) Large-scale identification of yeast integral membrane protein interactions. *Proceedings of the National Academy of Sciences of the USA, 102*(34), 12123-12128.

Miller, L. H., Baruch, D. I., Marsh, K., & Doumbo, O. K. (2002). The pathogenic basis of malaria. *Nature, 415*, 673-679.

Miller, S., Janin, J., Lesk, A. M., & Chothia, C. (1987). Interior and surface of monomeric proteins. *J Mol Biol, 196*, 641-656.

Milo, R., Itzkovitz, S., Kashtan, N., Levitt, R., & Alon, U. (2004, August). Response to comment on 'Network motifs: Simple building blocks of complex networks' and 'Superfamilies of evolved and designed networks'. *Science, 305*(5687), 1107d.

Milo, R., Itzkovitz, S., Kashtan, N., Levitt, R., Shen-Orr, S., Ayzenshtat, I., et al. (2004,March). Superfamilies of evolved and designed networks. *Science, 303*(5663), 1538–1542.

Milo, R., Kashtan, N., Itzkovitz, S., Newman, M. E. J., & Alon, U. (2003). On the uniform generation of random graphs with prescribed degree sequences. *cond-mat, 0312028*. Available from http://arxiv.org/abs/cond-mat/0312028

Milo, R., Shen-Orr, S., Itzkovitz, S., Kashtan, N., Chklovskii, D., & Alon, U. (2002, October). Network motifs: Simple building blocks of complex networks. *Science, 298*(5594), 824–827. (Software: http://www.weizmann.ac.il/mcb/UriAlon/groupNetworkMotifSW.html)

Milo, R., Shen-Orr, S., Itzkovitz, S., Kashtan, N., Chklovskii, D., & Alon, U. (2002). Network motifs: simple building blocks of complex networks. *Science, 298*, 824–827.

Mintseris, J., & Weng, Z. (2005). Structure, function, and evolution of transient and obligate protein-protein interactions. *Proc Natl Acad Sci U S A, 102*(31), 10930-10935.

Mintseris, J., & Weng, Z. (2005). Structure, function, and evolution of transient and obligate protein-protein interactions. *Proc Natl Acad Sci USA, 102*, 10930-10935.

Mintseris, J., Wiehe, K., Pierce, B., Anderson, R., Chen, R., Janin, J., & Weng, Z. (2005). Protein-Protein Docking Benchmark 2.0: an update. *Proteins, 60*, 214-216.

Mirny, L. A., & Shakhnovich, E. I. (1999). Universally conserved positions in protein folds: reading evolutionary signals about stability, folding kinetics and function. *J Mol Biol, 291*, 177-196.

Mishra, G. R., Suresh, M., Kumaran, K., Kannabiran, N., Suresh, S., Bala, P., et al. (2006). Human Protein Reference Database—2006 update. *Nucleic Acids Research, 34*(Database issue), D411-4.

Mitchell, T. (1997). *Machine Learning*. McGraw Hill.

Mootha, V. K., Lepage, P., Miller, K., Bunkenborg, J., Reich, M., Hjerrild, M., et al. (2003). Identification of a gene causing human cytochrome c oxidase deficiency by integrative genomics. *Proc Natl Acad Sci U S A, 100*(2), 605-610.

Mortazavi, A., Williams, B. A., McCue, K., Schaeffer, L., & Wold, B. (2008). Mapping and quantifying mammalian transcriptomes by RNA-Seq. *Nat Methods, 5*(7), 621-628.

Mrowka, R., Patzak, A., & Herzel, H. (2001). Is there a bias in proteome research? *Genome Research, 11*, 1971-1973

Mrowka, R., Patzak, A., & Herzel, H. (2001). Is there a bias in proteome research? *Genome Res, 11*(12), 1971-1973.

Mulder, N. J., Apweiler, R., Attwood, T. K., Bairoch, A., Bateman, A., Binns, D., et al. (2007). New developments in the InterPro database. *Nucleic Acids Res, 35*(Database issue), D224-228.

Murakami, Y., & Jones, S. (2006). SHARP2: protein-protein interaction predictions using patch analysis. *Bioinformatics, 22*, 1794-1795.

Murali, T. M., Wu, C. J., et al. (2006). The art of gene function prediction. *Nat Biotechnol, 24(*12), 1474-5; author reply 1475-6.

Murga, L. F., Ondrechen, M. J., & Ringe, D. (2008). Prediction of interaction sites from apo 3D structures when the holo conformation is different. *Proteins*.

Musso, G., Costanzo, M., Huangfu, M., Smith, A. M., Paw, J., San Luis, B. J., et al. (2008). The extensive and condition-dependent nature of epistasis among whole-genome duplicates in yeast. *Genome Res, 18*(7), 1092-1099.

Musso, G., Zhang, Z., & Emili, A. (2007). Retention of protein complex membership by ancient duplicated gene products in budding yeast. *Trends Genet, 23*(6), 266-269.

Nabieva, E., Jim, K., et al. (2005). Whole-proteome prediction of protein function via graph-theoretic analysis of interaction maps. *Bioinformatics, 21(*Suppl 1), i302-10.

Nacher, J. C., Hayashida, M., & Akutsu, T. (2006). Protein domain networks: Scale-free mixing of positive and negative exponents. *Physica A*, *367*, 538-552.

Nacher, J. C., Hayashida, M., & Akutsu, T. (2007). Topological aspects of protein networks. *Studies in Computational Intelligence*, *56*, 147-158.

Nacu, S., Critchley-Thorne, R., Lee, P., & Holmes, S. (2007). Gene expression network analysis and applications to immunology. *Bioinformatics*, *23*(7), 850-858.

Najafabadi, H. S, & Salavati, R. (2008). Sequence-based prediction of protein-protein interactions by means of codon usage. *Genome Biology*, *9*, R87.

Nash, R., Weng, S., Hitz, B., Balakrishnan, R., Christie, K. R., Costanzo, M. C., et al. (2007). Expanded protein information at SGD: new pages and proteome browser. *Nucleic Acids Res, 35*(Database issue), D468-471.

Neduva, V., & Russell, R. B. (2006). DILIMOT: discovery of linear motifs in proteins. *Nucleic Acids Res, 34*(Web Server issue), W350-355.

Neduva, V., & Russell, R. B. (2006). Peptides mediating interaction networks: new leads at last. *Curr Opin Biotechnol, 17*(5), 465-471.

Neduva, V., & Russell, R. B., (2006). Peptides mediating interaction networks: new leads at last. *Current Opinion in Biotechnology*, *17*(5), 465-471.

Negi, S. S., & Braun, W. (2007). Statistical analysis of physical-chemical properties and prediction of protein-protein interfaces. *Journal of molecular modeling, 13*, 1157-1167.

Negi, S. S., Schein, C. H., Oezguen, N., Power, T. D., & Braun, W. (2007). InterProSurf: a Web server for predicting interacting sites on protein surfaces. *Bioinformatics, 23*, 3397-3399.

Neuvirth, H., Heinemann, U., Birnbaum, D., Tishby, N., & Schreiber, G. (2007). ProMateus--an open research approach to protein-binding sites analysis. *Nucleic Acids Res, 35*, W543-W548.

Neuvirth, H., Raz, R., & Schreiber, G. (2004). ProMate: a structure based prediction program to identify the location of protein-protein binding sites. *J Mol Biol, 338*, 181-199.

Newman, A., Hunter, J., Li, Y. F., Bouton, C., & Davis, M. (2008). BioMANTA ontology: the integration of protein-protein interaction data. In *Interdisciplinary Ontology Conference*, Tokyo, Japan, 26-27 February, 2008.

Ng, S. K., et. Al. (2003). Integrative approach for computationally inferring protein domain interactions. *Bioinformatics, 19*, 923-929.

Ng, S-K., Zhang, Z., Tan, S-H., & Lin, K. (2003). InterDom: a database of putative interacting protein domains for validating predicted protein interactions and complexes. *Nucleic Acids Research, 31*, 251-254.

Nicola, J., et al. (2007). New developments in the InterPro database. *Nucleic Acids Research, 35*, D224-D228.

Nicolas, P. et. al. (1999). Discovering frequent closed itemsets for association rules. *Proceedings of the 7th International Conference on Database Theory*, Israel, (pp. 398–416).

Nijssen, S., & Kok, J. (2004). A quickstart in frequent structure mining can make a difference. *Proceedings of the 10th ACM SIGKDD International Conference on Knowledge Discovery and Data Mining* (pp. 647–652). ACM Press.

Nooren, I. M., & Thornton, J. M. (2003). Diversity of protein-protein interactions. *Embo J, 22*(14), 3486-3492.

Nooren, I. M., & Thornton, J. M. (2003). Structural characterisation and functional significance of transient protein-protein interactions. *J Mol Biol, 325*, 991-1018.

Nowozin, S., Bakir, G., & Tsuda, K. (2007). Discriminative subsequence mining for action classification. *Proceedings of the 11th IEEE International Conference on Computer Vision (ICCV 2007)*. IEEE Computer Society.

Nyfeler, B., Michnick, S.W., & Hauri, H.P. (2005). Capturing protein interactions in the secretory pathway of living cells. *Proc Natl Acad Sci U S A. 102*(18), 6350-6355.

Ofran, Y., & Rost, B. (2003). Analysing six types of protein-protein interfaces. *Journal of Molecular Biology, 325*(2), 377-387.

Oinn, T., Addis, M., Ferris, J., Marvin, D., Senger, M., Greenwood, M., Carver, T., Glover, K., Pocock, M. R., Wipat, A., & Li, P. (2004) Taverna: A tool for the composition and enactment of bioinformatics workflows. *Bioinformatics, 20*(17), 3045-54.

Oldfield, C. J., Meng, J., Yang, J. Y., Yang, M. Q., Uversky, V. N., & Dunker, A. K. (2008). Flexible nets: disorder and induced fit in the associations of p53 and 14-3-3 with their partners. *BMC genomics, 9*(Suppl 1), S1.

Oldham, M. C., Horvath, S., & Geschwind, D. H. (2006). Conservation and evolution of gene coexpression networks in human and chimpanzee brains. *Proc Natl Acad Sci U S A, 103*(47), 17973-17978.

Oltvai, Z. N., & Barabási, A. L. (2002). Life's complexity pyramid. *Science, 298*, 763-764.

Ooi, S. L., Pan, X., Peyser, B. D., Ye, P., Meluh, P. B., Yuan, D. S., et al. (2006). Global synthetic-lethality analysis and yeast functional profiling. *Trends Genet, 22*(1), 56-63.

Orchard, S., Salwinski, L., Kerrien, S., Montecchi-Palazzi, L., Oesterheld, M., Stümpflen, V., et al. (2007). The minimum information required for reporting a molecular interaction experiment (MIMIx). *Nature Biotechnology, 25*(8), 894-898.

Orlev, N., Shamir, R., & Shiloh Y. (2004). Pivot: Protein interaction visualization tool. *Bioinformatics, 20*(3), 424–425.

Oti, M., & Brunner, H. G. (2007). The modular nature of genetic diseases. *Clin Genet, 71*(1), 1-11.

Oti, M., Huynen, M. A., & Brunner, H. G. (2008). Phenome connections. *Trends Genet, 24*(3), 103-106.

Overbeek, R., Fonstein, M., et al. (1999). The use of gene clusters to infer functional coupling. *Proc Natl Acad Sci U S A, 96*(6), 2896-901.

Pache, R. A., Zanzoni, A., Naval, J., Mas, J. M., & Aloy, P. (2008). Towards a molecular characterisation of pathological pathways. *FEBS Lett, 582*(8), 1259-1265.

Pagel, P., Kovac, S., Oesterheld, M., Brauner, B., Dunger-Kaltenbach, I., Frishman, G., et al. (2005). The MIPS mammalian protein-protein interaction database. *Bioinformatics, 21*(6), 832-834.

Pal, C., Papp, B., & Hurst, L. D. (2003). Genomic function: Rate of evolution and gene dispensability. *Nature, 421*(6922), 496-7; discussion 497-8.

Pal, C., Papp, B., & Lercher, M. J. (2006). An integrated view of protein evolution. *Nat Rev Genet, 7*(5), 337-348.

Palla, G., Derényi, I., Farkas, I., & Vicsek, T. (2005). Uncovering the overlapping community structure of complex networks in nature and society. *Nature, 435*(7043), 814-818.

Pancholi, V., & Fischetti, V. (1992). A major surface protein on group A streptococci is a glyceraldehyde-3-phosphate-dehydrogenase with multiple binding activity. *Journal of Experimental Medicine, 176*, 415-426.

Pang, C. N., Krycer, J. R., Lek, A., & Wilkins, M. R. (2008). Are protein complexes made of cores, modules and attachments? *Proteomics, 8*(3), 425-434.

Paradesi, M. S. R. (2008). Graph-Based Protein-Protein Interaction Prediction in *Saccharomyces Cerevisiae*. Master's thesis, Kansas State University, Department of Computing and Information Sciences.

Paradesi, M. S. R., Caragea, D., & Hsu, W. H. (2007). Structural Prediction of Protein-Protein Interactions in *Saccharomyces cerevisiae*. *Proc. of IEEE 7th International Symposium on BioInformatics and BioEngineering*, 2, 1270-1274.

Paradesi, M. S. R., Wang, L., Brown, S. J., & Hsu, W. H. (2006). Mining Domain Association Rules From Protein-Protein Interaction data. *Intelligent Engineering Systems through Artificial Neural Networks, 16*, 213-218.

Parida, L. (2007). Discovering topological motifs using a compact notation. *Journal of Computational Biology, 14*(3), 300–323. (US Patent: 20070027662)

Pastor-Satorras, R., Smith, E., & Sole, R. V. (2003). Evolving protein interaction networks through gene duplication. *J Theor Biol, 222*(2), 199-210.

Patil, A., & Nakamura, H. (2005). Filtering high-throughput protein-protein interaction data using a combination of genomic features. *BMC Bioinformatics, 6*, 100.

Pawson, T., & Scott, J. D. (1997). Signalling through scaffold, anchoring, and adaptor proteins. *Science, 278*, 2075-2080.

Pazos, F., & Sternberg, M. J. (2004). Automated prediction of protein function and detection of functional sites from structure. *Proc Natl Acad Sci U S A, 101*(41), 14754-9.

Pazos, F., & Valencia, A. (2002). In silico two-hybrid system for the selection of physically interacting protein pairs. *Proteins, 47*, 219-227.

Pazos, F., Helmer-Citterich, M., Ausiello, G., & Valencia, A. (1997). Correlated mutations contain information about protein-protein interaction. *J Mol Biol, 271*, 511-523.

Pearson, H. (2008). Biologists initiate plan to map human proteome. *Nature, 452*(7190), 920-921.

Pearson, W. R., & Lipman, D. J. (1988). Improved tools for biological sequence comparison. *Proc Natl Acad Sci U S A, 85*(8), 2444-8.

Pei, J., Han, J., Mortazavi-Asl, B., Wang, J., Pinto, H., Chen, Q., Dayal, U., & Hsu, M.-C. (2004). Mining sequential patterns by pattern-growth: the prefixspan approach. *IEEE Trans. Knowledge and Data Engineering, 16*, 1424–1440.

Pei, P., & Zhang , A. (2005). A two-step approach for clustering proteins based on protein interaction profiles. In Proceedings of the *IEEE Int. Symposium on Bioinformatics and Bioengeneering (BIBE'2005)*, (pp. 201-209).

Pellegrini, M., Marcotte, E. M., et al. (1999). Assigning protein functions by comparative genome analysis: protein phylogenetic profiles. *Proc Natl Acad Sci U S A, 96*(8), 4285-8.

Pereira, J. B., Enright, A. J., & Ouzounis, C. A. (2004). Detection of functional modules from protein interaction networks. *Proteins: Structure, Fuctions, and Bioinformatics*, (20), 49-57.

Pereira-Leal, J. B., Enright, A. J., & Ouzounis, C. A. (2004). Detection of functional modules from protein interaction networks. *Proteins*, *54*(1), 49-57.

Pereira-Leal, J. B., Levy, E. D., Kamp, C., & Teichmann, S. A. (2007). Evolution of protein complexes by duplication of homomeric interactions. *Genome Biol, 8*(4), R51.

Perez-Iratxeta, C., Wjst, M., Bork, P., & Andrade, M. A. (2005). G2D: a tool for mining genes associated with disease. *BMC Genet, 6*, 45.

Peri, S. et al. (2003). Development of human protein reference database as an initial platform for approaching systems biology in humans. *Genome Research, 13*, 2363-2371.

Peri, S., Navarro, J. D., Amanchy, R., Kristiansen, T. Z., Jonnalagadda, C. K., Surendranath, V., et al. (2003). Development of human protein reference database as an initial platform for approaching systems biology in humans. *Genome Res, 13*(10), 2363-2371.

Pettit, F. K., Bare, E., Tsai, A., & Bowie, J. U. (2007). HotPatch: a statistical approach to finding biologically relevant features on protein surfaces. *J Mol Biol, 369*, 863-879.

Pevzner, P. A., & Sze, S-.H. (2000). Combinatorial Approaches to Finding Subtle Signals in DNA Sequences. *Proceedings of the International Conference on Intelligent Systems for Molecular Biology, 8*, 269-278.

Pinter, R. Y., Rokhlenko, O., Yeger-Lotem, E., & Ziv-Ukelson, M. (2005). Alignment of metabolic pathways. *Bioinformatics, 21*(16), 3401-3408.

Pitre, S., et. al. (2006). PIPE: a protein-protein interaction prediction engine based on the re-occurring short polypeptide sequences between known interacting protein pairs. *BMC Bioinformatics, 7*, 365.

Pizzuti, C., & Rombo, S. E. (2007). PINCoC: a Co-Clustering based Method to Analyze Protein-Protein Interaction Networks. In *the 8th International Conference on Intelligent Data Engineering and Automated Learning (IDEAL'07)* (pp. 821-830).

Pizzuti, C., & Rombo, S. E. (2008). Multi-functional Protein Clustering in PPI Networks. In *Proceedings of the 2nd International Conference on BioInformatics Research and Development (BIRD'08)*, Vienna, To appear.

Pletneva, E. V., Laederach, A. T., Fulton, D. B., & Kostic, N. M. (2001). The role of cation-pi interactions in biomolecular association. Design of peptides favoring interactions between cationic and aromatic amino acid side chains. *J Am Chem Soc, 123*, 6232-6245.

Porollo, A., & Meller, J. (2007). Prediction-based fingerprints of protein-protein interactions. *Proteins, 66*, 630-645.

Poyatos, J. F., & Hurst, L. D. (2004). How biologically relevant are interaction-based modules in protein networks? *Genome Biology*, *5*(11), R93.

Poyatos, J. F., & Hurst, L. D. (2006). Is optimal gene order impossible? *Trends Genet, 22*(8), 420-423.

Presser, A., Elowitz, M. B., Kellis, M., & Kishony, R. (2008). The evolutionary dynamics of the saccharomyces cerevisiae protein interaction network after duplication. *Proc Natl Acad Sci U S A, 105*(3), 950-954.

Presser, A., Elowitz, M. B., Kellis, M., & Kishony, R. (2008, January). The evolutionary dynamics of the saccharomyces cerevisiae protein interaction network after duplication. *Proc. National Academy of Sciences USA*, *105*(3), 950–954.

Prieto, C., & De Las Rivas, J. (2006). APID: Agile Protein Interaction DataAnalyzer. *Nucl. Acids Res., 34*, W298-W302.

Prieto, C., & De Las Rivas, J. (2006). APID: Agile Protein Interaction DataAnalyzer. *Nucleic Acids Res*, *34*(Web Server issue), W298-302.

Pržulj, N., Corneil, D. G., & Jurisica, I. (2006). Efficient estimation of graphlet frequency distributions in protein-protein interaction networks. *Bioinformatics*, *22*(8), 974–980.

Przulj, N., Wigle, D. A., & Jurisica, I. (2004). Functional topology in a network of protein interactions. *Bioinformatics*, *20*(3), 340-348.

Przulj, N., Wigle, D. A., & Jurisica, I. (2004). Functional topology in a network of protein interactions. *Bioinformatics*, *20*(3), 340–348.

Pu, S., Vlasblom, J., Emili, A., Greenblatt, J., & Wodak, S. J. (2007). Identifying functional modules in the physical interactome of Saccharomyces cerevisiae. *Proteomics*, *7*(6), 944-960.

Puig, O., Caspary, F., Rigaut, G., Rutz, B., Bouveret, E., Bragado-Nilsson, E., et al. (2001). The tandem affinity purification (TAP) method: a general procedure of protein complex purification. *Methods 24*(3), 218-229.

Pujana, M. A., Han, J. D., Starita, L. M., Stevens, K. N., Tewari, M., Ahn, J. S., et al. (2007). Network modeling links breast cancer susceptibility and centrosome dysfunction. *Nat Genet, 39*(11), 1338-1349.

Puntervoll, P., et. al. (2003). ELM server: a new resource for investigating short functional sites in modular eukaryotic proteins. *Nucleic Acids Research, 31*, 3625-3630.

Pupko, T., Bell, R. E., Mayrose, I., Glaser, F., and Ben-Tal, N. (2002). Rate4Site: An algorithmic tool for the identification of functional regions in proteins by surface mapping of evolutionary determinants within their homologues. *Bioinformatics,* (18 Suppl 1), S71-S77.

Qi, Y., Bar-Joseph, Z., & Klein-Seetharaman, J. (2006). Evaluation of different biological data and computational classification methods for use in protein interaction prediction. *Proteins, 63*(3), 490-500.

Qi, Y., Klein-Seetharaman, J., & Bar-Joseph, Z. (2005). Random forest similarity for protein-protein interaction prediction from multiple sources. *Pac Symp Biocomput*, 531-542.

Qi, Y., Klein-Seetharaman, J., & Bar-Joseph, Z. (2007). A mixture of feature experts approach for protein-protein interaction prediction. *BMC Bioinformatics, 8 Suppl 10*, S6.

Qin, S., & Zhou, H. X. (2007). meta-PPISP: A Meta Web Server for Protein-Protein Interaction Site Prediction. *Bioinformatics*.

Qiu, J., & Noble, W. S. Predicting co-complexed protein pairs from heterogeneous data. *PLoS Computational Biology*, *4*(4), e1000054.

Quinlan, J. R. (1986). Induction of decision trees. *Machine Learning*, *1*, 81-106.

Quinlan, J. R. (1993). *C4.5: Programs for Machine Learning*. Morgan Kaufmann.

Rain, J. C., Selig, L., De Reuse, H., Battaglia, V., Reverdy, C., Simon, S., et al. (2001). The protein-protein interaction map of Helicobacter pylori. *Nature, 409*(6817), 211-215.

Rain, J. C., Selig, L., De Reuse, H., Battaglia, V., Reverdy, C., Simon, S., et al. (2001). The protein-protein interaction map of helicobacter pylori. *Nature, 409*(6817), 211-215.

Rajagopalan, D., & Agarwal, P. (2005). Inferring pathways from gene lists using a literature-derived network of biological relationships. *Bioinformatics*, *21*(6), 788-793.

Ramani, A. K., Bunescu, R. C., Mooney, R. J., & Marcotte, E. M. (2005). Consolidating the set of known human protein-protein interactions in preparation for large-scale mapping of the human interactome. *Genome Biol*, *6*(5), R40.

Ramirez, F., Schlicker, A., Assenov, Y., Lengauer, T., & Albrecht, M. (2007). Computational analysis of human protein interaction networks. *Proteomics, 7*(15), 2541-2552.

Ramírez, F., Schlicker, A., Assenov, Y., Lengauer, T., & Albrecht, M. (2007). Computational analysis of human protein interaction networks. *Proteomics*, *7*(15), 2541-2552.

Reguly T. et al. (2006). Comprehensive curation and analysis of global interaction networks in Saccharomyces cerevisiae. *J. Biol, 5*(11).

Reguly, T., Breitkreutz, A., Boucher, L., Breitkreutz, B. J., Hon, G. C., Myers, C. L., et al. (2006). Comprehensive curation and analysis of global interaction networks in Saccharomyces cerevisiae. *J Biol, 5*(4), 11.

Reguly, T., Breitkreutz, A., Boucher, L., Breitkreutz, B. J., Hon, G. C., Myers, C. L., et al. (2006). Comprehensive curation and analysis of global interaction networks in Saccharomyces cerevisiae. *J Biol, 5*(4), 11.

Res, I., Mihalek, I., & Lichtarge, O. (2005). An evolution based classifier for prediction of protein interfaces without using protein structures. *Bioinformatics, 21*, 2496-2501.

Rhian, E. H., Joseph, L. D., Suad, A., David, C. K., Patrick, O., & Pradipsinh, K. R. (2000). Shotgun DNA microarrays and stage-specific gene expression in Plasmodium falciparum malaria *Molecular Microbiology, 35*(1), 6-14.

Rhodes, D. R., Tomlins, S. A., Varambally, S., Mahavisno, V., Barrette, T., Kalyana-Sundaram, S., et al. (2005). Probabilistic model of the human protein-protein interaction network. *Nat Biotechnol, 23*(8), 951-959.

Riechmann, L., & Winter, G. (2006). Early protein evolution: Building domains from ligand binding polypeptide segments. *Journal of Molecular Biology, 363*, 460-468.

Rigaut, G., Shevchenko, A., Rutz, B., Wilm, M., Mann, M., & Séraphin, B. (1999). A generic protein purification method for protein complex characterization and proteome exploration. *Nat Biotechnol. 17*(10), 1030-1032.

Risi Imre, K., & John, D. L. (2002). Diffusion Kernels on Graphs and Other Discrete Input Spaces. *Proceedings of the Nineteenth International Conference on Machine Learning.* Morgan Kaufmann Publishers Inc.

Rives, A. W., & Galitski, T. (2003). Modular organization of cellular. *Proceedings of the National Academy of Sciences of the USA*, *100*(3), 1128-1133.

Roch, K. G. L., Zhou, Y., Blair, P. L., Grainger, M., Moch, J. K., Haynes, J. D., et al. (2004). Discovery of Gene Function by Expression Profiling of the Malaria Parasite Life Cycle. *Science, 301*(5639), 1503-1508.

Rogozin, I. B., Makarova, K. S., Wolf, Y. I., & Koonin, E. V. (2004). Computational approaches for the analysis of gene neighbourhoods in prokaryotic genomes. *Brief Bioinform, 5*(2), 131-149.

Romero, P., Wagg, J., Green, M. L., Kaiser, D., Krummenacker, M., & Karp, P. D. (2005). Computational prediction of human metabolic pathways from the complete human genome. *Genome Biol, 6*(1), R2.

Rost, B. (1999). Twilight zone of protein sequence alignments. *Protein Eng, 12*(2), 85-94.

Rost, B., & Liu, J. (2003). The PredictProtein server. *Nucleic Acids Res, 31*(13), 3300-4.

Roth, E. J. (1990). Plasmodium falciparum carbohydrate metabolism: a connection between host cell and parasite. *Blood Cells, 16*(2-3), 453-460.

Roulet, E., Busso, S., Camargo, A. A., Simpson, A. J., Mermod, N., & Bucher, P. (2002). High-throughput SELEX SAGE method for quantitative modeling of transcription-factor binding sites. *Nat Biotechnol, 20*(8), 831-835.

Rual, J. F., Venkatesan, K., Hao, T., Hirozane-Kishikawa, T., Dricot, A., Li, N., et al. (2005). Towards a proteome-scale map of the human protein-protein interaction network. *Nature, 437*(7062), 1173-1178.

Rual, J. F., Venkatesan, K., Hao, T., Hirozane-Kishikawa, T., Dricot, A., Li, N., et al. (2005). Towards a proteome-scale map of the human protein-protein interaction network. Nature, 20, 437(7062), 1173-8.

Rual, J. F., Venkatesan, K., Hao, T., Hirozane-Kishikawa, T., Dricot, A., Li, N., et al. (2005). Towards a proteome-scale map of the human protein-protein interaction network. *Nature, 437*(7062), 1173-1178.

Rudolph, M.G., Bayer, P., Abo, A., Kuhlmann, J., Vetter I.R., & Wittinghofer, A. (1998). The Cdc42/Rac interactive binding region motif of the Wiskott Aldrich syndrome protein (WASP) is necessary but not sufficient for tight binding to Cdc42 and structure formation. *J. Biol. Chem. 273* (29), 18067-18076.

Rungsarityotin, W., Krause, R., Schödl, A., & Schliep, A. (2007). Identifying protein complexes directly from high-throughput TAP data with Markov random fields. *BMC Bioinformatics, 8*, 482.

Russell, R. B., Alber, F., Aloy, P., Davis, F. P., Korkin, D., Pichaud, M., et al. (2004). A structural perspective on protein-protein interactions. *Curr Opin Struct Biol, 14*(3), 313-324.

Saeed, R., & Deane, C. (2008). An assessment of the uses of homologous interactions. *Bioinformatics, 24*(5), 689-695.

Saha, S., & Heber, S. (2006). In silico prediction of yeast deletion phenotypes. *Genet Mol Res, 5*(1). 224-32.

Saigo, H., Nowozin, S., Kadowaki, T., Kudo, T., & Tsuda, K. (2008). gBoost: a mathematical programming approach to graph classification and regression. *Machine Learning*. to appear.

Saigo, H., Uno, T., & Tsuda, K. (2007). Mining complex genotypic features for predicting HIV-1 drug resistance. *Bioinformatics, 23,* 2455–2462.

Saito, R., Suzuki, H., et al. (2003). Construction of reliable protein-protein interaction networks with a new interaction generality measure. *Bioinformatics, 19*(6), 756-63.

Salama, J. J., et. al. (2001). Automatic annotation of BIND molecular interactions from three-dimensional structures. *Biopolymers, 61*, 111-120.

Salgado, H., Moreno-Hagelsieb, G., et al. (2000). Operons in Escherichia coli: genomic analyses and predictions. *Proc Natl Acad Sci U S A, 97*(12), 6652-7.

Salwinski, *et. al.* (2004). The database of interacting proteins: 2004 update. *Nucleic Acids Research, Database issue 32*, D449-D451.

Salwinski, L., & Eisenberg, D. (2003). Computational methods of analysis of protein-protein interactions. *Curr. Opin. Struct. Biol., 13*, 377-382.

Salwinski, L., Miller, C. S., Smith, A. J., Pettit, F. K., Bowie, J. U., & Eisenberg D. (2004). The Database of Interacting Proteins: 2004 update. *Nucleic Acids Research, 32*(Database Issue), D449-D451.

Salwinski, L., Miller, C. S., Smith, A. J., Pettit, F. K., Bowie, J. U., & Eisenberg, D. (2004, January). The database of interacting proteins: 2004 update. *Nucleic Acids Research, 32*, D449–D451. (URL: http://dip.doe-mbi.ucla.edu/)

Salwinski, L., Miller, C. S., Smith, A. J., Pettit, F. K., Bowie, J. U., & Eisenberg, D. (2004). The Database of Interacting Proteins: 2004 update. *Nucleic Acids Res, 32*(Database issue), D449-451.

Samanta, M. P., & Liang, S. (2003). Predicting protein functions from redundancies in large-scale protein interaction networks. *Proc Natl Acad Sci U S A, 100*(22), 12579-83.

Samantha, M. P., & Liang, S. (2003). Redundancies in large-scale protein interaction networks. *Proceedings of the National Academy of Science of the USA*, (pp. 12579-12583).

Sambrook, J., & Russell, D. W. (2006). Identification of Associated Proteins by Coimmunoprecipitation, *CSH Protocols*, doi:10.1101/pdb.prot3898.

Sandelin, A., Alkema, W., Engstrom, P., Wasserman, W. W., & Lenhard, B. (2004). JASPAR: an open-access database for eukaryotic transcription factor binding profiles. *Nucleic Acids Res, 32*(Database issue), D91-94.

Sander, C., & Schneider, R., (1991). Database of homology-derived protein structures and the structural meaning of sequence alignment. *Proteins, 9*, 56-68.

Sanderson, C. M. (2008). A new way to explore the world of extracellular protein interactions. *Genome Res, 18*(4), 517-520.

Saraiya, P., North C., & Duca, K. (2005). Visualizing biological pathways: requirements analysis, systems evaluation and research agenda. *Information Visualization* (pp. 1-15).

Sardiu, M. E., Cai, Y., Jin, J., Swanson, S. K., Conaway, R. C., Conaway, J. W., et al. (2008). Probabilistic assembly of human protein interaction networks from label-free quantitative proteomics. *Proc Natl Acad Sci U S A, 105*(5), 1454-1459.

Schachter, V. (2002). Construction and prediction of protein-protein interaction maps. *Ernst Schering Res Found Workshop,* (38), 191-220.

Scherens, B., & Goffeau, A. (2004). The uses of genome-wide yeast mutant collections. *Genome Biol, 5*(7), 229.

Schmith, J., Lemke, N., Mombach, J. C. M., Benelli, P., Barcellos, C. K., & Bedin, G. B. (2005). Damage, connectivity and essentiality in protein-protein interaction networks. *Physica A: Statistical Mechanics and its Applications, 349*(3-4), 675-684.

Schoelkopf, B., Tsuda, K., & Vert, J. (Eds.). (2004). *Kernel methods for computational biology.* MIT Press.

Scholkopf, B., Sung, K.-K., Burges, C. J. C., Girosi, F., Niyogi, P., Poggio, T., & Vapnik, V. (1997). Comparing support vector machines with gaussian kernels to radial basis function classiers. *IEEE Trans. on Signal Processing, 45*(11), 2758-65.

Scholtens, D., Chiang, T., Huber, W., & Gentleman, R. (2008). Estimating node degree in bait-prey graphs. *Bioinformatics, 24*(2), 218-224.

Scholtens, D., Vidal, M., & Gentleman, R. (2005). Local modeling of global interactome networks. *Bioinformatics, 21*(17), 3548-3557.

Schreiber, F., & Schwöbbermeyer, H. (2005a). Frequency concepts and pattern detection for the analysis of motifs in networks. *Transactions on Computational Systems Biology III, 3737*, 89–104.

Schreiber, F., & Schwöbbermeyer, H. (2005b). MAVisto: A tool for the exploration of network motifs. *Bioinformatics, 21*(17), 3572–3574. (Software: http://mavisto.ipk-gatersleben.de/)

Schuster-Böckler, B., & Bateman, A. (2007). Reuse of structural domain-domain interactions in protein networks. *BMC Bioinformatics, 8*, 259.

Schwikowski, B., Uetz, P., & Fields, S. (2000). A network of interacting proteins in yeast. *Nature Biotechnology, 18*(12), 1257-1261.

Schwock, J. o., Kirchberger, J. u., Edelmann, A., Kriegel, T. M., & Kopperschl¨ager, G. (2004). Interaction of 6-phosphofructokinase with cytosolic proteins of Saccharomyces cerevisiae. *Yeast, 21*, 483-494.

Scott, M. S., & Barton, G. J. (2007). Probabilistic prediction and ranking of human protein-protein interactions. *BMC Bioinformatics, 8*, 239.

Sebat, J. (2007). Major changes in our DNA lead to major changes in our thinking. *Nat Genet, 39*(7 Suppl), S3-5.

Sedgewick, R. (2002). *Algorithms in C, Part 5: Graph algorithms* (Third ed.). New York, USA: Addison-Wesley Publishing Co.

Segal, E., Wang, H., & Koller, D. (2003). Discovering molecular pathways from protein interaction and gene expression data. *Bioinformatics, 19* (Suppl 1), i264-i271.

Selenko, P., & Wagner, G. (2006). NMR mapping of protein interactions in living cells. *Nat Methods, 3*(2), 80-81.

Seringhaus, M., Paccanaro, A., Borneman, A., Snyder M., & Gerstein, M. (2006). Predicting essential genes in fungal genomes. *Genome Res, 16*(9), 1126-35.

Shannon, C. E. (2001). A mathematical theory of communication. *ACM SIGMOBILE Mobile Computing and Communications Review.*

Shannon, P. et al. (2003). Cytoscape: A software environment for integrated models of biomolecular interaction networks. *Genome Research, 13*(11), 2498-504.

Shannon, P., Markiel, A., Ozier, O., Baliga, N. S., Wang, J. T., Ramage, D., Amin, N., Schwikowski, B., & Ideker, T. (2003). Cytoscape: a software environment for integrated models of biomolecular interaction networks. *Genome Research, 13*(11), 2498-2504.

Shannon, P., Markiel, A., Ozier, O., Baliga, N. S., Wang, J. T., Ramage, D., Amin, N., Schwikowski, B., & Ideker, T. (2003). Cytoscape: a software environment for integrated models of bio-molecular interaction networks. *Genome Research. 13*(11), 2498-504.

Shannon, P., Markiel, A., Ozier, O., Baliga, N. S., Wang, J. T., Ramage, D., et al. (2003). Cytoscape: a software environment for integrated models of biomolecular interaction networks. *Genome Res, 13*(11), 2498-2504.

Shannon, P., Markiel, A., Ozier, O., Baliga, N. S., Wang, J. T., Ramage, D., et al. (2003). Cytoscape: a software environment for integrated models of biomolecular interaction networks. *Genome Res, 13*(11), 2498-2504.

Sharan, R., & Ideker, T. (2006). Modeling cellular machinery through biological network comparison. *Nature Biotechnology, 24*(4), 427-433.

Sharan, R., Suthram, S., Kelley, R. M., Kuhn, T., McCuine, S., Uetz, P., Sittler, T., Karp, R. M., & Ideker, T. (2005). Conserved patterns of protein interaction in multiple species. *Proceedings of the National Academy of Sciences USA, 102*(6), 1974-1979.

Sharan, R., Suthram, S., Kelley, R. M., Kuhn, T., McCuine, S., Uetz, P., Sittler, T., Karp, R. M., & Ideker, T. (2005). Conserved patterns of protein interaction in multiple species. *Proceedings of the National Academy of Sciences of the United States of America, 102*(6), 1974-1979.

Sharan, R., Suthram, S., Kelley, R. M., Kuhn, T., McCuine, S., Uetz, P., et al. (2005). Conserved patterns of protein interaction in multiple species. *Proc Natl Acad Sci U S A, 102*(6), 1974-1979.

Sharan, R., Ulitisky, I., & Shamir, R. (2007). Network-based prediction of protein function. *Molecular Systems Biology, 3*(88).

Shen, J., Zhang, J., Luo, X., Zhu, W., Yu, K., Chen, K., et al. (2007). Predicting protein-protein interactions based only on sequences information. *Proc Natl Acad Sci U S A, 104*(11), 4337-4341.

Shlomi, T., Segal, D., Ruppin, E., & Sharan, R. (2006). QPath: a method for querying pathways in a protein-protein interaction network. *BMC Bioinformatics, 7*(199)

Shoemaker, B. A., & Panchenko, A. R. (2007). Deciphering protein-protein interactions. Part I. Experimental techniques and databases. *PLoS Comput Biol, 3*(3), e42.

Shoemaker, B. A., & Panchenko, A. R. (2007). Deciphering Protein–Protein Interactions. Part II. Computational Methods to Predict Protein and Domain Interaction Partners. *PLoS Comput Biol 3*(4), e43.

Shoemaker, B. A., & Panchenko, A. R. (2007). Deciphering protein-protein interactions. Part I. Experimental techniques and databases. *PLoS Computational Biology, 3*(3), e42.

Shoemaker, B. A., & Panchenko, A. R. (2007). Deciphering protein-protein interactions. Part II. Computational methods to predict protein and domain interaction partners. *PLoS Computational Biology, 3*(4), e43.

Shoemaker, B. A., & Panchenko, A. R. (2007). Deciphering Protein–Protein Interactions. Part II. Computational Methods to Predict Protein and Domain Interaction Partners. *PLoS Comput Biol., 3*(3), e43.

Shoemaker, B., & Panchenko, A. (2007). Deciphering protein-protein interactions. Part I. Experimental techniques and databases. *PLoS Comput Biol., 3*(3), e42.

Shulman-Peleg, A., Shatsky, M., Nussinov, R., & Wolfson, H. J. (2007). Spatial chemical conservation of hot spot interactions in protein-protein complexes. *BMC Biol, 5*, 43.

Simonis, N., Gonze, D., Orsi, C., van Helden, J., & Wodak, S. J. (2006). Modularity of the transcriptional response of protein complexes in yeast. *Journal of Molecular Biology, 363*(2), 589-610.

Singer, G. A., Lloyd, A. T., Huminiecki, L. B., & Wolfe, K. H. (2005). Clusters of co-expressed genes in mammalian genomes are conserved by natural selection. *Mol Biol Evol, 22*(3), 767-775.

Singh, G. P., Ganapathi, M., & Dash, D. (2007). Role of intrinsic disorder in transient interactions of hub proteins. *Proteins, 66*, 761-765.

Singh, R., Xu, J., & Berger, B. (2006). Struct2net: integrating structure into protein-protein interaction prediction. *Pac Symp Biocomput*, 403-414.

Singh, R., Xu, J., & Berger, B. (2007). Pairwise global alignment of protein interaction networks by matching neighbourhood topology. *Proceedings of the Eleventh Annual International Conference on Research in Computational Molecular Biology (RECOMB 2007), 4453*, 16-31.

Singhal, M., & Resat, H. (2007). A domain-based approach to predict protein-protein interactions. *BMC Bioinformatics, 8*, 199.

Sirover, M. A. (1999). New insights into an old protein: the functional diversity of mammalian glyceraldehyde-3-phosphate dehydrogenase. *Biochemica et Biophysica Acta, 1432*, 159-184.

Skrabanek, L., et. al. (2008). Computational prediction of protein-protein interactions. *Molecular Biotechnology, 38*(1), 1-17.

Skrabanek, L., Saini, H. K., Bader, G. D., & Enright, A. J. (2008). Computational prediction of protein-protein interactions. *Molecular Biotechnology, 38*(1), 1-17.

Smalley, D. J., Whiteley, M., & Conway, T. (2003). In search of the minimal *Escherichia coli* genome. *Trends Microbiol, 11*(1), 6-8.

Soderberg, O., Gullberg, M., Jarvius, M., Ridderstrale, K., Leuchowius, K. J., Jarvius, J., et al. (2006). Direct observation of individual endogenous protein complexes in situ by proximity ligation. *Nat Methods, 3*(12), 995-1000.

Sohler, F., Hanisch, D., & Zimmer, R. (2004). New methods for joint analysis of biological networks and expression data. *Bioinformatics, 20*(10), 1517-1521.

Songyang, Z., et. al. (1993). SH2 domains recognize specific phosphopeptide sequences. *Cell, 72*(5), 767-778.

Soreq, L., Gilboa-Geffen, A., Berrih-Aknin, S., Lacoste, P., Darvasi, A., Soreq, E., et al. (2008). Identifying alternative hyper-splicing signatures in MG-thymoma by exon arrays. *PLoS ONE, 3*(6), e2392.

Spellman, P. T., Sherlock, G., Zhang, M. Q., Iyer, V. R., Anders, K., Eisen, M. B., et al. (1998). Comprehensive identification of cell cycle-regulated genes of the yeast Saccharomyces cerevisiae by microarray hybridization. *Mol Biol Cell, 9*(12), 3273-3297.

Spirin, V., & Mirny, L. A. (2003). Protein complexes and functional modules in molecular networks. *Proc Natl Acad Sci U S A, 100*(21), 12123-12128.

Splendiani, A. (2008). RDFScape: semantic Web meets systems biology. *BMC Bioinformatics, 9*(Suppl 4), S6.

Sporn, O., Honey, C. J., & Kötter, R. (2007). Identification and classification of hubs in brain networks. *Plos One, 10*, e1049.

Sprinzak, E, et. al. (2003). How reliable are experimental protein-protein interaction data? *Journal of Molecular Biology, 327*(5), 919-923.

Sprinzak, E., & Margalit, H. (2001). Correlated sequence-signatures as markers of protein-protein interaction. *Journal of Molecular Biology, 311*(4), 681-692.

Sprinzak, E., & Margalit, H. (2001). Correlated sequence-signatures as markets of protein-protein interaction. *Journal of Molecular Biology, 311*, 681-692.

Sprinzak, E., Altuvia, Y., & Margalit, H. (2006). Characterization and prediction of protein-protein interactions within and between complexes. *Proceedings of the National Academy of Sciences USA, 103*(40), 14718-14723.

Sprinzak, E., Altuvia, Y., & Margalit, H. (2006). Characterization and prediction of protein-protein interactions within and between complexes. *Proceedings of the National Academy of Sciences USA*, *103*, 14718-14723.

Sprinzak, E., Sattath, S., & Margalit, H. (2003). How reliable are experimental protein-protein interaction data? *J Mol Biol., 327*(5), 919-923.

Stagljar, I., & Fields, S. (2002). Analysis of membrane protein interactions using yeast-based technologies. *Trends Biochem Sci, 27*(11), 559-563.

Stamm, S. (2008). Regulation of alternative splicing by reversible protein phosphorylation. *J Biol Chem, 283*(3), 1223-1227.

Stark, C., Breitkreutz, B. J., Reguly, T., Boucher, L., Breitkreutz, A., & Tyers, M. (2006). BioGRID: A general repository for interaction datasets. *Nucleic Acids Research, 34*(Database Issue), D535-D539.

Stark, C., Breitkreutz, B. J., Reguly, T., Boucher, L., Breitkreutz, A., & Tyers, M. (2006). BioGRID: a general repository for interaction datasets. *Nucleic Acids Res, 34*(Database issue), D535-539.

Steinmetz, L. M., Scharfe, C., Deutschbauer, A. M., Mokranjac, D., Herman, Z. S., Jones, T., Chu, A. M., Giaever, G., Prokisch, H., Oefner, P. J., & Davis, R. W. (2002). Systematic screen for human disease genes in yeast. *Nat Genet, 31*(4), 400-4.

Stelzl, U., Worm, U., Lalowski, M., Haenig, C., Brembeck, F. H., Goehler, H., et al. (2005). A human protein-protein interaction network: A resource for annotating the proteome. *Cell* 2005, *122*(6), 957-68

Stephen, P. B. (2006). Identifying sets of key players in a social network. *Comput. Math. Organ. Theory, 12*(1), 21-34.

Strömbäck L, Jakoniene V, Tan H, & Lambrix P. (2006). Representing, storing and accessing molecular interaction data: a review of models and tools. *Briefings in Bioinformatics*, *7*(4), 331-338.

Stuart, J. M., Segal, E., Koller, D., & Kim, S. K. (2003). A gene-coexpression network for global discovery of conserved genetic modules. *Science, 302*(5643), 249-255.

Stumpf, M. P, Thorne, T., de Silva, E., Stewart, R., An, H. J., Lappe, M., & Wiuf, C. (2008). Estimating the size of the human interactome. *Proceedings of the National Academy of Sciences USA*, *105*(19), 6959-6964.

Stumpf, M. P., Thorne, T., de Silva, E., Stewart, R., An, H. J., Lappe, M., et al. (2008). Estimating the size of the human interactome. *Proc Natl Acad Sci U S A, 105*(19), 6959-6964.

Stumpf, M. P., Wiuf, C., & May, R. M. (2005). Subnets of scale-free networks are not scale-free: sampling properties of networks. *Proceedings of the National Academy of Sciences USA, 102*(12), 4221-4224.

Su, A. I., Wiltshire, T., Batalov, S., Lapp, H., Ching, K. A., Block, D., et al. (2004). A gene atlas of the mouse and human protein-encoding transcriptomes. *Proc Natl Acad Sci U S A, 101*(16), 6062-6067.

Su, Q. J., L. Lu, et al. (2005). eBLOCKs: enumerating conserved protein blocks to achieve maximal sensitivity and specificity. *Nucleic Acids Res, 33*(Database issue), D178-82.

Suderman, M., & Hallet, M. (2007). Tools for Visually Exploring Biological Networks. *Bioinformatics, 23*(20), 2651-9.

Sudol, M. (1998). From Src Homology domains to other signaling modules: Proposal of the 'protein recognition code'. *Oncogene, 17*, 1469-1474.

Sun, G., Liu, X., Cong, G., Zhou, M., Xiong, Z., Lin, C.-Y., & Lee, J. (2007). Detecting erroneous sentences using automatically mined sequential patterns. *ACL 2007, Proceedings of the 45th Annual Meeting of the Association for Computational Linguistics.*

Suthram, S., Sittler, T., & Ideker, T. (2005). The Plasmodium protein network diverges from those of other eukaryotes. *Nature, 438*(7064), 108-112.

Systems biologists hatch plan for virtual human. (2008, February 21) *Nature 451*(7181), 879.

Szilagyi, A., Grimm, V., Arakaki, A. K., & Skolnick, J. (2005). Prediction of physical protein-protein interactions. *Phys Biol, 2*(2), S1-16.

Tamura, M., & D'haeseleer, P. (2008). Microbial genotype-phenotype mapping by class association rule mining. *Bioinformatics*. to appear.

Tan, K., Shlomi, T., Feizi, H., Ideker, T., & Sharan, R. (2007). Transcriptional regulation of protein complexes within and across species. *Proceedings of the National Academy of Sciences USA, 104*(4), 1283-1288.

Tan, S.-H., et. al. (2004). Discovering Novel Interacting Motif Pairs from Large Protein-Protein Interaction Datasets. *4th IEEE International Symposium on BioInformatics and BioEngineering (BIBE), (pp.* 568-575).

Tan, S.-H., et. al. (2006). A correlated motif approach for finding short linear motifs from protein interaction networks. *BMC Bioinformatics, 7*, 502.

Tarassov, K., Messier, V., et al. (2008). An in vivo map of the yeast protein interactome. *Science, 320*(5882), 1465-70.

Tarassov, K., Messier, V., Landry, C. R., Radinovic, S., Molina, M. M., Shames, I., et al. (2008). An in vivo map of the yeast protein interactome. *Science, 320*(5882), 1465-1470.

Tarassov, K., Messier, V., Landry, C. R., Radinovic, S., Molina, M. M., Shames, I., et al. (2008). An in vivo map of the yeast protein interactome. *Science, 320*(5882), 1465-1470.

Tatusova, T. A., & Madden, T. L. (1999). Blast 2 sequences, a new tool for comparing protein and nucleotide sequences. *FEMS Microbiology Letters, 174*(2), 247-250.

Teichmann, S. A. (2002). The constraints protein-protein interactions place on sequence divergence. *J Mol Biol, 324*(3), 399-407.

Teichmann, S. A., & Veitia, R. A. (2004). Genes encoding subunits of stable complexes are clustered on the yeast chromosomes: An interpretation from a dosage balance perspective. *Genetics, 167*(4), 2121-2125.

Tew, K. L., Li, X. L., & Tan, S. H. (2007). Functional centrality: detecting lethality of proteins in protein interaction networks. *Genome Inform.,* ***19***, 166-77.

The Gene Ontology (GO) project in 2006. (2006). *Nucleic Acids Res, 34*(Database issue), D322-326.

The Gene Ontology Consortium (2001). Creating the gene ontology resource: design and implementation. *Genome Res, 11*(8), 1425-33.

The Gene Ontology Consortium. (2000). Gene Ontology: tool for the unification of biology. *Nature Genetics, 25*(1), 25–29. (URL: http://www.geneontology.org/)

The TAP method home page (2002). Retrieved August 30, 2008, from http://www-db.embl-heidelberg.de/jss/servlet/de.embl.bk.wwwTools.GroupLeftEMBL/ExternalInfo/seraphin/TAP.html.

The Tokyo Declaration. (2008). Retrieved August 15, 2008, from http://www.systems-biology.org/~myukiko/FCSB2008/lib/exe/fetch.php?id=workshop%3Astatement&cache=cache&media=workshop:fcsb_2008-tokyo-declaration.pdf.

Thomas, P. D., Campbell, M. J., Kejariwal, A., Mi, H., Karlak, B., Daverman, R., et al. (2003). PANTHER: a library of protein families and subfamilies indexed by function. *Genome Res, 13*(9), 2129-2141.

Thompson, J. D., Higgins, D. G., & Gibson, T. J. (1994). Clustal w: Improving the sensitivity of progressive multiple sequence alignment through sequence weighting, position-specific gap penalties and weight matrix choice. *Nucleic Acids Res, 22*(22), 4673-4680.

Thompson, W. et. al. (2003). Gibbs recursive sampler: Finding transcription factor binding sites. *Nucleic Acids Research, 31*, 3580-3585.

Tian, Y., McEachin, R. C., Santos, C., States, D. J., & Patel, J. M. (2007). SAGA: A subgraph matching tool for biological graphs. *Bioinformatics, 23*(2), 232-239.

Tibshirani, R. (1996). Regression shrinkage and selection via LASSO. J. R. *Stat. Soc. Ser. B Stat. Methodol., 58,* 267–288.

Tirosh, I., & Barkai, N. (2005). Computational verification of protein-protein interactions by orthologous co-expression. *BMC Bioinformatics*, *6*, 40.

Tjong, H., Qin, S., & Zhou, H. X. (2007). PI2PE: protein interface/interior prediction engine. *Nucleic Acids Res*, *35*, W357-W362.

Tong, A. H., Evangelista, M., Parsons, A. B., Xu, H., Bader, G. D., Pagé, N., Robinson, M., Raghibizadeh, S., Hogue, C.W., Bussey, H., Andrews, B., Tyers, M., & Boone, C. (2001). Systematic genetic analysis with ordered arrays of yeast deletion mutants. *Science*, *294*(5550), 2364-8.

Tong, A. H., Lesage, G., Bader, G. D., Ding, H., Xu, H., Xin, X., et al. (2004). Global mapping of the yeast genetic interaction network. *Science*, *303*(5659), 808-13.

Tornow, S., & Mewes, H. (2003). Functional modules by relating protein interaction networks and gene expression. *Nucleic Acid Research*, *31*(21), 6283-6289.

Tovchigrechko, A., & Vakser, I. A. (2005). Development and testing of an automated approach to protein docking. *Proteins*, *60*(2), 296-301.

Trinklein, N. D., Aldred, S. F., Hartman, S. J., Schroeder, D. I., Otillar, R. P., & Myers, R. M. (2004). An abundance of bidirectional promoters in the human genome. *Genome Res*, *14*(1), 62-66.

Troyanskaya, O. G., Dolinski, K., et al. (2003). A Bayesian framework for combining heterogeneous data sources for gene function prediction (in Saccharomyces cerevisiae). *Proc Natl Acad Sci U S A*, *100*(14), 8348-53.

Tsuda, K. (2007). Entire regularization paths for graph data. *Proceedings of the 24th International Conference on Machine Learning* (pp. 919–926).

Tsuda, K., Shin, H., et al. (2005). Fast protein classification with multiple networks. *Bioinformatics, 21(*Suppl 2), ii59-65.

Ucar, D., Asur, S. Catalyurek, U. V., & Parthasarathy, S. (2006). Improving functional modularity in protein-protein interactions graphs using hub-induced subgraphs. In *10th European Conference on Principles and Practice of Knowledge Discovery in Databases (PKDD)* (pp. 371-382).

Ucar, D., Parthasarathy, S., Asur, S., & Wang, C. (2005). Effective pre-processing strategies for functional clustering of a protein-protein interaction network. In *IEEE Int. Symposium on Bioinformatics and Bioengeneering (BIBE)*, (pp. 129-136).

Uetz, P. *et al.* (2000). A comprehensive analysis of protein-protein interactions in *Saccharomyces cerevisiae. Nature*, *403*, 623-627.

Uetz, P., & Goll, J. (2006). The elusive yeast interactome. *Genome Biology*, *7*, 223-229.

Uetz, P., Giot, L., Cagney, G., Mansfield, T. A., Judson, R. S., Knight, J. R., et al. (2000). A comprehensive analysis of protein-protein interactions in *Saccharomyces cerevisiae. Nature*, *403*(6770), 601-3.

Uetz, P., Giot, L., Cagney, G., Mansfield, T. A., Judson, R. S., Knight, J. R., et al. (2000). A comprehensive analysis of protein-protein interactions in Saccharomyces cerevisiae. *Nature*, *403*(6770), 623-627.

Uetz, P., Giot, L., Cagney, G., Mansfield, T. A., Judson, R. S., Knight, J. R., et al. (2000). A comprehensive analysis of protein-protein interactions in saccharomyces cerevisiae. *Nature*, *403*(6770), 623-627.

Ulitsky, I., & Shamir, R. (2007). Identification of functional modules using network topology and high-throughput data. *BMC Systems Biology*, *1*, 8.

Ulitsky, I., & Shamir, R. (2007). Pathway redundancy and protein essentiality revealed in the *Saccharomyces cerevisiae* interaction networks. *Mol Syst Biol*, *3*, 104.

Uno, T., Kiyomi, M., & Arimura, H. (2005). LCM ver.3: collaboration of array, bitmap and prefix tree for freque nt itemset mining. *OSDM '05: Proceedings of the 1st international workshop on open so urce data mining* (pp. 77–86). Chicago, Illinois.

Vagner, V., Dervyn, E., & Ehrlich, S. D. (1998). A vector for systematic gene inactivation in *Bacillus subtilis. Microbiology*, *144* (Pt 11), 3097-104.

Vaid, A., & Sharma, P. (2006). PfPKB, a protein kinase B-like enzyme from Plasmodium falciparum: II. Identification of calcium/calmodulin as its upstream activator and dissection of a novel signaling pathway. *J Biol Chem., 281*(37), 27126-27133.

Valaskovic, G. A. & Kelleher, N. L. (2002). Miniaturized formats for efficient mass spectrometry-based proteomics and therapeutic development. *Curr Top Med Chem., 2*(1), 1-12.

Valdar, W. S. (2002). Scoring residue conservation. *Proteins, 48*, 227-241.

Valencia, A., & Pazos, F., (2002). Computational methods for the prediction of protein interactions. *Current Opinion in Structural Biology, 12*, 368-373.

van de Vijver, M. J., He, Y. D., van't Veer, L. J., Dai, H., Hart, A. A., Voskuil, D. W., et al. (2002). A gene-expression signature as a predictor of survival in breast cancer. *The New England Journal of Medicine, 347*(25), 1999-2009.

van Driel, M. A., Bruggeman, J., Vriend, G., Brunner, H. G., & Leunissen, J. A. (2006). A text-mining analysis of the human phenome. *Eur J Hum Genet, 14*(5), 535-542.

Vapnik, V. (1998). *Statistical Learning Theory.* Wiley.

Vapnik, V. (1999). *The Nature of Statistical Learning Theory (Information Science and Statistics).* Springer.

Vastrik, I., D'Eustachio, P., Schmidt, E., Joshi-Tope, G., Gopinath, G., Croft, D., et al. (2007). Reactome: a knowledge base of biologic pathways and processes. *Genome Biol, 8*(3), R39.

Vazquez, A., Flammini, A., et al. (2003). Global protein function prediction from protein-protein interaction networks. *Nat Biotechnol, 21*(6), 697-700.

Vázquez, A., Flammini, A., Maritan, A., & Vespignani, A. (2003). Modeling of protein interaction networks. *ComPlexUs, 1*, 38-44.

Vestal, M. L. & Campbell, J. M. (2005). Tandem time-of-flight mass spectrometry. *Methods Enzymol.,* 402, 79-108.

Vidalain, P. O., Boxem, M., Ge, H., Li, S., & Vidal, M. (2004). Increasing specificity in high-throughput yeast two-hybrid experiments. *Methods, 32*(4), 363-370.

Vinayagam, A., del Val, C., et al. (2006). GOPET: a tool for automated predictions of Gene Ontology terms. *BMC Bioinformatics, 7*, 161.

von Mering, C. et al. (2007). STRING 7 – recent developments in the integration and prediction of protein interactions. *Nucleic Acids Research, 35*(Database issue), D358-D362.

von Mering, C., & et al. (2002). Comparative assessment of a large-scale data sets of protein-protein interactions. *Nature, 417*(6887), 399-403.

von Mering, C., Jensen, L. J., Kuhn, M., Chaffron, S., Doerks, T., Kruger, B., et al. (2007). STRING 7--recent developments in the integration and prediction of protein interactions. *Nucleic Acids Res, 35*(Database issue), D358-362.

von Mering, C., Krause, R., Snel, B., Cornell, M., Oliver, S. G., Fields, S., et al. (2002). Comparative assessment of large-scale data sets of protein–protein interactions. *Nature, 417*, 399-403.

von Mering, C., Krause, R., Snel, B., Cornell, M., Oliver, S. G., Fields, S., et al. (2002). Comparative assessment of large-scale data sets of protein-protein interactions. *Nature, 417*(6887), 399-403.

Wach, A., Brachat, A., Pohlmann R., & Philippsen, P. (1994). New heterologous modules for classical or PCR-based gene disruptions in *Saccharomyces cerevisiae. Yeast, 10*(13), 1793-808.

Wachi, S., Yoneda, K., & Wu, R. (2005). Interactome-trancriptome analysis reveals the high centrality of genes differentially expressed in lung cancer tissues. *Bioinformatics, 21*(23), 4205-4208.

Wagner, A. (2001). The yeast protein interaction network evolves rapidly and contains few redundant duplicate genes. *Mol Biol Evol, 18*(7), 1283-1292.

Wagner, A. (2002). Asymmetric functional divergence of duplicate genes in yeast. *Mol Biol Evol, 19*(10), 1760-1768.

Walhout, A. J., Sordella, R., Lu, X., Hartley, J. L., Temple, G. F., Brasch, M. A., et al. (2000). Protein interaction mapping in c. Elegans using proteins involved in vulval development. *Science, 287*(5450), 116-122.

Wang, H., et. al. (2007). InSite: a computational method for identifying protein-protein interaction binding sites on a proteome-wide scale. *Genome Biology, 7*, R192.

Wang, K., & Samudrala, R. (2005). FSSA: a novel method for identifying functional signatures from structural alignments. *Bioinformatics, 21*(13), 2969-77.

Wang, T., Touchman, J. W., Zhang, W., Suh, E. B., & Xue, G. (2005, October). A parallel algorithm for extracting transcriptional regulatory network motifs. In *Proc. 5th IEEE Symposium on Bioinformatics and Bioengineering* (pp. 193–200). Washington, DC, USA: IEEE Computer Society Press.

Wang, Y., Klijn, J.G., Zhang, Y., Sieuwerts, A. M., Look, M. P., Yang, F., et al. (2005). Gene-expression profiles to predict distant metastasis of lymph-node-negative primary breast cancer. *Lancet*, *365*, 671–679.

Wang, Z., & Zhang, J. (2007). In search of the biological significance of modular structures in protein networks. *PLoS Computational Biology*, *3*(6), e107.

Watt, D. J. (1999) *Small worlds*. Princeton University Press.

Watts, D. J., & Strogatz, S. H. (1998). Collective dynamics of 'small-world' networks. *Nature, 393*(6684), 440-2.

Watts, D. J., & Strogatz, S. H. (1998). Collective dynamics of 'small-world' networks. *Nature, 393*(6684), 440-442.

Wei, P., & Pan, W. (2008). Incorporating gene networks into statistical tests for genomic data via a spatially correlated mixture model. *Bioinformatics*, *24*(3), 404-411.

Wei, Z., & Li, H. (2007). A Markov random field model for network-based analysis of genomic data. *Bioinformatics*, *23*(12), 1537-1544.

Wernicke, S. (2006). Efficient detection of network motifs. *IEEE/ACM Transactions on Computational Biology and Bioinformatics*, *3*(4), 347–359.

Wernicke, S., & Rasche, F. (2006). FANMOD: A tool for fast network motif detection. *Bioinformatics*, *22*(9), 1152–1153. (Software: http://www.minet.uni-jena.de/~wernicke/motifs/)

Wernicke, S., & Rasche, F. (2007). Simple and fast alignment of metabolic pathways by exploiting local diversity. *Bioinformatics, 23*(15), 1978-1985.

West, A. G., Gaszner, M., & Felsenfeld, G. (2002). Insulators: Many functions, many mechanisms. *Genes Dev, 16*(3), 271-288.

Wheeler, D. L., Chappey, C., Lash, A. E., Leipe, D. D., Madden, T. L., Schuler, G. D., et al. (2000, January). Database resources of the National Center for Biotechnology Information. *Nucleic Acids Research*, *28*(1), 10–14.

Wilkinson, D. M., & Huberman, B. A. (2004). A method for finding communities of related genes. *Proc. Natl. Acad. Sci., 101*(1), 5241-5248.

Wilson, A. C., Carlson, S. S., & White, T. J. (1977). Biochemical evolution. Annu Rev Biochem, *46*, 573-639.

Winter, C., et. al. (2006). SCOPPI: A structural classification of protein protein interfaces. *Nucleic Acids Research, Database issue 34,* D310-D314.

Witten, I. H., & Frank, E. (2005). *Data Mining: Practical machine learning tools and techniques*. 2nd Edition. Morgan Kaufmann.

Wojcik, J., Boneca, I. G., & Legrain, P. (2002). Prediction, assessment and validation of protein interaction maps in bacteria. *J Mol Biol, 323*(4), 763-770.

Wolf, D. M., & Arkin, A. P. (2003, April). Motifs, modules and games in bacteria. *Current Opinion in Microbiology*, *6*(2), 125–134.

Wolfe, K. H., & Shields, D. C. (1997). Molecular evidence for an ancient duplication of the entire yeast genome. *Nature, 387*(6634), 708-713.

Wu, C. H., Apweiler, R., et al. (2006). The Universal Protein Resource (UniProt): an expanding universe of protein information. *Nucleic Acids Res, 34*(Database issue), D187-91.

Wu, H.-Y., Tomizawa, K., Oda, Y., Wei, F.-Y., Lu, Y.-F., Matsushita, M., et al. (2004). Critical Role of Calpain-mediated Cleavage of Calcineurin in Excitotoxic Neurodegeneration. *J. Biol. Chem., 279*(6), 4929-4940.

Wu, J., Kasif, S., et al. (2003). Identification of functional links between genes using phylogenetic profiles. *Bioinformatics, 19*(12), 1524-30.

Wu, T. D. et. al. (1996). Discovering Empirically Conserved Amino Acid Substitution Groups in Databases of Protein Families. *Proceedings of the International Conference on Intelligent Systems for Molecular Biology, 4*, 230-240.

Wu, X., & Brutlag, D. L. (2006). Prediction of yeast protein-protein interaction network: insights from the gene ontology and annotations. *Nucleic Acids Research, 34*, 2137-2150.

Wu, X., Jiang, R., Zhang, M. Q., & Li, S. (2008). Network-based global inference of human disease genes. *Mol Syst Biol, 4*, 189.

Wu, X., Zhu, L., Guo, J., Zhang D. Y., & Lin, K. (2006). Prediction of yeast protein-protein interaction network: insights from the Gene Ontology and annotations. *Nucleic Acids Res, 34*(7), 2137-50.

Wuchty, S. (2004). Evolution and topology in the yeast protein interaction network. *Genome Res, 14*(7), 1310-1314.

Wuchty, S. (2007). Rich-club phenomenon in the interactome of P. falciparum--artifact or signature of a parasitic life style? *PLoS ONE, 2*(3), e335.

Wuchty, S., Barabasi, A. L., & Ferdig, M. T. (2006). Stable evolutionary signal in a yeast protein interaction network. *BMC Evol Biol, 6*, 8.

Wuchty, S., Oltvai, Z. N., & Barabási, A.-L. (2003, October). Evolutionary conservation of motif constituents in the yeast protein interaction network. *Nature Genetics, 35*(2), 176–179.

Xenarios, I., & Eisenberg, D. (2001). Protein interaction databases. *Current Opinion in Biotechnology, 12*, 334-339.

Xenarios, I., Rice, D. W., Salwinski, L., Baron, M. K., Marcotte, E. M., & Eisenberg, D. (2000). DIP: the database of interacting proteins. *Nucleic Acids Res, 28*(1), 289-291.

Xenarios, I., Rice, D. W., Salwinski, L., Baron, M. K., Marcotte, E. M., & Eisenberg, D. (2000) DIP: The Database of Interacting Proteins. *NAR, 28*, 289-91.

Xenarios, I., Salwnski, L., Duan, X. J., Higney, P., Kim, S. M., & Eisenberg, D. (2002). DIP, the database of interacting proteins: a research tool for studying cellular networks of protein interactions. *Nucleic Acids Res, 30*, 303–5.

Xia, K., Dong, D., & Han, J. D. (2006). IntNetDB v1.0: an integrated protein-protein interaction network database generated by a probabilistic model. *BMC Bioinformatics, 7*, 508.

Xiong, H., He, X., Ding, C., Zhang, Y., Kumar, V., & Holbrook, S. R. (2005) Identification of Functional Modules in Protein Complexes via Hyperclique Pattern Discovery. *Pacific Symposium on Biocomputing, 10* 221-232.

Xiong, J., Rayner, S., et al. (2006). Genome wide prediction of protein function via a generic knowledge discovery approach based on evidence integration. *BMC Bioinformatics, 7*, 268.

Xu, J., & Li, Y. (2006). Discovering disease-genes by topological features in human protein-protein interaction network. *Bioinformatics, 22*, 2800-2805.

Yan, X., & Han, J. (2002). gSpan: Graph-based substructure pattern mining. In *Proc. 2002 IEEE International Conference on Data Mining* (pp. 721–724). Washington, DC, USA: IEEE Computer Society Press.

Yan, X., & Han, J. (2002). gSpan: graph-based substructure pattern mining. *Proceedings of the 2002 IEEE International Conference on Data Mining* (pp. 721–724). IEEE Computer Society.

Yang, J., Gu, Z., & Li, W. H. (2003). Rate of protein evolution versus fitness effect of gene deletion. *Mol Biol Evol, 20*(5), 772-4.

Yang, Q., & Sze, S.-H. (2007). Path Matching and Graph Matching in Biological Networks. *Journal of Computational Biology, 14*(1), 56-67.

Yao, Z., Weinberg, Z., & WL, W. R. (2005). CMfinder–a covariance model based RNA motif finding algorithm. *Bioinformatics, 22*, 445–452.

Yeger-Loten, E., Sattah, S., Kashtan, N., Itzkovitz, S., Milo, R., Pinter, R., Alon, U., & Margalit, H. (2004). Networks motifs in integrated cellular networks of transcription-regulation and protein-protein interactions. *PNAS, 101*(16), 5934-5939.

Yeo, G., & Burge, C. B. (2004). Maximum entropy modeling of short sequence motifs with applications to RNA splicing signals. *J Comput Biol, 11*(2-3), 377-394.

Yook, S. H., Oltvai, Z. N., & Barabasi, A. L. (2004). Functional and topological characterization of protein interaction networks. *Proteomics, 4*(4), 928-942.

Yoon, J., Blumer, A., & Lee, K. (2006). An algorithm for modularity analysis of directed and weighted biological networks based on edge-betweenness centrality. *Bioinformatics, 22*(24), 3106-3108.

Young, K. H. (1998). Yeast two-hybrid: so many interactions, (in) so little time. *Biol Reprod, 58*(2), 302-11.

Yu, H., Greenbaum, D., Xin Lu, H., Zhu, X., & Gerstein, M. (2004). Genomic analysis of essentiality within protein networks. *Trends Genet, 20*(6), 227-231.

Yu, H., Kim, P. M., Sprecher, E., Trifonov, V., & Gerstein, M. (2007). The importance of bottlenecks in protein networks: correlation with gene essentiality and expression dynamics. *PLoS Computational Biology, 3*(4), e59.

Zabka, M., Les'niak, W., Prus, W., nicki, J. K., & Filipek, A. (2008). Sgtl has co-chaperone properties and is upregulated by heat shock. *Biochemical and Biophysical Research Communications, 370*, 179-183.

Zaki, M., Parthasarathy, S., Ogihara, M., & Li, W. (1997). New algorithms for fast discovery of association rules. *Proc. KDD 1997* (pp. 283–286).

Zanzoni, A., Montecchi-Palazzi, L., Quondam, M., Ausiello, G., Helmer-Citterich, M., & Cesareni, G. (2002). MINT: a Molecular INTeraction database. *FEBS Lett, 513*(1), 135-140.

Zeeberg, B. R., Feng, W., Wang, G., Wang, M. D., Fojo, A. T., Sunshine, M., Narasimhan, S., Kane, D. W., Reinhold, W.C., Lababidi, S. et al. (2003). GoMiner: A resource for biological interpretation of genomic and proteomic data. *Genome Biol, 4*, R28.

Zehetner, G. (2003). OntoBlast function: From sequence similarities directly to potential functional annotations by ontology terms. *Nucleic Acids Res, 31*(13), 3799-803.

Zhang, B., & Horvath, S. (2005). A general framework for weighted gene co-expression network analysis. *Stat Appl Genet Mol Biol, 4*(17).

Zhang, B., Park, B. H., Karpinets, T., & Samatova, N. F. (2008). From pull-down data to protein interaction networks and complexes with biological relevance. *Bioinformatics, 24*(7), 979-986.

Zhang, B., Schmoyer, D., Kirov, S., & Snoddy, J. (2004). GOTree Machine (GOTM): a web-based platform for interpreting sets of interesting genes using Gene Ontology hierarchies. *BMC Bioinformatics, 5*(16).

Zhang, C., Zhao, X. & Zhang, M.Q. (2007). Functional in silico analysis of gene regulatory polymorphism. In M. R. Barnes (Ed.), *Bioinformatics for Geneticists: A Bioinformatics Primer for the Analysis of Genetic Data* (2nd ed., pp. 281-310). England: Wiley.

Zhang, J., & He, X. (2005). Significant impact of protein dispensability on the instantaneous rate of protein evolution. Mol Biol Evol, *22*(4), 1147-55.

Zhang, L. V., King, O. D., Wong, S. L., Goldberg, D. S., Tong, A. H. Y., Lesage, G., Andrews, B., Bussey, H., Boone, C., & Roth, F. P. (2005). Motifs, themes and thematic maps of an integrated Saccharomyces cerevisiae interaction network. *Journal of Biology, 4*, 6.

Zhang, M. Q., & Marr, T. G. (1993). A weight array method for splicing signal analysis. *Comput Appl Biosci, 9*(5), 499-509.

Zhang, S., Jin, G., Zhang, X. S., & Chen, L. (2007). Discovering functions and revealing mechanisms at molecular level from biological networks. *Proteomics, 7*(16), 2856-2869.

Zhang, S., Zhang, X.-S., & Chen, L. (2008). Biomolecular network querying: a promising approach in systems biology. *BMC Systems Biology, 2*(5).

Zhao, X., Huang, H., & Speed, T. P. (2005). Finding short DNA motifs using permuted Markov models. *J Comput Biol, 12*(6), 894-906.

Zhou, H. X., & Qin, S. (2007). Interaction-site prediction for protein complexes: a critical assessment. *Bioinformatics, 23*, 2203-2209.

Zhou, H. X., & Shan, Y. (2001). Prediction of protein interaction sites from sequence profile and residue neighbor list. *Proteins, 44*, 336-343.

Zhou, M., & Cui, Y. (2004). GeneInfoViz: constructing and visualizing gene relation networks. *In Silico Biol, 4*(3), 323-333.

Zhou, X., Kao, M. C., et al. (2002). Transitive functional annotation by shortest-path analysis of gene expression data. *Proc Natl Acad Sci U S A, 99*(20), 12783-8.

Zhu, H., Bilgin, M., Bangham, R., Hall, D., Casamayor, A., Bertone, P., et al. (2001). Global analysis of protein activities using proteome chips. *Science, 293*(5537), 2101-2105.

Zotenko, E., Guimarães, K. S., Jothi, R., & Przytycka, T. M. (n/d). Decomposition of Overlapping Protein Complexes: A Graph Theoretical Method for Analyzing Static and Dynamic Protein Associations. *Algorithms for Molecular Biology, 1*(7).

About the Contributors

Xiao-Li Li is currently a principal investigator at the Data Mining Department, Institute for Infocomm Research, A*Star. He also holds an appointment of adjunct assistant professor at the School of Computer Engineering, Nanyang Technological University. Xiao-Li received his PhD degree in computer science from Chinese Academy of Sciences in 2001. He was with National University of Singapore (School of Computing/Singapore-MIT Alliance) as a research fellow from 2001 to 2004 before he joined the Institute for Infocomm Research. His research interests include bioinformatics, data mining and machine learning. He has been serving as the members of editorial boards or technical program committees in numerous bioinformatics, data mining and machine learning related conferences and journals. In 2005, he received best paper award in the 16th International Conference on Genome Informatics (GIW 2005). In 2008, he received the best poster award in the 12th Annual International Conference Research in Computational Molecular Biology (RECOMB 2008). Dr. Li is also co-editor-in-chief of the *International Journal of Knowledge Discovery in Bioinformatics.*

See-Kiong Ng is currently the department head of the Data Mining Department at Institute for Infocomm Research. He is also an adjunct associate professor at the School of Computer Engineering, Nanyang Technological University. See-Kiong obtained his PhD in computer science from Carnegie Mellon University. When he was a graduate student at CMU, See-Kiong wrote the TrueAllele software which was used by biotech and pharmaceutical companies to fully automate the analysis of data from large-scale genotyping studies. Since then, See-Kiong has continued his journey into the exciting field of genomics as a computer scientist, publishing award-winning papers and serving on the editorial boards or technical program committees in major bioinformatics conferences and journals. See-Kiong's current research focuses on unraveling the underlying functional mechanisms of protein interaction networks as well as other real-world networks. His continuing and emerging diverse and cross-disciplinary research interests include bioinformatics, text mining, social network mining, and privacy-preserving data mining. Dr. Ng is also co-editor-in-chief of the *International Journal of Knowledge Discovery in Bioinformatics.*

* * *

Tero Aittokallio received his PhD in applied mathematics from the University of Turku in 2001. He is Research Fellow of the Academy of Finland at the Department of Mathematics, University of Turku, Finland. He has authored more than 50 articles in international peer-reviewed journals. His main research field is model-based analysis of biological systems, with special focus on developing mathematical and statistical data mining approaches to address concrete biomedical problems.

Tatsuya Akutsu received BEng and MEng in aeronautics and DEng in information engineering from University of Tokyo, 1984, 1986 and 1989, respectively. From 1989 to 1994, he was with Mechanical Engineering Laboratory, Japan. From 1994 to 1996, he was an associate professor in the Department of Computer Science at Gunma University. From 1996 to 2001, he was an associate professor in Human Genome Center, Institute of Medical Science, University of Tokyo. Since 2001, he has been a professor in Bioinformatics Center, Institute for Chemical Research, Kyoto University. His research interests include bioinformatics and the design and analysis of algorithms.

Doina Caragea is an assistant professor at Kansas State University. Her research interests include artificial intelligence, machine learning, data mining, information integration and information visualization, with applications to bioinformatics. Doina received her PhD in computer science from Iowa State University in August 2004 and was honored with the Iowa State University Research Excellence Award for her work. She has published more than 20 refereed conference and journal articles and contributed to the design and implementation of INDUS (an open source system for Intelligent Data Understanding). Doina is teaching machine learning, data mining and bioinformatics courses.

Hon Nian Chua is a research engineer at the Data Mining Department of the Institute for Infocomm Research, A*STAR. He obtained his PhD degree in bioinformatics from the National University of Singapore in 2008, under the support of the A*STAR Graduate Scholarship. His current research interest is in the application of machine learning and graph-based techniques in biological and medical research.

Joaquin Dopazo is the head of the Department of Bioinformatics and Genomics at the CIPF (Valencia). In previous appointments he was responsible of Bioinformatics units at the CNIO (Madrid) and at GlaxoWellcome SA (Madrid). He has supervised several large scale projects of software development, as the GEPAS (http://www.gepas.org) or the Babelomics (http://www.Babelomics.org) where more than 1000 microarray experiments are daily analysed. J. Dopazo has a master degree in chemistry (Universidad de Valencia) and a PhD in biology (Universidad de Valencia). He has more than one hundred papers published in international peer reviewed journals and has edited a book on genomic data analysis. His main interests include functional and comparative genomics.

Valeria Fionda is currently attending the second year of the "Dottorato in Matematica e Informatica" (Doctorate in Mathematics and Computer Science) at the Department of Mathematics, Università della Calabria. Her research interests include issues related to bioinformatics, computational biology and graph matching.

Sirisha Gollapudi is a PhD student at the University of Nottingham, within the Multidisciplinary Centre for Integrative Biology (MyCIB). She obtained a BSc in computer science from the University of Nottingham, and a postgraduate certificate in bioinformatics from the University of Manchester. Her PhD research focuses on the use of web service technology to expose software for the analysis of biological networks, including holistic models of metabolism and protein-protein interactions. Her main research interest is the combination of such web services into scientific workflows to automate analyses and query heterogeneous data resources.

Morihiro Hayashida received BSci and MSci in information science from University of Tokyo, 2000 and 2002, and DInf in informatics from Kyoto University, 2005. Since 2005, he has been an assistant professor in Bioinformatics Center, Institute for Chemical Research, Kyoto University. His research interests include issues related to protein function prediction and bioinformatics.

Charlie Hodgman dates his first bioinformatics project to 1980. During the 1980s and 90s, he acquired an international reputation for elucidating biomolecular function using informatics approaches. Since the mid-90s, his attention has turned to interaction networks and dynamic modelling of multi-scale systems. His research career began in ICI Ltd (UK) in the 1970s. Following fifteen years in Cambridge and nine in GlaxoSmithKline, he is now Professor of Bioinformatics and Systems Biology and director of the Multidisciplinary Centre for

Integrative Biology in the University of Nottingham (UK). He was also an associate editor of Bioinformatics from 2000 to 2006.

William H. Hsu is an associate professor in the Department of Computing and Information Sciences at Kansas State University. He received a BS in mathematical sciences and computer science and an MSEng in computer science from Johns Hopkins University in 1993, and a PhD in computer science from the University of Illinois at Urbana-Champaign in 1998. His research interests include machine learning and probabilistic reasoning, with applications to information extraction, time series prediction, link mining, and bioinformatics (especially computational genomics and proteomics).

Lilia M. Iakoucheva is a research assistant professor at the Rockefeller University in New York, NY, USA. She holds a PhD degree in molecular biology and immunology, which she received in Moscow, Russia. Her research interests are focused on understanding how alterations in protein structure, function and interactions contribute to human disease. Dr. Iakoucheva is interested in investigating this question from two different levels, molecular and systems biology. She uses both computational and experimental approaches to investigate these questions. Another aspect of her scientific interests includes investigation of functional properties of intrinsically disordered proteins. Dr. Iakoucheva's research was published in a number of high-impact journals including *Bioinformatics, PLoS Computational Biology, Nucleic Acids Research, Journal of Molecular Biology, Protein Science, Journal of Proteome Research,* and *Biochemistry*. Her research was supported by the National Science Foundation (NSF) and the National Institutes of Health (NIH).

Daisuke Kihara PhD is an assistant professor in the Departments of Biological Sciences and Computer Science at Purdue University. He received his PhD degree from Kyoto University in 1999. His research projects include protein function and structure prediction and protein surface shape searching for function prediction and docking. His research projects are funded by the National Institutes of Health and the National Science Foundation of the USA.

David La is a PhD candidate in the Department of Biological Sciences at Purdue University. Prior to starting his graduate studies at Purdue University, he received his MS degree from the California State Polytechnic University at Pomona where his thesis work involved the computational prediction of protein functional sites and their application at a proteome-scale. Currently, his research is focused on gaining greater insight to the evolution and prediction of protein-protein binding regions.

Takashi Makino is the research fellow in the Genetics Department at Trinity College Dublin. His research focuses on evolution of protein interaction networks and comparative genomics of vertebrates. He studied molecular evolution under his supervisor who is prof. Takashi Gojobori in National institute of genetics, Mishima, Japan. His work has been published in international journals including *PLoS Biology, Molecular Biology and Evolution, Nucleic Acids Research and GENE.*

Hiroshi Mamitsuka received the BS in biochemistry and biophysics, the ME in information engineering and the PhD in information sciences from the University of Tokyo in Japan, in 1988, 1991 and 1999, respectively. He worked with NEC Corporation in Japan on developing machine learning and data mining techniques and their applications to problems in industry and the sciences for eleven years since 1991. He joined the Institute for Chemical Research (ICR) of Kyoto University in 2002, focusing on research matters in molecular biology. Currently he is a professor of ICR, being jointly appointed as a faculty of the School of Pharmaceutical Sciences of the same university. His present research interests include bioinformatics, data mining, machine learning and information retrieval.

After graduating in 2003 at the "Sapienza" University of Rome in Control Engineering, **Paolo Marcatili** directed his interest to the study of system level description and modeling of biological data, with a specific focus on protein-protein interaction in immunology and pathology. Currently he is working for his PhD at the "Sapienza" University under the supervision of Anna Tramontano.

Alex Marshall is a systems biologist with a biology background, specialising in bioinformatics and network biology in plants. Alex recently gained an MSc in bioinformatics at Cranfield University (UK) and that has sparked her interest in systems-based research. Currently, Alex is studying a PhD in plant cross-species transcriptomics and systems biology at the Multidisciplinary Centre for Integrative Biology at the University of Nottingham (UK).

Aoife McLysaght is the principal investigator of the Molecular Evolution research group in the Genetics Department at Trinity College Dublin. Her research focuses on comparative genomics of vertebrates and virus genome evolution. Dr. McLysaght is an associate editor for the *Journal of Molecular Biology and Evolution*, the leading specialist journal in her field. Her work has been published in major international journals including Nature, Nature Genetics, and the Proceedings of the National Academy of Sciences, USA and Trends in Genetics. In 2005 she was awarded a prestigious President of Ireland Young Researcher Award by Science Foundation Ireland.

Pablo Minguez is a PhD student at the Bioinformatics Department of the Centro de Investigación Principe Felipe, Valencia, Spain. He has been working in functional genomics since 2004 developing methodologies for the functional profiling of genome-scale experiments augmenting the discovery possibilities of the field by introducing new sources of annotation such as bioentities associated to genes by co-occurrences in the scientific literature, phenotype specific expression levels and protein-protein interactions. He is the developer of several modules in the Babelomics suite for functional profiling of genome-scale experiments including SNOW, a program to infer functional modules from transcriptomic and protein-protein interaction data and a method to evaluate its significance in cellular processes.

Smita Mohanty is currently pursuing her PhD in the field of computational biology from Molecular Biophysics Unit of the Indian Institute of Science, Bangalore. Her research focuses on understanding protein-protein interactions at the level of domain families and applying this knowledge into effective prediction of protein-protein interactions. The major thrust of her work is in understanding the biology of the malarial parasite P.falciparum, especially focusing on the role of protein-protein interactions in the parasite's life-cycle.

Luigi Palopoli is a professor of computer science at DEIS of Università della Calabria, where he chairs the School of Computer Engineering, and research associate with ICAR-CNR. He was a visiting scholar at UCLA (Los Angeles, USA), AT&T Laboratories (Murray Hills, USA), Oxford University (Oxford, UK) and Technical University of Wien (Wien, Austria). His research interests are in the areas of information systems and databases, artificial intelligence, game theory and bioinformatics. He authored about 150 papers appearing in refereed journals and conference proceedings. He serves in the editorial board of AI Communications – The European J. on AI , and in several conference program committees.

Shashi Bhushan Pandit completed his PhD at the Indian Institute of Science Bangalore. During his Ph.D, he worked on sequence analysis of proteins with emphasis on the function prediction at the genomic scale. After his Ph.D he joined the research group of Dr. Skolnick at Center of Excellence in Bioinformatics, Buffalo, USA to pursue his research in the field of protein tertiary structure prediction. Presently, he is working at Georgia Institute of Technology, Atlanta, USA. His current research focuses on the development of tools for the protein tertiary and quaternary structure modeling and its application at the genomic scale.

Martin S.R. Paradesi graduated with a master's degree in computer science from Kansas State University in 2008. Prior to that, he obtained a bachelor's degree in computer science and engineering from Jawaharlal Nehru Technological University in 2005. His research interests include bioinformatics, machine learning, artificial intelligence and data mining. This book chapter is a partial result of his Master's thesis that investigated the effect of accuracy and other statistical metrics caused by the use of graph features in predicting protein interactions.

Clara Pizzuti received the Laurea degree in Mathematics from the University of Calabria, Italy. She is a senior researcher at the Institute of High Performance Computing and Networking (ICAR) of the Italian National Research Council (CNR). Since 1995 she is also a contract professor in the department of Computer Science at the University of Calabria. In the past, she worked in the research division of a software company on deductive databases, advanced logic based systems, and abduction. She has published more than seventy papers in conference proceedings and journals. Her research interests include knowledge discovery in databases, data mining, bioinformatics, evolutionary computation, genetic algorithms, and genetic programming. She is serving as program committee member of international conferences, and as reviewer for several international journals.

Simona Ester Rombo received the Laurea degree in electronic engeneering and the PhD in computer science from the University "Mediterranea"of Reggio Calabria, Italy. She is a post-doc fellow at DEIS, Università della Calabria. Since 2006 she is also a contract professor at the University of Calabria. Her research interests data mining, bioinformatics, combinatorial algorithms. She is serving as program committee member of international conferences, and as reviewer for international journals.

Christian Schönbach's research expertise lies in computational and functional genomics. Currently, he is an associate professor at Division of Genomics and Genetics, School of Biological of Biological Sciences, Nanyang Technological University (Singapore). Previous positions included team leader at RIKEN Genomic Sciences Center (Japan), principal investigator at Kent Ridge Digital Labs (Singapore), and research scientist at Chugai Research Institute for Molecular Medicine, Inc. (Japan). He obtained his doctor degree from Eberhard Karls University Tübingen (Germany) through immunogenetics research at Max-Planck Institute for Biology, and did his post-doctoral training in experimental and computational immunology at The University of Tokyo.

Narayanaswamy Srinivasan obtained his PhD at the Indian Institute of Science and had three postdoctoral stints in the UK: Birkbeck College, London, Ludwig Institute for Cancer Research, London and Department of Biochemistry, University of Cambridge. He is currently leading a research team working on protein structure, function and interactions at the Molecular Biophysics Unit, Indian Institute of Science. He is in the editorial boards of a few journals including *Bioinformatics*. He is a visiting professor at the Reunion university, France, senior fellow of the Manchester University, UK and a fellow of two of the Academies of Sciences in India.

Kar Leong Tew graduated from University of Exeter with a bachelor in computer science (Hons) in 2005. He is current working as a research officer in the Data Mining Department at Institute for Infocomm Research (I2R) which is one of the research institute under Agency for Science, Technology and Research (A*Star). His current research works involves networking mining, data mining, bioinformatics, and social networks.

Anna Tramontano graduated in physics at the University of Naples, Italy. She was a post-doc at UCSF and a staff scientist in the Biocomputing programme of the EMBL in Heidelberg. In 1990, she went back to Italy to work at the Merck Research Laboratories near Rome. In 2001, she returned to academia as Chair Professor of Biochemistry at the "Sapienza" University where she works on protein structure prediction and analysis. She is a member of the European Molecular Biology Organization, the Scientific Council of Institute Pasteur-Fondazione Cenci Bolognetti, and of the organizing committee of the Critical Assessment of Techniques for Protein Structure

Prediction (CASP) initiative. She is also the director of the Bioinformatics Center of CRS4 in Pula, Sardinia and of a Master program in bioinformatics.

Koji Tsuda is project leader at Max Planck Institute for Biological Cybernetics, Tuebingen, Germany. In 2001-2002 and 2005, he worked as a research scientist in AIST Computational Biology Research Center, Tokyo, Japan. In 2000-2001, he was a visiting scholar at GMD FIRST, Berlin, Germany. He received doctor of engineering degree from Kyoto University in 1998. His scientific interests are in the fields of machine learning, data mining and bioinformatics.

Raymond Wan received the BSc degree in computer science from the University of British Columbia, Canada, in 1997 and the PhD degree from the University of Melbourne, Australia, in 2004. His doctoral dissertation, "Browsing and Searching Compressed Documents", describes a method for browsing the phrases of a compressed document collection through minimal decompression. Since graduation, he has been a post-doctoral fellow in bioinformatics at Kyoto University, Japan, working on microarray analysis and text retrieval from biomedical texts. His current research interests include bioinformatics, graph theory, machine learning, information retrieval, and text compression.

Hugo Willy received his bachelor of computing degree from the National University of Singapore (NUS), in 2003. He is currently pursuing his study as a PhD candidate under the joint research programme between the School of Computing, National University of Singapore and the Institute of Infocomm Research, Singapore. His early research was on the structure-templated RNA secondary structure inference by arc-annotated sequence model. Currently, he is working on mining sequence and structural motifs implicated in Protein-Protein Interaction.

Limsoon Wong is a professor in the School of Computing and the School of Medicine at the National University of Singapore. He is currently working mostly on knowledge discovery technologies and is especially interested in their application to biomedicine. He serves on the editorial boards of *Journal of Bioinformatics and Computational Biology* (ICP), *Bioinformatics* (OUP), and *Drug Discovery Today* (Elsevier).

After completing a BSc in applied biology at the University of Bath and an MSc in bioinformatics at Sheffield Hallam University, **Daniel Zadik** is working towards completing a PhD at the Multidisciplinary Centre for Integrative Biology (MyCIB) at the University of Nottingham. His PhD involves the construction of a network model of dog metabolism and its control mechanisms, with the aim of better understanding the relationships between genetics, nutrition and health, and elucidating the specific dietary requirements of dog breeds. In the past, he has worked as a laboratory scientist at University College London, and a bioinformatician at the University of Nottingham.

Michael Q. Zhang is professor and the director of the Bioinformatics Core of Cancer Center at CSHL. He is also an adjunct professor at Stony Brook University and a Chang Jiang Chair (Guest) Professor at Tsinghua University. He holds degrees in mechanical engineering (BS), and physics (PhD). Michael has co-edited, *Current Topics in Computational Molecular Biology*. Michael has published his research in a number of high-impact journals including *Cell, Nat. Genet.s, Nat. Cell Biol., Gene & Dev., Mol. Sys.. Biol., PNAS, Genome Res., TIG, PLoS Comp. Biol., Nucl. Acid. Res.* and *Bioinformatics*. associate editor of the BMC journals, and editor for *Bioinformatics, J Bioinfo. Comp. Biol., Systems and Synthetic Biol., Comparative and Func. Genomics*. His research interests include computational biology and bioinformatics.

Xiaoyue Zhao is research scientist at Bionovo Inc., located in Emeryville, CA, USA. She received PhD degree in statistics from the University of California, Berkeley and did postdoc training at Cold Spring Harbor Laboratory. Her research interests include developing and applying statistical methods for understanding biology and human diseases, specifically related to women's health and neurological disorders. She focuses on addressing statistical challenges from high-throughput genomic data, such as protein-protein interaction, microarray and sequencing data. Dr. Zhao's research was published in a number of high-impact journals including *Proceedings of National Academy of Sciences*, *Genome Biology*, *Journal of Computational Biology*, *Annals of the Institute of Statistical Mathematics*, *Molecular Endocrinology*, and *Endocrinology*. She also contributed in the book *Bioinformatics for Geneticists: A Bioinformatics Primer for the Analysis of Genetic Data*

Index

A

B

C

D

H

I

J

K

L

M

N

O

P

Q

R

S

T